Evolutionary Processes and Theory

Evolutionary Processes and Theory

EDITED BY

SAMUEL KARLIN
Department of Mathematics
Stanford University
Stanford, California

EVIATAR NEVO
Institute of Evolution
University of Haifa
Mount Carmel, Haifa
Israel

1986

ACADEMIC PRESS, INC.
Harcourt Brace Jovanovich, Publishers
Orlando   San Diego   New York   Austin
London   Montreal   Sydney   Tokyo   Toronto

ACADEMIC PRESS, INC.
Orlando, Florida 32887

United Kingdom Edition published by
ACADEMIC PRESS INC. (LONDON) LTD.
24–28 Oval Road, London NW1 7DX

Library of Congress Cataloging in Publication Data

Evolutionary processes and theory.

   "Based on a workshop held in Israel in March 1985"—
Pref.
   Includes index.
   1. Evolution—Congresses.   I. Karlin, Samuel,
Date   . II. Nevo, Eviatar.
QH359.E938   1986   575   86-47517
ISBN 0–12–398760–1 (hardcover) (alk. paper)
ISBN 0–12–398761–X (paperback) (alk. paper)

PRINTED IN THE UNITED STATES OF AMERICA

86 87 88 89   9 8 7 6 5 4 3 2 1

Contents

Preface

These proceedings, based on a workshop held in Israel in March 1985, are a natural sequel to the proceedings of a workshop held in Israel 10 years earlier and published as *Population Genetics and Ecology* (Academic Press, New York, 1976). The 1975 workshop discussed evolutionary problems in molecular and organismal biology and focused specifically on problems concerning spatial and temporal allozyme (electromorph) frequency patterns in natural populations and their associations with ecological parameters.

Over the past decade advances in our knowledge of differentiation, a greater understanding of metabolic and immunological mechanisms, and the development of the powerful recombinant DNA technology have provided new data and perspectives on the molecular biology of the genome. This new information needs to be incorporated into our thinking about evolutionary processes. Challenging new problems have emerged with the discovery of interrupted genes, pseudogenes, transposable elements, eukaryotic virus-mediated transmission, nonhomologous recombination, sequence expansion, correction and transposition, high orders of repeated tandem and interspersed DNA segments, an abundance of multigene families as distinguished from unique genes, and the paradox that 80–90% of the genome is not transcribed in mammals and other eukaryotes.

On the organismal level, scientists are learning more about paleobiology, micro- and macroevolution, and processes of speciation, and continue to search for ways to quantify the evolution of behavioral traits based on genic, individual, group, and population fitness interactions. Among the specific issues being debated are the origin and evolution of sexual systems, the genetics of altruism, and general forms and levels of social evolution. Social behavior in the animal world manifests virtually a continuous spectrum from asocial to advanced eusocial expression. Its many forms range from the striking phenomenon of sterile castes in certain Hymenoptera species, through hierarchical status in groups, cooperative foraging strategies, and intra- and intergenerational conflicts, to a vast array of recognition and communication systems in animal groups that extend to cultural relations in man.

Other currently active issues of evolutionary theory bear on population growth and extinction patterns, the mode and tempo of evolutionary change, the description and analysis of sexual selection and nonrandom mating structures, the resolution of the components of polygenic inheritance, the merits of the neutralist explanation of molecular evolution as opposed to the selectionist explanation, and the role of sex and recombination in evolution.

A primary objective of this conference has been to bring together experts from many fields—molecular biologists, evolutionists, population geneticists and ecologists, naturalists, ethologists, and mathematical biologists—to assess the implications of contemporary molecular and organismal biology for evolutionary processes and theory, and to examine recent developments relating to genetic and environmental factors that contribute to ecological and behavioral forms in natural populations.

The contents divide into six parts: I. Evolutionary Problems of Molecular Biology; II. Tempo and Mode of Molecular Evolution; III. Comparative Analysis of DNA and Protein Sequences; IV. Models and Evidence of Speciation; V. Population Genetics: Observation, Experiment, and Theory; VI. Population Genetics of Ecological and Behavioral Interactions. The abstracts accompanying each paper highlight the key results. Many of the papers report experimental and field data with discussions, others propose models, hypotheses, and speculations pertinent to the evolution of molecular, morphological, physiological, and behavioral mechanisms.

The peripatetic workshop was conducted at five locations in Israel. The opening three sessions at the Hebrew University in Jerusalem were devoted to topics in molecular biology and molecular evolution; the fourth considered a variety of models and data on the population genetics of behavioral traits. The conference convened next at Haifa University to focus on modes and mechanisms of speciation and sexual selection, stochastic selection effects, and other topics of population genetics. The sixth session, at Tel Aviv University, considered issues of human molecular evolution; and the seventh, at Beer Sheva University in the Negev Desert, concentrated on the mix of population ecology and genetics. The concluding two sessions at the Weizmann Institute returned to problems of molecular biology and molecular evolution.

We are indebted to the European Molecular Biology Organization (EMBO), Haifa University, Hebrew University in Jerusalem, Tel Aviv University, Weizmann Institute, Ben-Gurion University of the Negev, The Israel National Academy of Sciences and Humanities, The Israel National Council for Research and Development, and the Dobrin Center for Nutrition and Plant Research at the Weizmann Institute of Science for providing financial support for the Conference. Finally, the first editor acknowledges continuing support from the U.S. NIH grant GM10452-22, NSF grant MCS 82-15131, and a grant from the Sloan Foundation to the Population Group at Stanford University.

Samuel Karlin
Eviatar Nevo

PART I. EVOLUTIONARY PROBLEMS OF MOLECULAR BIOLOGY

GENE REGULATION AND ITS ROLE IN EVOLUTIONARY PROCESSES

Kenneth Paigen

Department of Genetics
University of California, Berkeley
Berkeley, California 94720

ABSTRACT

It is thought that changes in the regulation of gene activity are as important in evolution as are modifications of protein structure. However, in contrast to the extensive body of information available on the evolution of protein sequences, there is relatively little experimental data on regulatory adaptations in evolution. For this reason, evolutionary models are largely derived from our knowledge of regulatory polymorphisms within contemporary species.

Regulatory polymorphisms affect enzyme levels by altering either the degradation or synthesis of specific proteins. Among the polymorphisms affecting enzyme synthesis are *systemic* mutations, altering enzyme levels in all cells and developmental stages equivalently; *effector response* mutations affecting the interaction of structural genes (or their products) with regulatory signals; and *temporal* mutations which cause tissue and stage specific changes in protein levels. Many such mutations are gene specific in their effects but at least two distinct types of pleiotropic regulators have been described in which mutation affects the level of many proteins. One type is represented by changes in hormone/receptor systems that activate many genes. The other type are "inverse" regulators that appear to be involved in phenomena of dosage compensation. The properties of many, but not necessarily all, regulatory polymorphisms are explicable in terms of the known molecular biology of transcription and translation.

3

When considered in detail, several conclusions can be drawn from our knowledge of regulatory polymorphism: a) the magnitude of a phenotypic change is often not correlated with the magnitude of the underlying genetic change; b) relatively short lengths of DNA are sufficient to accomplish ostensibly complex regulatory tasks; c) transposons then become a potentially important source of regulatory diversity; d) structural gene duplication allows the evolution of two independent regulatory systems for the same protein; e) we expect regulatory genes with pleiotropic effects to be under tighter selection than gene specific regulators; f) the close proximity of regulatory and structural DNA sequences means that the two will cosegregate and be coselected; and g) because close proximity greatly reduces the chances of recombination, even small differences in fitness will cause marked linkage disequilibrium among various combinations of regulatory and structural sequences.

There are also some gaps in our knowledge. In particular: 1) we lack a substantial body of experimental information on regulatory polymorphism in natural populations; 2) very little is known about the successive regulatory changes in the evolution of new phenotypes; 3) there is no objective notation system or terminology to facilitate the comparison of regulatory phenotypes; 4) no data is available on the DNA changes underlying examples of regulatory evolution; and 5) we do not understand the relative evolutionary significance of pleiotropic and gene specific regulatory elements.

I. INTRODUCTION

It is likely that evolution proceeds as much by changing the relative amounts of various proteins as by changing their amino acid sequences. This importance of regulatory as distinct from structural changes in proteins during evolution has been emphasized by several workers (Zuckerkandl, 1963; Wallace, 1963; Wilson, 1975; Wilson *et al.*, 1974, 1979; King and Wilson, 1975; Fischer and Whitt, 1978; Ferris and Whitt, 1979; Klose, 1982; MacIntyre, 1982; Laurie-Ahlberg, 1985), and attempts have been made to take the role of regulatory genes into account in developing theoretical models of population genetics (Hedrick and McDonald, 1980). There is, however, a paucity of direct evidence on the evolution of regulatory processes (see below). Consequently, attention has been directed at extrapolating our knowledge of gene regulation in living organisms to develop evolutionary models. To review this approach, an example of a major physiological change in evolution is first presented as a means of defining some of the issues that require explanation. This is followed by a summary of the experimental work on regulatory evolution that is available; a description of the regulatory polymorphism seen in contemporary species; and finally, a summary of current evidence on how regulatory information is organized in DNA. Considering this material together allows some conclusions to be drawn and raises a series of questions for the future.

II. PHYSIOLOGICAL EVOLUTION AND REGULATORY CHANGE

The transition from aquatic to terrestrial life required the evolution of new pathways for excretion of nitrogenous wastes, especially the amino nitrogen coming from protein catabolism. This physiological adaptation was almost certainly an example of regulatory evolution, and it raises a series of issues that provide a context for considering the experimental evidence on how such regulatory changes come about.

Aquatic organisms characteristically use ammonium ion as the end product of nitrogen metabolism, but this is only feasible when large volumes of water are available for excretion since it is a relatively toxic substance. Land vertebrates, whose water intake is limited, have evolved several solutions to the problem. Mammals, some reptiles and most amphibians synthesize urea, which is both non-toxic and water soluble, by greatly increasing the flow of material through the metabolic pathway for biosynthesis and degradation of arginine that preexisted in their aquatic ancestors. Birds, many reptiles and some desert amphibians, whose problems of water balance are more severe, use urate as the end product because its low water solubility means it can be conveniently excreted in solid form. For this they emphasize the metabolic pathway for purine biosynthesis and degradation that preexisted in ancestral species. In each case a set of enzymes that already served another metabolic function was brought into play by greatly increasing their concentrations in liver.

The urea cycle, for example, uses one ammonia molecule from oxidation of amino acids, one molecule of carbon dioxide, and one amino group from aspartate (which can be replenished by transamination from other amino acids), together with a supply of energy to synthesize urea in a series of five reactions, each catalyzed by a different enzyme (Figure 1). The first four steps of the cycle were originally the biosynthetic pathway for arginine, and the last step was the first reaction in arginine

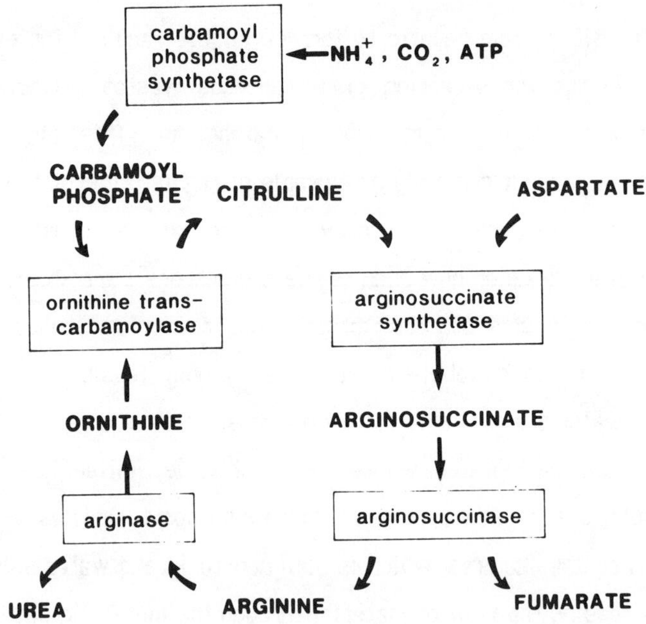

Figure 1

catabolism. The genes for these five enzymes are quite distinct, and those whose chromosomal locations are known are not genetically linked (Sparkes *et al.,* 1984). During the adaptation to terrestrial life, it is possible that a cataclysmic mutation somehow resulted in the up regulation of all of these enzymes to very high levels in liver, and for a few of them to high levels in other tissues as well. Alternatively, each individual enzyme may have been affected by separate mutations, producing incremental increases in enzyme activity. In this latter case, steady selection would drive the level of each enzyme up in small mutational jumps, with the possibility of some overshoot at each jump. Each overshoot would create a new genotype that resulted in new selective pressures on the other enzymes of the cycle.

From the standpoint of regulatory evolution, the experimental problems inherent in explaining the evolution of new pathways of nitrogen excretion are to understand the regulatory mechanisms controlling tissue levels of these and other enzymes, and to describe their phylogenetic differences. The conceptual problem is then to assemble these ideas into models of regulatory evolution that can be tested using observational data. As the following sections illustrate, many of these tasks still lie ahead. What is available at the present time, and provides much of the information that can be drawn upon, are studies of regulatory polymorphisms within species. Such polymorphisms presumably serve as the raw material of evolutionary selection, and their properties are suggestive of how regulatory evolution might occur, but the evidence they provide is necessarily indirect.

III. PHYLOGENETIC STUDIES OF REGULATORY EVOLUTION

Comparative studies of specific regulatory processes of the type that allow reconstruction of past events are still few in number. Some of the most informative of these have come from the work of Dickinson and colleagues (Dickinson and Carson, 1979; Dickinson, 1980a,b,c) on the rapidly evolving species of the Hawaiian picture-winged *Drosophila*. He has compared the levels of expression of 23 enzymes in a variety of tissues among these species. A number of cases were observed where an enzyme was expressed at high level in a particular tissue of one species and was virtually absent from the same tissue in another closely related species. This was not due to a general defect in enzyme production since the two species had equivalent levels of enzyme in other tissues. Analysis of F_1 hybrids indicated that in a few cases the regulation was dominant or recessive, suggesting differences in diffusible regulatory factors, but in many cases it showed co-dominant inheritance suggesting differences in cis-acting elements linked to the structural gene. At the molecular level Rabinow and Dickinson (1981) showed that one of these evolutionary regulatory changes involving alcohol dehydrogenase involves the control of mRNA levels. Among the Hawaiian *Drosophila*, then, dramatic changes in tissue specific enzyme regulation are common, suggesting that the rate of regulatory evolution in this group has kept pace with its rapid evolution. Dickinson and co-workers (1984) have also detected differences in cis-acting regulatory elements for alcohol de-hydrogenase among the sibling species *D. melanogaster* and *D. simulans* which appear to have adaptive significance in determining resistance to ethanol toxicity and thus the range of habitats occupied by the two species.

Another body of information on the evolution of regulatory controls comes from the study of duplicated genes. Gene duplication to produce isozymic forms of the same protein are common in vertebrates, probably reflecting a tetraploidization event early in vertebrate evolution. Studies of duplicated genes, including the mammalian globins (Bunn and Forget, 1984), creatine kinase in fishes (Fisher and Whitt, 1978) and lactate dehydrogenase in a variety of vertebrates (Markert *et al.*, 1975), have emphasized the importance of gene duplication in evolving isozymes with distinct levels of tissue and stage specific regulation as well as different catalytic properties. It has been difficult, however, to demonstrate experimentally the physiological advantages associated with performing the same metabolic function using different isozymes in separate tissues, where each isozyme is presumably adapted to its particular environment. One graphic example where this has been possible is the stomach lysozyme *c* of ruminants which has undergone both structural and regulatory evolution to allow it to function in the digestion of the large quantities of bacteria produced in the foregut of these animals. The structural changes that occurred allow this isozyme to function at the low pH of the stomach and in the presence of pepsin. The regulatory changes have vastly increased the concentration of this enzyme in the stomach. Together, they have provided an important physiological adaptation.

For large gene families, such as the histones, where many gene copies are present, the possibilities for divergent regulatory evolution are even greater. New regulatory programs can evolve among some members of the family while old ones still continue among other members. For the histone gene family, a study of extant sea urchin species suggests that just such a major regulatory change occurred in this gene family coincident with a macroevolutionary radiation 190-200 Myr ago (Raff *et al.*, 1984). Cidaroid sea urchins, along with most other echinoderms, lack maternally synthesized α-histone mRNA in oocytes. In contrast, the advanced echinoids, which separated in that radiation, synthesize maternal α-histone mRNA whose translation is delayed until after fertilization.

Whether large gene families also evolve to a significant extent by changing the number of gene copies, and hence the quantity of gene product, is an open question. It is not clear whether gene amplification akin to the process seen under selective conditions in cell culture (Schimke, 1982) also operates in natural populations. There are many structural proteins (such as actins and myosins) that are coded for by multi-gene families, and in some cases the multiple genes do function within a single cell type (for example, in the synthesis of chorion proteins, *Drosophila* salivary glue proteins, and the major urinary proteins (MUPs)) produced in mouse hepatocytes. Whether the presence of multiple genes reflects the advantages of producing a diversity of structural forms to perform some repertoire of related function, or whether it represents selection for increased production capacity is uncertain. That the former is more likely to be the case is suggested by the fact that so many examples of multiple genes for the same protein involve structural proteins, and that gene duplication of catalytic proteins, even those present in high concentrations, is rare. Certainly, amplification *per se* is not required to produce large amounts of protein, as witness the massive production of protein from single structural genes in the case of serum albumin in mammalian liver or silk fibroin in *Bombyx* larvae.

As brief as the experimental examples of regulatory evolution may be, they illustrate the possibilities of obtaining useful data from the comparative analysis of presently extant species, much as the analysis of amino acid and nucleotide sequences from the same sources has proved so powerful. Unlike sequence analyses, however, one unfortunate limitation in comparative studies of regulation is the difficulty of determining the underlying genetic and molecular bases of the regulatory differences that do evolve. Interspecific genetic crosses are only possible between closely related species, and even then the hybrid progeny are usually sterile, making conventional genetic analysis impossible. On the molecular side, many changes in DNA sequences and organization arise between species, and it will require detailed DNA transformation experiments to assign specific regulatory changes to particular DNA sequences. Nevertheless, this information is of such importance that considerable effort is justified in overcoming these difficulties.

IV. CONTEMPORARY REGULATORY POLYMORPHISM

Studies of regulatory variation among contemporary organisms have provided information about which aspects of protein function are polymorphic, how the heriditary elements determining regulatory processes are organized, and what steps in the pathways of protein synthesis and degradation are changed to produce these differences. Especially important are observations on the relative roles of regulatory sequences that are near structural genes and those that are located at a distance. Adjacent sequences are likely to act cis and show co-dominant inheritance; distant sequences act trans and almost certainly function by producing molecular signals that go from one gene to another.

Among the higher eukaryotes, most of the regulatory polymorphisms that have been described come from studies of a relatively few species. These have been primarily two mammals (mouse and human), an insect (*Drosophila*), and a higher plant (maize). Fortunately, the genetic data are consistent among these organisms, giving some confidence in generalizing from this small sample. Moreover, surveys of related species, such as that of the Hawaiian *Drosophila* by Dickinson, show regulatory differences that parallel the polymorphisms seen within a species, encouraging us to believe that contemporary regulatory polymorphisms do reflect the genetic changes attending speciation. Indeed, in at least one case (McDonald *et al.,* 1977) it has been possible to show that laboratory selection of *Drosophila* with increased tolerance to ethanol occurs by regulatory changes increasing levels of alcohol dehydrogenase and not by changes in the structure of this enzyme, a result parallel to the interspecific differences between *D. melanogaster* and *D. simulans* described above (Dickinson *et al.,* 1984). It is for these reasons, then, that the study of contemporary polymorphism has provided a framework for evolutionary modeling.

A. Unicellular and Multicellular Organisms

Although the regulatory parameters of rather diverse multicellular organisms appear quite similar, that does not extend to free-living unicellular organisms. Unicellular organisms are characterized by populations of phenotypically equivalent cells whose major regulatory challenge is responding to a fluctuating environment; in large part they do so by altering their metabolic capabilities, especially in response to changes in external nutrient supply. Multicellular organisms on the other hand must produce a diversity of cell types in a common internal environment, where it is the entire organism, and not the individual cell, that must adapt to environmental fluctuations. In effect, multicellular organisms have given up individual cell responses to environmental change in order to allow differentiation and development of specialized tissues. Their cells are largely shielded from the external world and much internal signaling is achieved by hormonal and neuronal pathways. Because the regulatory needs of unicellular and multicellular organisms are rather different, it is likely that they have evolved distinct regulatory mechanisms, and the data that are available for comparisons suggests that this may indeed be the case (Paigen, 1979a).

B. Gene Specific Regulators

Many of the regulatory polymorphisms discovered so far are gene-specific, and appear to modulate the concentration of a single protein species. When the DNA sequence determining the regulatory phenotype is a cis acting element closely linked to a structural gene, it is likely that this specificity is absolute. For regulatory loci that are located at some distance from the structural gene they control, absolute specificity can never be proven.

Regulation of macromolecular concentrations reflect the steady-state metabolism characteristic of higher organisms. Cellular constituents such as enzymes and structural proteins are constantly being degraded and replaced, and the relative concentration of any particular protein depends upon its rates of both synthesis and degradation. Following the earliest work on protein turnover (Price *et al.*, 1962; Segal and Kim, 1963; Schimke *et al.*, 1964), many radio-labeling experiments have shown that for each protein a fixed number of molecules are synthesized per day and a constant fraction of the amount present is degraded in the same time. In these circumstances $d(\text{Protein})/d(t) = K_s - K_d (\text{Protein})$, where K_s is the number of protein molecules synthesized per day and K_d is the fraction degraded each day. Because the concentration of a protein does not change under steady state conditions, $d(\text{Protein})/d(t) = 0$, and $(\text{Protein}) = K_s / K_d$. Consequently, the concentration of a given protein is proportional to its rate of synthesis and inversely proportional to its rate of degradation, so that the steady state concentration is equally sensitive to a regulatory change in the rate of synthesis or degradation.

Mutations affecting rates of degradation are known for at least three enzymes--catalase in mouse (Rechcigl and Heston, 1967; Ganschow and Schimke, 1969), alcohol dehydrogenase (ADH) in maize (Lai and Scandalios, 1980), and sn-glycerophosphate dehydrogenase (GPDH) in *Drosophila* (King and McDonald, 1983). The three are similar in that the mutations do not map near the corresponding structural genes, the inheritance pattern is recessive/dominant, and the differential phenotype is only expressed in some tissues.

Mutations affecting the synthesis of specific proteins are more varied in phenotype, falling into several apparently distinct classes (Paigen, 1979a). Many, like the mutations affecting mouse β-galactosidase (Felton *et al.*, 1974) and Drosophila GPDH (Shaffer and Bewley, 1983), are *systemic*. That is, enzyme levels are altered by an apparently constant factor in all cells at all stages of development. These regulatory mutations typically are located at or near the corresponding structural gene and are cis-acting. These properties suggest the mutations do not alter any diffusible regulatory molecule and that the DNA sequence changes probably alter either the transcription rate or the functional properties of the mRNA produced.

A second group of mutants, typified by mouse β-glucuronidase (Swank *et al.*, 1973, 1978) and mouse renin (Wilson *et al.*, 1977, Wilson and Taylor, 1982), are altered in the ability of these genes to respond to hormonal stimulation. These are *effector response* mutants, where the ability of the structural gene (or its mRNA product) to interact with regulatory signals is changed. These mutations are characteristically located at or near the responsive structural gene and are cis-acting. The β-glucuronidase polymorphism affects the time course and extent of androgen induction of the single structural gene for this enzyme in kidney proximal tubule epithelial cells (Watson *et al.*, 1981; Pfister *et al.*,

1984). In the case of renin, the polymorphism involves whether or not a tandemly duplicated copy of the renin structural gene is present (Piccini *et al.*, 1982; Mullins *et al.*, 1982; Panthier *et al.*, 1982). It is this second copy of the renin gene that has acquired the ability to respond to androgen stimulation.

Changes in tissue specific regulation of enzyme activity are an important component of regulatory evolution. In this context, the tissue specificity of enzyme induction by hormones has some puzzling features that are almost certainly of evolutionary importance. Typically, a steroid hormone induces a different set of proteins in each responsive tissue, although it is presumably the same receptor acting on the same genome in the various tissues. For example, many murine tissues contain androgen receptor protein, and a single mutation that inactivates this androgen receptor ablates all androgen responses (Ohno and Lyon, 1970). Nevertheless, each androgen sensitive tissue induces a different set of proteins in response to androgen stimulation (Paigen and Peterson, 1981). Why a given gene is responsive in one cell type and not another, when both contain functional androgen receptor protein, is not known. There are obviously additional components to the system, where evolutionary change could have novel physiological effects. Recently, De Franco *et al.* (1985) have suggested from the results of *in vitro* experiments that the source of tissue specificity in hormonal responses might lie in a synergistic interaction between hormone dependent promoter elements for transcription and tissue specific enhancer sequences.

The third class of regulatory changes are developmental in character and cause tissue and stage specific changes in protein levels. Characteristically, the activity of an enzyme is altered in some tissues and stages of development, but not in others. Because tissue specific as well as stage specific alterations must involve a prior developmental switch at some time in the developing organism, this class of regulators has been referred to as *temporal* regulatory genes (Paigen, 1979b). In contrast to the prior two classes, in which the regulatory sequences are located at or near the structural gene they control, temporal regulators may be distant from the structural gene or close to it. For a few structural genes, notably mouse β-galactosidase (Berger *et al.*, 1979) and *Drosophila* amylase (Abraham and Doane, 1978), both adjacent cis-acting and distant trans-acting regulatory loci have been identified. In these cases the final phenotype appears to result from interaction of the two types of loci, suggesting that the distant loci are generating molecular regulatory signals that act through the adjacent sites. An evolutionarily significant feature of these systems is that they show additive inheritance so that the phenotype of F_1 hybrids is intermediate between the two homozygous parents. This is true not only of the cis-acting sites, for which this is expected, but for the distant trans-acting sites, for which it is something of a surprise. Typically, mutants of trans-acting regulatory genes in unicellular organisms show recessive or dominant expression consistent with their known modes of action. The observation of intermediate phenotypes in heterozygotes of regulatory mutations in multicellular organisms suggests that many act by rather different mechanisms.

C. Pleiotropic Regulators

In contrast to these specific systems, some regulatory loci have been found that are highly pleiotropic; mutation in these genes affects the regulation of a number of structural genes. Included in these pleiotropic systems are the genes for the synthesis of various hormones and hormone receptors and a group of "inverse" regulators that are at least partially responsible for the phenomenon of chromosomal dosage compensation.

Among the most dramatic receptor mutants are the *Tfm* (testicular feminization) mutants of mouse and man that lose functional androgen receptor protein (Attardi and Ohno, 1974; Bullock and Bardin, 1974; Gehring and Tomkins, 1974). As a result they lack all androgen responsiveness, including the ability of androgen to induce specific proteins (Ohno and Lyon, 1970), and they develop externally as phenotypic females. Other mutants that have lost enzymes required for the metabolic synthesis of androgens show related phenotypes reflecting their inability to generate an internal hormonal signal (Gutai *et al.*, 1977). Polymorphisms affecting hormone and receptor function are not limited to the steriods. Mutants such as *dwarf* and *little* in mice are unable to produce a normal complement of pituitary peptide hormones and have a drastically altered body type (Green, 1981). More subtle is the *Ah* polymorphism in mice, which affects the relative affinity of the aryl hydrocarbon receptor protein for ligand and hence its ability to induce xenobiotic metabolizing enzymes (Poland and Glover, 1975). In this case the change is not all or none, but instead affects the relative sensitivity of the system to incoming signals.

Receptor mutations present possiblities for evolutionary change that are rather distinct from changes in the regulation of single structural genes. Potentially, entire new systems involving many proteins, can be brought into play, eliminated, or modulated at one step. In a sense, the differences between receptor and structural gene mutations parallel the conceptual differences between macro and micro evolution. It is to be expected, however, that because receptor changes involve multiple proteins with manifold phenotypic changes, there is likely to be strong selection against most receptor mutations.

The other class of pleiotropic regulatory elements that have been identified have rather different properties. Following on earlier studies in *Drosophila* (Rawls and Lucchesi, 1974a,b), studies of chromosome dosage effects in maize (Birchler, 1979; Birchler and Newton, 1981) have revealed the existence of a set of "inverse" regulators that act to down regulate the expression of structural genes according to the dosage of the regulatory locus.   That is, the level of expression of an affected structural gene is inversely proportional to the dosage of the regulatory locus.   These loci are pleiotropic and affect more than one enzyme or protein in a tissue, but do not affect all (Birchler and Newton, 1981). Their action appears to be important in chromosome dosage compensation. When a chromosome carries both a structural gene and an inverse regulator for that gene, the level of gene expression becomes independent of chromosome dosage.   The effects of changing structural and regulatory gene dosage act in opposite directions to cancel each other out.   Similar effects appear to occur in other plant species (Birchler, 1983).

How inverse regulators participate in evolutionary change is uncertain. Little is known about the number of such genes, whether or not they have overlapping specificities in modulating structural genes, and whether or not they interact with each other. Their mechanism of action is also unknown. These uncertainties make extrapolation difficult.

D. Levels of Regulation

Three interlocking and regulated cycles are involved in the expression of any gene coding for a protein.   The first is the activation and deactivation of chromatin, which may well involve different biochemical reactions in the two directions. Levels of active chromatin determine the rate of the second cycle involving a series of reactions of polynucleotide metabolism. Here, primary transcripts are first assembled, then processed to mature mRNA by reactions that can include capping, polyadenylation and splicing, and finally degraded at a rate specific to each mRNA by a process whose details are unknown. In turn, levels of mRNA determine the rate of the third cycle of polypeptide metabolism which included polypeptide synthesis, processing to mature enzyme, and loss via degradation or secretion. Because of these catalytic effects, the three cycles are hierarchical (Figure 2). Altering either the rate of synthesis or degradation of a protein changes its concentration. Similarly, changing either the rate of synthesis or breakdown of the two

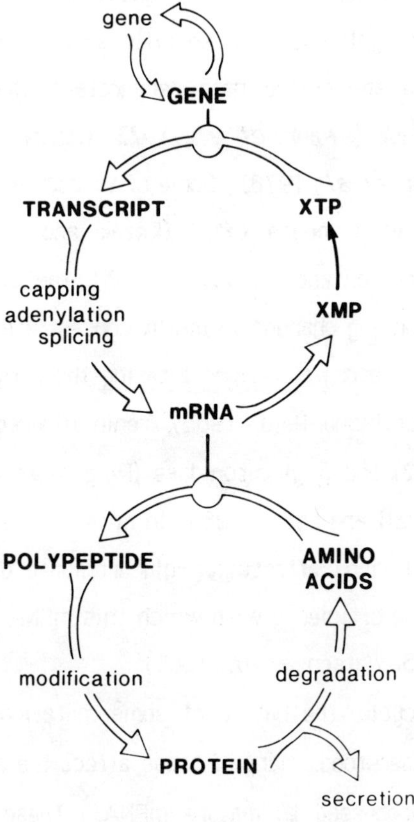

Figure 2

catalytic components, active chromatin and mRNA, changes their concentrations, which in turn alters the outcome by changing the rate of protein synthesis. Potentially, mutations can affect the final outcome by changing the concentration, the efficiency, or the specificity of any critical component.

Many of these possible changes have been detected. In the third cycle of polypeptide metabolism, several mutations increase the rate of degradation of specific proteins and thereby reduce their concentration (see above). Within the same cycle many mutations are known that affect the relative rate of synthesis of a specific enzyme or protein. These include, for example amino-levulinate dehydratase (Doyle and Schimke, 1969), β-glucuronidase (Swank *et al.*, 1973; Ganschow, 1975) and β-galactosidase (Berger *et al.*, 1978). Some of these have been analyzed at the mRNA level. With mouse GPDH (Kozak and Ratner, 1980) and Drosophila ADH (Anderson and McDonald, 1983), the changes in enzyme derive from corresponding changes in mRNA concentration. Such changes also characterize regulatory mutations altering the hormonal responsiveness of MUP (Sampsell and Held, 1985), renin (Rougeon *et al.*, 1981; Piccini *et al.*, 1982) and β-glucuronidase (Paigen *et al.*, 1979; Palmer *et al.*, 1983; Catterall and Leary, 1983) in mice. However, at least two other regulatory mutations affecting β-glucuronidase do not change the amount, but rather the efficiency with which this mRNA can be translated (Pfister *et al.*, 1985; Watson *et al.*, 1985).

In the second cycle, the nature of some mutations changing mRNA concentration are understood. Many of these affect the ability of primary transcripts to be processed to mature mRNA. These have been most extensively analyzed in the globin system (reviewed by Antonarakis *et al.*, 1985) where defects in both splicing and polyadenylation have been reported. To my knowledge, no mutations affecting mRNA stability have yet been identified, but this is probably only a matter of time. From an evolutionary standpoint, one of the more interesting polymorphisms to be analyzed in terms of mRNA is that determining androgen inducibility of renin in the submaxillary gland of mice. The results reiterate the role of gene duplication in evolution. All mouse strains carry the *Ren-1* gene which is active at a high level only in kidney where it produces the characteristic renin found there. Some mouse strains also carry a second renin gene, *Ren-2*, in the form of a gene duplication (Piccini *et al.*, 1982; Mullins *et al.*, 1982; Panthier *et al.*, 1982). The duplicated gene is only expressed at high level in submaxillary gland after androgen stimulation and produces a second renin isozyme. The result is that production of renin in kidney is monomorphic, whereas the production of renin in submaxillary tissue is polymorphic, depending on the presence of the gene duplication. At the DNA level the two genes show high homology except for a region extending upstream from a point about 150 base pairs before the coding sequence. There, 163 bases of *Ren-1* sequence are replaced in *Ren-2* by a 376 base sequence that includes an *Alu* type repetitive element; it is possible that sequences within this replacement provide the new tissue specific hormonal response (Fields *et al.*, 1984).

Within the first cycle, mutations are known that alter sequences required for the initiation of transcription. For globin (Antonarakis *et al.*, 1985), these involve both the TATA and CCAAT boxes (see below). As yet, no mutations are known that modify the activation/deactivation of chromatin. However, the properties of some β-glucuronidase regulatory mutants suggests that they may involve changes in the ability of chromatin to be activated by a steriod receptor protein (Watson *et al.*, 1981; Pfister *et al.*, 1984).

V. STRUCTURAL ASPECTS OF DNA REGULATION

As the previous sections illustrate, the regulation of protein levels can be polymorphic and these polymorphisms involve many of the possible regulatory steps in the chain of cycles that connects a gene with its protein product. A major contemporary concern is the location, identity, and function of the cis-acting DNA sequences in and around each structural gene that are essential for gene expression and regulation. The bulk of the available information has come from the identification of consensus sequences, in vitro mutagenesis combined with DNA transformation systems, and the study of enhancer-like sequences, and is derived from a wider array of species than has been analyzed for regulatory polymorphism. Collectively, these experiments define some of the system components where regulatory polymorphism could arise by mutation and selection, and as such they represent potential sites of regulatory evolution. What is still lacking at the molecular level is an extensive description of trans-acting regulatory elements that presumably determine diffusible regulatory signals.

Among the consensus sequences participating in the initiation of transcription are the TATA and CCAAT "boxes" about 30 and 80 bases, respectively, upstream of the start site for initiation of transcription, as well as sequences further upstream in the -100 region. Additionally, intron-exon junction sequences are required for splicing, and other sequences appear to determine capping, termination, and polyadenylation of the transcript. Because these are consensus sequences, they are common to many genes and so do not suggest specificity. They may, however, be important in natural polymorphisms exactly because they are "consensus" and not "rigorous" sequences. That is, because related but different sequences can satisfy a requirement, and these sequences are not equally efficient, they represent logical sites at which systemic regulatory variation could occur. These sites are also potentially important in the tissue specific or stage specific expression of genes. The genes for mouse amylase (Young *et al.*, 1981) and Drosophila alcohol dehydrogenase (Benyajati *et al.*, 1983) each have two promoter sequences. In the case of amylase the alternate promoters are used in different tissues; in the case of alcohol dehydrogenase the alternate promoters are used at larval v. adult stages of development. Depending on which promoter is used, different primary transcripts are produced. For both genes the alternative transcripts are spliced to produce functional mRNAs that synthesize identical polypeptides. What appears to be a further step in the evolution of stage specific regulation has occurred in some species of *Drosophila* where the structural gene for alcohol dehydrogenase has become tandemly duplicated with one copy utilized in larval stages and the other in adults (Batterham *et al.*, 1983).

Experiments using DNA transformation into cultured cells in developing embryos have made it clear that a great deal of regulatory information is located close to the ends of structural genes, often within a few hundred base pairs (see, for example, Palmiter *et al.*, 1983). Importantly, these regions include sequences required for tissue specificity of gene expression; for example, the sequences required for correct expression of pancreatic insulin and chymotrypsin, which are synthesized in two distinct cell types in that organ, are located 200-300 bases upstream of the transcription start site (Walter *et al.*, 1983). This clustering of regulatory sequences so close to each other and to the structural gene they control makes recombination between these sequences rare, creating circumstances favoring the occurrence of some combination of sequences in preference to others (see Conclusions).

The discovery of viral enhancer sequences (Benoist and Chambon, 1981; Gruss *et al.*, 1981; Banerji *et al.*, 1981; Moreau *et al.*, 1981) has introduced a new element into speculation on the evolution of regulatory controls. Their prototypic structural pattern is a 72 base pair repeat. Functionally, enhancers strongly increase the rate of transcription from adjacent genes. Enhancers are unique in that this ability does not depend on their exact location or orientation; they can be moved and/or inverted and still function. Subsequently, non-viral, organismal enhancer-like elements were detected in the introns of immunoglobulin genes (Neuberger, 1983; Gillies *et al.*, 1983; Banerji *et al.*, 1983; Queen and Baltimore, 1983). The specificity, small size and relaxed locational requirement of enhancers make it tempting to consider that the transposition of these sequences plays a significant role in regulatory evolution.

VI. CONCLUSIONS FROM THE DATA

Several insights are suggested by the available facts.

(a) The magnitude of a phenotypic change is often not correlated with the magnitude of the underlying genetic change. Small genetic changes can have large phenotypic consequences. This is apparent at the metabolic level when considering the urea cycle. Among the thousands of metabolic reactions that occur, it was a change in the regulation of five pre-existing enzymes that brought a new pathway of nitrogen excretion into being and removed an important limitation of terrestrial existence. This does not imply that changing the regulation of five enzymes by itself was sufficient to initiate a major step in vertebrate evolution, but it does make clear that only small genetic changes were required to allow a major physiological adaptation. The same is true at the molecular level.

A change of only a few base pairs can significantly alter the expression of a gene, and a gene can acquire or lose an entire regulatory control by a single insertion or deletion.

(b) Relatively short lengths of DNA are sufficient to accomplish ostensibly complex regulatory tasks, such as deciding the relative expression of a protein in different tissues. These lengths are measured in tens or hundreds of base pairs and are small enough to be readily carried by transposons.

(c) Transposons thus become a potentially important source of regulatory diversity in evolutionary processes by providing a means of shuttling entire regulatory sequences around the genome. Experience with the *white* locus in *Drosophila* bears this out; many of the mutations at this locus that change the level of gene expression are due to the insertion or removal of transposable elements in non-coding regions (Collins and Rubin, 1982; Zachar and Bingham, 1982). McClintock (1984) has elaborated this into a model of speciation by suggesting that when organisms are exposed to an environmental stress beyond their capacity to adapt they may enter a state of "genome shock" which initiates active transposition of regulatory elements and brings about rapid evolutionary change. Experience with maize alcohol dehydrogenase is also thought-provoking in this regard. Extensive efforts to mutagenize the gene for this enzyme in the laboratory have failed to produce regulatory changes analogous to those seen in natural polymorphisms, suggesting that the natural polymorphisms represent DNA rearrangements other than the simple sequence changes induced by conventional mutagens (Freeling and Bennett, 1986).

(d) Duplication of structural genes provides an oportunity to evolve two independent regulatory systems for the same functional protein. This can be coupled with changes in the structural sequences of the two genes to produce proteins that are adapted to function optimally in different cellular environments.

(e) We would expect regulatory genes with pleiotropic effects to be under tighter selection than specific regulators, and hence probably less polymorphic. The more phenotypic consequences a mutational change has, the more likely it is that some of them will be deleterious. Conversely, however, because the phenotypic consequences of a single mutation in a pleiotropic regulator can be manifold, such genes present the possibility of large evolutionary change from a single genetic event.

(f) The presence of regulatory elements in close proximity to a structural gene means that the two will tend to cosegregate and hence be coselected. For higher organisms, where the overall recombination frequency is of the order of 10^{-5} per kilobase of DNA, these effects can be very strong. In effect, it is the entire complex of closely linked structural and regulatory sequences that segregates in populations, and strong hitch-hiking between structural and regulatory alleles is expected. Many structural polymorphisms are probably driven by selection of linked regulatory elements.

(g) Because of the short distances and low recombination frequencies involved, even small differences in fitness among various combinations of alleles at adjacent regulatory sequences will be sufficient to produce very strong linkage disequilibrium with the occurrence of some allelic combinations at frequencies much higher or lower than expected from the product of the frequencies of the individual alleles. When recombination between linked regulatory elements is of the order of 10^{-5}, a reduction in fitness for some combinations as small as 10^{-3} to 10^{-2}, is enough to remove them from the population, leaving the remaining combinations to predominate.

VII. OPEN QUESTIONS

Our knowledge of genetic regulatory systems is obviously rudimentary compared to the level of insight required for analyzing evolutionary problems. Several issues attract attention.

1) There is not yet a substantial body of information on the frequency and extent of regulatory polymorphism in natural populations. Among mammals, most of the available information is derived from clinical studies of human disease or from surveys of inbred laboratory strains. In *Drosophila*, surveys of chromosomes extracted from wild populations have provided some estimate of the range of enzyme activity possible. These surveys have also demonstrated the existence of genetic modifiers of enzyme activity unlinked to enzyme structural genes (c.f. Laurie-Ahlberg *et al.*, 1982), but in no case do we know how many loci are involved or what they do.

2) Very little is known about the sequential changes that actually occurred in regulatory evolution. Returning to the example of the urea cycle, it is not known how many evolutionary steps were involved, whether the changes in regulation of each enzyme were the result of independent mutations, what the order of events was, or how long it took. In no case has it yet been possible to reconstruct similar processes from observations on contemporary species in the same way that protein lineages are reconstructed. It is not known whether these changes come clustered in time, and whether they are associated with rapid evolution of other biochemical or morphological features. In searching for answers it is conceivable that regulatory ontogeny might provide clues to regulatory phylogeny. Amphibian metamorphosis, for example, mimics evolution in that a switch in nitrogen catabolism from ammonium ion to urea as the

end product accompanies the transition from aquatic tadpole to terrestrial adult.

3) There is no objective notation system or terminology to describe regulatory evolution, one capable of categorizing or quantitating the kind and magnitude of the changes that occur. Morphological evolution can be described by quantitating changes in size and shape, especially using allometric equations and topological transformations. Protein and nucleic acid evolution can be described in terms of numbers and types of amino acid and nucleotide substitutions. As yet, however, there is no unit component of regulatory change, and often uncertainty about how to quantitate the phenotype. The enzyme level of an organism is not a single parameter, but depends in physiologically significant ways on the state of development, tissue examined, and response to external signals. These dimensions must be clearly distinguished. At the present time the most meaningful statements about regulatory changes in evolution are qualitative ones, and it may well be that experimental systems where qualitative statements are relevant will be especially useful in the immediate future.

4) Data is not yet available on the DNA changes underlying regulatory evolution. It is not known what sequence changes occur, and to what extent they come about by processes of transposition rather than base substitution and deletion.

5) It is not known what kinds of evolutionary changes are possible through mutation of pleiotropic regulatory genes, and whether they are indeed under tighter selection because mutation in such elements alters so many enzymes. It is also not clear to what extent functionally related sets of structural genes are under the control of common regulators, so that saltatory or macroevolution could occur as the consequence of single

mutation. Conversely, it is possible that pleiotropic regulatory networks are so complex that evolution can only occur by small incremental steps.

VIII. COMMENTARY

The study of gene regulation in evolutionary processes is relatively recent. There has not yet been time to acquire a substantial body of fact, and theory is largely derived by extrapolation from the molecular biology of transcription/translation and the genetics of intraspecific regulatory polymorphisms. As our understanding of the underlying molecular biology and genetics progresses, it should become increasingly possible to devise useful parameters for describing regulatory evolution and to carry out phylogenetic studies. It is fascinating to speculate about, but difficult to foresee, how our models of evolutionary processes will be affected by the results of those efforts.

ACKNOWLEDGMENTS

I am particularly grateful to my colleagues Allan C. Wilson and Leonard J. Rabinow for their constructive comments in the final preparation of this manuscript.

REFERENCES

Abraham, I., and Doane, W. W. (1978). Genetic regulation of tissue-specific expression of amylase structural genes in *Drosophila melanogaster*. *Proc. Natl. Acad. Sci. USA* **75**, 4446-4450.

Anderson, S. M., and McDonald, J. F. (1983). Biochemical and molecular analysis of naturally occurring Adh variants in *Drosophila melanogaster*. *Proc. Natl. Acad. Sci. USA* **80**, 4798-4802.

Antonarakis, S. E., Kazazian, H. H. Jr., and Orkin, S. H. (1985). DNA polymorphism and molecular pathology of the human globin gene clusters. *Hum. Genet.* **69**, 1-14.

Attardi, B., and Ohno, S. (1974). Cytosol androgen receptor from kidney of normal and testicular feminized (Tfm) mice. *Cell* **2**, 205-212.

Banerji, J., Rusconi, S., and Schaffner, W. (1981). Expression of a B-globin gene is enhanced by remote SV40 DNA sequences. *Cell* **27**, 299-308.

Banerji, J., Olson, L., and Schaffner, W. (1983). A lymphocyte-specific cellular enhancer is located downstream of the joining region in immunoglobulin heavy chain genes. *Cell* **33**, 729-740.

Batterham, D., Lovett, J. A., Starmer, W. T., and Sullivan, D. T. (1983). Differential regulation of duplicate alcohol dehydrogenase genes in *Drosophila mojavensis*. *Dev. Biol.* **96**, 346-354.

Benoist, C., and Chambon, P. (1981). *In vivo* sequence requirements of the SV40 early promoter region. *Nature* **290**, 304-310.

Benyajati, C., Spoerel, N., Haymerle, H., and Ashburner, M. (1983). The messenger RNA for alcohol dehydrogenase in *Drosophila melanogaster* differs in its S^1 end in different developmental stages. *Cell* **33**, 125-133.

Berger, F. G., Paigen, K., and Meisler, M. H. (1978). Regulation of the rate of β-galactosidase synthesis by the *Bgs* and *Bgt* loci in the mouse. *J. Biol. Chem.* **253**, 5280-5282.

Berger, F. G., Breen, G. A. M., and Paigen, K. (1979). Genetic determination of the developmental program for mouse liver β-galactosidase: Involvement of sites proximate to and distant from the structural gene. *Genetics* **92**, 1187-1203.

Birchler, J. (1979). A study of enzyme activities in a dosage series of the long arm of chromosome one in maize. *Genetics* **92**, 1211-1229.

Birchler, J. (1981). The genetic basis of dosage compensation of alcohol dehydrogenase-1 in maize. *Genetics* **97**, 625-637.

Birchler, J., and Newton, K. J. (1981). Modulation of protein levels in chromosomal dosage series of maize: The biochemical basis of aneuploid syndromes. *Genetics* **99**, 247-266.

Birchler, J. A. (1983). Allozymes in gene dosage studies. *In* "Isozymes in Plant Genetics and Breeding" (S. D. Tanksley and T. J. Orton, eds.). Elsevier Scientific Publ. Co., Amsterdam, The Netherlands.

Bullock, L. P., and Bardin, C. W. (1974). Androgen receptors in mouse kidney: A study of male, female and androgen insensitive (Tfm/Y) mice. *Endocrinology* **94**, 746-756.

Bunn, H. F., and Forget, B. G. (1984). "Hemoglobin: Molecular, Genetic, and Clinical Aspects." W. B. Saunders, Philadelphia, PA.

Catterall, J. F., and Leary, S. L. (1983). Detection of early changes in androgen-induced mouse renal β-glucuronidase messenger ribonucleic acid using cloned complementary deoxyribonucleic acid. *Biochemistry* **22**, 6049-6053.

Collins, M., and Rubin, G. M. (1982). Structure of *Drosophila* mutable allele, *white-crimson,* and its *white-ivory* and wild-type derivatives. *Cell* **30**, 71-79.

DeFranco, D., Wrange, O., Merryweather, J., and Yamamoto, K. R. (1985). Biological activity of a glucocorticoid regulated enhancer: DNA sequence requirements and interactions with other transcriptional enhancers. *In* "Genome Rearrangement," UCLA Symposium on Molecular and Cellular Biology, New Series (I. Herskowitz and M. Simon, eds.), pp. 305-321. Alan Liss, New York.

Dickinson, W. J. (1980a). Evolution of patterns of gene expression in Hawaiian picture-winged Drosophila. *J. Mol. Evol.* **16**, 73-94.

Dickinson, W. J. (1980b). Complex cis-acting regulatory gene demonstrated in Drosophila hybrids. *Dev. Gen.* **1**, 229-240.

Dickinson, W. J. (1980c). Tissue specificity of enzyme expression regulated by diffusible factors: Evidence from Drosophila hybrids. *Science* **207**, 995-997.

Dickinson, W. J., and Larson, H. L. (1979). Regulation of the tissue specificity of enzyme expression by a cis-acting genetic element: Evidence from *Drosophila* hybrids. *Proc. Natl. Acad. Sci. USA* **76**, 4559-4562.

Dickinson, W. J., Rowan, R. G., and Brennan, M. D. (1984). Regulatory gene evolution: Adaptive differences in expression of alcohol dehydrogenase in *Drosophila melanogaster* and *Drosophila simulans.* *Heredity* **52**, 215-225.

Dobson, D. E., Prager, E. M., and Wilson, A. C. (1984). Stomach lysozymes of ruminants: I. Distribution and catalytic properties. *J. Biol. Chem.* **259**, 11607-11616.

Dobson, D. E., Prager, E. M., and Wilson, A. C. (1984). Stomach lysozymes of ruminants: II. Amino acid sequence of cow lysozyme 2 and immunological comparisons with other lysozymes. *J. Biol. Chem.* **259**, 11617-11625.

Doyle, D., and Schimke, R. T. (1969). The genetic and developmental regulation of hepatic F-amino-levulinate dehydrotase in mice. *J. Biol. Chem.* **244**, 5449-5459.

Felton, J., Meisler, M., and Paigen, K. (1974). A locus determining β-galactosidase activity in the mouse. *J. Biol. Chem.* **249**, 3267-3272.

Ferris, S., and Whitt, G. (1979). Evolution of the differential regulation of duplicate genes after polyploidization. *J. Mol. Evol.* **12**, 267-317.

Field, L. J., Philbrick, W. M., Howles, P. N., Dickinson, D. P., McGowan, R. A., and Gross, K. W. (1984). Expression of tissue-specific *Ren-1* and

Ren-2 genes of mice: Comparative analysis of 5'-proximal flanking regions. *Molec. Cell Biol.* **4**, 2321-2331.

Fischer, S. E., and Whitt, G. S. (1978). Evaluation of isozyme loci and their differential tissue expression. *J. Mol. Evol.* **12**, 25-55.

Freeling, M., and Bennett, D. C. (1986). Maize *Adh1*. *Ann. Rev. Genet.* (in press).

Ganschow, R. E. (1975). Simultaneous genetic control of the structure and rate of synthesis of murine glucuronidase. *In* "Isozymes: Genetics and Evolution," Vol. 4 (C. L. Markert, ed.), pp. 633-647. Academic Press, New York.

Ganschow, R. E., and Schimke, R. T. (1969). Independent genetic control of the catalytic activity and the rate of degradation of catalase in mice. *J. Biol. Chem.* **264**, 4649-4658.

Gehring, U., and Tomkins, G. M. (1974). Characterization of a hormone receptor defect in the androgen-insensitivity mutant. *Cell* **3**, 59-64.

Gillies, S. C., Morrison, S. L., Oi, V. T., and Tonegawa, S. (1983). A tissue-specific transcription enhancer element is located in the major intron of a rearranged immunoglobulin heavy chain gene. *Cell* **33**, 717-728.

Green, M. C. (1981). "Genetic Variants and Strains of the Laboratory Mouse." Gustav Fischer Verlag, New York.

Gruss, P., Dhar, R., and Khoury, G. (1981). Simian virus 40 tandem repeated sequences as an element of the early promoter. *Proc. Natl. Acad. Sci. USA* **78**, 943-947.

Gutai, J. P., Kowarski, A. A., Migeon, C. J. (1977). The detection of the heterozygous carrier for congenital virilizing adrenal hyperplasia. *J. Pediatr.* **90**, 924.

Hedrick, P. W., and McDonald, J. F. (1980). Regulatory gene adaptation: An evolutionary model. *Heredity* **45**, 83-97.

King, J. J., and McDonald, J. F. (1983). Genetic localization and biochemical characterization of a trans-acting regulatory effect in *Drosophila*. *Genetics* **105**, 55-69.

King, M. C., and Wilson, A. C. (1975). Evolution at two levels in humans and chimpanzees. *Science* **188**, 107-116.

Klose, J. (1982). Genetic variability of soluble proteins studied by two-dimensional electrophoreses on different inbred mouse strains and on different mouse organs. *J. Mol. Evol.* **18**, 315-328.

Kozak, L. P., and Ratner, P. L. (1980). Genetic regulation of translatable mRNA levels for mouse sn-glycerol 3-phosphate dehydrogenase during development of the cerebellum. *J. Biol. Chem.* **255**, 7589-7594.

Lai, Y. K., and Scandalios, J. G. (1980). Genetic determination of the developmental program for maize scutellar alcohol dehydrogenase:

Involvement of a recessive, trans-acting, temporal-regulatory gene. *Dev. Genet.* **1**, 311.

Laurie-Ahlberg, C. C. (1985). Genetic variation affecting the expression of enzyme coding genes in *Drosophila* : An evolutionary perspective. *In* "Isozymes: Current Topics in Biology and Medical Research," Vol. 12, pp. 33-88.

Laurie-Ahlberg, C. C., Wilton, A. N., Curtsinger, J. W., and Emigh, T. H. (1982). Naturally occurring enzyme activity variation in *Drosophila melanogaster.* I. Sources of variation for 23 enzymes. *Genetics* **102**, 191-206.

McClintock, B. (1984). The significance of responses of the genome to challenge. *Science* **226**, 792-801.

MacDonald, J. F., Chambers, G. K., David, J., Ayala, F. J. (1977). Adaptive response due to changes in gene regulation: A study with *Drosophila. Proc. Natl. Acad. Sci. USA* **74**, 4562-4566.

MacIntyre, R. J. (1982). Regulatory genes and adaptation-past, present, and future. *Evol. Biol.* **15**, 247-285.

Markert, C. L., Shaklee, J., and Whitt, G. (1975). Evolution of a gene. *Science* **189**, 102-114.

Moreau, P., Hen, R., Everett, R., Gaub, M. P., and Chambon, P. (1981). The SV40 72 base repair (sic) repeat has a shrinking effect on gene expression both in SV40 and other chimeric recombinants. *Nucleic Acids Res.* **9**, 6047-6068.

Mullins, J. J., Burt, D. W., Windass, J. D., McTurk, P., George, H., and Brammar, W. J. (1982). Molecular cloning of two distinct renin genes from the DBA/2 mouse. *EMBO J.* **1**, 1461-1466.

Neuberger, M. S. (1983). Expression and regulation of immunoglobulin heavy chain gene transfected into lymphoid cells. *EMBO J.* **2**, 1373-1378.

Ohno, S., and Lyon, M. F. (1970). X-linked testicular feminization in the mouse as a non-inducible regulatory mutation of the Jacob-Monod type. *Clin. Genet.* **1**, 121-127.

Paigen, K. (1979a). Acid hydrolases as models of genetic control. *Ann. Rev. Genet.* **13**, 417-466.

Paigen, K. (1979b). Genetic factors in developmental regulation. *In* "Physiological Genetics" (J. Scandalios, ed.), pp.1-61. Academic Press.

Paigen, K., and Peterson, J. (1981). The developmental appearance of androgen receptor protein and androgen inducibility of the *Gus* gene complex in mouse kidney. *Develop. Genet.* **2**, 269-278.

Palmer, R., Gallagher, P. M., Boyko, W. L., and Ganschow, R. E. (1983). Genetic control of levels of murine kidney glucuronidase mRNA in response to androgen. *Proc. Natl. Acad. Sci. USA* **80**, 7596-7600.

Panthier, J. J., Holm, T., and Rougeon, F. (1982). The mouse Rn locus: S allele of the renin regulator gene results from a single structural gene duplication. *EMBO J.* **1**, 1417-1421.

Palmiter, R. D., Norstedt, G., Gelinas, R. E., Hammer, R. E., and Brinster, R. L. (1983). Metallothionein-human GH fusion genes stimulate growth of mice. *Science* **228**, 809-814.

Pfister, K., Watson, G., Chapman, V., and Paigen, K. (1984). Kinetics of β-glucuronidase induction by androgen: Genetic variation in the first order rate constant. *J. Biol. Chem.* **259**, 5816-5820.

Pfister, K., Chapman, V., Watson, G., and Paigen, K. (1985). Genetic variation for enzyme structure and systemic regulation in two new haplotypes of the β-glucuronidase gene of *Mus musculus castaneus*. *J. Biol. Chem.* **260**, 11588-11594.

Piccini, N., Knopf, J. L., and Gross, K. W. (1982). A DNA polymorphism, consistent with gene duplication, correlates with high renin levels in the mouse submaxillary gland. *Cell* **30**, 205-213.

Poland, A., and Glover, E. (1975). Genetic expression of arylhydrocarbon hydroxylase by 2,3,7,8-tetrachlorodibenzo-p-dioxin: Evidence for a receptor mutation in genetically non-responsive mice. *Mol. Pharmacol.* **11**, 389-398.

Price, V. E., Sterling, W. H., Tarantola, V. A., Hartley, R. W. Jr., Rechcigl, M. Jr. (1962). The kinetics of catalase synthesis and destruction *in vivo*. *J. Biological Chem.* **237**, 3468-3475.

Queen, C., and Baltimore, D. (1983). Immunoglobulin gene transcription is activated by downstream sequence elements. *Cell* **33**, 741-748.

Rabinow, L., and Dickinson, W. J. (1981). A cis-acting regulator of enzyme tissue specificity in *Drosophila* is expressed at the RNA level. *Mol. Gen. Genet.* **183**, 264-269.

Raff, R. A., Anstrom, J. A., Huffman, C. J., Leaf, D. S., Loo, J. H., Showman, R. M., and Wells, D. E. (1984). Origin of a gene regulatory mechanism in the evolution of echinoderms. *Nature* **310**, 312-314.

Rawls, J. M., Lucchesi, J. (1974a). Regulation of enzyme activity in *Drosophila*: I. The detection of regulatory loci by dosage responses. *Genet. Res.* **24**, 59-72.

Rawls, J. M., Lucchesi, J. (1974b). Regulation of enzyme activity in *Drosophila*: II. Characterization of enzyme responses in aneuploid flies. *Genet. Res.* **24**, 73-80.

Rechcigl, M. Jr., and Heston, W. E. (1967). Genetic regulation of enzyme activity in a mammalian system by the alteration of the rate of enzyme degradation. *Biochem. Biophys. Res. Commun.* **27**, 119-124.

Rougeon, F., Chambrand, B., Foote, S., Panthier, J., Nateotte, R., Corvol, P. (1981). Molecular cloning of a mouse submaxillary gland renin cDNA fragment. *Proc. Natl. Acad. Sci. USA* **78**, 6367-6371.

Sampsell, B. M., and Held, W. A. (1985). Variation in the major urinary protein multigene family in wild-derived mice. *Genetics* **109**, 549-568.

Schaffer, J. B., and Bewley, G. C. (1983). Genetic determination of sn-glycerol 3-phosphate dehydrogenase in *Drosophila melanogaster*: A cis-labeled controlling element. *J. Biol. Chem.* **258**, 10027-10033.

Schimke, R. T., Sweeney, E. W., and Berlin, C. M. (1974). An analysis of the kinetics of rat liver trypophan pyrolase induction: The significance of both enzyme synthesis and degradation. *Biochem. Biophys. Res. Commun.* **15**, 214-219.

Segal, H. L., Kim, Y. S. (1963). Glucocorticoid stimulation of the biosynthesis of glutamic-alanine transaminase. *Proc. Natl. Acad. Sci. USA* **50**, 912-918.

Sparkes, R. S., Berg, K., Evans, H. J., and Klinger, H. P. (1984). Human Gene Mapping 7. "Seventh International Workshop on Human Gene Mapping." *Cytogenet. Cell Genet.* **37**, 1-666.

Swank, R. T., Paigen, K., and Ganschow, R. (1973). Genetic control of glucuronidase induction in mice. *J. Mol. Biol.* **81**, 225-243.

Swank, R. T., Paigen, K., Davey, R., Chapman, V., Labarca, C., Watson, G., Ganschow, R., Brandt, E. J., and Novak, E. K. (1978). Genetic regulation of mammalian glucuronidase. *Rec. Prog. Horm. Res.* **34**, 401-436.

Wallace, B. (1963). Genetic diversity, genetic uniformity and heterosis. *Canad. J. Genet. Cytol.* **5**, 239-253.

Watson, G., Davey, R. A., Labarca, C., and Paigen, K. (1981). Genetic determination of kinetic parameters in β-glucuronidase induction by androgen. *J. Biol. Chem.* **256**, 3005-3011.

Watson, G., Felder, M., Rabinow, L., Moore, K., Labarca, C., Tietze, C., VanderMolen, G., Bracey, L., Brabant, M., Cai, J., and Paigen, K. (1985). Properties of rat and mouse β-glucuronidase mRNA and cDNA, including evidence for sequence polymorphism and genetic regulation of mRNA levels. *Gene* (in press).

Wilson, A. C. (1975). Evolutionary importance of gene regulation. *Stadler Symp.* **7**, 117-133.

Wilson, A. C., Carlson, S. S., and White, T. J. (1977). Biochemical evolution. *Ann. Rev. Biochem.* **46**, 573-639.

Wilson, A. C., Maxson, L. R., and Sarich, V. M. (1974). Two types of molecular evolution, evidence from studies of interspecific hybridization. *Proc. Natl. Acad. Sci. USA* **71**, 2843-2847.

Wilson, C. M., Erdos, E. G., Dunn, J. F., and Wilson, J. D. (1977). Genetic control of renin activity in the submaxillary gland of the mouse. *Proc. Natl. Acad. Sci. USA* **74**, 1185-1189.

Wilson, C. M., and Taylor, B. A. (1982). Genetic regulation of thermostability of mouse submaxillary gland renin. *J. Biol. Chem.* **257**, 217-223.

Young, R. A., Hagenbuchle, O., and Schibler, U. (1981). A single mouse a-amylase gene specifies two different tissue-specific mRNAs. *Cell* **23**, 451-458.

Zachar, Z., and Bingham, P. M. (1982). Regulation of *white* locus expression: The structure of mutant alleles at the *white* locus of *Drosophila melanogaster. Cell* **30**, 529-541.

Zuckerkandl, E. (1963). Perspectives in molecular anthropology. *In* "Classification and Human Evolution" (S. L. Washburn, ed.), pp. 243-272. Aldine Publ. Co., Chicago.

THE EVOLUTION OF TRANSCRIPTIONAL CONTROL SIGNALS: COEVOLUTION OF RIBOSOMAL GENE PROMOTER SEQUENCES AND TRANSCRIPTION FACTORS

Norman Arnheim[1]

Biochemistry Department and Molecular Biology Program
State University of New York
Stony Brook, NY 11794

ABSTRACT

Three RNA polymerases are involved in the transcription of eucaryotic genes. One of them, RNA polymerase I, has a unique relationship with its templates in that the latter are members of a single multigene family which codes for the 18S and 28S ribosomal RNAs. We have carried out experiments on the control of ribosomal RNA transcription in two mammalian species. Our data showed that mice and humans have unique transcription factors involved in the expression of their ribosomal genes and that these factors, unlike those involved in mRNA and tRNA synthesis, are species-specific. The development of species-specificity appears to involve the coevolution of the ribosomal gene promoter sequences with the transcription factors that interact at the promoter.

In higher organisms, three different RNA polymerases function in transcription. RNA polymerase I is responsible for transcribing the ribosomal gene family (rDNA) which codes for the 18S and 28S ribosomal

---

[1]Present address: Biology Department, University of Southern California, Los Angeles, CA 90089-0371.

RNAs (rRNAs). rRNA polymerase II transcribes all of the messenger RNA coding genes while RNA polymerase III's major role is to produce transcripts of the 5S RNA and tRNA genes. Both *in vitro* and *in vivo* studies have revealed a good deal about the nucleotide sequence signals responsible for the selective transcription by these polymerases (Shenk, 1981). These signals can be divided into at least three categories: those required for defining the precise origin of transcription, those that modulate the absolute levels of transcription, and signals involved in cell-specific regulation.   In the following discussion, we will focus on the controlling elements (promoters) involved in the first two processes.   In the case of RNA polymerase II, elements important to transcription initiation are found 5' of the actual transcription start site. The TATA and CCAAT boxes are examples of such consensus sequences which are required for this function and have been found in approximately the same position relative to the transcription start site in virtually all eucaryotic genes. Consensus sequences responsible for the transcription initiation of 5S and tRNA genes by RNA polymerase III have also been identified in all eucaryotic species 3' of the transcription start site. Direct experimental evidence supporting the idea that these conserved RNA polymerase II and III transcriptional control signals serve a common function in all eucaryotes also exists. Thus cloned protein coding genes as well as tRNA and 5s RNA sequences can be faithfully transcribed in a wide variety of heterologous cell-free systems. For example, the silk fibroin gene from *Bombyx mori* is faithfully transcribed in a human *in vitro* transcription system (Tsujimoto *et al.*, 1980) and sea urchin histone gene fragments can also be correctly expressed when injected into *Xenopus* oocytes (Grosschedl and Birnstiel, 1980). Finally, interspecific cell hydrids provide another kind of evidence for the evolutionary conservation of required RNA polymerase II and III transcriptional control signals.

Mouse-human hybrid cells are known to be capable of expressing protein coding sequences of both parental species, and this fact has been used to map human protein coding genes to specific chromosomes (Ruddle and Creagan, 1975).

On the contrary, the transcription initiation signals for ribosomal RNA synthesis directed by RNA polymerase I do not appear to have been conserved during eucaryotic evolution. Comparisons of the nucleotide sequences around the origins of transcription of human (Miesfeld and Arnheim, 1982), mouse (Miller and Sollner-Webb, 1981), rabbit (Seperack and Arnheim, unpublished data), *Xenopus* (Sollner-Webb and Reeder, 1979), *Drosophila* (Long et al., 1981), and yeast rDNA (Swanson and Holland, 1983) show little, if any sequence homology. More significant, however, are the observations on the lack of functional equality of these ribosomal RNA genes in *in vitro* expression assays and cell hybrids. For example, in contrast to the experiments with genes transcribed by RNA polymerases II and III, cloned rRNA sequences from human, rabbit, or *Drosophila* cannot be transcribed in a mouse cell-free extract nor can mouse or rabbit sequences be transcribed in a human cell-free extract (Grummt *et al.*, 1982; Miesfeld and Arnheim, 1984; Seperack and Arnheim, unpublished data). In addition, hybrid cell lines containing all of the mouse chromosomes and only a few human ones do not transcribe the available human rDNA sequences nor do hybrid cells with all of the human chromosomes but only a few mouse chromosomes produce any mouse ribosomal RNA when the rodent rDNA sequences are present (Eliceiri and Green, 1969; Marshall *et al.*, 1975; Miller, D A., *et al.*, 1976; Perry *et al.*, 1979; Miesfeld *et al.*, 1984; Croce *et al.*, 1977; Miller, O. J. *et al.*, 1976). Other examples of this phenomenon, known as "nucleolar dominance," have been identified in a variety of other plant and animal groups (see review by Rieger *et al.*, 1979). In summary, species-independent RNA polymerase II and III transcription reflects an evolutionary conservation of fundamental transcription initiation signals whereas it appears as if the analogous rDNA control signals have evolved under less severe constraints resulting in species-specificity.

We have carried out experiments designed to examine the molecular basis of species-specificity in rDNA transcription. Several years ago our lab (Miesfeld and Arnheim, 1982) and others (Learned and Tijian, 1982; Grummt, 1982) developed mammalian cell-free transcription systems dependent on RNA polymerase I along the lines of those devised for studying RNA polymerase II (Manley *et al.*, 1980; Weil *et al.*, 1979). Using this system we were able to identify the sequences surrounding the origin of transcription for human ribosomal DNA. We and others also used the system to examine cloned ribosomal genes from other species and found, for example, that mouse rDNA was not transcribed in a human cell-free extract nor could human rDNA be transcribed in a mouse cell-free system. We then set out to examine the molecular basis for this example of species-specificity. The results (Miesfeld and Arnheim, 1984) are reviewed below. We first fractionated a human cell extract by phosphocellulose chromatography into four fractions; A, B, C, and D. By recombining only three of these fractions (A, C, and D), we were able to regain the ability to specifically transcribe cloned human ribosomal genes. This reconstituted extract also exhibited species-specificity in that it was incapable of transcribing a mouse ribosomal gene clone which was itself active in a mouse cell-free extract. We reasoned that there might be a factor in the human fractions A, C, or D which specifically recognized nucleotide sequences present in the human ribosomal gene but, because of the sequence differences that exist between the human and mouse genes, was incapable of stimulating transcription of the latter clone. Our next step was to study fractions A, C, and D individually. We asked whether there might exist in one of these fractions a factor or factors that formed a stable complex with the human rDNA template itself.   In these experiments we used two slightly diffrent human rDNA clones whose RNA transcripts could be distinguished from each other due to a size difference. When the two clones were added to a mixture of fractions A, C, and D, both RNA transcripts were made in equal amounts. In the next experiment one of the clones was added to a mixture of A, C, and D under conditions where transcription initiation could not take place. The second clone was subsequently added and then transcription was started. When the RNA products were analyzed, we observed that there was a strong preferential transcription of the clone which was added first.   This suggested that some factor or factors in either fraction A, C, or D formed a stable complex with the first added human DNA template which allowed for its preferential transcription. We then carried out a series of similar studies to determine which one of the three fractions contained such a factor(s). Our results (Miesfeld and Arnheim, 1984) suggested that material in fraction D was primarily responsible for the formation of this complex.

Studies by another group (Mishima *et al.*, 1982) using fractions prepared from mouse and human extracts by a protocol similar to ours had shown that fractions A, B, and C from human, when combined with a mouse fraction equivalent to human D, was capable of transcribing a mouse but not a human rDNA clone. Conversely, the analogous mouse fractions a, b, and c would transcribe a human rDNA clone if supplemented with human fraction D. This localization of a species-specific factor or factors to the fraction that we found was involved in complex formation with the ribosomal gene led naturally to the following hypothesis: species-specificity results from the presence of a factor (or factors) in human cells which is required for transcription and which is capable of binding to human rDNA but not mouse rDNA. Mouse cells would be expected to have a factor(s) with the analogous species-specific binding properties. If this hypothesis were true, we might expect that the regions of the ribosomal DNA involved in the formation of the complex with the factor(s) would be those already known to be directly involved in the initiation of RNA synthesis. We therefore examined which human rDNA sequences, in fact, were involved in the binding of the human species-specific material in fraction D. In this experiment (Miesfeld and Arnheim, 1984) various fragments of the human rDNA clone were purified after restriction enzyme digestion. Included in these fragments was one that contained sequences far upstream of the origin of transcription, one encompassing DNA between 170 base pairs upstream of the origin and 80 base pairs downstream, and finally one containing sequences greater than 80 base pairs downstream of the start site. Each fragment was tested individually for its ability to form a complex with human fraction D. Our results showed that only the fragment containing sequences between 170 base pairs upstream and 80 base pairs downstream of the origin of transcription could form a stable complex with fraction D (Miesfeld and Arnheim, 1984). Unfortunately, the material is not of sufficient purity to allow a more precise definition of the binding site by DNA footprinting.

Our binding data provides significant information when viewed in light of other studies on human rDNA expression. Using deleted human ribosomal gene clones in *in vitro* assays, it has been established that sequences involved in transcription initiation of human rRNA synthesis lie between 250 base pairs upstream and 10 base pairs downstream of the start site (Learned *et al.*, 1983). This overlaps the region we defined as being involved in the formation of the complex. We carried out analogous studies with a mouse rDNA clone and a fractionated mouse cell-free extract with similar results (Miesfeld and Arnheim, 1984). A mouse phosphocellulose fraction (d), with the same chromatographic characteristics of the human fraction D, formed a stable complex with a mouse rDNA clone.   The specific region of the mouse clone which was responsible for the formation of the complex was also right around the origin of transcription in sequences defined by others using deletion mutants as being required for transcription (Yamamoto *et al.,* 1984).

We then tested the hypothesis that complex formation of the factors in fraction D (human) and d (mouse) with the rDNA elements containing sequences required for transcription was species-specific. Human fraction D was assayed for its ability to form a complex with the mouse rDNA fragment containing promoter sequences that we had shown to interact with mouse fraction d. No complex could be observed nor in a test with mouse fraction d were we able to observe the formation of a complex with the human rDNA fragment containing sequences required for transcription (Miesfeld and Arnheim, 1984).

In another line of experiments (Miesfeld *et al.,* 1984), we extended these observations to the "nucleolar dominance" observed in mouse-human hybrid cells. We were able to show in this study that the inability of cells with all of the mouse chromosomes but only a few human chromosomes to produce human rRNA (even though the human rDNA sequences were present) is due to the absence of material found in human fraction D. Thus cell-free extracts from these hybrid cells can transcribe a mouse rDNA clone but are incapable of transcribing a human rDNA clone unless the extract is supplemented with purified human fraction D. In summary, both mouse and human cells contain a specific factor or factors which are required for transcription of rDNA but can interact only with transcriptionally significant rDNA sequence of the homologous species.

The data on the control of RNA polymerase I transcription in different species reveals several interesting features. First of all, the nucleotide sequences in and around the origin of transcription have evolved considerably with respect to the signals responsible for transcription initiation whereas regions of the gene which code for the 18S and 28S ribosomal RNAs are highly conserved. Second, the experimental data showing the lack of functional equivalence of rDNA's from different animal groups in cell-free extracts further suggests that the sequence differences that have accumulated during evolution have transcriptional consequences. Finally, different species have transcription factors that interact only with the ribosomal genes of that species and not those of more distantly related groups. Thus, different animal groups have diverged from one another both with respect to the nucleotide sequences of their ribosomal genes and the factors involved in controlling transcription initiation from these templates such that they are no longer functionally compatible. However, this process of evolution must have involved changes in the templates and factors such that within any one evolutionary lineage they remained functionally compatible; they must have coevolved.

Why has coevolution leading to species-specific transcription been observed for the RNA polymerase I transcription system but not in the case of RNA polymerase II or III? The answer to this question may lie in two fundamental differences between the former and two latter transcription machineries. In the first place, the transcription templates for RNA polymerase I are the members of a single multigene family and members of this family undergo concerted evolution. (For reviews, see Arnheim, 1983; Dover, 1982.) That is, as a consequence of genetic exchanges (gene conversion or unequal crossing over), the members of the family are kept from drifting apart in the face of constant nucleotide substitutions due to spontaneous mutation. This does not lead necessarily to an unchanging sequence during evolution but allows for the gradual alteration (through genetic drift or natural selection) in the sequence of each family member within the unit as a whole. Thus in different evolutionary lineages the nucleotide sequences of those regions responsible for transcriptional control can become distinct although within a species the different members of the family will, on the average, have the same nucleotide sequence. For example, mutations that advantageously affect rRNA synthesis can be propagated throughout the whole rDNA family even if the mutation initially occurred only in one of its members. It is extremely difficult to imagine this occurring, for example, in the case of genes transcribed by the RNA polymerase II system. Nucleotide substitutions in the controlling region of one protein coding gene that provide an advantage towards initiating the synthesis of its product cannot be propagated throughout all of the protein coding genes of the cell by genetic exchanges due to the lack of sequence homology among them. For all protein coding genes to acquire such a mutation, the same substitution would have to occur independently in the control regions of each one. This fact would place severe constraints upon the capability of control signals involved in the fundamental aspects of transcriptional initiation to undergo evolutionary change.

The second significant difference between ribosomal RNA expression and the synthesis of other cellular RNAs concerns the requirements of the latter transcription factors to recognize diverse templates. Because the templates for ribosomal RNA synthesis are members of a single multigene family, the transcription factors in the cell that are specific for rRNA production have a relatively homogeneous set of DNA sequences with which to interact. On the contrary, transcription factors which are involved exclusively in the initiation of messenger RNA production have a variety of transcription templates with which they are required to be functionally compatible.

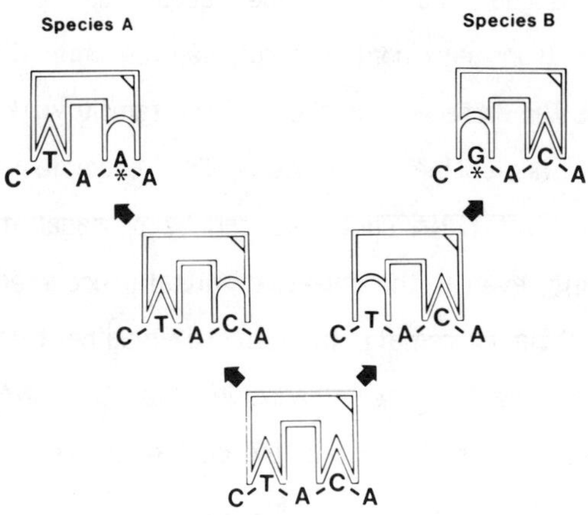

Figure 1. Hypothetical Scheme for the Coevolution of Ribosomal Gene Transcription Factors and Templates. The common ancestor of species A and B has the sequence CTACA in the promoter region. Two bases (T and C) interact with a transcription factor which has two pockets (Λ) capable of binding pyrimidines but not purines. In the lineage leading to species A, one binding pocket is altered by mutation in the gene coding for the transcription factor so that it can accommodate either a purine or pyrimidine (∩) and this is followed by fixation of a C to A nucleotide substitution (*) in the rDNA promoter region. In the lineage leading to species B, the other binding pocket is altered by random mutation in the same way and a T to G substitution is eventually fixed in the promoter. When compared to the common ancestor, the structures of the promoter regions and transcription factors have been altered in both species. Within a lineage, the fixation of mutations in factor and template might be due to natural selection or to the fixation of neutral mutations by drift along lines suggested by Fitch and Markowitz (1970). Although the control signals and factors have evolved within a lineage so as to remain functionally compatible, these transcription components are no longer able to function directly with heterologous factors or templates. Thus the factor in species A will not fit the promoter sequence of species B due to the presence of the guanine residue nor will species B's factor accommodate the presence of an adenine in the promoter of species A.

These two different aspects of rRNA and messenger RNA synthesis go a long way towards explaining their different potentials for the molecular coevolution of their transcription factors and templates. Consider, for example, the RNA polymerase I transcription system. If a nucleotide substitution occurs in a gene coding for a required transcription factor and the change in the amino acid sequence of the factor is compatible with the rDNA controlling elements, the mutation has a chance of being fixed. Its interaction with any one family member will, in all likelihood, be the same as with all the other members due to the relative homogeneity in the nucleotide sequence of the rDNA family (see Ohta and Dover, 1983). Likewise, a mutation that occurs in the controlling region of one rDNA gene that enhances its interaction with the transcription factor has a chance of being fixed in all rDNA family members due to the genetic exchanges that occur among them (concerted evolution). In this way both factors and templates can coevolve with one another (Miesfeld and Arnheim, 1984; Dover and Flavell, 1984). Since the direction of coevolutionary change can be different in independent evolutionary lineages, functional incompatibility between the transcription factors of one species and the templates of another can arise (Figure 1) and lead to species-specificity (Arnheim, 1983; Miesfeld and Arnheim, 1984; Dover and Flavell, 1984) although this is not obligatory (see Wilkinson and Sollner-Webb, 1982).

On the contrary, the possibility of coevolutionary change in the factors and templates involved in messenger RNA synthesis by RNA polymerase II would be expected to be severely constrained. Mutations that occurred in the transcriptional control region of one protein coding gene (even if advantageous for the transcription of that gene) could not be fixed in all the species' protein coding genes due to the lack of genetic interactions among them. Similarly, a mutation in a transcription factor which enhanced its interaction with a newly mutant protein coding gene promoter would, in all likelihood, be detrimental to the interaction of the factor with most of the other genes in the genome which lack this mutation and, thus, would have little chance of being fixed in the population. These facts would suggest that coevolution leading to species-specificity in fundamental aspects of RNA polymerase II or III transciptional control would be rare. On the other hand, the phenomenon of coevolution between transcription factors and the promoter sequences of a multigene family coding for protein should not be considered to be excluded all together. If a protein coding multigene family, for example, depends for its cell specific expression pattern on a transcription factor whose only function is to regulate the tissue-specificity of that family of proteins then the principal of coevolution exemplified by the ribosomal gene system might apply here as well, as long as the gene family evolved in a concerted fashion. Coevolution in such cases could result in species-specificity with respect to tissue-specific gene regulation and might contribute to the characteristic differences in biochemistry, physiology, and morphology that have accumulated among species during evolution.

ACKNOWLEDGMENTS

I would like to thank Kathy Levenson at Cetus Corporation for typing the first draft of this manuscript and the Cetus Graphic Services Department for preparation of the figure.

REFERENCES

Arnheim, N. (1983). Concerted evolution of multigene families. *In* "Evolution of Genes and Proteins (R. Koehn and M. Nei, eds.), pp. 38-61. Sinauer, Sunderland, MA.

Croce, C. M., Talavera, A., Basilico, C., and Miller, O. J. (1977). Suppression of production of mouse 28S ribosomal RNA in mouse-human hybrids segregating mouse chromosomes. *Proc. Natl. Acad. Sci. USA* **74**, 694-697.

Dover, G. A. (1982). Molecular drive. *Nature* **299**, 111-117.

Dover, G. A., and Flavell, R. B. (1984). Molecular coevolution: DNA divergence and the maintenance of function. *Cell* **38**, 622-623.

Eliceiri, G. L., and Green, H. (1969). Ribosomal RNA synthesis in human-mouse hybrid cells. *J. Mol. Biol.* **41**, 253-260.

Fitch, W. M., and Markowitz, E. (1970). Improved method for determining codon variability in a gene and its application to the rate of fixation of mutations in evolution. *Biochemical Genetics* **4**, 579-593.

Grosschedl, R., and Birnstiel, M. L. (1980). Identification of regulatory sequences in prelude sequences of an H2A histone gene by the study of specific deletion mutants *in vivo. Proc. Natl. Acad. Sci. USA* **78**, 727-731.

Grummt, I. (1982). Nucleotide sequence requirements for specific initiation of transcription by RNA polymerase I. *Proc. Natl. Acad. Sci. USA* **79**, 6908-6911.

Grummt, I., Roth, E., and Paule, M. R. (1982). Ribosomal RNA transcription *in vitro* is species-specific. *Nature (London)* **296**, 173-174.

Learned, R. M., Smale, S., Haltiner, M. M., and Tjian, R. (1983). Regulation of human ribosomal RNA transcription. *Proc. Natl. Acad. Sci. USA* **80**, 3558-3562.

Learned, R. M., and Tjian, R. (1982). *In vitro* transcription of human ribosomal RNA genes by RNA polymerase I. *J. Mol. Appl. Genet.* **1**, 575-584.

Long, E., Rebbert, M. L., and Dawid, J. B. (1981). Nucleotide sequence of the initiation site for ribosomal RNA transcription in *Drosophila melanogaster*: Comparison of genes with and without insertions. *Proc. Natl. Acad. Sci. USA* **78**, 1513-1517.

Manley, J., Fire, A., Cano, A., Sharp, P., and Gefter, M. (1980). DNA-dependent transcription of adenovirus genes in a soluble, whole-cell extract. *Proc. Natl. Acad. Sci. USA* **77**, 3855-3859.

Marshall, C. J., Handmaker, S. D., and Bramwell, M. E. (1975). Synthesis of ribosomal RNA in synkaryons and heterokaryons formed between human and rodent cells. *J. Cell Sci.* **17**, 307-325.

Miesfeld, R., and Arnheim, N. (1982). Identification of the *in vivo* and *in vitro* origin of transcription in human rDNA. *Nucleic Acids Res.* **10**, 3933-3949.

Miesfeld, R., and Arnheim, N. (1984). Species-specific rDNA transcription is due to promoter-specific binding factors. *Mol. Cell. Biol.* **4**, 221-227.

Miesfeld, R., Sollner-Webb, B., Croce, C., and Arnheim, N. (1984). The absence of a human-specific ribosomal DNA transcription factor leads to nucleolar dominance in mouse human-hybrid cells. *Mol. Cel. Biol.* **4**, 1306-1312.

Miller, D. A., Dev, V. G., Tantravahi, R., and Miller, O. J. (1976). Suppression of human nucleolus organizer activity in mouse-human somatic hybrid cells. *Exp. Cell Res.* **101**, 235-243.

Miller, K. G., and Sollner-Webb, B., (1981). Transcription of mouse rRNA genes by RNA polymerase I: *In vitro* and *in vivo* initiating and processesing sites. *Cell* **27**, 165-174.

Miller, O. J., Miller, D. A., Dev, V. G., Tantravahi, R., and Croce, C. M. (1976). Expression of human and suppression of mouse nucleolus organizer activity in mouse-human somatic cell hybrids. *Proc. Natl. Acad. Sci. USA* **73**, 4531-4535.

Mishima, Y., Financsek, L., Kominami, R., and Muramatsu, M. (1982). Fractionation and reconstitution of factors required for accurate transcription of mammalian ribosomal RNA genes; identification of a species-dependent factor. *Nucleic Acids Res.* **10**, 6659-6670.

Ohta, T., and Dover, G. A. (1983). Popualtion genetics of multigene families that are dispersed on two or more chromosomes. *Proc. Natl. Acad. Sci. USA* **80**, 4079-4083.

Perry, R. P., Kelley, D. E., Schibler, U., Huebner, K., and Croce, C. M. (1979). Selective suppression of the transcription of ribosomal genes in mouse-human hybrid cells. *J. Cell. Physiol.* **98**, 553-560.

Rieger, R., Nicoloff, H., and Anastassova-Kristeva, M. (1979). "Nucleolar dominance" in interspecific hybrids and translocation lines - a review. *Biol. ABL* **98**, 385-398.

Ruddle, F. H., and Creagan, R. P. (1975). Parasexual approaches to the genetics of man. *Ann. Rev. Genet.* **9**, 407-486.

Shenk, T. (1981). Transcriptional control regions: Nucleotide sequence requirements for initiation by RNA polymerase II and III. *In* "Initiation Signals in Viral Gene Expression" (A. Shatkin, eds.), pp. 25-45. Springer Verlag, New York.

Sollner-Webb, B., and Reeder, R. H. (1979). The nucleotide sequence of the initiation and termination sites for ribosomal RNA transcription in *X. laevis. Cell* **18**, 485-499.

Swanson, M. E., and Holland, M. J. (1983). RNA polymerase I-dependent selective transcription of yeast ribosomal DNA. *J. Biol. Chem.* **258**, 3242-3250.

Tsujimoto, Y., Hirose, S., Tsuda, M., and Suzuki, Y. (1981). Promoter sequence of fibroin gene assigned by *in vitro* transcription system. *Proc. Natl. Acad. Sci. USA* **78**, 4838-4842.

Wilkinson, J., and Sollner-Webb, B. (1982). Transcription of *Xenopus* ribosomal RNA genes by RNA polymerase I *in vitro*. *J. Biol. Chem.* **257**, 14375-14383.

Weil, P. A., Luse, D. S., Segal, J., and Roeder, R. (1979). Selective and accurate initiation of transcription at the Ad2 major late promoter in a soluble system dependent on purified RNA polymerase II and DNA. *Cell* **18**, 469-494.

Yamamoto, O., Jakakusa, N., Misima, Y., Kominami, R., and Muramatsu, M. (1984). Determination of the promoter region of mouse ribosomal RNA gene by an *in vitro* transcription system. *Proc. Natl. Acad. Sci. USA* **81**, 299-303.

RAPID EVOLUTION OF HUMAN INFLUENZA VIRUSES

Peter Palese

Department of Microbiology
Mount Sinai School of Medicine
New York, New York 10029

ABSTRACT

The advent of genetic engineering methods has made it possible to study the evolution of animal viruses on a molecular level. The most extensively studied of these sytems are the influenza viruses, which represent an interesting model of rapid evolution in nature. Among the three influenza virus types (A, B and C), influenza A viruses change most dramatically with time. Influenza A virus isolates obtained since 1933 exhibit an accumulation of 0.22 – 0.34% nucleotides changes per year in their NS genes (and comparable changes in the other viral genes). This number translates to an evolutionary rate for influenza A virus genes of $2.2 - 3.4 \times 10^{-3}$ substitutions/site/year, and it is thus 10^6-fold higher than the values observed for eukaryotic genes such as globin. The phenomenon of rapid sequential change in influenza A viruses is the single most important factor in explaining the occurrence of influenza A epidemics in humans. In contrast, other viruses, including influenza C viruses, do not evolve in such a rapid fashion and show different evolutionary patterns.

INTRODUCTION

Influenza viruses are classified into three types: A, B and C. Strains of the same type possess internal proteins which are immunologically

53

related. In contrast, the proteins of viruses belonging to different types do not readily cross-react in serologic tests. There are, however, many structural and biochemical features shared by type A, B and C viruses (for review, see Palese and Young, 1982). For example, all influenza viruses possess a segmented genome consisting of single stranded RNAs of negative polarity. Furthermore, influenza viruses are characterized by possessing glycoprotein spikes which are embedded into a host-derived lipid membrane. Type A, B and C viruses have hemagglutinin (HA) spikes, which are required for the attachment of the viruses to cells. These HA spikes also carry the antigenic determinants responsible for inducing neutralizing antibodies in infected cells. (A schematic diagram of the proteins associated with mature influenza virus particles is shown in Fig. 1.) Another interesting hallmark of influenza viruses is the ability of these viruses to readily exchange RNA segments. Two different influenza viruses belonging to the same type may undergo reassortment of genes

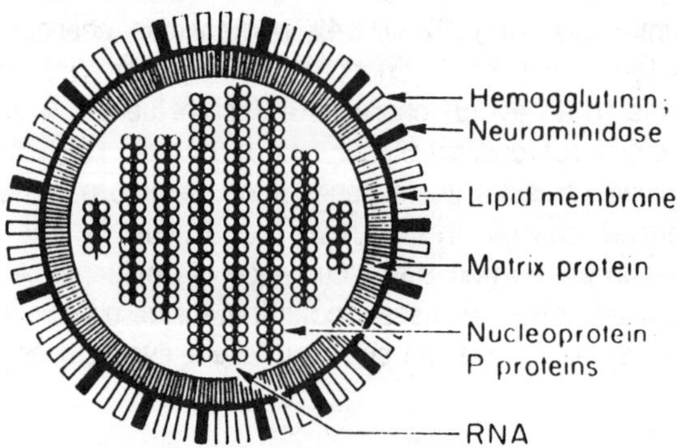

Figure 1. Schematic diagram of influenza viruses. The spherical particles are characterized by spikes consisting of hemagglutinin (open bars) and neuraminidase (closed bars), a lipid membrane, the underlying M (matrix) protein and the internal ribonucleoprotein complex. Influenza C viruses are most likely similar in structure to influenza A and B viruses, but they lack a gene coding for neuraminidase.

following coinfection of a single cell. If reassortment takes place between two viruses containing 8 RNA segments each, 254 different reassortant viruses may be formed (i.e., 2^8 viruses minus the two parents).

Much has been learned about the molecular biology of influenza viruses. Genes of influenza A, B and C virus have been cloned into plasmids and sequences of many influenza virus genes have been obtained (for review, see Lamb, 1983). Furthermore, the mechanisms of RNA transcription and replication have been extensively explored (for review see Krug, 1983) and X-ray crystallographic studies have provided detailed information on the structure of the influenza viral proteins, hemagglutinin and neuraminidase (Wilson *et al.*, 1981; Varghese *et al.*, 1983). This vast amount of information has made influenza viruses one of the best known viral agents.

Table I

	A	B	C
Severity of disease	+++	++[*]	+
Frequency of isolation	+++	++	+
Human reservoir	yes	yes	yes
Animal reservoir	yes	no	no
Appearance of subtypes	yes	no	no
RNA segments	8	8	7
Variation	+++	++	+

It should be noted that one of the serious complications of influenza B (and to a lesser extent of influenza A) virus infections is Reye's syndrome. Although fewer than 3,000 cases of Reye's syndrome occur annually in the U.S., this disease is associated with a high degree (20-50%) of fatality.

II. EPIDEMIOLOGY OF INFLUENZA VIRUSES

In terms of a disease-causing entity, influenza virus continues to attract a lot of attention (for review, see Palese and Young, 1983). Although influenza virus vaccines have been available for the last 40 years, the disease has not been eradicated nor has it been possible to significantly curb major epidemic outbreaks over the last several decades. The single major factor contributing to this unchanging pattern appears to be the phenomenon of genetic variation associated with influenza viruses. Figure 2 reveals some of the features which have been associated with the epidemiology of influenza virus infections. In this century at least three different subtype strains of influenza A viruses were introduced into the human population, resulting in widespread disease. The "swine" influenza outbreak of 1918 caused by an H1 (hemagglutinin subtype 1) virus is

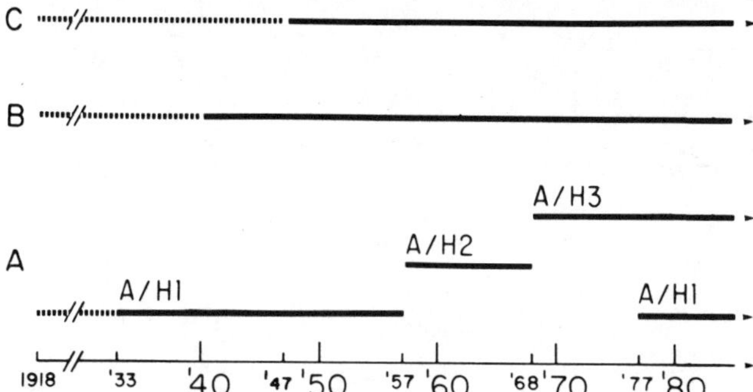

Figure 2. Circulation of human influenza viruses. Three different A virus subtypes, A/H1, A/H2 and A/H3, defined by the hemagglutinins of the viruses, have been observed in this century. In addition to changes associated with subtype "shifts", A viruses belonging to the same subtype also undergo antigenic drift. For example, H3 viruses isolated in 1985 vary considerably from H3 isolates observed in 1980. Subtypes of influenza B and C viruses have not been found, but it should be noted that anitgenic drift also occurs among these viruses. The earliest dates for the isolation of influenza A, B and C viruses were 1933, 1940 and 1947, respectively. Broken lines indicate possible circulation before these dates.

estimated to have caused two million deaths in the U.S. and 20 million worldwide. This subtype virus remained in the population until 1956. In 1956 the Asian, or H2 (hemagglutinin subtype 2) influenza virus strain travelled from China to the rest of the world and replaced the earlier H1 strains for a period of eleven years. The emergence of the new H2 subtype strains were clearly demonstrated since the HAs of the new strains did not cross-react serologically with those of the earlier H1 strains (antigenic shift). The Asian strains then disappeared in 1968 following the introduction of an H3 virus (Hong Kong strain). At the present time the H3 viruses continue to circulate in the human population. Interestingly, a second subtype strain, again of the H1 subtype, was found to be reintroduced in 1977. Studies in our own laboratory (Nakajima *et al.*, 1978) and by other groups (Scholtissek *et al.*,1978; Kendal *et al.*, 1978; Webster *et al.*, 1979) have demonstrated that the 1977 H1 strains are very closely related to the H1 strains circulating in 1950. Since in the past strains belonging to a single subtype were prevalent, the present situation--with strains of the H1 as well as the H3 subtype cocirculating --represents an anomaly. Although it is not possible to precisely define the circumstances of the reemergence of the H1 strains in 1977, an accidental introduction of laboratory strains cannot be excluded. It should be noted that--in addition to subtype shift--antigenic drift is always observed among influenza A viruses isolated in successive years (see below).

Influenza B and C viruses have not been demonstrated to occur in the form of subtypes; rather, influenza B virus strains undergo only slow antigenic drift in their hemagglutinins and influenza C viruses appear to lack any great degree of antigenic variability. The epidemiologic situation with influenza B and C viruses is thus very different from that observed for influenza A viruses.

III. EMERGENCE OF SUBTYPE STRAINS OF HUMAN INFLUENZA A VIRUSES

Cloning and sequencing of hemagglutinin genes derived from human influenza viruses has revealed that subtype A hemagglutinins differ by more than 50% on the nucleotide level. Similar differences are observed when the deduced amino acid sequences are compared (Table II). Since changes of that extent are unlikely to occur by simple mutation within a year's time, other explanations have been forwarded to explain the emergence of new influenza A subtype strains. The most widely accepted theory suggests the involvement of animal and human strains through reassortment of genes (Laver and Webster, 1972). In support of this theory, evidence for reassortment among different human influenza A viruses in nature has been obtained (Young and Palese, 1979). Also, reassortment of genes among animal influenza viruses has been observedin nature to generate strains with novel surface antigen combinations

Table II. Sequence homologies of HAs from different human influenza viruses.

Comparison	Amino acid conservation, %	
	HA1	HA2
H1 vs. H2	58	79
H1 vs. H3	35	53
H2 vs. H3	36	50
B vs. H1	24	39

Legend. H1, A/PR/8/34; H2, A/Jap/305/57; H3, A/Aichi/2/68; and B, B/Lee/40. The sequences of the H1, H2, H3 and B virus HAs are from Winter *et al.*,1981; Gething *et al.*, 1980; Verhoeyen *et al.*,1980; Krystal *et al.*, 1982.

Homologies were calculated as numbers of homologous residues/total number of residues. The total number of residues is the average of the length of the two sequences. The signal peptide sequences are not used for this comparison (from Palese and Young, 1983).

(Desselberger *et al.*, 1978; Hinshaw *et al.*, 1980). Thus, there is general agreement that similar reassortment events may occur between a human and an animal influenza virus strain to give rise to a new subtype strain. Such a strain would derive one or both surface antigen genes from an animal virus and most or all of its other genes would be retained from the human virus.

IV. EMERGENCE OF VARIANT INFLUENZA A VIRUSES: SELECTION OF DOMINANT VARIANTS

Several lines of evidence including epidemiologic data suggest that dominant influenza A viruses emerge with time. For example, strains belonging to the same influenza A virus subtype have been demonstrated to reinfect patients following a 2-5 year interval. Serologic analysis reveals that influenza A viruses undergo rapid antigenic drift, which changes the antigenic makeup of the virus sufficiently to allow reinfection of patients. It should be noted that this antigenic drift is different from a subtype change (antigenic shift) described above. The advent of sequencing techniques subsequently permitted determination of whether antigenic (genetic) drift occurs as the result of selection of successive variants or through the random emergence of variants in a pool of cocirculating strains. In the latter case successive dominant variants would not belong to a common evolutionary tree. Based on several lines of evidence it is suggested that the first mechanism is responsible for the observed antigenic (genetic) drift of influenza A viruses. First, it was demonstrated by comparative sequence analysis of H3 hemagglutinins isolated between 1968 and 1980 that these molecules share a common lineage (Both *et al.*, 1983; Skehel *et al.*, 1983). Second, the largest data base was obtained by examining the evolution of the gene coding for

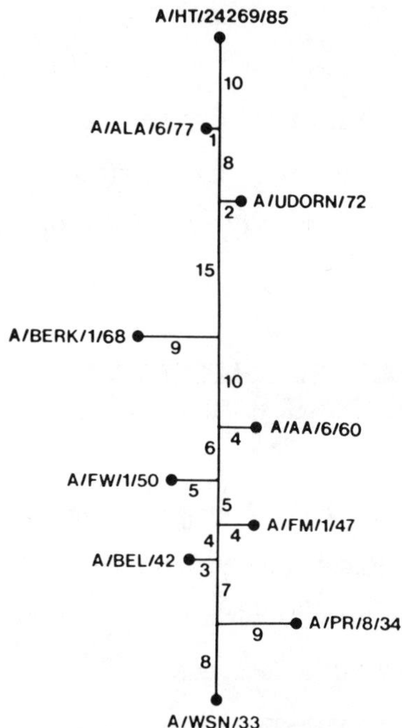

Figure 3. Nucleotide differences among NS genes. Evolutionary tree depicting the nucleotide relationship among NS genes of human influenza viruses. The NS gene of the oldest strain, A/WSN/33, which was isolated in 1933, is used as baseline. (The last number in the strain designation refers to the year of isolation.) The numbers on the evolutionary tree indicate single base changes observed among the different genes. All NS genes are 890 nucleotide long. Nucleotide positions in which more than one change occurred were omitted from analysis to allow for construction of an evolutionary tree with additive genetic distances. Data is from Krystal *et al.,* (1983a), Buonagurio *et al.* (1984, and unpublished).

the NS1 and NS2 proteins of influenza A viruses (Krystal *et al.,* 1983a; Buonaguirio *et al.,* 1984) (Fig. 3). Sequencing analysis of this gene derived from strains isolated as far back as 1933 shows a linear accumulation of nucleotide changes. This study clearly reveals that all of these NS genes share a common evolutionary tree. It should be noted that the NS genes derive from strains belonging to different subtypes, i.e., the H1, H2 and H3 subtypes. However, as discussed above, the genes determining the subtype (the HA genes) are introduced into the virus through reassortment, leaving the other genes unchanged. Thus, the evolutionary tree of the NS genes gives a legitimate picture of the evolutionary changes of influenza A viruses over a period of more than five decades (Fig. 3). The amino acid changes of the polypeptides NS1 and NS2, both coded for by the NS gene, show the same evolutionary relationship as that observed by comparing the nucleotide sequences (data not shown). Finally, the same phenomenon was studied by sequence analysis of the M genes of different influenza viruses (Ortin *et al.*,1983). Again it was found that successive variants of influenza A viruses emerge over time.

V. GENETIC DRIFT AMONG INFLUENZA B VIRUSES

As seen with influenza A virus HAs, variation of influenza B virus HAs also proceeds through the accumulation of nucleotide changes with time (Fig. 4). Although fewer data are available for the viruses, it appears that the overall rate of change among influenza B virus HAs is slower than that among the A virus HAs (Krystal *et al.*, 1983b). It should also be remembered that influenza B viruses do not appear as subtypes and that the only significant degree of change is associated with antigenic (genetic) drift. On a molecular level, the changes observed are the result of point mutations and short deletions and insertions (Krystal *et al.*, 1983b).

VI. VARIATION AMONG INFLUENZA C VIRUSES

In contrast to influenza A and B virus HAs, sequencing and comparison of different influenza C virus HAs revealed the noncumulative nature of changes in this system (Buonagurio *et al.*,1985) (Fig. 5).

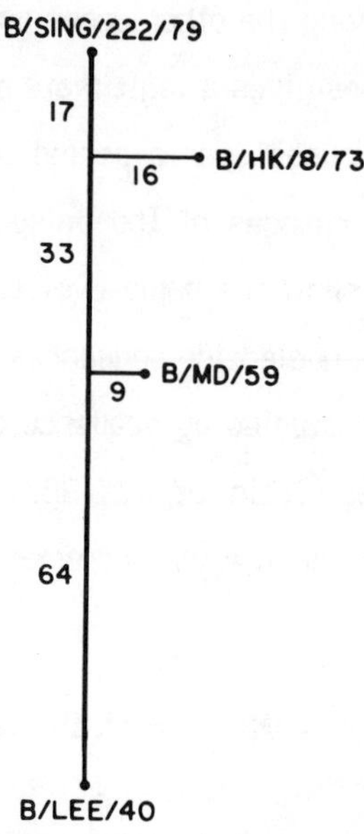

Figure 4. Evolutionary tree of influenza B viruses. Nucleotide differences among the HA genes of four different influenza B viruses are used in the analysis. The viruses were isolated in 1940, 1959, 1973, and 1979, respectively. The sequences of the entire HA genes (each is approximately 1880 nucleotides in length) were used for comparison. Numbers on the evolutionary tree indicate nucleotide changes among the genes. Data are from Krystal *et al.* (1983b), and Verhoeyen *et al.* (1983).

Sequence analysis of eight different influenza C virus HAs derived from viruses isolated between 1947 and 1983 suggested that different strains cocirculate and that these strains have different evolutionary relationships. Similar evidence is obtained when the NS genes of these strains are compared (Buonagurio *et al.*, unpublished). Since these viruses are not related through a direct evolutionary tree in which changes accumulate, at any one time strains lying on different evolutionary trees can be sampled. It appears that this situation is similar to that found for herpes or rhinoviruses. In each of these cases single epidemiologic

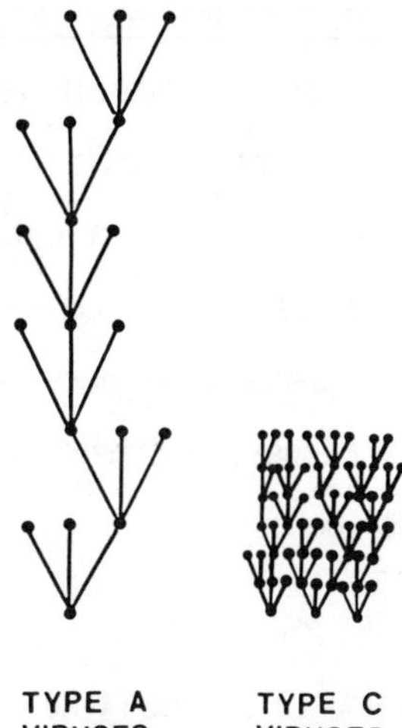

Figure 5. Evolutionary model for the propagation of influenza A and influenza C viruses. The length of the branches indicates relative genetic distances. Dots lying on a horizontal line represent influenza virus isolates obtained in the same season (year). The left part of the cartoon shows the emergence of influenza A virus variants directly belonging to the same evolutionary tree. Dominant variants appear to emerge which show an accumulation of changes. The right part of the diagram depicts the cocirculation of influenza C virus variants derived from multiple evolutionary pathways. Variation appears to be slower than in the case of influenza A viruses, as demonstrated by the shorter length of the branches on the different evolutionary trees. For both influenza A and influenza C viruses an arbitrary number of seven seasonal cycles is shown on the diagram.

variants are not prevalent at one particular time; rather, many variants cocirculate at the same time. Furthermore, it should be noted that the variation rate in the influenza C virus system must be slower than that observed for influenza A and B viruses. For example, the HA genes of the C/AA/50 and C/YA/81 strains, which were isolated 31 years apart, differ by only two nucleotides (Buonagurio *et al.,* 1985). This finding suggests a relative stability of the influenza C virus genome over the last three decades (Fig. 5).

Despite the observed conservation of sequences, the genome of influenza C viruses does not appear to be evolutionarily fixed. This can be shown by comparing the HA genes of influenza A, B an C viruses. Although the A and B virus HAs show an overall amino acid homology of approximately 30% (Table II) there is little or no primary amino acid sequence homology between A and C virus or between B and C virus HAs. Specifically, several cysteine residues are conserved among the different HA species, but the overall sequence divergence between A and C or B and C HAs (Nakada *et al.*, 1984) suggests a considerable genetic flexibility in influenza A, B and C virus HA genes.

VII. MOLECULAR CLOCK: RAPID EVOLUTION OF INFLUENZA VIRUSES

Analysis of the data shown in Figures 3 and 4 and of other results suggests an extremely high rate of nucleotide substitutions in the genomes of influenza viruses (Nei, 1983; Hayashida *et al.*, 1985; Buonagurio *et al.*, 1984). This phenomenon was first extensively studied by sequence analysis of six different NS genes derived from influenza viruses isolated between 1934 and 1977 (Buonagurio *et al.*, 1984). Based on this data a variation rate of 2.2−3.4% per 10 years was calculated. For third position changes a rate of 3.1−4.2% was obtained. Since that time more data have been collected and values have been obtained to indicate that the influenza A and B virus genes change at a rate of 1.8−5.3 × 10^{-3} substitutions/site/year (Table III). Interestingly, these values are very different from those obtained for eukaryotic genes. As an exmple, the data calculated by Li and Gojobori (1983) for changes in globin genes are included in the table. Comparison shows a 10^{6}-fold difference in rate between the influenza virus genes and those observed in eukaryotic genomes. Similar data for high evolutionary rates of

Table III. Rate of nucleotide substitutions

Genes	% change/10 yrs. (nucleotides)	Subtitutions/site/yr. (nucleotides)
Influenza A and B virus HA	1.8 – 5.3	$1.8 - 5.3 \times 10^{-3}$
Influenza A virus NS	2.2 – 3.4	$2.2 - 3.4 \times 10^{-3}$
Globin*	N.A.	$2-5 \times 10^{-9}$

*From W.-H. Li and T. Gojobori (1983).

influenza virus genes by Nei (1983) and extensive analyses using different methodologies (for review see Holland, 1982) appear to support the remarkable difference observed in the variation rate of influenza viruses and other organisms. This difference has called forth a variety of explanations. One suggestion is that RNA viruses, lacking a replicating system with proofreading capacity (Holland, 1982), may be subject to a higher mutation rate than are other viral or eukaryotic systems. Furthermore, the generation time of RNA viruses is short, which suggests a large number of replication cycles. In conjunction with the high frequency of influenza A and B virus infections, these may be the factors responsible for the high variation rate observed with influenza A or B viruses.

Experiments are now underway to determine the effect of mutation rates on viral evolution by directly measuring these rates for influenza viruses and other animal viruses. It should be remembered, however, that mutation rate may only be one of the components that determines the remarkable evolutionary success of influenza viruses. Future studies will be necessary to explore all the devices used by the virus (or the host) which result in the remarkable changes that are observed during propagation of these viruses in nature. Because influenza viruses are simple organisms with a relatively low degree of complexity, and because of the availability of strains that have changed dramatically during the last 50 years, this system may provide an excellent model to study evolution in nature and to test the validity of different evolutionary theories.

REFERENCES

Both, G., Sleigh, M., and Cox, N. (1983). Antigenic drift in influenza virus H3 hemagglutinin from 1968 to 1980. Multiple evolutionary pathways and sequential amino acid changes at key antigenic sites. *J. Virol.* **48**, 52-60.

Buonagurio, D. A., Krystal, M., Palese, P., DeBorde, D. C., and Maasab, H. F. (1984). Analysis of an influenza A virus mutant with a deletion in the NS segment. *J. Virol.* **49**, 418-425.

Buonagurio, D. A., Nakada, s., Desselberger, U., Krystal, M., and Palese, P. (1985). Non-cumulative sequence changes in the hemagglutinin genes of influenza C virus isolates. *Virology* **146**, 221-232.

Desselberger, U., Nakajima, K., Alfino, P., Pedersen, F. S., Haseltine, W., Hannoun, C., and Palese, P. (1978). Biochemical evidence that "new" influenza virus strains in nature may arise by recombination (reassortment). *Proc. Natl. Acad. Sci. USA* **75**, 3341-345.

Gething, M. J., Bye, J., Skehel, J. J., Waterfield, M. (1980). Cloning and DNA sequence of double stranded copies of hemagglutinin genes from H2 and H3 strains elucidate antigenic shift and drift in human influenza virus. *Nature* **287**, 301-306.

Hayashida, H., Toh, H., Kikuno, R., Miyata, T. (1985). Evolution of influenza virus genes. *Mol. Biol. and Evol.* **2**, 289-303.

Hinshaw, V. S., Bean, W. J., Webster, R. G., Sriram, G. (1980). Genetic reassortment of influenza A viruses in the intestinal tract of ducks. *Virology* **102**, 412-419.

Holland, J., Spindler, K., Horodyski, F., Grabav, E., Nichol, S., and Van de Pol., S. (1985). Rapid evolution of RNA genomes. *Science* **215**, 1577-1585.

Kendal, A. P., Noble, G. R., Skehel, J. J., and Dowdle, W. R. (1978). Antigenic similarities of influenza A/H1N1 viruses from epidemics in 1977-1978 to "Scandinavian" strains isolated in epidemics of 1950-1951. *Virology* **89**, 632-636.

Krug, R. M. (1983). Transcription and replication of influenza viruses. *In* "Genetics of Influenza Viruses" (P. Palese and D. W. Kingsburgy, eds.), pp. 70-98. Springer Verlag, New York.

Krystal, M., Elliott, R. M., Benz, E. W., Young, J. F., and Palese, P. (1982). Evolution of influenza A and B viruses: Conservation of structural features in the hemagglutinin genes. *Proc. Natl. Acad. Sci. USA* **79**, 4800-4804.

Krystal, M., Buonagurio, D., Young, J. F., and Palese, P. (1983a). Sequential mutations in the NS genes of influenza virus field strains. *J. Virol.* **45**, 547-554.

Krystal, M., Young, J. F., Palese, P., Wilson, I. A., Skehel, J. J., and Wiley, D. C. (1983b). Sequential mutations in the hemagglutinins of influenza B virus isolates: Definition of antigenic domains. *Proc. Natl. Acad. Sci. USA* **80**, 4527-4531, including correction *PNAS* **81**, 1261.

Lamb, R. A. (1983). The influenza virus RNA segments and their encoded proteins. *In* "Genetics of Influenza Viruses" (P. Palese and D. W. Kingsburgy, eds.), pp. 21-69. Springer-Verlag, New York.

Laver, W. G., and Webster, R. G. (1972). Studies on the origin of pandemic influenza. *Virology* **48**, 445-455.

Li, W.-H., and Gojobori, T. (1983). Rapid evolution of goat and sheep globin genes following gene duplication. *Mol. Biol. and Evol.* **1**, 94-108.

Nakada, S., Creager, R. S., Krystal, M., Aaronson, R. P., and Palese, P. (1984). Influenza C virus hemagglutinin: Comparison with influenza A and B virus hemagglutinins. *J. Virol.* **50**, 118-124.

Nakajima, K., Desselberger, U., and Palese, P. (1978). Recent human influenza A viruses are closely relatd genetically to strains isolated in 1950. *Nature* **274**, 334-339.

Nei, M. (1983). Genetic polymorphism and the role of mutation in evolution. *In* "Evolution of Genes and Proteins" (M. Nei and R. K. Koehn, eds.), pp. 165-190. Sinauer Assoc., Inc., Massachusetts.

Palese, P., and Young, J. F. (1982). Variation of influenza A, B and C viruses. *Science* **215**, 1468-1474.

Palese, P., and Young, J. F. (1983). Molecular epidemiology of influenza virus. *In* "Genetics of Influenza Viruses" (P. Palese and D. W. Kingsburgy, eds.), pp. 321-336. Springer-Verlag, New York.

Scholtissek, C., von Hoyningen, V., and Rott, R. (1978). Genetic relatedness between the new 1975 epidemic strains (H1N1) of influenza and human influenza strains isolated between 1947 and 1957 (H1N1). *Virology* **89**, 613-617.

Skehel, J. J., Daniels, R. S., Douglas, A. R., and Wiley, D. C. (1983). Antigenic and amino acid sequence variations in the hemagglutinins of type A influenza viruses recently isolated from human subjects. *Bull. W.H.O.* **61**, 671-676.

Varghese, J. N., Laver, W. G., an Colman, P. M. (1983). Structure of the influenza virus glycoprotein antigen neuraminidase at 2.9. A resolution. *Nature* **303**, 35-40.

Verhoeyen, M., Fang, R., Min Jou, W., Devos, R., Huylebroeck, D., Saman, E., and Fiers, W. (1980). Antigenic drift between the hemagglutinin of the Hong Kong influenza strains A/Aichi/2/68 and 75. *Nature* **286**, 771-776.

Verhoeyen, M., van Rompuy, L., Jou, W. M., Huylebroek, D., and Fiers, W. (1983). Complete nucleotide sequence of the influenza B/Singapore/222/79 virus hemagglutinin gene and comparison with the B/Lee/40 hemagglutinin. *N.A.R.* **11**, 4703-4712.

Webster, R. G., Kendal, A. P., and Gerhard, W. (1979). Analysis of antigenic drift in recently isolated influenza A (H1N1) viruses using monoclonal antibody preparations. Virology **96**, 258-264.

Wilson, I. A., Skehel, J. J., and Wiley, D. C. (1981). Structure of the hemagglutinin membrane glycoprotein of influenza virus at 3 A resolution. *Nature* **289**, 366-373.

Winter, G., Fields, S., and Brownlee, G. G. (1980). Nucleotide sequence of the hemagglutinin gene of a human influenza virus H1 subtype. *Nature* **292**, 72-75.

Young, J. F., and Palese, P. (1979). Evolution of human influenza A viruses in nature: recombination contributes to genetic variation of H1N1 viruses. *Proc. Natl. Acad. Sci. USA* **76**, 6547-6551.

STRUCTURAL DIVERSITY AND EVOLUTION OF INTERMEDIATE FILAMENT PROTEINS[1]

Israel Hanukoglu

Department of Biology
Technion-Israel Institute of Technology
Haifa 32000, Israel

Elaine Fuchs

Department of Molecular Genetics and Cell Biology
The University of Chicago
Chicago, Illinois 60637 USA

ABSTRACT

The cytoskeletal network of most mammalian cells includes three major types of filamentous systems: microfilaments, intermediate filaments (IF) and microtubules with respective diameters of 6nm, 8-10nm and 25nm. Each of these filamentous systems is assembled from only one or two different subunits. The proteins that form these filaments are all encoded by multigene families, the members of which are differentially expressed in different tissues. While the protein sequences of actins and tubulins (which form microfilaments and microtubules, respectively) are highly conserved, the IF proteins show a

---

[1] Our work reviewed here was supported by a U.S. National Institutes of Health grant. I. H. was the recipient of a U.S. National Cancer Institute National Research Service Award. E. F. is the recipient of a National Institutes of Health Career Development Award and a Presidential Young Investigator Award.

69

much higher degree of diversity both in their sequences and the length of the polypeptide chains (Mr = 40-140K). Despite their diversity, the ultrastructures of the filaments formed by the different IF proteins are highly similar and these all resemble the structure of the microfibrils which form the backbone of such epidermal appendages as wool and hair. Analyses of the sequences of IF proteins and the microfibrillar α-keratins indicate that within the central 300 residues of these proteins there is a remarkable conservation of α-helical structural domains despite the divergence of sequences in this region. The IF proteins can be categorized into three major families on the basis of sequence homology in this central region: 1) type I keratins, 2) type II keratins, and 3) desmin, vimentin, glial filament protein and neurofilament proteins. The sequence homology within each group is > 50%, while between groups it is 24-35%. The terminal sequences on both sides of the central region are even more variable and the size heterogeneity among IF proteins is a result of the differences in the length of these terminal regions. The conservation of structural features despite diversity of the IF protein sequences and the juxtaposition of hyper-variable terminal sequences with relatively constant domains provide valuable data and pose intriguing questions about the mechanisms of evolutionary change.

I. INTRODUCTION

The cytoskeletal network of most mammalian cells includes three major types of filamentous systems: microfilaments, intermediate filaments (IF) and microtubules with respective diameters of 6nm, 8-10nm and 25nm. Each of these filamentous sytems is assembled from only one or two different subunits. The proteins that form these filaments are all encoded by multigene families, the members of which are differentially expressed in different tissues. While the protein sequences of actins and tubulins (which form microfilaments and microtubules, respectively) are highly conserved, the IF proteins show a much higher degree of diversity both in their sequences and the length of the polypeptide chains (M_r 40,000-140,000). Here, we shall present an

overview of our current knowledge of the sequence and structural relatedness of this most divergent group of cytoskeletal proteins and their genes.

II. MULTIPLICITY OF IF PROTEINS AND THEIR STRUCTURAL RELATIONSHIP

In mammalian organisms the total number of distinct proteins
capable of forming IF in different tissues and at different stages of
development may be at least 20-30 (Lazarides, 1982). The expression of
subsets of this group of proteins is generally limited to specific types of
tissues: keratins are expressed in epithelial cells, desmin in muscle
cells, glial filament protein (GFP) in glial cells, neurofilament proteins
(NFP) in neurons, and vimentin in mesenchymal cells and many other
tissues. However, the IF proteins cannot all be grouped into strict
categories based on their tissue specificity of expression as some of the
IF proteins (especially vimentin together with others) coexist in different
tissues, and the spectrum of IF proteins in a specific tissue may change
during development and differentiation (Gard *et al.*, 1979; Franke *et al.*,
1982; Lazarides, 1982; Sharp *et al.*, 1982; Granger and Lazarides, 1983;
Lane *et al.*, 1983; Ben-Zeev, 1984; Holthofer *et al.*, 1984). Proteins and
genomic sequences that represent all of these five groups of IF proteins
have been detected in all vertebrates examined to date (Fuchs and Marchuk,
1983; Quax *et al.*, 1984; Lewis *et al.*, 1984; Lewis and Cowan, 1985).
There is immunological evidence that vimentin related polypeptides may
also exist in invertebrates (Walter and Biessmann, 1984). Using cDNA
probes, sequences homologous to the mRNAs of some IF proteins have been
detected in organisms lower than the vertebrates; however, until the
nature of the proteins coded by these sequences is established, these

remain as tentative indications for the presence of IF proteins in invertebrates (Fuchs and Marchuk, 1983).

Among the IF proteins, the keratins represent the largest and the most diverse group. The keratins which form IF are highly homologous to the keratins which form the backbone of such epidermal appendages as hair and wool. Thus, the multigene families of IF proteins also include these keratins. While in epithelial cells keratin filaments are organized in seemingly irregular patterns of a cytoskeletal network (henceforth named as cytoskeletal keratins), in epidermal appendages such as hair and wool the keratins form very orderly arrays of microfibrils (henceforth named as microfibrillar keratins) embedded in a rigid matrix of other proteins (much like steel rods embedded in concrete!) (Jones, 1975; for reviews see Crewther *et al.*, 1965; Fraser *et al.*, 1972; Fuchs and Hanukoglu, 1986). The proteins that constitute the matrix in epidermal appendages are also called keratins. However, as these "matrix keratins" are completely different from the microfibrillar α-keratins, both in terms of their sizes (6-20K vs. 40-70 K) and their primary and secondary structures, we shall not cover studies on these keratins here.

In vertebrates the total number of different keratins varies between 2-20 across species (Moll *et al.*, 1982; Fuchs and Marchuk, 1983). Most, if not all, keratins (both cytoskeletal and microfibrillar) can be grouped into two distinct classes of sequences (Crewther *et al.*, 1978, 1980; Fuchs *et al.*, 1981; Hanukoglu and Fuchs, 1982, 1983; Dowling *et al.*, 1983; Fuchs and Marchuk, 1983; Steinert *et al.*, 1983, 1984). These two classes were named as type I and type II keratins, respectively (Hanukoglu and Fuchs, 1983) extending a nomenclature used for wool keratin fragments (Crewther *et al.*, 1978). The type I keratins are relatively acidic (isoelectric pH 4.5-5.5) and small (M_r 40,000-59,000), whereas the type II keratins are more basic (isoelectric pH 6.5-7.5) and larger (M_r

53,000-67,000). In contrast to the other IF proteins, a single keratin polypeptide cannot polymerize into filaments by itself. The formation of keratin filaments requires a pairwise association and polymerization of at least two different keratins (Lee and Baden, 1976; Steinert *et al.*, 1976, 1982). *In vitro* polymerization studies indicate that the type I and type II keratins may be these necessary building blocks of keratin filaments (Franke *et al.*, 1983; Quinlan *et al.*, 1984). This hypothesis is further strengthened by the observations that at least one member of each of these two classes of keratins is present in all epithelial cells (Kim *et al.* 1983; Eichner *et al.*, 1984), and that type I and type II keratins and their corresponding genes are found in all vertebrates (Fuchs *et al.*, 1981; Fuchs and Marchuk, 1983).

The category of neurofilament proteins includes three distinct polypeptides (M_r 62,000, 107,000, and ~140,000) named, respectively, as NFP-L, NFP-M, and NFP-H (Geisler *et al.*, 1984, 1985). These three NFP, and the other non-keratin IF proteins, desmin, vimentin, and GFP represent single proteins and appear to be encoded by genes that are present in a single copy in the genome of all vertebrate species examined to date (Capetanaki *et al.*, 1983; Lewis and Cowan, 1985; Quax *et al.*, 1984; Zehner and Paterson, 1983; Lewis *et al.*, 1984). Consistent with the singular presence of desmin, vimentin, and GFP in most tissues, these proteins can assemble into IF by self-polymerization (homopolymer formation) (Steinert *et al.*, 1982; Geisler, Kaufmann and Weber, 1982). Desmin, vimentin and GFP can also form heteropolymer IF by copolymerization *in vitro* (Steinert *et al.*, 1982), as well as *in situ* in tissues and cells where these proteins coexist (Quinlan and Franke, 1982; Wang *et al.*, 1984). NFP-L can form filaments by itself but NFP-M, and NFP-H also participate in the formation of neurofilaments *in situ*

(Liem and Hutchison, 1982; Sharp *et al.*, 1982; Steinert *et al.*, 1982; Hirokawa *et al.*, 1984).

Despite major differences among the various IF proteins (especially the M_r range: 40,000-70,000, excluding NFP-M and NFP-H), the ultrastructures of the filaments formed by these proteins are highly similar (Kallman and Wessells, 1967; Henderson *et al.*, 1982; Milam and Erickson, 1982). Prior to the determination of the sequences of IF proteins and microfibrillar keratins, a number of biochemical, immunological and physiochemical studies indicated that both of these two groups of proteins are structurally related, and that both contain long regions of mainly α-helical conformation in the center, and a staggered conformation of unknown structure at the amino and carboxy terminal ends (Crewther and Harrap, 1967; Fraser *et al.*, 1972, 1976; Skerrow *et al.*, 1973; Jones, 1975; Steinert, 1978; Steinert, Idler and Goldman, 1980; Weber, Osborn and Franke, 1980). Since the early studies of Pauling and Corey (1953) and Crick (1953) it is thought that the helical regions of these proteins intertwine around one another to form a coiled-coil rod and that these rods are linked end-to-end to form a rope-like filament that constitutes a protofibril. A bundle of about 10 such protofibrils constitutes the microfibrils and IF. Although some early studies suggested that in each protofibril the number of polypeptide chains aligned side-by-side is three (Crewther and Harrap, 1967; Skerrow *et al.*, 1973; Steinert, 1978; Steinert, Idler and Goldman, 1980), more recent studies indicate that this number is two (Gruen and Woods, 1981; Geisler and Weber, 1982; Woods and Gruen, 1983; Quinlan *et al.*, 1984). Model building studies and analyses of the sequences of microfibrillar keratins also indicate that the basic unit of protofibrillar structure is made up of two intertwined α-helical polypeptide chains (Parry *et al.*, 1977; McLachlan, 1978).

The positioning of the non-helical terminal regions of the IF proteins in the overall structure of the protofibril is not known yet. The terminal sequences of IF proteins can be extensively digested without disrupting the ability of these proteins to assemble into filaments (Steinert *et al.,* 1983; Lu and Johnson, 1983). Yet, the helical segments without the terminal extensions cannot assemble into filaments by themselves (Geisler *et al.,* 1982). Thus, it is currently thought that the terminal domains largely project from the central coiled-coil rod and contribute to both end-to-end linkage of IF proteins within the protofibril and the interactions of these proteins with other cellular molecules.

III. COMPARISON OF THE SEQUENCES AND PREDICTED SECONDARY STRUCTURES OF THE IF PROTEINS

Currently, the complete or nearly complete sequences of over 10 different IF proteins and microfibrillar keratins are known (Fig. 1). Analysis of these sequences indicate two major structural motifs: 1) A central region of about 300 residues predominantly in α-helical conformation, and 2) amino and carboxy terminal regions that show an amazing variability in their lengths and sequences across different IF proteins subclasses, but which appear to be conserved across species for each subclass. These results are in general consistent with the observations noted in the previous section. However, the analyses based primarily on sequence data present a more precise model for the structure of the IF proteins that is, in some important respects, different from those of some of the earlier studies. (For a detailed discussion see Geisler and Weber, 1982). A comparison of the sequences and the

Amino terminus
```
MT-I  :   1 SVLYCSSSKQFSSSRSGGGGGGGSVRVSSTRGSLGGGLSSGGFSGGSFSRGSSGGGCFGGSSGGYGGFGGGGSFGGGYGGSSFGGGYGGSSFGGGYGGSS
        101 FGGGSFGGGSFGGGSFGGGGCGGGFGGGGFGGDGGGLLSGNFIDKVRF
HT-I  :   1 MTTCSRQFTSSSSMKGSCGIGGGIGAGSSRISSVLAGGSCRAPNTYGGGLSVSSSRFSSGGAYGLGGGYGGGFSSSSSSFGSGFGGGYGGGLGTGLGGGF
        101 GGGFAGGDGLLVGS
HT-II :   ...
MT-II :   1 STKTTIKSQTSHRGYSASSARVLGLNRSGFSSVSVCRSRGSGGSSAMCGGAGFGSRSLYGVGSSKRISIGGGSCGIGGGYGSRFGGSFGIGGGAGSGFGF
        101 GGGAGFGGGYGGAGFPVCPLGGIQEVTINQSLLTPLNLQIDPTIQRVRTE
CDES  :   1 SQSYSSSQRVSSYRRTFGGGTSPVFPRASFGSRGSGSSVTSRVYQVSRTSAVPTLSTFRTTRVTPLRTYGSAYQGAGELL   DFSLADAMNQEFLQTRTN
HVIM  :   1   STRSVSSSSYRRMFGGPGTSNRQSSNRSYVTTSTRTYSLGSLRPSTSRSLYSSSPGGAYVTRSSAVRLRSSMPGVRLLQDSVDFSLADAINTEFKNTRTN
MGFP  :   1                                 ...LGTMPRFSLSRMTPPLPARVDFSLAGALNAGFKETRAS
PNFP  :   1 SYTLDSLGNPSSAYRRVTETRSSFSRVSGSPSSGFRSQSWSRGSPSTVSSSYKRSALAPRLTYSSAMLSSAESSLDFSQSSSLLDGGSGPGGDYKLSRSN
```

Helical domain I
```
MT-I  : 142 GRVTMRNLNDRLASYMDKVRALEESNYELEGKIKEVVREARQLKPREPRDYS
HT-I  :  52 EKVTMQNLNDRLASYLDKVRALEEANADLEVKIRDWYQRQR   PAEIKDYS
HT-II :   1 .................................................QNLEP
MT-II : 151 EREQIKTLNNKFASFIDKVRFMERQNKVMDTKWALLREQDTKTVRQNMEP
CDES  :  99 EKVELQELNDRFANYIEKVRFLEQQNALMVAEVNRLRGKQPTR   VAE
HVIM  : 101 EKVELQELNDRFANYIDKVRFLEQQNKILLAELEQLKGQGKSR   LGD
MGFP  :  39 ERAEMMELNDRFASYIEKVRFLEQQNKALAAELNQLRAKEPTK   LAD
PNFP  : 101 EKEQIQGLNDRFAGYIEKVHYLEQQNKEIEAEIQALRQKQASHAQLGD
```

Helical domain II
```
MT-I  : 194 KYYKTIECLKGQILTLTTDNANVLLQIDNARLAADDFRLKYENEVTLRQSVEADINGLRRVLDELTLSQSVLELQIESLNEELAYLKKNLEEEMRDLQN
HT-I  : 101 PYFKTIEDLRNKILTATVDNANVLLQIDNARLAADDFRTKYETELNLRMSVEADINGLRRVLDELTLARADLEMQIESLKEELAYLKKNHEEEMNALRG
HT-II :   6 LFEQYINNLRRQLDSIVGERGRLDSELRGMQDLVEDFKNKYEDEINKRTAAENEFVTLKKDVDAAYMNKVELQAKADTLTDEINFLRALYDAELSQMQT
MT-II : 201 MFEQYISNLRRQLDSIIGERGRMNSELRNMQELVEELRNKYEDEINKRTDAENEFVTLKKDVDAAYMNKAELQAKADSLTDDINFLRALYEAELSQMQT
CDES  : 145 MYEEELRELRRQVDALTGQRARVEVERDNLLDNLQKLKQKLQEEIQLKQEAENNLAAFRADVDAATLARIDLERRIESLQEEIAFLKKVHEEEIRELQA
HVIM  : 147 LYEEEMRELRRQVDQLTNDKARVEVERDNLAEDIMRLREKLQEEMLQREEAESTLQSFRQDVDNASLARLDLERKVESLQEEIAFLKKLHDEEIQELQA
MGFP  :  85 VYQAELRELRLRLDQLTANSARLEVERDNFAQDLGTLRQKLQDETNLRLEAENNLAAYRQEAHEATLARVDLERKVESLEEEIQFLRKIYEEEVRDLRE
PNFP  : 149 AYDQEIRELRATLELVNHEKAQVQLDSDHLEEDIHRLKERFEEEARLRDDTEAAIRALRKDIEEASLVKVELDKKVQSLQDEVAFLRSNHEEEVADLLA
```

```
MT-I  : 293 VSTGD VNVE MNAAPGV
HT-I  : 200 QVGGD VNVE MDAAPGV
HT-II : 105 HISDTSVVLS MDNNRNL
MT-II : 300 HISDTSVVLS MVNNRSL
CDES  : 244 QLQEQHIQVE MDISKP
HVIM  : 246 QIQEQHVQID VDVSKP
MGFP  : 184 QLAQQQVHVE MDVAKP
PNFP  : 248 QIQASHITVERKDYLKT
```

Helical domain III
```
MT-I  : 309 DLTQLLNNMRNQYEQLAEKNRKDAEEWFNQKSKELTTEIDSNIAQMSSHKS
HT-I  : 216 DLSRILNEMRDQYEKMAEKNRKDAEEWFFTKTEELNREVATNSELVQSGKS
HT-II : 122 DLDSIIAEVKAQYEEIAQRSRAEAESWYQTKYEELQVTAGRHGDDLRNTKQ
MT-II : 317 VLDSIIAEVKAQFEVIAQRSRAEAESLYQTKYEELQVTAGRHGDDLRNTKQ
CDES  : 260 DLTAALRDVRQQYESVAAKNLQEAEEWYKSKVSDLTQAANKNNDALRQAKQ
HVIM  : 262 DLTAALRDVRQQYESVAAKNLQEAEEWYKSKFADLSEAANRNNDALRQAKQ
MGFP  : 200 DLTAALREIRTQYEAVATSNMQETEEWYRSKFADLTDAASRNAELLRQAKH
PNFP  : 265 DISSALKEIRSQLECHSDQNMAQAEEWFKCRYAKLTEAAQENKEAIRSAKE
```

Helical domain IV
```
MT-I  : 360 EITELRRTVQGLEIELQSQLALKQSLEASLAETVESLLRQLSQIQSQISALEEQLQQIRAETECQNAEYQQLLDIKTRLENEIQTYRSLLEGEGSSS
HT-I  : 267 EISELRRTMQNLEIELQSQLSMKASLENSLEETKGRYCMQLAQIQEMIGSVEEQLAQLRCEMEQQNQEYKILLDVKTRLEQEIATYRRLLEGEDAHL
HT-II : 173 EIAEINRMIQRLRSESDHVKKQCANLQAAIADAEQRGEMALKDAKNKLEGLEDALQKAKQDLARLLKEYQELMNVKLALDVEIATYRKLLEGEECRL
MT-II : 368 EIAEINRMIQRLRSEIDHVKKQCANLQAAIADAEQRGEITLKDARGKLEGLEDALQKAKQDMAMLLKEYRELMNVKLALDVEIATYRKLLEGEECRL
CDES  : 311 EMLEYRHQIQSYTCEIDALKGTNDSLMRQMREMEERFAGEAGGYQDTIARLEEEIRHLKDEMARHLREYQDLLNVKMALDVEIATYRKLLEGEENRI
HVIM  : 313 ESNEYRRQVQSLTCEVDALKGTNESLERQMREMEENFALEAANYQDTIGRLQDEIQNMKEEMARHLREYQDLLNVKMALDIEIATYRKLLEGEESRI
MGFP  : 251 EANDYRRQLQALTCDLESLRGTNESLERQMREQEERHARESASYQEALARLEEEGQSLKEEMARQLQEYQDLLNVKLALDIEIATYRKLLEGEENRI
PNFP  : 316 EIAEYRRQLQSKSIELESVRGTKESLERQLSDIEERHNHDLSSYQDTIQQLENELRGTKWEMARHLREYQDLLNVKMALDIEIAAYRKLLEGEETRF
```

Carboxy terminus
```
MT-I  : 457 GGGGGRRGGSGGGSYGGSSGGGSYGGSSGGGGSYGGSSGGGGSYGGSSGCGGRGGGSGGGYGGGSSSGGAGGRGGGSGGGYGGGGSSGRRGGSGGF 553
        554 SGTSGGGDQSSKGPRY 569
HT-I  : 364 SSSQFSSGSQSSRDVTSSSRQI       RTKVMDVH     DGKVVSTHEQVLRTKN 409
HT-II : 270 NGEGVGQVNISVVQSTVSSGYGGASGVGSGLGLGGGSSYSYGSGLGVGGGFSSSSGRAIGGGLSSVGGGSS TIKYTTTSSSSRKSYKH
MT-II : 465 NGEGVGPVNISVVQSTVSSGYGSAGGASSSLGMGGGSSYSYSSSHGLGGGFSAGSGRAIGGGLSSSGGLSSSTIKYTTTSSS KKSYRQ 552
CDES  : 408 SIRMHQTFASALNFRETSPDQRGS EVHTKKTVMIKTIETRDGEVVSEATQQQHEVL 463
HVIM  : 410 SLPLPN FSS LNLRETNLESLPLVDTHSKRTLLIKTVETRDGQVINETSQHHDDLE 464
MGFP  : 348 TIPV QTF SNLQIRETSLDTKSVSEGHLKRNIVVKTVEMRDGEVIKDSKQEHKDVVM 403
PNFP  : 413 STFAGSITGPLYTHRQPSITISSK...FVEEIIEETKVEDEKSEM...KEEVKEEEAEEKEEKQEAEEVEAAKK...
```

Figure 1

Figure 1 Legend. The amino acid sequences of IF proteins and the common location of the predicted helical domains. The sequences are from the following sources: 1) MT-I: Type I cytoskeletal keratin from mouse epidermis (Mr = 59K) (Steinert *et al.*, 1983); 2) HT-I: Type I cytoskeletal keratin from human epidermal cells (Mr = 50 K) (Hanukoglu and Fuchs, 1982; Marchuk *et al.*, 1984); 3) HT-II: Type II cytoskeletal keratin from human epidermal cells (Mr = 56 K) (Hanukoglu and Fuchs, 1983); 4) MT-II: Type II cytoskeletal keratin from mouse epidermal cells (Mr = 60K) (Steinert *et al.*, 1984); 5) CDES: Desmin from chicken muscle (Geisler and Weber, 1982); 6) HVIM: Vimentin from hamster lens (Quax *et al.*, 1983); 7) MGFP: Glial fibrillary protein from mouse brain (Lewis *et al.*, 1984); 8) PNFP: Neurofilament protein-M from porcine spinal cord (Geisler *et al.*, 1984). Table I includes additional sequence comparisons with wool microfibrillar keratins: WT-I: Type I microfibrillar keratin from sheep wool (Crewther *et al.*, 1980; Gough *et al.*, 1978; Dowling *et al.*, 1983); WT-II: Type II microfibrillar keratin from sheep wool (Crewther *et al.*, 1978; 1980; Gough *et al.*, 1978; Sparrow and Inglis, 1980; Dowling *et al.*, 1983). The dots (.) indicate missing sequence information. In cases where the sequence of the amino terminus is not known the numbering on the left side of the sequences start with the first known amino terminal residue of the longest segment. The gaps in the sequences indicate gaps introduced in order to align the sequences for optimal homology. The position of the helical domains are based on computerized secondary structure prediction analyses using the Chou and Fasman (1978, 1979) and Garnier, Osguthorpe and Robson (1978) methods as previously described (Hanukoglu and Fuchs, 1982; 1983).

predicted secondary structures of these proteins indicate the following major similarities and differences among the IF proteins (for references see legend of Fig. 1).

A. The Central Helical Region Is Structurally Conserved

The central approximately 300 residue long portion of all IF protein sequences can be aligned for optimal homology without any or only a few (1-6) gaps. In this central region different IF proteins share 24% to 90% homology with each other (Fig. 1, Table I). The IF proteins can be

Table I. Percent homology between the central regions of IF proteins (see Fig. 1).

	HT-I	MT-I	WT-I	HT-II	MT-II	WT-II	CDES	HVIM	MGFP	PNFP
HT-I		66.0	56.8	26.6	25.9	29.7	35.2	34.6	37.1	35.0
MT-I			50.2	24.4	23.7	27.6	32.9	32.9	34.2	32.2
WT-I				26.2	24.7	28.0	30.8	30.5	33.3	32.3
HT-II					88.8	58.6	34.9	34.9	32.7	31.9
MT-II						54.9	34.4	34.4	32.8	32.1
WT-II							35.9	37.1	37.1	32.9
CDES								74.1	63.4	47.8
HVIM									63.1	50.0
MGFP										48.1
PNFP										

categorized into three major families on the basis of sequence homology in this central region: 1) type I keratins; 2) type II keratins; and 3) desmin, vimentin, GFP, and neurofilament proteins. The sequence homology within each group is > 50%, while between groups it is 24-35%.

The presently available few sequences of type I and type II keratin suggest that within both of these groups cytoskeletal and microfibrillar keratins constitute distinct subfamilies with a higher degree of homology by at least 10% (Table I). However, it remains to be seen whether this homology difference will hold when more sequences from both putative subfamilies will be known.

Even though there is only a low degree of homology between some of the IF protein sequences, the secondary structures of the central regions of all IF proteins seem to be remarkably conserved. Secondary structure prediction analyses using the methods of Chou and Fasman (1978, 1979) and Garnier *et al.* (1978) indicate that within this central region, there are four richly α-helical domains which are demarcated from one another by three regions for which β-turns are predicted with a high degree of probability in all sequences. The first two of these β-turn regions contain proline(s) in some, but not all sequences. The four helical domains (marked as I, II, III, and IV in Fig. 1) are predicted to be nearly constant in size in all IF proteins and they are approximately 30-40, 100, 35-40, and 100 residues long, respectively. Different secondary structure prediction analyses yield slightly different lengths for the helical domains (see references in Fig. 1 legend). Thus, the predictions of the points of initiation and ends of the helical domains are not precise, and these remain to be determined by physicochemical structural analyses.

Within the predicted helical domains both charged and hydrophobic residues display specific periodicities. The hydrophobic residues predominantly occupy positions **a** and **d** in consecutive heptade repeats of a-b-c-d-e-f-g. This periodicity is characteristic of coiled-coil α-helical super-secondary structures and thus provides further support for the validity of the predicted positions of the helical domains (Fraser *et al.*, 1976; Parry *et al.*, 1977; Elleman *et al.*, 1978; McLachlan, 1978; McLachlan, and Stewart, 1982; references in Fig. 1 legend). In these domains both the hydrophobic and charged residues are conserved more frequently and many substitutions for them represent conservative replacements, e.g., Asp (-) for Glu (-), or Arg (+) for Lys (+).

Although among all IF proteins amino acid sequence homology is higher within the predicted helical domains, it is especially prominent in domain III and in the carboxy terminal end of domain IV.

Despite evolutionary divergence of sequences, the amino acid compositions of the total central region and those of the individual helical domains of IF proteins have remained remarkably similar. The conservation of the helical secondary structure of IF proteins, despite the divergence of their sequences, may be partly a result of this conservation of amino acid compositions compatible with α-helicity. For example, for each IF protein 25-30% of the residues in this region are Glu and Leu, both

Amino terminal
AA Frequency (%)
Ala Ile Leu Met Phe Val Cys Asn Gln Gly Pro Ser Thr Trp Tyr Asp Glu His Arg Lys
Central region
AA Frequency (%)
Ala Ile Leu Met Phe Val Cys Asn Gln Gly Pro Ser Thr Trp Tyr Asp Glu His Arg Lys
Carboxy terminal
AA Frequency (%)
Ala Ile Leu Met Phe Val Cys Asn Gln Gly Pro Ser Thr Trp Tyr Asp Glu His Arg Lys
T-I   T-II   GFP   NFP   DESMIN   VIMENTIN

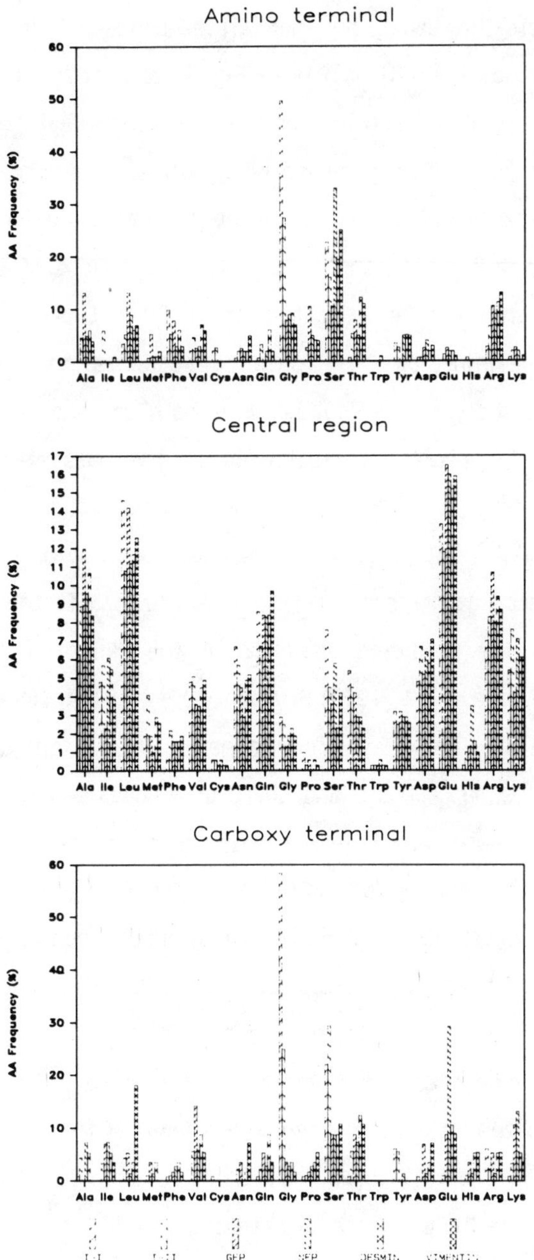

Figure 2

Figure 2 Legend.   The amino acid composition of the amino terminal, central, and carboxy terminal regions of IF proteins. The IF protein sequences included in this figure are shown in Figure 1. The keratins, marked as T-I and T-II, are mouse keratins. The "central region" extends from the first residue of the first helical domain to the last residue of the fourth helical domain. The terminal regions are as shown in Figure 1.

of which highly favor α-helical structures (Fig. 2).   It should be noted that the significant homologies among IF proteins do not result from this conserved amino acid composition, as a shift of the optimal alignment of the sequences by a single residue reduces the homologies to random levels.

B. The Amino and Carboxy Terminal Regions are Extremely Variable Across Subclasses of IF Proteins

The amino and carboxy terminal regions of different IF proteins show an extreme degree of heterogeneity in their sizes, sequences, and amino acid compositions (Figs. 1 and 2). The size differences among IF proteins is a result of differences in the length of these terminal portions rather than the structurally conserved central region. In terms of amino acid composition, the only conspicuous feature common to *all* IF proteins appears to be a relatively high ratio of serines in the amino and, to a lesser degree, carboxy terminal regions (Fig. 2). The secondary and higher orders of structure of these regions are not known yet, and structure prediction analyses cannot be applied to them because of their unusual sequences and highly skewed amino acid compositions.   Thus, future structural information may disclose novel features at the ends of IF proteins.

Although the terminal regions show a great diversity, the members of each of the three major families (type I keratins, type II keratins, and non-keratins) can be further subgrouped on the basis of sequence homology of these terminal regions as noted below. The available sequences of IF proteins from different species indicate that for each subclass of IF protein the sequences of these regions are conserved across species. However, the extent of the evolutionary conservation of terminal region sequences of each subclass remains to be further defined by determination of sequences from a greater number of species.

1. Type I and type II keratins

a) *Within both type I and type II keratin families, the cytoskeletal and microfibrillar keratins can be distinguished by their terminal sequences.* Although the central helical domains of the cytoskeletal and microfibrillar keratins share 50-60% homology within each type, the known terminal sequences of these two groups of proteins show no homology (Fuchs and Hanukoglu, 1985). In the terminal regions both wool microfibrillar type I and type II keratins are rich in cysteine and proline (Crewther *et al.*, 1980; Sparrow and Inglis, 1980). The cysteine richness of these regions readily suggests that these residues are involved in S-S covalent bond formation between individual keratins contributing greatly to the special characteristics and strength of the microfibrils. In contrast, the terminal regions of cytoskeletal type I and type II keratins contain no or only a few cysteines or prolines, but instead they are highly rich in glycine as noted below. However, as the sequences of only a few members of these subfamilies are known, their characteristics cannot be outlined fully yet. In addition, the possibility of the existence of keratins that share properties of both forms also remains.

b) *The cytoskeletal keratins can be further subgrouped on the basis of evolutionarily conserved differences in the terminal sequences.* The amino acid compositions of both cytoskeletal type I and type II keratins are unusually rich in glycine. The great majority of these glycines appear in the terminal regions of the keratins as glycine is not compatible with α-helix. The amino termini of all cytoskeletal keratin sequences known to date include irregularly spaced short repeats of three or four glycines separated by other residues or clusters of serines, whereas the carboxy termini of some but not all cytoskeletal keratins include GGGX type repeats (Fig. 1).

At present, the partial sequences of only two different type I keratins from one species are known (Jorcano *et al.*, 1984). The central helical regions of these bovine keratins are highly homologous with one another (68% homology) as well as with those of other type I keratins from human, mouse and Xenopus (Hanukoglu and Fuchs, 1982; Steinert *et al.*, 1983; Hoffman and Franz, 1984). However, the carboxy termini of these two bovine keratins do not show even a vestige of sequence homology. On the basis of a comparison of these two sequences with the sequences of type I keratins from other species, one subfamily of type I keratins may be characterized by a carboxy terminus that includes GGGXYGG type repeats, whereas the other by a non-glycine rich carboxy terminus with a specific conserved sequence. Most interestingly, the 3' non-coding regions of the mRNAs of these subfamilies are also conserved (Jorcano *et al.*, 1984). Most probably, as more keratin sequences from different species are determined, other similar evolutionarily conserved distinct variations will be observed in both type I and type II keratin families.

As noted above, in contrast to non-keratins, the assembly of a keratin filament requires the pairwise association of a type I keratin with a type II keratin. The microfibrillar keratins include specific structural pairs of type I and type II keratins. Whether similarly some or all of the cytoskeletal type I keratin subfamilies structurally match specific type II keratin subfamilies remains to be determined. This subject can be examined from two perspectives: Are specific type I and type II subfamilies expressed together in different tissues? Which specific type I and type II subfamilies can copolyermize *in vitro*?

2. Non-keratins

a) *Desmin, vimentin and GFP which can form both homopolymers and heteropolymers with one another, share similar terminal sequences.* Although desmin, vimentin and GFP share a lower homology with one another in their terminal sequences than in their central regions, these terminal regions are much less divergent than those of the keratin subfamilies (Quax *et al.*, 1984). As the terminal sequences of the IF proteins appear to be involved in end-to-end linkage of protofibrillar subunits, the similarity of these sequences may partly explain the ability of these proteins to copolymerize both *in vitro* and *in situ* (Steinert *et al.*, 1982; Wang *et al.*, 1984).

b) *Among IF proteins the most extreme size variation is observed in the terminal regions of NFP.* As noted above, in vertebrates the NF are constituted from three different proteins NFP-L, NFP-M, and NFP-H, with respective M_r of 62,000, 107,000 and ~140,000. Despite this extreme size variation, all three NFP share the same α-helical central region common to all IF proteins, and the size differences result from the different sizes of the carboxy terminal regions (Geisler *et al.*, 1984, 1985). The partial sequences and amino acid compositions of fragments isolated from the terminal regions indicate that they are highly variable in sequence but generally share a high percentage of glutamate (20-30%) and lysine (16-25%) (Geisler *et al.*, 1984, 1985).

IV. COMPARISON OF THE GENE STRUCTURES OF IF PROTEINS

Recently, the complete sequences of a hamster vimentin and a human type I keratin genes have been determined (Quax *et al.*, 1983; Marchuk *et al.*, 1984, 1985). These two sequences as well as mapping of intron positions in bovine type I and type II keratin genes (Lehnert *et al.*, 1984) reveal the following major similarities and differences among IF protein genes:

The positions of introns are remarkably well conserved among all IF protein genes. Both human and bovine type I keratin genes characterized to date have seven introns, while bovine type II keratin genes and the hamster vimentin gene have eight introns. In each gene all but one of the introns appear in the central helical region of the coded protein. Six of these introns occupy identical positions in this structurally conserved region.

In many genes intron positions have been observed to mark the boundaries of structural or functional domains of the coded proteins (Craik, 1983; Go, 1983; Lonberg and Gilbert, 1985). However, in IF protein genes most of the highly conserved intron positions do not appear to border the structural domains of the proteins. Only one intron (intron 6 in type I keratin and vimentin, and intron 7 in type II keratin) occurs at the end of the central helical region. The conserved presence of this intron at the border of the conserved central region and the highly variable carboxy terminal region raises the possibility that intron mediated gene fragment shuffling might have resulted in the juxtaposition of a structurally conserved region with highly variable segments. However, the other introns in the central region are all located within the helical domains. As noted above, both the sequence and the structure of the helical domains are conserved much more than other regions of IF proteins. Introns appear neither at the border of the first helical domain and the highly divergent amino terminal region, nor at the borders of the helical domains and the proline and glycine rich non-helical linker regions that connect these domains (Fig. 1; Quax *et al.,* 1983; Marchuk *et al.,* 1984).   The conservation of the positions of the introns suggests that these may be of some functional significance.   Yet, with our present knowledge of the structure of the IF proteins if these positions have a structural significance it eludes us.

Despite the strict conservation of intron position, across and within species comparisons of the intron sequences indicate that neither the sizes nor the sequences of introns are conserved among IF protein genes.

The regulatory sequences 5' upstream from the type I keratin and the vimentin genes appear to be very divergent.   Furthermore, the type I keratin gene, which is highly expressed in some epidermal tissues, contains sequences that share significant homology with enhancer elements found in other genes, whereas vimentin gene does not appear to have similar sequences in the corresponding gene segments.   These differences in the 5' regulatory regions may ultimately explain the differential regulation of expression of IF genes in different tissues (Marchuk *et al.,* 1985).

V. SEQUENCE BASED CLASSIFICATION OF IF PROTEINS

The IF proteins have been sometimes referred to as a "family of proteins." Yet, as detailed above, sequence homology between *some* IF proteins can be about 10% over the complete lengths of the proteins and as low as 24% in the central structurally conserved region (Table I).   This degree of diversity of sequences mandates the use of a larger set name for the IF proteins.   Thus, consistent with nomenclature used for other multigene families (e.g., Regier *et al.*,1983), we suggest that the group of IF proteins be referred to as a "*superfamily*".

To summarize the sequence comparisons presented above, the 20-30 proteins which constitute the IF protein superfamily can be grouped in three distinct families based on their sequence homologies in the structurally conserved central region: 1) type I keratins, 2) type II keratins, 3) non-keratins. The first two families include about 10 members each in mammals, whereas the third includes desmin, vimentin, GFP, NFP-L, NFP-M, and NFP-H. The degree of central region homology between proteins from different families is 24-35% and between proteins within the same family it is > 50% (Table I). The proteins in each of the three families, but especially the multi-membered keratin families, can be further grouped into subfamilies primarily by the similarities and differences of the sequences of their amino and carboxy terminal regions.

This classification of IF proteins based strictly on sequence similarities also reflects developmental, structural, and functional differences among the proteins. The keratins are expressed in epithelial tissues and apparently can form filaments only by association of specific type I and type II pairs, whereas the non-keratins are expressed predominantly in non-epithelial tissues and can form filaments by both self-polymerization and, in tissues and cells where they co-exist, by heteropolymer formation in association with another non-keratin IF protein.

VI. THE EVOLUTION OF IF PROTEINS

The characteristics common to all IF proteins and their genes provide strong evidence that all IF proteins have evolved from a common ancestral protein. Thus, we can surmise that the IF proteins that exist today have

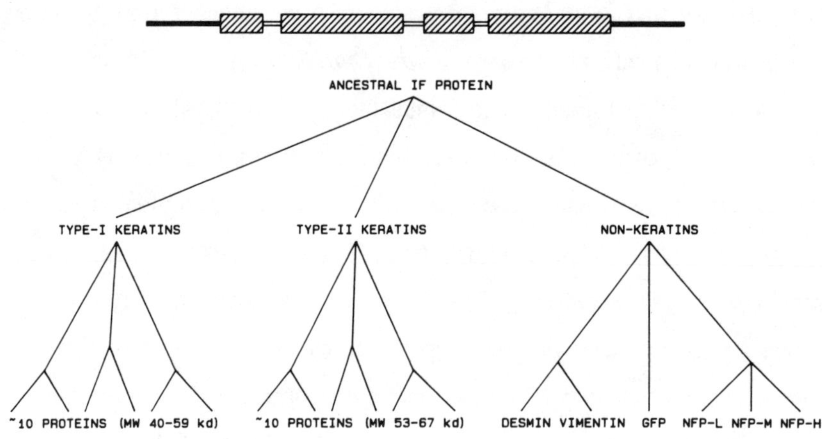

Figure 3. A hypothetical scheme for the evolution of the IF proteins. The top portion of the figure shows a secondary structural model for the ancestral IF protein based on Figure 1. In this model the bars with the parallel lines represent the four α-helical domains; the empty bars represent the short β-turn regions that link the α-helical domains; and the black bars represent the non-helical terminal regions that highly differ in length and sequence among the different IF proteins (Fig. 1). The lengths of the domains within the central region are depicted in proportion to the lengths of the respective domains (Fig. 1). The nodes from which three lines emanate indicate uncertainty in the respective order of emergence of the different IF protein classes during evolution. The branches of keratin subfamilies are hypothetical and as only a few sequences of keratins are known the number of subfamilies cannot be estimated at present.

probably evolved from an ancestral protein which also possessed a predominantly α-helical central region (Fig. 3). Yet the nature of the ancestral terminal sequences on either side of this helical region cannot be conjectured because of the extreme variability of these in present day IF proteins. The possibility that the ancestral protein had no, or very short, terminal sequences also exists, as one cytoskeletal keratin has an M_r of 40,000, which indicates that its terminal sequences are very short (Fuchs and Marchuk, 1983; Kim *et al.*, 1984).

The groupings of IF proteins by central and terminal region sequence homologies indicate that the evolution of IF proteins probably proceeded in two major stages: 1) Generation of the prototypes of the three major families; 2) The evolution of subfamilies of proteins from these prototypes mainly by diversification of the sequences of the terminal regions, and to a minor degree by divergence of the sequence of the central region (Fig. 3). The evolutionary paths that have led to both of these stages most probably consisted of a series of events that included gene duplications, and subsequent point mutations, insertions and deletions, as previously suggested for other multigene families (e.g., Markert *et al.*, 1975; Wahli *et al.*, 1981; Jones and Kafatos, 1982).

The great differences in the types of evolutionary modifications observed in the central versus terminal regions of IF proteins represent outstanding examples to illustrate the interdependence of evolutionary changes and structural and functional roles of protein segments. While, during the course of evolution, very few insertions and deletions have been accepted in the central region, the terminal regions have been completely redesigned and structured for different IF proteins.

The conservation of the size and the secondary structural organization of the central region during evolution was probably necessary to maintain the ability of these proteins to assemble into coiled-coil filaments. Between some IF proteins the overall sequence of this region shows up to 76% divergence. Hence, structural conservation cannot be ascribed solely to the maintenance of a similar sequence, but also to the maintenance of (1) an α-helix compatible amino acid composition, and (2) the specific periodicities of charged and hydrophobic residues which delineate points of interaction between helices of polypeptide chains forming coiled-coils (see Section III.A)

The high degree of divergence of the sequences of the central regions indicates that many details of the surface topology of the IF proteins (the R groups of amino acid residues project outwards from the main axis of an α-helix) are not crucial for defining the similarity of their secondary and super-secondary structures. This idea is strengthened further by the observation that IF can be formed *in vitro* and *in situ* by combinations of several different IF proteins. This degree of freedom in the choice of amino acids that make up the sequences of IF proteins greatly contrasts with the strict conservation of the sequences of globular proteins that form cytoskeletal filaments -- actin and tubulin. For example, between human and *Drosophila* cytoplasmic actin sequences, there is only a 2% divergence (Hanukoglu *et al.*, 1983). This difference suggests that the globular shapes of actins and tubulins may be disrupted more readily by single residue alterations and that their surface topology is much more precisely architectured for their many structural interactions, whereas the long helical segments of IF proteins appear to tolerate substitutions easily as long as specific residues at points of interaction between different helical domains are maintained.

The amino and carboxy terminal regions of IF proteins are extremely diversified across subfamilies, but the presently available few sequences from different species suggest that the representative of each subfamily is conserved across species (see Section III.B.1 and III.B.2). If these observations are generalizable to a larger number of species, then the across-species conservation of the unusual terminal sequences would suggest that they fulfill indispensable roles in the structure of IF proteins within the protofibril, and the interactions of these proteins with other cellular molecules. Thus, the different terminal sequences may be important in selective interactions with specific molecules in different tissues. Yet, the question that remains is to what degree the complete sequences of the terminal regions are necessary for these specific interactions? The possibility can be raised that these terminal regions include segments without structural importance on the basis of the finding that partial digestion of these do not seem to disrupt the ability of IF proteins to form IF (Section II.B). But, in the absence of good understanding of the structural organization and interactions of these regions and the degree of their across-species conservation for each subfamily, these possibilities remain speculations and the posed question cannot be answered reliably at present.

Understanding the mechanisms and selective pressures that have led to the generation and diversification of the unusual amino and carboxy terminal sequences pose some of the biggest puzzles in the evolution of IF proteins. The following questions represent some of these: How were the extremely variable amino and carboxy terminal regions juxtaposed with the conserved central region? How did the glycine rich inexact repeats at the amino and carboxy terminal regions of cytoskeletal keratins arise? Do they share a common origin? Judging from their conserved central region sequences, the microfibrillar and cytoskeletal keratin subfamilies are much more closely related within each type I and type II family. But, while the terminal sequences of these subfamilies appear to be completely unrelated, cytoskeletal keratins from different families show similar glycine rich repeats. The sequences of these terminal regions cannot be aligned with a degree of certainty, but do their glycine rich repeats nonetheless indicate a common origin, or represent a case of convergent evolution serving the need of these proteins to assemble together? How did the carboxy terminal portions of NFP-M and NFP-H reach their gigantic sizes and sequences different from all other IF proteins?

In conclusion, the many IF proteins with their segmented architecture of a conserved central α-helical region flanked by extremely variable amino and carboxy terminal sequences provide valuable data and pose intriguing questions about the mechanisms of evolutionary change.

REFERENCES

Ben-Zeev, A. (1984). Control of intermediate filament protein synthesis by cell-cell interaction and cell configuration. *FEBS Lett.* **171**, 107-110.

Chou, P. Y., and Fasman, G. D. (1978). Prediction of the secondary structure of proteins from their amino acid sequence. *Adv. Enzymol.* **47**, 45-148.

Chou, P. Y., and Fasman, G. D. (1979). Prediction of β-turns. *Biophys. J.* **16**, 367-383.

Craik, C. S., Rutter, W. J., and Fletterick, R. (1983). Splice junctions: Association with variation in protein structure. *Science* **220**, 1125-1129.

Capetanaki, Y. G., Ngai, J., Flytzanis, C. N., and Lazarides, E. (1983). Tissue-specific expression of two mRNA species transcribed from a single vimentin gene. *Cell* **35**, 411-420.

Crewther, W. G., and Harrap, B. S. (1967). The preparation and properties of a helix-rich fraction obtained by partial proteolysis of low sulfur S-carboxymethyl-kerateine from wool. *J. Biol. Chem.* **242**, 4310-4319.

Crewther, W. G., Inglis, A. S., and McKern, N. M. (1978). Amino acid sequences of α-helical segments from S-carboxymethylkerateine-A. *Biochem. J.* **173**, 365-371.

Crewther, W. G., Dowling, L. M., and Inglis, A. S. (1980). Amino acid sequence data from a microfibrillar protein of α-keratin. *In* "The Structure and Chemical Reactions of Keratins," Vol. 2, pp. 79-91.

Crick, F. H. C. (1953). The packing of α-helices: simple coiled-coils. *Acta Cryst.* **6**, 689-697.

Dowling, L. M., Parry, D. A. D., and Sparrow, L. G. (1983). Structural homology between hard α-keratin and the intermediate filament proteins desmin and vimentin. *Biosci. Rep.* **3**, 73-78.

Eichner, R., Bonitz, P., and Sun, T.-T. (1984). Classification of epidermal keratins according to their immunoreactivity, isoelectric point and mode of expression. *J. Cell Biol.* **98**, 1388-1396.

Elleman, T. C., Crewther, W. G., and Touw, J. V. D. (1978). Amino acid sequences of α-helical segments from S-carboxymethylkerateine-A: Statistical analysis. *Biochem. J.* **173**, 387-391.

Franke, W. W., Schmid, E., Schiller, D. L., Winter, S., Jarasch, E. D., Moll, R., Denk, H., Jackson, B. W., and Illmensee, K. (1982). Differentiation-related patterns of expression of protein of intermediate-size filament in tissues and cultured cells. *Cold Spring Harbor Symp. Quant. Biol.* **46**, 431-474.

Franke, W. W., Schiller, D. L., Hatzfeld, M., and Winter, S. (1983). Protein complexes of intermediate-sized filaments: melting of cytokeratin complexes in urea reveals different polypeptide separation characteristics. *Proc. Natl. Acad. Sci. USA* **80**, 7113-7117.

Fraser, R. D. B., MacRae, T. P., and Rogers, G. E. (1972). "Keratins: Their Composition, Structure and Biosynthesis. C. C. Thomas, Springfield, USA.

Fraser, R. D. B., MacRae, T. P., and Suzuki, E. (1976). Structure of the α-keratin microfibril. *J. Mol. Biol.* **108**, 435-452.

Fuchs, E. V., Coppock, S. M., Green, H., and Cleveland, D. W. (1982). Two distinct classes of keratin genes and their evolutionary significance. *Cell* **27**, 75-84.

Fuchs, E., and Marchuk, D. (1983). Type I and type II keratins have evolved from lower eukaryotes to form the epidermal intermediate filaments in mammalian skin. *Proc. Natl. Acad. Sci. USA* **80**, 5857-5861.

Fuchs, E., and Hanukoglu, I. (1986). Epidermal α-keratins: Structural diversity and changes during tissue differentiation. *In* "The Biology of the Integument" (in press).

Gard, D. L., Bell, P. B., and Lazarides, E. (1983). Coexistence of desmin and the fibroblastic intermediate filament subunit in muscle and nonmuscle cells: Identification and comparative peptide analysis. *Proc. Natl. Acad. Sci. USA* **76**, 3894-3898.

Garnier, J., Osguthorpe, D. J., and Robson, B. (1978). Analysis of the accuracy and implications of simple methods for predicting the secondary structure of globular proteins. *J. Mol. Biol.* **120**, 97-120.

Geisler, N., and Weber, K. (1982). The amino acid sequence of chicken muscle desmin provides a common structural model for intermediate filament proteins. *EMBO J.* **1**, 1649-1656.

Geisler, N., Kaufmann, E., and Weber, K. (1982). Protein chemical characterization of three structurally distinct domains along the protofilament unit of desmin 10 nm filaments. *Cell* **30**, 277-286.

Geisler, N., Fischer, S., Vandekerckhove, J., Plessman, U., and Weber, K. (1984). Hybrid character of a large neurofilament protein (NF-M):

intermediate filament type sequence followed by a long and acidic carboxy-terminal extension. *EMBO J.* **4**, 57-63.

Geisler, N., Fischer, S., Vandekerckhove, J., Van Damme, J., Plessman, U., and Weber, K. (1985). Protein-chemical characterization of NF-H, the largest mammalian neurofilament component; intermediate filament-type sequences followed by a unique carboxy terminal extensions. *EMBO J.* **4**, 57-63.

Go, M. (1983). Modular structural units, exons, and function in chicken lysozyme. *Proc. Natl. Acad. Sci. USA* **80**, 1964-1968.

Gough, K. H., Inglis, A. S., and Crewther, W. G. (1978). Amino acid sequences of α-helical segments from S-carboxymethylkerateine-A. *Biochem. J.* **173**, 373-385.

Granger, B. L., and Lazarides, E. (1983). Expression of major neurofilament subunit in chicken erythrocytes. *Science* **221**, 553-556.

Gruen, L. C., and Woods, E. F. (1983). Structural studies on the microfibrillar proteins of wool. *Biochem. J.* **209**, 587-595.

Hanukoglu, I., and Fuchs, E. (1982). The cDNA sequence of a human epidermal keratin: divergence of sequence but conservation of structure among intermediate filament proteins. *Cell* **31**, 243-252.

Hanukoglu, I., and Fuchs, E. (1983). The cDNA sequence of a type II cytoskeletal keratin reveals constant and variable structural domains among keratins. *Cell* **33**, 915-924.

Hanukoglu, I., Tanese, N., and Fuchs, E. (1983). Complementary DNA sequence of a human cytoplasmic actin. *J. Mol. Biol.* **163**, 673-678.

Hirokawa, N., Glicksman, M. A., Willard, M. B. (1984). Organization of mammalian neurofilament polypeptides within the neuronal cytoskeleton. *J. Cell Biol.* **98**, 1523-1536.

Henderson, D., Geisler, N., and Weber, K. (1982). A periodic ultrastructure in intermediate filaments. *J. Mol. Biol.* **155**, 173-176.

Hoffman, W., and Franz, J. K. (1984). Amino acid sequence of the carboxy-terminal part of an acidic type I cytokeratin of molecular weight 51000 from Xenopus laevis epidermis as predicted from the cDNA sequence. *EMBO J.* **3**, 1301-1306.

Holthofer, H., Miettinen, A., Lehto, V.-P., Lehtonen, E., and Virtanen, I. (1984). Expression of vimentin and cytokeratin types of intermediate filament proteins in developing and adult human kidneys. *Lab. Invest.* **50**, 552-559.

Jones, C. W., and Kafatos, F. C. (1982). Accepted mutations in a gene family: Evolutionary diversification of duplicated DNA. *J. Mol. Evol.* **19**, 87-103.

Jones, L. N. (1975). The isolation and characterization of α-keratin microfibrils. *Biochim. Biophys. Acta.* **412**, 91-98.

Jorcano, J. L., Rieger, M., Franz, J. K., Schiller, D. L., Moll, R., and Franke, W. W. (1984). Identification of two types of keratin polypeptides within the acidic cytokeratin subfamily I. *J. Mol. Biol.* **179**, 257-281.

Kallman, F., and Wessells, N. K. (1967). Periodic repeat units of epithelial cell tonofilaments. *J. Cell. Biol.* **32**, 227-231.

Kim, K. H., Rheinwald, J. G., and Fuchs, E. (1983). Tissue specificity of epithelial keratins: Differential expression of mRNAs from two multigene families. *Mol. Cell. Biol.* **3**, 495-502.

Kim, K. H., Marchuk, D., and Fuchs, E. (1984). Expression of unusually large keratins during terminal differentiation: Balance of type I and type II keratins is not disrupted. *J. Cell Biol.* **99**, 1872-1877.

Kim, K. H., Schwartz, F., and Fuchs, E. (1984). Differences in keratin synthesis between normal epithelial cells and squamous cell carcinomas are mediated by vitamin A. *Proc. Natl. Acad. Sci. USA* **81**, 4280-4284.

Lazarides, E. (1982). Intermediate filaments; A chemically heterogeneous, developmentally regulated class of proteins. *Ann. Rev. Biochem.* **51**, 219-250.

Lane, E. B., Hogan, B. L. M., Kurkinen, M., and Garrels, J. I. (1983). Co-expression of vimentin and cytokeratins in parietal endoderm cells of early mouse embryo. *Nature* **303**, 701-704.

Lee, L. D., and Baden, H. P. (1976). Organisation of the polypeptide chains in mammalian keratin. *Nature* **264**, 377-379.

Lehnert, M. E., Jorcano, J. L., Zentgraf, H., Blessing, M., Franz, J. K., and Franke, W. W. (1984). Characterization of bovine keratin genes: similarities of exon patterns in genes coding for different keratins. *EMBO J.* **3**, 3279-3287.

Lewis, S. A., Balcarek, J. M., Krek, V., Shelanski, M., and Cowan, N. J. (1984). Sequence of a cDNA clone encoding mouse glial fibrillary acidic protein: structural conservation of intermediate filaments. *Proc. Natl. Acad. Sci. USA* **81**, 2743-2746.

Lonberg, N., and Gilbert, W. (1985). Intron/exon structure of the chicken pyruvate kinase gene. *Cell* **40**, 81-90.

Lu, Y.-J., and Johnson, P. (1983). The N-terminal domain of desmin is not involved in intermediate filament formation: evidence from thrombic digestion studies. *Int. J. Biol. Macromol.* **5**, 347-350.

Marchuk, D., McCrohon, S., and Fuchs, E. (1984). Remarkable conservation of structure among intermediate filament genes. *Cell* **39**, 491-498.

Marchuk, D., McCrohon, S., and Fuchs, E. (1985). Complete sequence of a gene encoding a human type I keratin: Sequences homologous to enhancer elements in the regulatory region of the gene. *Proc. Natl. Acad. Sci. USA* **82**, 1609-1613.

Markert, C. L., Shaklee, J. B., and Whitt, G. S. (1975). The evolution of a gene. *Science* **189**, 102-114.

McLachlan, A. D. (1978). Coiled coil formation and sequence regularities in the helical regions of α-keratin. *J. Mol. Biol.* **124**, 297-304.

McLachlan, A. D., and Stewart, M. (1982). Periodic charge distribution in the intermediate filament proteins desmin and vimentin. *J. Mol. Biol.* **162**, 693-698.

Milam, L., and Erickson, H. P. (1982). Visualization of a 21-nm axial periodicity in shadowed keratin filaments and neurofilaments. *J. Cell. Biol.* **94**, 592-596.

Moll, R., Franke, W. W., Schiller, D. L., Geiger, B., and Krepler, R. (1982). The catalog of human cytokeratins: Patterns of expression in normal epithelia, tumors and cultured cells. *Cell* **31**, 11-24.

Parry, D. A. D., Crewther, W. G., Fraser, R. D. B., and MacRae, T. P. (1977). Structure of α-keratin: Structural implication of the amino acid sequences of the type I and type II chain segments. *J. Mol. Biol.* **113**, 449-454.

Pauling, L., and Corey, R. B. (1953). Compound helical configurations of polypeptide chains: structure of proteins of the α-keratin type. *Nature* **171**, 59-61.

Quax, W., Egberts, W. V., Hendrisk, W., Quax-Jeuken, Y., and Bloemendal, H. (1983). The structure of the vimentin gene. *Cell* **35**, 215-223.

Quax, W., Heuvel, R.V.D., Egberts, W.V., Quax-Jeuken, Y., and Bloemendal H. (1984). Intermediate filament cDNAs from BHK-21 cells: demonstration of distinct genes for desmin and vimentin in all vertebrate classes. *Proc. Natl. Acad. Sci. USA* **81**, 5970-5974.

Quinlan, R. A., and Franke, W. W. (1982). Heteropolymer filaments of vimentin and desmin in vascular smooth muscle tissue and cultured baby hamster kidney cells demonstrated by chemical crosslinking. *Proc. Natl. Acad. Sci. USA* **79**, 3452-3456.

Quinlan, R. A., Cohlberg, J. A., Schiler, D. L., Hatzfeld, M., and Franke, W. W. (1984). Heterotypic tetramer (A2D2) complexes of non-epidermal keratins isolated from cytoskeletons of rat hepatocytes and hepatoma cells. *J. Mol. Biol.* **178**, 365-388.

Regier, J. C., Kafatos, F. C., and Hamodrakas, S. J. (1983). Silkmoth chorion multigene families constitute a superfamily: Comparison of C and B family sequences. *Proc. Natl. Acad. Sci. USA* **80**, 1043-1047.

Sharp, G., Osborn, M., and Weber, K. (1982). Occurrence of two different intermediate filament protein in the same filament *in situ* within a human glioma cell line. *Ex. Cell Res.* **141**, 385-395.

Skerrow, D., Matoltsy, G., and Matoltsy, M. (1973). Isolation and characterization of the helical regions of epidermal prekeratin. *J. Biol. Chem.* **248**, 4820-4826.

Steinert, P. M. (1978). Structure of the three-chain unit of the bovine epidermal keratin filaments. *J. Mol. Biol.* **123**, 49–70.

Steinert, P. M., Idler, W. W., and Zimmerman, S. B. (1976). Self-assembly of bovine epidermal keratin filaments *in vitro. J. Mol. Biol.* **108**, 547–467.

Steinert, P. M., Idler, W. W., and Goldman, R. D. (1980). Intermediate filaments of baby hamster kidney (BHK-21) cells and bovine epidermal keratinocytes have similar ultrastructures and subunit domain structures. *Proc. Natl. Acad. Sci. USA* **77**, 4534–4538.

Steinert, P., Idler, W., Aynardi-Whitman, M., Zackroff, R., and Goldman, R. D. (1982). Heterogeneity of intermediate filaments assembled *in vitro. Cold Spring Harbor Symp. Quant. Biol.* **46**, 465–474.

Steinert, P. M., Rice, R. H., Roop, D. R., Trus, B. L., and Steven, A. C. (1983). Complete amino acid sequence of a mouse epidermal keratin subunit and implications for the structure of intermediate filaments. *Nature* **302**, 794–800.

Steinert, P. M., Parry, D. A. D., Racoosin, E. L., Idler, W. W., Steven, A. C., Trus, B. L., and Roop, D. R. (1984). The complete cDNA and deduced amino acid sequence of a type II mouse epidermal keratin of 60,000 Da: analysis of sequence differences between type I and type II keratins. *Proc. Natl. Acad. Sci. USA* **81**, 5709–5713.

Sun, T.-T., Shih, C., and Green, H. (1979). Keratin cytoskeletons in epithelial cells of internal organs. *Proc. Natl. Acad. Sci. USA* **76**, 2813–2817.

Wahli, W., Dawid, I. B., Ryfell, G. U., and Weber, R. (1981). Vitellogenesis and vitellogenin gene family. *Science* **212**, 298–304.

Walter, M. F., and Biessmann, H. (1984). A monoclonal antibody that detects vimentin-related proteins in invertebrates. *Mol. Cell. Biochem.* **60**, 99–108.

Wang, E., Cairncross, J. G., and Liem, R. K. H. (1984). Identification of glial filament protein and vimentin in the same intermediate filament system in human glioma cells. *Proc. Natl. Acad. Sci. USA* **81**, 2102–2106.

Weber, K., Osborn, M., and Franke, W. W. (1980). Antibodies against merokeratin from sheep wool decorate cytokeratin filaments in non-keratinizing epithelial cells. *Eur. J. Cell. biol.* **23**, 110–114.

Woods, E. F., and Gruen, L. C. (1983). Sturctural studies on the microfibrillar proteins of wool: characterization of the α-helix particle produced by chymotryptic digestion. *Aust. J. Biol. Sci.* **34**, 515–526.

Zehner, Z. E., and Paterson, B. M. (1983). Characterization of the chicken vimentin gene: single copy gene producing multiple mRNAs. *Proc. Natl. Acad. Sci. USA* **80**, 911–915.

Note added in proof:

A recent paper has reported partial sequence homologies between some IF proteins and oncogene proteins (Crabbe, 1985). In addition, the recently determined sequence of the Scrapie protein reveals Gly-Gly-X type repeats that are also found in cytoskeletal keratins (Oesch *et al.*, 1985).

Crabbe, M. J. C. (1985). Partial sequence homologies between cytoskeletal proteins, c-myc, Rous sarcoma virus and adenovirus proteins, transducin, and β- and γ- crystallins. *Biosci. Rep.* **5**, 167-174.

Oesch, B., Westaway, D., Walchli, M., McKinley, M. P., Kent, S. B. H., Aebersold, R., Barry, R. A., Tempst, P., Teplow, D. B., Hood, L. E., Prusiner, S. B., and Weissmann, C. (1985). A cellular gene encodes scrapie PrP 27-30 protein. *Cell* **40**, 735-746.

After the completion of this manuscript additional sequences of intermediate filament protein genes have been reported (see references below). The characteristics of these sequences are consistent with the conclusions noted in this review.

Geisler, N. Plessmann, U., and Weber, K. (1985). The complete amino acid sequence of the major mammalian neurofilament protein (NF-L). *FEBS Lett.* **182**, 475-478.

Johnson, D. L., Idler, W. W., Zhou, X.-M., Roop, D. R., and Steinert, P. M. (1985). Structure of a gene for the human epidermal 67-kDa keratin. *Proc. Natl. Acad. Sci. USA* **82**, 1896-1900.

Jorcano, J. L., Franz, J. K., and Franke, W. W. (1984). Amino acid sequence diversity between bovine epidermal cytokeratin polypeptides of the basic (type II) subfamily as determined from cDNA clones. *Differentiation* **28**, 155-163.

Krieg, T. M., Schafer, M. P., Cheng, C. K., Filpula, D., Flaherty, P., Steinert, P. M., and Roop, D. R. (1985). Organization of a type I keratin gene. *J. Biol. Chem.* **260**, 5867-5870.

Steinert, P. M., Parry, D. A. D., Idler, W. W., Johnson, D. L., Steven, A. C., and Roop, D. R. (1985). Amino acid sequences of mouse and human epidermal type II keratins of Mr 67,000 provide a systematic basis for the structural and functional diversity of the end domains of keratin intermediate filament subunits. *J. Biol. Chem.* **260**, 7142-7149.

Tyner, A. L., Eichman, M. J., and Fuchs, E. (1985). The sequence of a type II keratin gene expressed in human skin: Conservation of structure among all intermediate filament genes. *Proc. Natl. Acad. Sci. USA* **82**, 4683-4687.

EVOLUTIONARY ORIGIN OF ANTIGEN-BINDING POCKETS[1]

Susumu Ohno

Beckman Research Institute of The City of Hope
Duarte, CA 91010

ABSTRACT

The adaptive immune system unique to vertebrates apparently owes its existence to the β_2-microglobulin superfamily of genes. Single domain monomeric members of this superfamily are β_2-microglobulin itself and Thy-1 T cell antigen. Two domain, monomeric members are Lyt-1, Lyt-2,3 antigens of helper and cytotoxic T cells, whereas class II MHC (major histocompatibility) antigens of macrophages and B cells as well as T cell receptor constitute the two domain, dimeric memberships. Ubiquitously expressed class I MHC antigens are comprised of three β_2-microglobulin-like domains while immunoglobulin light chains are two domain and immunoglobulin heavy chains are four domain structures.

It is granted that all β_2-microglobulin-like domains are alike being comprised of two β-sheet structures held together by a single intrachain disulfide bridge, yet there occurred, within this superfamily, a remarkably novel evolutionary innovation. V_L and V_H domains of immunoglobulins as well as $V_{\alpha T}$ and $V_{\beta T}$ cell receptors acquired an ability to form dimeric antigen-binding pockets. Needless to say, this innovation was the corner stone in development of the adaptive immune system. In our previous two papers (Ohno *et al.*, 1984; and Ohno *et al.*, 1985), this innovation was

---

[1]Abbreviations. V_{KL}, $V_{\lambda L}$: Immunoglobulin κ and λ-type light chain variable region. C_L : Immunoglobulin light chain constant region. V_H : Immunoglobulin heavy chain variable region. C_H : Immunoglobulin heavy chain constant region. CDR regions: Complementarity determining regions. MHC antigens: Major histocompatibility antigens.

identified as development of the tryptophane loop within the second β-sheet structure. In this paper, it will be shown that the tryptophane loop is already developed in the first domain of T8 antigen (human equivalent of Lyt-2,3) as well as in the first domain of poly-Ig A transepithelial transport receptors. Thus, it is implied that the innovation of the tryptophane loop by certain β_2-microglobulin-like domains must have occurred in the immediate predecessor of vertebrates such as *Cephalochordata* or even *Urochordata*.

I. INTRODUCTION

In evolution, truely novel innovations occurred but seldom. Accordingly, when it occurred, it was plagiarized in every conceivable way during the subsequent period which might have encompassed tens of millions of years. For example, body forms of all vertebrates, from limbless snakes to winged birds, from ground crawling lizards to bipedal man, are but various plagiarisms of the one basic body form; a skull connected to a vertebral column to which pectoral and pelvic girdles are attached. This basic body form already evident in jawless ostracoderm fish of Silurian time harks back to the free swimming tadpole-like larva of a sessile filter feeder which resembled certain modern tunicates and its predecessor was already in existence at the beginning of Cambian times, some 500 million years ago. In such free-swimming larva, the gill apparatus is situated in its swollen "head" region. Behind, there is a muscular swimming tail, strengthened by a stout, but flexible, longitudinal notochord; predecessor of the vertebral column. There exists a longitudinal dorsal nerve cord which, in the head region, receives sensory information from rudimentary sense organs. By the process of Paedomorphosis, the adult stage as a Sessile filter feeder was eliminated and the larval form became sexually mature and reproduced. Thus, the seed of all vertebrates was sown.

The adaptive immune system characterized by the ability to generate specific antibodies directed against any and all antigens of the universe is another truely novel innovation that occurred only in the phylum *Chordata* at about the same time as the basic body form of vertebrates was set forth by tadpole-like larva of a sessile filter-feeder. Defensive mechanism against viral and bacterial infections employed by various invertebrates are apparently based on entirely different principles. Their body fluids contain a variety of lectins and foreign materials aggregated by lectins are actively phagocitized by macrophages (Komano *et al.*, 1980). Their body fluids also appear to contain various forms of wide spectrum bacteriocidal polypeptides. There is little doubt that the adaptive immune system developed by extensive plagiarization of the β_2-microglobulin-like domain. Yet during this plagiarization, there was a truely novel innovation which would be the subject of this paper.

II. A DIFFERENCE BETWEEN NON-ANTIGEN-BINDING AND ANTIGEN-BINDING DOMAINS OF THE β_2-MICROGLOBULIN SUPERFAMILY

Both X-ray crystallographic studies (Saul *et al.*, 1978) and the simulation studies by computer graphics (Travers *et al.*, 1984) have revealed the essential characteristics of all β_2-microglobulin-like domains to be two separate β-sheet structures held close together by a single intrachain disulfide bridge. In the schematic drawing of two domain, dimeric T cell receptor shown in Figure 1, components of the first β-sheet structure are drawn solid black, while those of the second β-sheet structure are outlined, and three connecting pieces are shaded. Figure 1 also reveals a distinct difference that separates antigen-binding domains from non-antigen-binding domains. The first β-sheet structures

1
$V_\alpha T$
S
TRP
110
TRP
S
$C_\alpha T$
S
1
$V_\beta T$
TRP
S
110
S
TRP
$C_\beta T$
S

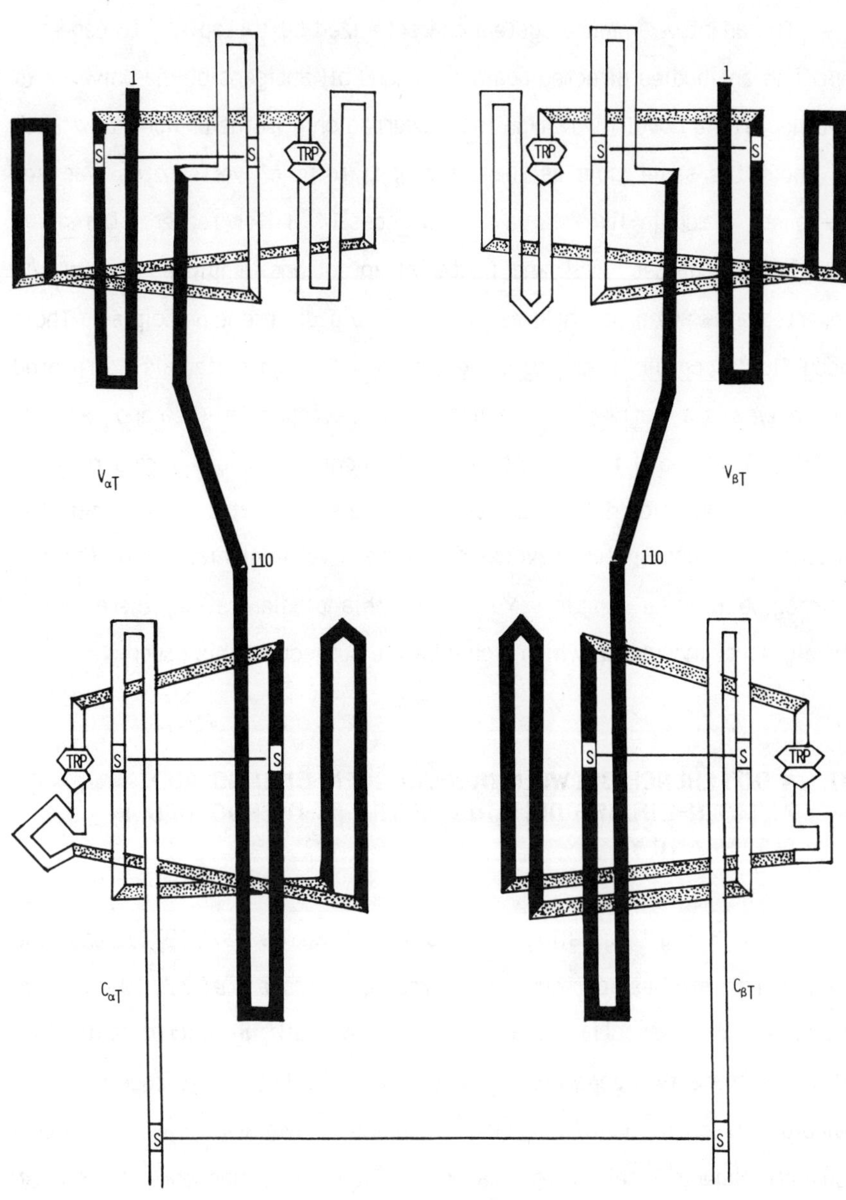

Figure 1

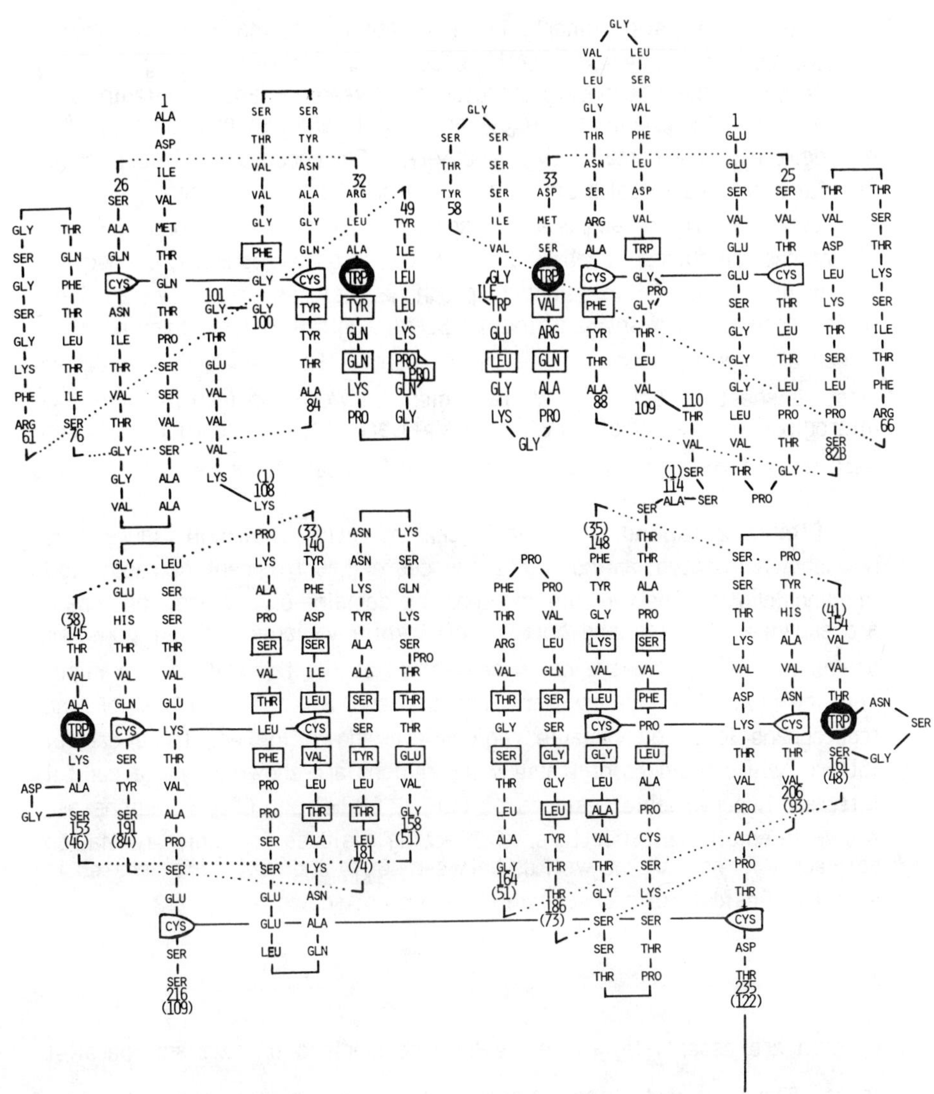

Figure 2

Figure 1 legend. Dimeric T-cell receptor. Schematic demonstration of two domain, dimeric T cell receptor shown emphasizing striking differences in their secondary structures between non-antigen-binding β_2-microglobulin-like domains ($C_{\alpha T}$ and $C_{\beta T}$) and antigen-binding β_2-microglobulin-like domains ($V_{\alpha T}$ and $V_{\beta T}$). Components of first β-sheet structure are drawn solid black, while those of second β-sheet structure are outlined. Three segments connecting first and second β-sheet structures are shaded. Each disulfide bridge is shown as a line connecting a pair of S, and the invariant tryptophane residues are shown as TRP. Dimer between non-antigen-binding β_2-microglobulin-like domains (between $C_{\alpha T}$ and $C_{\beta T}$) are formed by the contact between their respective first β-sheet structures. While dimer between antigen-binding β_2-microglobulin-like domains ($V_{\alpha T}$ and $V_{\beta T}$) are made by contacts between each respective tryptophane loop of second β-sheet structure.

Figure 2 legend. Schematic demonstration shown in Figure 1 is now shown in actual amino acid sequences of Fab fragment of rabbit IgG immunoglobulin. The variable and constant domains of κ-class light chain are shown at the top and bottom left, while variable and C_{H1} constant domains of γ-class heavy chain are shown at the top and bottom right. The invariant TRP of constant regions and all the residues of the tryptophane loops of variable regions are shown in very large capital letters, while residues of frame-work regions are shown in large capital letters. Hypervariable residues of CDR-1, CDR-2 and CDR-3 regions are shown in small capital letters. Contacting residues for dimer formation between $V_{\kappa L}$ and V_H as well as between $C_{\kappa L}$ and $C_{\lambda H1}$ are encased in squares. Residues of connecting pieces are not shown.

of both are essentially the same being comprised of two anti-parallel loops. The first loop is comprised of 26 to 35 residues from the amino terminus, while the second loop is comprised of 51st to 72nd or 73rd residues in the case of non-antigen-binding domains, 61st to 76th (V_L of immunoglobulins) or 66th to 82nd residues (V_H of immunoglobulins and $V_{\alpha}T$ and $V_{\beta}T$ of T cell receptors) in the case of antigen-binding domains as shown in Figure 2 (Ohno *et al.*, 1984). Whether or not a given

β_2-microglobulin-like domain can serve as a component of the dimeric antigen-binding pocket is solely determined by the construction of its second β-sheet structure and the key to it is the absolutely invariant tryptophane residue occupying 41st or 44th position in the case of non-antigen-binding domains and 35th or 36th residues in the case of antigen-binding domains (Figs. 1 and 2). This tryptophane residue is missing only from single domain, monomeric β_2-microglobulin and Thy-1 antigen, but present in every other member of the vast β_2-microglobulin superfamily (Kabat *et al.*, 1983). In the case of all non-antigen-binding domains, the second β-sheet structure remains under-developed because the invariant tryptophane residue is included only in an 8 to 9 residues long stand, thus, the second β-sheet structure being comprised merely of this strand and a single loop made of the carboxyl terminal 26 or so residues (Figs.1 and 2). In the case of all antigen-binding domains (V_L and V_H of immuno-globulins as well as $V_\alpha T$ and $V_\beta T$ of T cell receptors), on the other hand, this tryptophane is now included in a loop, thus, their second β-sheet structure is comprised, at the minimum as in the case of V_L , of two anti-parallel loops (Figs. 1 and 2). Furthermore, a third loop is formed within the second β-sheet structure by residues of CDR-2 region in the case of V_H as well as in the case of both types of V_T's as discussed in Ohno *et al.* (1984), and in Figures 1 and 2.

Whether or not the second β-sheet structure contains the tryptophane loop determines the manner of dimer formation between two β_2-microglobulin domains that come face to face. Dimers between two non-antigen-binding domains are made by the contact between 12 or so residues of each respective first β-sheet structure as schematically illustrated at the bottom of Figure 1. This occurs between first and second domains of α- and β-subunits of class II MHC antigens (Travers *et al.*, 1984) and between $C_\alpha T$ and $C_\beta T$ of T cell receptors (Ohno *et al.*,

1984) as well as between C_L and C_{H1} of immunoglobulins (Saul *et al.*, 1978). In Figure 2, 12 contacting residues each residing in first β-sheet structure of $C_{\kappa}L$ (bottom left) and in that of $C\gamma_{H1}$ (bottom right) of rabbit immunoglobulins are encased in squares. Thus, the invariably present tryptophane residue of the second β-sheet structure swings outside (Figs. 1 and 2). In Sharp contrast, the dimer between two antigen-binding domains such as between V_L and V_H of immoglobulins as well as between $V_{\alpha}T$ and $V_{\beta}T$ of T cell receptors are formed by contact between each tryptophane loop of second β-sheet structure. At the top of Figure 2, contacting residues residing within the tryptophane loop of rabbit $V_{\kappa L}$ (Fig. 2, top left) as well as in that of rabbit V_H (Fig. 2, top right) are encased in squares. A few residues within the carboxyl terminal loop of the second β-sheet structure also make contact with each other (Fig. 2, top).

It thus became clear that the true innovation in development of the adaptive immune system was the creation of the tryptophane loop within the second β-sheet structure of β_2-microglobulin-like domains, for the floor of dimeric antigen-binding pockets of immunoglobulins as well as T cell receptors are formed by a pair of contacting tryptophane loops furnished by V_L and V_H or $V_{\alpha T}$ and $V_{\beta T}$. The roof of the dimeric antigen-binding pocket on the other hand is furnished by hypervariable residues of CDR-1 and CDR-3 regions in the case of V_L ; hypervariable residues of CDR-2 also contributing in the case of V_H and V_T's (Ohno *et al.*, 1984; Ohno *et al.*, 1985).

III. THE ORIGIN OF TRYPTOPHANE LOOPS: MONOSPECIFIC, MONOMERIC COMBINING SITES OF POLY-IgA TRANSEPITHELIAL TRANSPORT RECEPTOR AND T CELL LYT-1, LYT-2,3 ANTIGENS

As with many an innovative thought in genetics, J. B. S. Haldane was apparently the first to realize that each organism or rather each population of organisms has to pay a toll for the privilege of maintaining each gene locus in the genome in the form of a deleterious mutation rate (Haldane, 1957). Because of this mutation load, there exists an inverse correlation between the spontaneous mutation rate primarily determined by the accuracy of copying by DNA polymerase as well as by the efficacy of repair mechanisms and the total number of gene loci in the genome. Viruses being endowed with but a few gene loci of their own can afford to be venturesome, thus, being equipped with mistake prone nucleic acid polymerases and extremely high spontaneous mutation rate estimated to be of the order of 10^{-3}/base pair/year. The total number of gene loci in the vertebrate genome approaches 10^5. Accordingly, conservation of the gains acquired in the past at the expense of venturesomeness became the rule to abide by; the spontaneous mutation rate becoming of the order of 10^{-9}/base pair/year. Because of this million-fold difference in evolutionary rates between parasitic pathogens and the host, the adaptive immune system of vertebrates had no choice but to generate specific antibodies before the actual encounter with antigens. Herein is found the Kafakaesque absurdity of the adaptive immune system in that it is prepared to confront the universe generating specific antibodies directed against any and all antigens; although some of them were present only in the remote past and no longer relevant today, while others may not come into being even in the future. It has been estimated that the repertoire of antibody diversity approaches 10^7 (Jerne, 1971). Such diversity is created in parts by the possession in the genome of 100 or so different V_H genes, 50 or so V_L genes as well as 20 or so D_H coding segments and five each of J_H and J_L coding segments; D and J contributing hypervariable residues of CDR-3 region.   Somatic mutations make further and very significant contributions for the expansion of the repertoire.   The recent study on the primitive shark, *Heterodontus,* indicated that this capacity of the adaptive immune system to confront the universe was already acquired in early vertebrates soon after its inception (Litman, *et al.,* 1985).   It follows then that the invention of the tryptophane loop must necessarily have preceded the birth of variable region genes; V_L, V_H and V_T .   In short, there had to be β_2-microglobulin-like domains with the tryptophane loop that were neither variable regions of immunoglobulins nor those of T cell receptors.

Of various classes of immunglobulins, only polymeric forms of IgM and IgA are secreted to saliva, guts, milk, etc.   This secretion is due to the presence of Poly-Ig receptor that engages in transepithelial transport of polymeric IgA and IgM. Recently, the entire amino acid sequence of this poly-Ig receptor of the rabbit was deduced (Mostov *et al.,* 1984).   It consists of six β_2-microglobulin-like domains and amino acid sequences of its six domains were shown to share an average of 28% identical residues with the consensus sequence of mammalian V_{KL} (Mostov *et al.,* 1984).   The first domain of the poly-Ig receptor sharing 36% homology with the V_{KL} was of particular interest, for it was clearly endowed with the tryptophane loop.   At the top of Figure 3 the consensus amino acid sequence of rabbit V_{KL} is aligned in the manner to reveal its first and second β-sheet structures. At the bottom of Figure 3, first domain of the rabbit poly-Ig receptor is aligned in the similar way.   Comparing the top with the bottom, it would be noted, many of the residues occupying critical positions for the maintenance of similar secondary structures are

94
SER
95
SER
TYR THR
ASN VAL
ALA VAL
LEU GLY GLY
ALA GLN PHE
32
ARG
49
TYR
ILE
LEU TRP CYS GLY
LEU TYR TYR
LYS GLN GLY GLY
101
PRO GLN TYR
PRO THR
GLN LYS ALA
84
GLY PRO
41 40
1
ALA
ASP
ILE
VAL
MET
THR
GLN
GLY
THR
PRO
GLU
SER
VAL
VAL
SER
LYS
LYS
107
26
SER
ALA THR
GLN GLN
CYS PHE
ASN THR
ILE LEU
THR THR
SER VAL ILE
VAL THR SER
SER GLY
ALA GLY
ALA VAL
14 15
69 68
THR GLY
GLN SER
PHE GLY
THR SER
THR LYS
ILE PHE
SER ARG
76 61
RABBIT V_{κL} CONSENSUS

1
GLN
SER
SER
LEU
LEU SER
GLY THR
PRO THR
SER PRO
SER TYR GLY
ILE TYR
PHE CYS PHE
GLY THR
PRO ILE VAL
GLY SER
VAL GLU
VAL SER
ASP
84
34
77 76
LYS ASP
PRO
PHE
ASP
THR
LEU
LYS
GLY
ARG
68
103
GLY
ASN ARG
VAL GLY
GLY LEU
VAL ASP
PHE GLY PHE
TRP CYS GLY
CYS CYS LYS
57 56 40
SER ALA ARG
THR LEU LYS
GLY THR PHE GLY
TYR VAL
THR CYS
61
ARG ARG
GLY GLU SER
SER GLU
GLU
48
111
VAL ASN
TYR VAL
ILE
VAL
SER
GLN
LYS VAL
PRO LEU
118 19
GLU
20
RABBIT POLY-IG RECEPTOR FOR TRANSEPITHELIAL
TRANSPORT OF POLY-IGM AND POLY-IGA

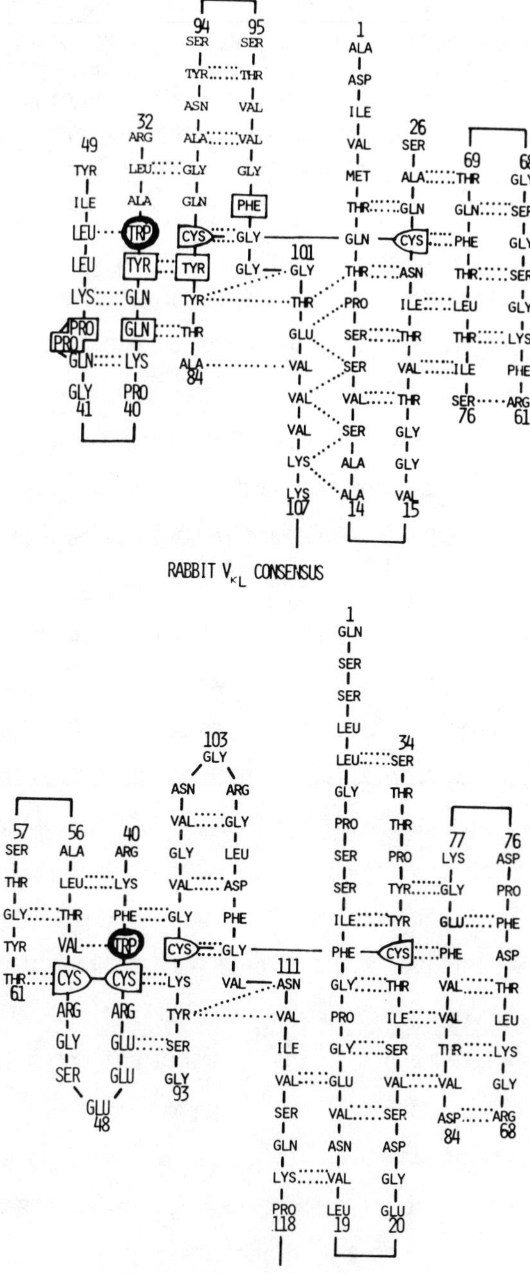

Figure 3

Figure 3 Legend.   The secondary structure assumed by the consensus amino acid sequence of rabbit V_{KL} is shown at the top, and the deduced secondary structure assumed by the first domain of the rabbit poly-Ig receptor for transepithelial transport of poly-IgM and poly-IgA is shown at the bottom.   Again, residues of connecting pieces are omitted. Hydrogen bonds to be formed between certain pairs of residues facing each other are indicated by broken lines.

identical. The transformation from the secondary structure characteristic of non-antigen-binding β_2-microglobulin-like domains (bottom of Figs. 1 and 2) to the one whose second β-sheet structure is endowed with the tryptophane loop was not likely to have been accomplished by mere random amino acid substitutions.   For the particular secondary structure once established acquires its own inertia and tends to perpetuate itself.   For example, loops of first and second β-sheet structures of β_2-microglobulin-like domains often contain penta to heptapeptidic segments which would, when left alone, rather form α-helix.   Nevertheless, they are incorporated into loops of β-sheet structures because of Ser, Thr, Val, Leu residues in their neighborhoods that by hydrogen bonding with each other stabilize the loop structures.   Thus, it was of extreme interest to note the presence of a pair of cysteine residues within the tryptophane loop of the first domain of poly-Ig receptor (Fig. 3, bottom).   The disulfide bridge formation between this pair would have sufficed to break the inertia and transformed the secondary structure characteristic of non-antigen-binding β_2-microglobulin-like domains to that characteristic of antigen-binding domain by the forced formation of the tryptophane loop within the second β-sheet structure.   I found the fifth domain of rabbit poly-Ig receptor also to be capable of the tryptophane loop formation in its second β-sheet.   Whether or not first and fifth domains of poly-Ig receptor form a single dimeric binding-pocket specific for poly-IgM and IgA remain to be seen.   Conversely, two monomeric domains may independently recognize different parts of poly-IgM and poly-IgA. Nevertheless, the point is that aside from V_L and V_H of immunoglobulins and $V_{\alpha T}$ and $V_{\beta T}$ of T cell receptors, β_2-microglobulin-like domains with the tryptophane loop in their second β-sheet structures have been utilized as monospecific ligand binding sites.

While the above was very instructive as to possible evolutionary steps that culminated in the creation of tryptophane loop within β_2-microglobulin-like domain, the poly-Ig receptor is an obvious late comer in development of the adaptive immune system.

Fortunately, there is yet another example of the above, and this example can indeed be considered as the precursor of antigen-binding domains of T-cell receptors as well as immunoglobulin.

Unlike humoral antibodies produced by B cells, membrane-bound receptors of T cells do not recognize alien antigens *per se* but monitor conformational perturbation of self MHC antigens caused by alien antigens; the altered self recognition mechanism of Zinkernagel and Doherty (1974). While receptors of cytotoxic T cells monitor altered self class I MHC antigens, those of helper T cells monitor altered self class II MHC antigens. Yet so far as $V_{\alpha T}$ and $V_{\beta T}$ subunits that comprise dimeric antigen-binding pockets of T cell receptors are concerned, there is no evidence of functional specialization, rather both cytotoxic and helper T cells apparently draw their $V_{\alpha T}$ and $V_{\beta T}$ from common pools (Ohno *et al.*, 1984; Litman *et al.*, 1985). Thus, it appears that it is the coexisting cell-type specific plasma membrane antigen that causes functional specialization. Indeed, in the mouse, there exist helper T cell specific Lyt-1 antigen and cytotoxic T cell specific Lyt-2,3 antigen. Recently the deduced amino acid sequence of human cytotoxic T cell specific T8 antigen (human homologue of mouse Lyt-2,3) have been determined and the remarkable sequence homology between the first

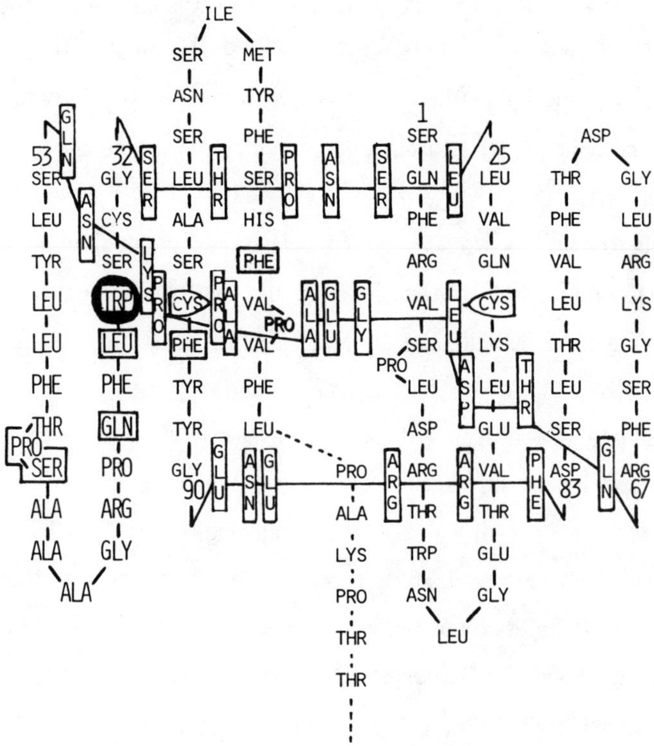

Figure 4. Tryptophane loop of human cytotoxic T cell T8 antigen (mouse Lyt-2,3 equivalent) autodimer for class I MHC antigen binding? The deduced secondary structure assumed by the first domain of human cytotoxic T cell specific T8 plasma membrane antigen is shown. This time, residues of three connecting pieces bridging first and second β-sheet structures are also shown being encased in elongated square.

β_2-microglobulin-like domain of T8 antigen and V_{KL} was noted (Litman *et al.*, 1985). The deduced secondary structure of the first domain shown in Figure 4 and rabbit V_{KL} shown at the top of Figure 3 is very evident. Thus, it would appear that either as a monomer or as an autodimer, first domain of human T8 antigen, therefore, Lyt-2,3 antigen of the mouse is endowed with the monospecific binding affinity toward the evolutionary conserved portion of class I MHC antigens.

All in all, there is little doubt that vertebrates became endowed with the adaptive immune system solely because of the previous innovation that created the tryptophane loop within the second β-sheet structure of certain β_2-microglobulin-like domains.

REFERENCES

Haldane, J. B. B. (1957). The cost of natural selection. *J. Genet.* **55**, 511-524.

Jerne, N. K. (1971). Somatic generation of immune recognition. *Europ. J. Immnol.* **1**, 1-9.

Kabat, E. A., Wu, T. T., Bilofsky, H., Reid-Miller, M., and Perry, H. (1983). "Sequences of Proteins of Immunological interest." U.S. Dept. of Health and Human Services, Natl. Inst. of Health, Bethesda, Maryland.

Komano, H., Mizuno, D., and Natori, S. (1980). Purification of lectin induced in the hemolymph. *J. Biol. Chem.* **255**, 2912-2924.

Litman, G. W., Berger, L., Murphy, K., Litman, R., Hinds, K., and Erickson, B. W. (1985). Immunoglobulin V_H gene structure and diversity in *Heterodontus*, a phylogenetically primitive shark. *Proc. Natl. Acad. Sci. USA* **82**, 844-848.

Litman, D. R., Thomas, Y., Maddon, P. J., Chess, L., and Axel, R. (1985). The isolation and sequence of the gene encoding T8: A molecule defining functional classes of T lymphocytes. *Cell* **40**, 237-246.

Mostov, K. E., Friedlander, M., and Blobel, G. (1984). The receptor for transepithelial transport of IgA and IgM contains multiple immunoglobulin-like domains. *Nature* **308**, 37-43.

Ohno, S., Matsunaga, T., and Lee, A. D. (1984). The invariably present Tryptophan loop as the core of all divergent antigen-binding pockets. *Scand. J. Immnol.* **20**, 377-388.

Ohno, S., Mori, N., and Matsunaga, T. (1985). Antigen-binding specificities of antibodies are primarily determined by seven residues of V_H. *Proc. Natl. Acad. Sci. USA* **82**, 2760-2764.

Patten, P., Yokota, T., Rothbard, J., Chien, Y.-H., Arai, K., and Davis, M. M. (1984). Structure, expression and divergence of T-cell receptor β-chain variable regions. *Nature* **312**, 40-46.

Saul, F. A., Amzel, L. M., and Poljak, R. J. (1978). Preliminary refinement and structural analysis of the Fab fragments from human immunoglobulin New at 2.0 A resolution. *J. Biol. Chem.* **253**, 585-597.

Travers, P., Blundell, T. L., Sternberg, M. J. E., and Bodmer, W. F. (1984). Structural and evolutionary analysis of *HLA-D*-region products *Nature* **310**, 235-238.

Zinkernager, R. M., and Doherty, P. D. (1974). Immunological surveillance against altered self-components by sensitized T-lymphocytes in lymphocytic choriomeningitis. *Nature* **251**, 547-549.

ARE THE MAJOR HISTOCOMPATIBILITY COMPLEXES OF THE MOUSE t-HAPLOTYPES DISTINCTIVE GENE POOLS?

Gabriel Gachelin
Christiane Delarbre

Unité de Biologie Moléculaire du Gène
U. 277 I.N.S.E.R.M., U.A. 535
C.N.R.S. – Institut Pasteur
25 rue du Dr. Roux
75724 Paris Cédex 15, France

Takashi Morita

Research Institute for Microbial Disease
Osaka University
Osaka, Japan

ABSTRACT

The t-haplotypes are natural polymorphisms of the mouse chromosome 17, found in up to 50% of the wild mouse populations. They are characterized by several intriguing genetic features, and especially by their inability to recombine with normal chromosome 17, over a chromosomal segment which appears to include the Major Histocompatibility Complex (MHC). In order to get some insight into the relationships that could exist between the MHC's of t- and non-t haplotypes, we have examined by the Southern blotting technique, the H-2K, D, L alleles of 10 independent t haplotypes, and determined the complete nucleotide sequence of the H-2K gene of the t^{w32} haplotype. All these genes appear to be extremely homologous to the corresponding alleles in non-t haplotypes. Our results, taken along with the numerous data reported by others concerning the MHC of t haplotypes and re-examined here, suggest the occurrence of gene exchanges, between the MHC's of t and non-t haplotypes. Moreover, the t haplotypes might behave as a "reservoir" for at least some of the genes found in the MHC of normal, non-t mice.

I. INTRODUCTION

The chromosome 17 of the mouse harbors two complex and polymorphic genetic systems, which have attracted for years the attention of mammalian geneticists: first, genes located within the T/t complex, which exert a wide range of effects on the embryonic development and on the reproduction of the affected animals (reviewed in Bennett, 1975; Klein, 1977; Lyon, 1981; Silver, 1981); second, the genes of the Major Histocompatibility Complex (MHC) which participate to most functions of the immune system (reviewed in Dorf, 1981; Hood *et al.*, 1983; Klein, 1975; Klein *et al.*, 1983). The genetic map of the chromosome 17 of the mouse is given in Figure 1.

T-mutations, also located on chromosome 17, are responsible for the characteristic short-tailed phenotype of T/+ animals. All original t haplotypes have been identified on the basis of the interaction of their *tct* gene (allelic to T and present in all t haplotypes) with T, which leads to the tailless phenotype of T/t animals. t haplotypes have been identified in several outbred laboratory stocks (Klein and Hammerberg, 1977; Silver *et al.*, 1984); however, most of them have been recovered from wild mouse populations (Klein, 1977; Klein *et al.*, 1984) in which they are present at a high frequency, up to 50% in certain areas (Klein *et al.*, 1984). The t haplotypes are thus major polymorphisms of the chromosome 17 in natural populations of mice.

All complete t haplotypes share the same basic properties. 1) Embryos homozygous for two t haplotypes unable to complement each other, die during their development. 2) Animals heterozygous for two t haplotypes belonging to different complementation groups are viable: 16 complementation groups have been identified so far, defining at least as many different "lethality factors" in the t region (Klein *et al.*, 1984). In addition, t^x/t^y males are sterile, whereas t^x/t^y females are fertile. 3) +/t^x males transmit the t chromosome at a high frequency; distortion of segregation can reach up to 99% in certain t haplotypes and is associated with the functioning of distorter and responder genes (Lyon, 1984). 4) Recombination between a T (or +) chromosome, and most t chromosomes, is dramatically reduced over at least 15cM from the centromere. This property has prevented for long the detailed genetic analysis of the t region. The recent finding that recombination occurs in t^x/t^y females, has allowed a tentative mapping of the several traits defining t haplotypes (Artzt, 1984; Artzt *et al.*, 1982; Pla and Condamine, 1984; Shin *et al.*, 1984). The entire tf-MHC segment appears likely to be inverted (Shin *et al.*, 1983) in t chromosomes, a chromo-somal rearrangement which can account for the inhibition of the recombination. Finally, careful analysis of several gene products of the t region (the Tcp proteins) has led to the conclusion that all t haplotypes share the same electromorphs, some of them different from those found in classical laboratory animals (Silver *et al.*, 1983). These and other findings (Shin *et al.*, 1982; Silver, 1982; Sturm *et al.*, 1982) suggest strongly that t haplotypes derive all from a common ancestor animal, the origin of which is unknown, which could have "borrowed" the entire t region from a different mouse semi-species (Silver, 1981) or was already present at the time of the split between *Mus musculus musculus* and *M. m. domesticus* (Figueroa *et al.*, 1985).

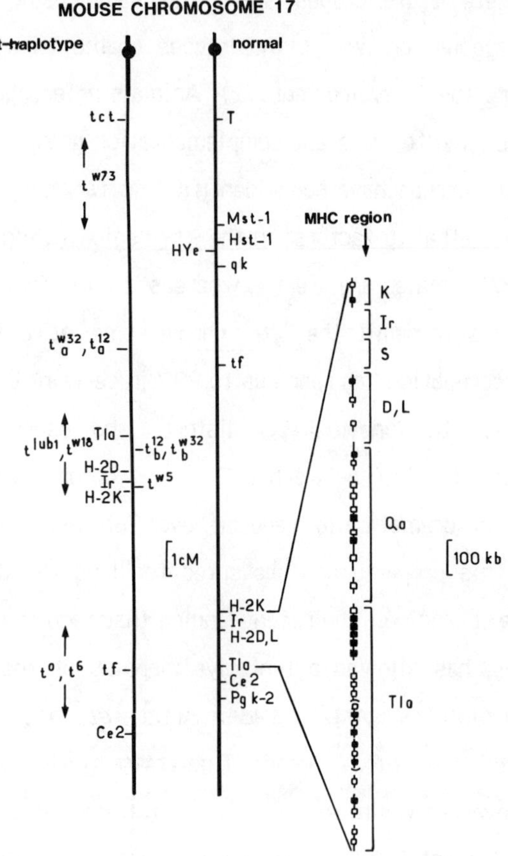

Figure 1. Genetic map of normal and t-bearing mouse chromosome 17. Gene symbols are defined in Artzt *et al.* (1984), Artzt (1984), and Shin *et al.* (1984) for t-haplotypes, and in Forejt (1981) and Klein *et al.* (1983) for the normal chromosome. The molecular map of class I genes of the MHC region is from Hood *et al.* (1983). Open boxes are for genes which can be expressed; black ones for genes the expression of which could not be detected.

An account of the present knowledge of the molecular structure of the normal murine MHC is given elsewhere (this volume; Hood *et al.*, 1983; Weiss *et al.*, 1984). Briefly, the MHC is a 1.5 – 2 cM long portion of chromosome 17, located at about 15cM from the centromere on the normal chromosome 17 (Fig. 1). It contains a cluster of genes controlling the efficiency of immune responses, some genes of the complement cascade, and the multigene family of the 25-35 class I genes encoding class I antigens and related molecules (Hood *et al.*, 1983; Weiss *et al.*, 1984; Fig. 1). Although some are secreted, most class I gene products are cell surface antigens which are distributed into two main classes: the highly polymorphic classical histocompatibility antigens, named H-2K, D and L antigens, encoded each by a unique gene located in the corresponding H-2 locus (Goodenow *et al.*,1982; Weiss *et al.*,1984); oligomorphic and monomorphic "differentiation" antigens, encoded by genes located in the Qa-TLa region of the MHC (Hood *et al.*, 1983; Weiss *et al.*, 1984). Most of the class I genes are found located in the latter region. Because of their extensive antigenic polymorphism, the different H-2 antigens can be easily distinguished one from each other by serotyping. On the other hand, with the recent availability of DNA probes, Class I genes can be studied now at the genomic level.

A close association exists between t haplotypes and their MHC: on the basis of the results of a serological analysis, the inhibition of recombination displayed by most t chromosomes has been found to involve the entire MHC of t haplotypes (Hammerberg *et al.*, 1976; Klein and Hammerberg, 1977). In other words, the 15cM long region of t chromosomes which harbors the two complexes, is considered as behaving as a "genetic unit" essentially undisturbed by recombination. Because of the extremely high linkage disequilibrium between MHC and "t genes," MHC antigens and genes were indeed taken as markers for t chromatin, in studies aimed at the understanding of the origin of t haplotypes as well as on the evolution of the MHC itself.   On the basis of the complex restriction patterns yielded by the whole class I multigene family, it was concluded that the MHC's of the various t haplotypes studied were closely related one to each other (Shin *et al.,* 1982; Silver, 1982). A refined serological analysis (Sturm *et al.,* 1982) led independently to a similar conclusion, thus in favor of a common origin of the t-haplotypes. However, the complexity of the results published later on, has become such, that, although the existence of a common ancestor to t haplotypes remains the most likely hypothesis (Dembic *et al.,* 1984, 1985); Figueroa *et al.,* 1982; Klein *et al.,* 1983; Klein *et al.,* 1984; Nizetic *et al.,* 1984) the existence of a clear-cut relationship between a given t haplotype (as defined by its complementation group) and its MHC haplotype was clearly to be abandoned. We have been interested for long in the exchanges of MHC genes occurring among wild mice, and especially between t and non-t animals. The increasing complexity and diversity of the results obtained by us and many others is such that a re-evaluation of the relationships existing between the MHC of t and of non-t animals could not be avoided. On our side, we have compared the restriction patterns of the H-2K, D, L genes present in non-t and t chromosomes (Delarbre *et al.,* 1985), and sequenced the H-2K gene of the t^{w32} haplotype (Morita *et al.,* 1985).   In the present review, we have undertaken the re-examination of these results and those published by others (Dembic *et al.,* 1984, 1985; Figueroa *et al.,* 1982; Hammerberg *et al.,* 1976; Klein *et al.,*1983; Klein *et al.,* 1984; Nizetic *et al.,* 1984). The conclusion is the following: the genetic distance between the MHC of t and non-t animals appears to be most often negligible. Irrespectively of the origin of t-chromosomes, we conclude that at least some t and non-t MHC's have shared until recently the same evolutionary history, and that, in spite of the nearly complete inhibition of recombination, some gene exchanges should occur between the MHC's of the two mouse populations.

II. RFLP ANALYSIS OF THE H-2K, D, L GENES IN LABORATORY MICE AND IN t-HAPLOTYPES

Most of the southern blot analyses of class I genes of t-haplotypes have been carried out using probes able to hybridize with a large number of, if not all, class I genes (Shin *et al.*, 1982; Silver, 1982). This has prevented the assignment of the rare polymorphisms observed, to any specific class I gene, and thus precluded the extensive use of these results in a study of the evolution of the MHC. We have, therefore, decided to focus on the relationships between alleles. Because of the availability of a cDNA probe specific for H-2K genes and of a genomic probe specific for H-2K, D, L genes (Delarbre *et al.*, 1985), we have been able to carry out RFLP analysis of the H-2K, D, L genes without any significant contribution of the other class I genes of the multigene family. The details of this analysis have been published elsewhere (Delarbre *et al.*, 1985). The probes used have been described elsewhere (Delarbre *et al.*, 1985; Morita *et al.*, 1985). Briefly, one probe is a 165 bp long *Pst* I-*Hinf* I fragment, derived from the 3' end of an H-2K^d cDNA, which hybridizes solely with H-2K genes (Delarbre *et al.*,1985). The second, a 950 bp long *Pst* I-*Pst* I fragment derived from the H-2K^d gene (Kvist *et al.*, 1983), contains the 3' end and some of its flanking sequence: it was found to recognize the H-2K, D, L genes of all MHC haplotypes tested so far, as well as the pseudogene K', associated with the H-2^{b,d,k} genes (Bernd *et al.*,1984; Delarbre *et al.*, 1985; Hood *et al.*, 1983; Weiss *et al.*,1984). No cross-hybridizing DNA sequence has been found elsewhere in the genome and especially in the genes believed to act as donor of sequences in conversion-like events of H-2K genes (Mellor *et al.*, 1983). The evolution of the 3' part of the K, D, L genes should not therefore be significantly contributed to by the latter mechanism. The size of the restriction fragments and their assignment to the H-2K, D, L genes, as deduced from the analysis of the DNA of recombinant mice and of H-2^b and H-2^d class I cosmid clones have been published elsewhere (Delarbre *et al.*, 1985). The same and typical restriction patterns for the H-2K, D, L genes have been found for animals sharing the same serologically defined MHC, but assumed to be of different origins, or to have been separated for long one from each other (Klein, 1975). The possibility of exact gene duplications could not be considered in such a RFLP analysis.

The same probes and the same restriction enzymes, have been used in a survey of the 10 independent t haplotypes maintained at the Pasteur Institute as the progeny of (T/t^x × T/t^x) crosses. With the exception of t^{WPA1} and t^{WPA2} haplotypes (Guenet *et al.*, 1980), all the t-animals studied here are derived from the original stocks kept in Dr. Bennett's laboratory. A brief description of the t haplotypes studied here is given in Table I. We have decided to investigate t haplotypes in their "original" genetic configuration as well as in some proper F1 animals, rather than in nearly inbred animals on the 129/Sv background, because possible recombination involving the MHC has not been checked for, neither during the inbreeding process, nor after.

A summary of the data concerning the 3' end of the K^t genes is presented in Table II. Since we do not possess recombinants separating the K and the D, L loci in the MHC of t haplotypes, the genetic assignment of the restriction fragments remains only tentative and based upon a comparison with the restriction patterns of non-t animals, except when the nucleotide sequence of an H-2K^t gene is available (Morita *et al.*,

TABLE I. The geographic origin and the main features of the different t-haplotypes used in the present studies are given in the table. A more complete description of these natural polymorphisms of the chromosome 17 can be found in Bennett and Dunn (1964), Guenet *et al.* (1980), Klein, (1977), Klein *et al.* (1984), Lyon (1981), Pla and Condamine (1984), Silver *et al.* (1984).

t-haplotype	Complementation group	H-2 haplotype	Death stage at	Geographic origin
t^{w32}	t^{12}	W28	Morula 2-3	Clinton (Montana) (1957)
t^{w2}	Semi-lethal	W29	Semi-lethal forebrain 14-21	New York (1956)
t^0	t^0	W29	Inner cell mass 6-7	Paris (1936)
t^{w5}	t^{w5}	W31	Embryonic ectoderm 8-10	New York (1956)
t^{Lub1}	t^{Lub1}	W33	Day 7	Alpic Drobic (Italy) (1978)
t^{w73}	t^{w73}	W32	Trophectoderm 4-5	Jutland (Danmark)(1973)
t^{w1}	t^{w1}	W30	Neural tube 11-21	New York (1956)
t^{w18}	t^9	(recombines)	Primitive streak 8-11	Storrs (Conn.) (1960)
t^{wPA1}	t^{wPA1}	nd	Day 7	M.m. brevirostris (Mus I) Villeneuve/Lot (France) (1980)
t^{wPA2}	Viable	nd	Viable	M.m. brevirostris (Mus I) Galeria (Corsica) (1980)

Table II. DNAs were prepared from T/t* heterozygous animals and digested to completion with *Bgl* II, *Xba* I or *Eco* RI restriction enzymes. The resulting fragments were analyzed by Southern blotting technique, using K and KDL specific probes. After deduction of the fragments derived from the T chromosomes (described in Delarbre *et al.*, 1985) and by comparison with the size of the fragments generated by the K, D, L genes of classical H-2 haplotypes, the K region of the 10 t-haplotypes can be classified into six groups. Letters within the table refer to the assignment of the observed fragment with a restriction fragment identified with the same probe and the same enzyme in a classical laboratory haplotype (b is for H-2^b, etc.). Under the same analysis, the D, L region appears very monomorphic and related to H-2^b or H-2^q haplotypes.

Groups	Names of t-haplotypes	K REGION		
		BglII	XbaI	EcoRI
1	t^0	d	d	d (s,q)
	t^{w2}	d	d	d (s,q)
	t^{wPA1}	d	d	d (s,q)
2	t^{w18}	b	b	b (k)
3	t^{w5}	k (s,q)	k	k (b)
4	t^{wPA2}	k (s,q)	k	d (s,q)
5	t^{w32}	q (s,k)	q	q (s,d)
6	t^{Lub1}	k (s,q)	d	k (b)
	t^{w73}	k (s,q)	d	k (b)
	t^{w1}	k (s,q)	d	k (b)

1985). The primary conclusion is that the 3' ends of the presumptive H-2K genes present in the MHC of the 10 independent t haplotypes display the very same restriction sites, which are found also at the 3' end of the H-2K genes of classical H-2 haplotypes. The restriction patterns of H-2K^t genes fall into six groups, four of them assignable to "classical" H-2K genes, which, interestingly enough, are among those most frequently found in the survey of random wild mice, checked for the absence of   t mutations (Nadeau *et al.*, 1981). It is interesting to note that the K^b gene was not found on any of the t chromosomes we have studied, with the exception of t^{w18}, a t haplotype known to recombine at a normal frequency: this finding is of interest, since the T chromosome used to maintain the t^{w18} haplotype possesses the $H-2^b$ haplotype (Delarbre *et al.*, 1985).

On the basis of the restriction patterns of the 3' end of the H-2K genes, the group comprising t^0, t^{w2} and t^{wPA1} haplotypes would possess an $H-2K^d$ like gene; group 2, t^{w18} is $H-2^b$ (as it is over the entire MHC, because it has recombined with a T ($H-2^b$) chromosome); group 3, t^{w5}, appears to be $H-2^k$ like; group 5 (t^{w32}) possesses an $H-2K^q$ gene. Groups 6 (t^{Lub-1}, t^{w73} and t^{w1}) and group 4 (t^{wPA2}) remain unassigned to any classical H-2K gene studied so far (b, d, f, k, p, q and s) but some restriction sites of those genes are retained. The pseudogene present in the vicinity of the $H-2K^{b,d \text{ and } k}$ genes is also present in the MHC of the t^{w32} haplotype. It is of the $H-2^q$ type (Delarbre *et al.*, 1985) suggesting that the K region of the t^{w32} haplotype, is organized as in $H-2^q$ haplotype.

By contrast, the D, L region of all t haplotypes studied was found to display a very low diversity. The restriction fragments that the $H-2D^t$, L^t genes generate, could not be distinguished from those generated from the corresponding genes in the b haplotype, with two exceptions: a) the $H-2D$, L region of the t^{w32} was of the $H-2^q$ type, and b) that of t^0 was compatible with both the H-2D, L genes of the $H-2^d$ and $H-2^q$ haplotypes (Delarbre *et al.*, 1985).

Results of RFLP can be considered as being preliminary and only indicative. Anyhow, the results obtained on H-2 genes proper suggest the existence of a close parenthood between the H-2K, D, L genes of t and of non-t MHC. In addition, a preculiarity of the t haplotypes we have studied, would be the low polymorphism of their D, L regions, as if the K and the DL loci of the MHC of t animals, were evolving at different rates, or along different rules. This last conclusion is to be compared with the recent finding that A_β, E_β and E_α genes, of t haplotypes, fall into only three main categories (Dembic *et al.*, 1984, 1985; Figueroa *et al.*, 1985; Mathis *et al.*, 1983a,b). Then, in t-haplotypes, the entire region from Ir genes to D, L genes appears to be much less variable than the H-2K region.

To which extent does the extensive serotyping analysis of the t-haplotypes, carried out by J. Klein's group, support the tentative conclusion that $H-2K^t$ genes share much homology with classical H-2 genes? As mentioned above, our study of the $H-2K^t$, D^t, L^t genes, at the level of genomic DNA is based upon restriction mapping of the non-coding 3' ends of the genes. Because of the lack of DNA probes unique to the 5' part of the H-2K genes, we could not perform a similar RFLP analysis on the 5' end of the $H-2K^t$ genes. However, the 5' part of the genes encodes the most variable part of H-2 antigens, and all of the antigenic polymorphism is associated to it. Serotyping of H-2 antigen can be considered as detecting differential short amino acid stretches encoded in the 5' part of the H-2K, D, L genes. In the following comparison, we have used data published by J. Klein's group. It should be noted, however, that most of the serotyping of the MHC of t haplotypes reported in the literature has not been performed in order to assess the parenthood of t and non-t class I genes, but rather to identify those antigenic specificities which could be unequivocally associated to the different t complementation groups (Hammerberg *et al.*, 1976; Klein and Hammerberg, 1977; Nizetic *et al.*, 1984; Sturm *et al.*, 1982). However, enough data have been obtained to allow a comparison between the H-2 antigens of t and non-t animals. We have listed the antigenic specificities of classical H-2 antigens which have been also detected on H-2^t antigens (Table III, left panel). Two conclusions can be reached. Firstly, the various t haplotypes can be sorted out into several groups sharing the same antigenic specificities: t^{w32}; t^0 and t^{w2}, compatible with the expression of an H-2K^d like antigen; t^{Lub1}, t^{w73} and t^{w1}, slightly different from each other, but compatible with the expression of an antigen bearing H-2K^f antigenic specificities. Finally, t^{w5} is compatible with the expression of an H-2K gene displaying H-2K^k antigenic specificities. No serotyping is available as yet neither for t^{wPA1} (although an unpublished report indicates it is similar to t^0) nor for t^{wPA2}. These groups are the same as those defined by RFFP analysis: there exists an excellent correlation between groups defined by serotyping, and groups defined by RFLP. It is therefore tempting to conclude that the same classification of the H-2K^t genes can be reached through restriction mapping of the 3' end of the K genes, and through serotyping of the protein region encoded by the 5' end of the H-2K genes. Secondly, H-2^t antigens display a strong reactivity towards classical anti-H-2 reagents, as if sharing extensive homology with the antigenic determinants recognized by the latter antibodies. Several antigenic specificities have been found, in initial studies as being associated exclusively to t haplotypes (Hammerberg *et al.*, 1976; Klein and Hammerberg, 1977) (Table III, right panel). Further serological analysis of the H-2 antigens of wild mouse populations (Nadeau *et al.*, 1981) has led to the conclusion that most of these antigenic specificities can be found also in wild non-t animals, although some of them, such as specificities (107, 108, 117) appear to be more preferentially associated with t chromosomes. The conclusion is that there would be no genuine t-specific antigenic specificities, although overall antigenic profiles

Table III. This table is a summary of the data collected over 10 years through serotyping of class I antigens of the MHC of t-haplotypes, and taken from Hammerberg (1976), Klein *et al.* (1983), Klein and Hammerberg (1977), Nadeau *et al.* (1981), Nizetic *et al.* (1984), Sturm *et al.* (1982). The names of the t-haplotypes are listed on the left of the table. On the left panel, we have gathered data obtained through the use of antisera (Bernd *et al.*, 1984; Nizetic *et al.*, 1984) reacting with clearly identified MHC antigenic determinants (d, k, Kd, etc.). On the right panel, the antigenic specificities (105 to 146), once assumed to be t-specific, are given on line with the corresponding t-haplotypes. Most, if not all of them, have been found expressed at the surface of cells of non-t, wild mice.

Haplotype	Reactions (antiserum: determinant)	Specificities
t^{12}, t^{w32}	5: b,k,p,q,r,s,u,v,z; m153: D^V; cm2: D^b	106,107; 146
t^0, t^{w2}	3: d,k,p,q,r,s,u,v; 6: all but k; 7: j,f,p,s; 8: d,f,k,p,r; 31: K^d; m11: K^q,r; c16: K^p; c33: K^b; m153: D^V	105; 107; 121
t^{Lub1}	m11: K^q,r; c14: D^V; c26: K^f; m36: b,d,s,u; m89: $D^{d,v}$; m153: D^V	107; 117; 126
t^{w73}	c14: D^V; c26: K^f; m36: b,d,s,u; m89: $D^{d,v}$	108,109,117; 126
t^{w1}	5: b,k,p,q,r,s,u,v,z; 13: d,q; c14: D^V; c26: K^f; m36: b,d,s,u; m89: $D^{d,v}$	108; 117; 126
t^{w5}	1: k,p,q,r,s,u,v; 5: b,k,p,q,r,s,u,v,z; m11: K^q,r; 23: K^k; m24: $K^{b,f,k,q,s}$,r,z; 25: $K^{k,r}$; m90: D^k,r	109

might be associated to the t-haplotypes. Thus, all the data concerning H-2K genes, whatever obtained through the use of DNA probes or antisera, point to the resemblance of the K genes of t and of non-t mice, with no distinctive feature associated to $H-2K^t$ genes. Would such conclusions be generalized, the frequency of the different K^t genes in natural mouse populations is expected to parallel that of the different H-2K genes in the same non-t populations.

III. NUCLEOTIDE SEQUENCE OF AN $H-2K^t$ GENE

Conclusions from results of RFLP, as well as serotyping (which in *ultima fine* detects products associated to localized gene conversion) events are only tentative. In order to substantiate our conclusion, we have taken advantage of the availability of a K specific probe, and of the isolation of viable partial t haplotypes derived from t^{w32} (which could be made homozygous for the MHC of the t^{w32} haplotype ($H-2^{w28}$ haplotype)), to isolate by molecular cloning a gene (P5) from $H-2^{w28}$ haplotype (Morita *et al.*, 1985). Upon sequencing of the entire gene, P5 turned out to possess indeed all traits associated with an H-2K gene, and was subsequently named $H-2K^{w28}$ gene (Morita *et al.*, 1985). The restriction map of the 3' part of $H-2K^{w28}$ is compatible with that of the $H-2K^q$ gene. The nucleotide sequence of this latter gene is not available as yet. Only a nearly complete cDNA encoding for an $H-2K^q$ antigen has been sequenced (Mellor *et al.*, 1983). Upon comparison of this sequence with that of the putative transcript of the $H-2K^{w28}$ gene, we found only five differences along a 1435 bp transcript, all of them silent. In other words, with the possible exception of nucleotide differences in the coding region of $H-2K^{w28}$ and $H-2K^q$ genes occurring in the missing 30 first nucleotides of the second exon of the K^q cDNA, the $H-2K^{w28}$ gene has to be considered as encoding an H-2K^q antigen. Such a nearly complete identity was surprising and raised questions concerning the origin of the t^{w32} haplotype. The t^{w32} chromosome maintained at the Pasteur Institute behaves as a complete t haplotype, lethal at the morula stage, transmitted at a high frequency and showing a marked inhibition of recombination. It also fails to complement t^{12}. However, its origin is nebulous: it has appeared in 1957, in the progeny of a Ttf/t^{w10} lethal cross, in which t^{w10} (nowadays extinct) belonged to the t^{w5} complementation group (Bennett and Dunn, 1964). One possibility for the generation of the t^{w32} haplotype could have been that the t^{w10} chromosome has recombined with a T-tf (H-2^q chromosome such as the TF strain (Hammerberg *et al.*, 1976)). As discussed elsewhere (Morita *et al.*, 1985), this would make it a unique example of a complete t haplotype generated by recombination with a wild type chromosome (Silver, 1983). Indeed, the restriction patterns of the H-2K, D, L^{w28} genes could not be distinguished from the corresponding genes in H-2^q haplotype (Delarbre *et al.*, 1985). Would such a recombination have occurred, this could be considered as meaning that gene exchange between t and non-t MHC has not impaired the overall properties of a t-haplotype. However, no genetic evidence of such recombination was published (Bennett and Dunn, 1964) and restriction pattern analysis of the entire MHC's of both w28 and q haplotypes has shown that the genes in the MHC, outside the H-2K-DL region, are significantly different (C. Delarbre *et al.*, unpublished). Thus, the t^{w32} haplotype is not likely to have been generated by simple recombination of t^{w10} with a T-tf (H-2^q) chromosome, and the H-2K, DL^q genes were probably present on the original t^{w10} chromosome. Assuming this conclusion is correct, H-2K^{w28} differs from H-2K^q by few silent mutations, which appear to reflect a divergence merely caused by mutational drift. Would one assume molecular clock for the evolution of the MHC, a distance could be computed. Whatever calculation hypothesis is used, K^{w28} and K^q gene should have split rather recently.

The nucleotide sequences of the H-2K^b, K^d, K^k genes and of several other class I genes, is now available (Bernd *et al.*, 1984; Evans *et al.*, 1982; Kvist *et al.*, 1983; Mellor *et al.*, 1984; Steinmetz *et al.*, 1981; Weiss *et al.*, 1983). Upon alignment of these sequences with that of H-2K^{w28}, several conclusions can be drawn which have been discussed in detail elsewhere (Morita *et al.*, 1985) and are in good agreement with the conclusions of others (Hayashida and Miyata, 1983; Weiss *et al.*, 1983). Briefly, the non-coding sequences are more conserved than the coding sequences (Delarbre *et al.*, 1985). Most of the nucleotide changes are clustered and lead to clustered amino acid changes (Fig. 2). The clustered nucleotide changes seem to operate mostly in exons, although some can be observed within the first introns of the gene. They are obviously reminiscent of highly localized gene conversion like events, presumably responsible for the generation of some of the polymorphism (Hayashida and Miyata, 1983; Mellor *et al.*, 1983; Nairn *et al.*, 1980; Weiss *et al.*, 1983). These alterations occur along a gradient from the 5' end to the transmembrance region, none of them being observed downstream exon 4. The clustered changes occur at very specific locations, most of them already assigned to the mutants of the bm series (Nairn *et al.*, 1980) as if these genetic exchanges which are fixed, have not occurred randomly along the entire gene but rather at some specific places. Finally, and most important here, the very same donor sequences appear to have been used to generate the variable regions of all K genes presently available to study, including the H-2K^{w28} gene itself (Morita *et al.*, 1985). This can be taken as meaning that the same donor genes have been present, at some time of their evolutionary history, in t and non-t animals. It

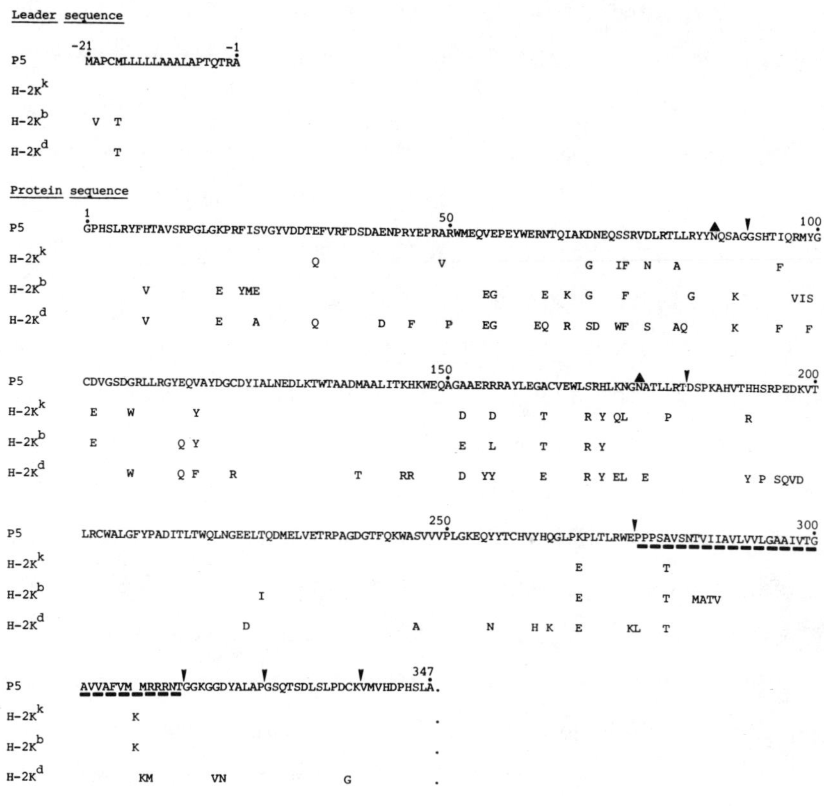

Figure 2. We have aligned the amino acid sequences deduced from the nucleotide sequences of H-2K^w28 (P5), H-2K^k, H-2K^b and H-2K^d genes, assuming a normal exon distribution of the fully processed transcript. With very few exceptions, the amino acid changes are clustered; the H-2K^d genes appears to be the most distant. The heavy line indicates the transmembrane region.

Table IV. This table illustrates the total nucleotide changes (T) and those leading to amino acid changes (c) existing between the various exons of the transcribed region of H-2Kw28 gene and of the other class I gene (listed on the left). U.T. is for untranslated region. (2) indicates the presence of a frame shift resulting in a greater number of amino acid changes than of nucleotide changes.

	5' U.T. 25 n	Exon 1 63 n		Exon 2 270n		Exon 3 276n		Exon 4 276n		Exon 5 117n		Exons 6,7,8 105n		3' U.T. 425n	Nucleotide sequence homology	Amino acid sequence homology
	T	T	C	T	C	T	C	T	C	T	C	T	C	T		
H-2K^q				3 over 240n	0	0	0	0	0	0	0	0	0	2	99.7	100
H-2K^k	1		0	13	7	24	12	2	2	5	2	0	0	4	96.9	94
H-2K^b	2	4	2	20	13	26	11	2	2	10	6	0	0	2	95.8	91
H-2K^d	4	2	1	37	20	28	18	17	12	7	5	3	3	20	92.4	84
H-2L^d	9	8	9^a	26	17	24	13	11	9	19	11	8 over 78 n	5	144	83.7	82
Q-10	> 5	10	3	40	22	38	20	17	12	45 over 81 n	18			>112	81.1	77
27-1	10	13	7	36	19	43	25	29	15	20 over 111n	12			94	83.1	76

should be kept in mind that these exchanges seem to occur most exclusively in females (Loh and Baltimore, 1984), an interesting point in view of the fact that t^x/t^y females are fertile and can recombine freely, as they do presumably also exchange short stretches of DNA by conversion-like events. A semi-quantitative comparison (treating nucleotide changes over the transcribed regions of the genes as if they were randomly distributed) of the coding regions of the class I genes and cDNA sequenced, shows that, $H-2K^{w28}$ is 99.8% homologous ot $H-2K^q$. But $H-2K^{w28}$ is closer to $H-2K^k$ and $H-2K^b$ than those two latter genes are to $H-2K^d$ (Tables IV and V).

Table V. Because of active gene conversion-like events among class I genes, the frequency of nucleotide changes cannot be taken as indicative of genetic distance. However, it can help in the definition of "families" among H-2K alleles. We have plotted the total number of nucleotide alterations between each couple of H-2K gene, and computed their relative homology (numbers underlined). The $H-2K^q$ (or $H-2K^{w28}$) appears to be much closer to $H-2K^k$ and $H-2K^b$ than these two are to $H-2K^d$.

$H-2K^{w28}$	--	3	49	66	125
		_99.8_	_96.9_	_95.8_	_92.0_
$H-2K^q$	3	--	49	59	110
	_99.8_		_96.6_	_95.9_	_92.4_
$H-2K^k$	49	49	--	65	106
	_96.9_	_96.6_		_95.8_	_93.2_
$H-2K^b$	66	59	65	--	122
	_95.8_	_95.9_	_95.8_	--	_92.2_
$H-2K^d$	125	110	106	122	--
	_92.0_	_92.4_	_93.2_	_92.2_	

IV. MHC GENES MUST BE EXCHANGED BETWEEN t AND NON-t HAPLOTYPES

An increasing body of evidence suggests the absence of specific features associated to the MHC of t-haplotypes. This included restriction pattern analysis of class I (Delarbre *et al.*, 1985) and class II (Dembic *et al.*, 1984, 1985; Figueroa *et al.*, 1985; Mathis *et al.*, 1983a,b), abnormalities of the E_α gene (the same deletion found in t^{12} and t^{w5} haplotypes are found also in the $H-2^b$ and $H-2^s$ haplotypes and in wild mice (Dembic *et al.*, 1984, 1985); the unstability of the E_α mRNA in the $H-q^q$ and $H-2^f$ haplotypes is also found in the t^0 and t^{w2} haplotypes whereas the other t haplotypes have a normal E^α gene (Dembic *et al.*, 1984, 1985; Nizetic *et al.*, 1984). No antigenic specificity of the class I and class II antigens have been found associated exclusively to t haplotypes: those which are usually considered as being t-specific have been detected occasionally in some non-t animals, sampled out of wild mouse populations. We add to this list of data, restriction enzyme analysis of K, DL genes (Delarbre *et al.*, 1985) and the sequence of the $H-2K^{w28}$ gene (t^{w32} haplotype) (Morita *et al.*, 1985), which point to the close homology of at least certain class I genes, in t and non-t animals. The frequency of the alleles of H-2K, D, L genes in the MHC of t haplotypes might, therefore, reflect in some way the distribution of the corresponding genes in non-t, wild, populations of mice, with no real distinctive trait being associated to the MHC of the t chromosomes. Finally, the MHC of t-animals appears to be a mosaic, all components of which have been already identified either in wild mice, or in classical laboratory haplotypes. As an example, the MHC of t^{w5} appears to be $H-2K^k$, $E_\alpha^{b,s}$, DL^b, etc. Similar conclusions have been reached using A_β , E_β and C4 probes (Figueroa *et al.*, 1985).

One of the distinctive features of complete t haplotypes is the inhibition of recombination, which limits the exchanges of genes between t and non-t chromosomes and extends over the MHC. How to reconcile this fact with the exchange of genes between the MHC of t and non-t animals, which must exist, or have existed, as the most likely explanation to account for the above reported observations? First, it should be kept in mind that intrachromosomal gene conversion is likely to occur among class I (and class II) genes of t chromosomes, even in the absence of recombination (Klein, 1984). The probable identity of the t with non-t H-2K genes could then be merely due to active conversion-like processes within an MHC otherwise "locked" by the inhibition of recombination. However, such a mechanism cannot account simultaneously for the homology of H-2K genes, the persistence of E_α gene anomalies and the relative oligomorphism of the Ir-DL region. A possibility never ruled out so far, would be the continuous production of t haplotypes from non-t, wild, animals. Such an hypothesis is consistent with the fact that the frequency of the genes of the MHC in t haplotypes seems to correlate with the frequency of the corresponding genes in natural, non-t, mouse populations, but cannot easily account for the persistence of E_α mutations. Alternatively, and more provocative, the MHC of some non-t chromosomes could be generated from that of some t-chromosomes. In the frame of such an hypothesis recombination between t and non-t chromosomes is necessary to occur in order to allow the separation of the various MHC genes, and their combination into different haplotypes. Is such a recombination a rare event? At least under laboratory conditions, the respective organizations of the MHC of t haplotypes appear to be very stable. Nevertheless, the occurrence of partial t haplotypes derived by recombination, have been observed at several times. The MHC of non-t animals can be derived from the MHC of t chromosomes after the occurrence of similar rare recombination events yielding partial, viable and able to recombine, t haplotypes. This has already been suggested by J. Klein's group, to account for the persistence of E_α genes anomalies in wild mouse populations (Dembic *et al.,* 1984, 1985). Then, t-chromosomes would behave as a reservoir for classical MHC genes, as well as, perhaps, for the many other genes involved in the male reproduction and which are located in the same region of chromosome 17 (Forejt, 1981). This conclusion is not conflicting with the hypothesis by Artzt *et al.* according to which the MHC of t haplotype is inverted and transposed (Shin *et al.,*1983) and that lethality factors are intermingled with class I genes (Artzt, 1984; Shin *et al.,* 1984), provided the inverted MHC^t is located distally to the tf marker of the T chromosome. Indeed, we have observed, most likely by double recombination between a T and a t^{w32} chromosome, the occurrence of partial, viable t haplotypes, which have retained most of the MHC of the t^{w32} chromosome with the probable exception of the Tla region. These MHC genes are now free to be separated one from each other by normal recombination processes an to diffuse into normal mouse population. This could be one of the mechanisms through which genes of the MHC of t-animals pass into non-t mouse populations, and is suggestive of the possibility of exchanges between genes of the MHC in t and non-t mice, at least in wild mouse populations.

ACKNOWLEDGMENTS

We are grateful to Drs. H. Condamine and P. Kourilsky for stimulating discussions, and to Mrs. V. Caput for preparing the manuscript. The work carried out at the Pasteur Institute was supported by grants from the Institut Pasteur, C.N.R.S., I.N.S.E.R.M. and F.R.M.F.

REFERENCES

Artzt, K. (1984). Gene mapping within the T/t complex of the mouse. III) lethal genes are arranged in three clusters on chromosomes 17. *Cell* **39**, 565-572.

Artzt, K., Shin, H. S., and Bennett, D. (1982). Gene mapping within the T/t complex of the mouse. II) Anomalous position of the H-2 complex in t-haplotype. *Cell* **28**, 471-476.

Bennett, D. (1975). The T-locus of the mouse. *Cell* **6**, 441-454

Bennett, D., and Dunn, L. C. (1964). Repeated occurrences in the mouse of lethal alleles of the same complementation groups. *Genetics* **49**, 949-958.

Bernd, A., Burgert, H. G., Archibald, A. L., and Kvist, S. (1984). Complete nucleotide sequence of the murine H-2K^k gene. Comparison of three H-2k locus alleles. *Nucl. Acids. Res.* **12**, 9473-9487.

Delarbre, C., Morita, T., Kourilsky, P., and Gachelin, G. (1985). Evolution within the multiple family coding for the class I histocompatibility antigenes: the case of the mouse t-haplotypes. *Ann. Inst. Pasteur, Immunol.* **136C**, 51-70.

Dembic, Z., Ayane, M., Klein, J., Steinmetz, M., Benoist, C. O., and Mathis, D. J. (1985). Inbred and wild type mice carry identical deletions in their E_α MHC genes. *EMBO J.* **4**, 127-131.

Dembic, Z., Singer, P. A., and Klein, J. (1984). $E_\alpha{}^o$: a history of a mutation. *EMBO J.* **3**, 1647-1654

Dorf, M. E. (1981). "The Role of the Major Histocompatibility Complex in Immunobiology." Garland Press.

Evans, G. A., Margulies, D. H., Camerini-Otero, R. D., Ozata, K., and Seidman, J. G. (1982). Structure and expression of a mouse major histocompatibility antigen gene, H-2K^d. *Proc. Natl. Acad. Sci. USA* **79**, 1994-1998.

Figueroa, F., Golubic, M. L., Nizetic, D., and Klein, J. (1985). Evolution of mouse major histocompatibility complex genes borne by t-chromosomes. *Proc. Natl. Acad. Sci. USA* **82**, 2819-2823.

Forejt, J. (1981). Hybrid sterility gene, located in the T/t supergene on chromosome 17. *In* "Current Trends in Histocompatibility" (Reisfeld *et al.*, eds.). Plenum Press, New York.

Goodenow, R. S., McMillan, M., Nicolson, M., Sher, B. T., Eakle, K., Davidson, N., and Hood, L. (1982). Identification of the class I genes of the mouse major histocompatibility complex by DNA-mediated gene transfer. *Nature* **300**, 231-237.

Guenet, J. L., Condamine, H., Gaillard, J., and Jacob, F. (1980). t^{wPa-1}, t^{wPa-2}, t^{wPa-3} : three new t-haplotypes in the mouse. *Genet. Res. Comb.* **36**, 211-217.

Hammerberg, C., Klein, J., Artz, K., and Bennett, D. (1976). Histocompatibility-2 system in wild mice. II) H-2 haplotypes of t-bearing mice. *Transplantation* **21**, 199-212.

Hayashida, H., and Miyata, T. (1983). Unusual evolutionary conservation and frequent DNA segment exchange in class I genes of the major histocompatibility complex. *Proc. Natl. Acad. Sci. USA* **80**, 2671-2675.

Hood, L., Steinmetz, M., and Malissen, B. (1983). Genes of the major histocompatibility complex of the mouse. *Ann. Rev. Immunology* **1**.

Klein, H. L. (1984). Lack of association between intrachromosomal gene conversion and reciprocal exchanges. *Nature* **310**, 748-752.

Klein, J. (1975). "Biology of the Mouse Histocompatibility-2 Complex." Springer-Verlag, pp. 620.

Klein, J., Figueroa, F., and Davis, C. S. (1983). H-2 haplotypes, genes and antigens: second listing. II) the H-2 complex. *Immunogenetics* **17**, 553-596.

Klein, J., Figueroa, F., and Nagy, Z. A. (1983). Genetics of the major histocompatibility complex; the final act. *Ann. Rev. Immunol.* **1**, 119-142.

Klein, J., and Hammerberg, C. (1977). The control of differentiation by the T complex. *Immunol. Rev.* **33**, 70-104.

Klein, J., Sipos, P., and Figueroa, F. (1984). Polymorphism of t-complex genes in European wild mice. *Genet. Res. Camb.* **44**, 39-46.

Kvist, S., Roberts, L., and Dobberstein, B. (1983). Mouse histocompatibility genes: structure and organization of a K^d gene. *EMBO J.* **2**, 245-254.

Loh, D. Y., and Baltimore, D. (1984). Sexual preference of apparent gene conversion events in MHC genes of mice. *Nature* **309**, 639.

Lyon, M. F. (1981). The t-complex and the genetical control of development. *In* "Biology of the House Mouse" (R. J. Berry, ed.), pp. 455-477. Academic Press, London.

Lyon, M. F. (1984). Transmission ratio distortion in mouse t-haplotypes is due to multiple distorter genes acting on a responder locus. *Cell* **37**, 621-628.

Mathis, D. J., Benoist, C. O., Williams, V. E. II, Kanter, M. R., and McDevitt, H. O. (1983). Several mechanisms can account for defective E_α gene expression in different mouse haplotypes. *Proc. Natl. Acad. Sci. USA* **80**, 273-277.

Mathis, D. J., Denoist, C. O., Williams, V. E. II, Kanter, M. R., and McDevitt, H. O. (1983). The murine E_α immune response gene. *Cell* **32**, 745-754.

Mellor, A. L., Weiss, E. H., Kress, M., Jay, G., and Flavell, R. A. (1984). A non-polymorphic class I gene in the murine major histocompatibility complex. *Cell* **36**, 139-144.

Mellor, A. L., Weiss, E. H., Ramachandran, K., and Flavell, R. A. (1983). A potential donor gene for the bml gene conversion event in the C57BL mouse. *Nature* **306**, 792-795.

Morita, T., Delarbre, C., Kress, M., Kourilsky, P., and Gachelin, G. (1985). An H-2K gene of the t^{w32} mutant at the T/t complex is closely related to H-2K^b or H-2K^q genes. *Immunogenetics* **21**, 367-384.

Nadeau, J. H., Wakeland, E. K., Götze, D., and Klein, J. (1981). The population genetics of the H-2 polymorphism in European and North African populations of the mouse (Mus musculus L). *Genet. Res. Camb.* **37**, 17-31.

Nairn, R., Yamaya, K., and Nathenson, S. G. (1980). Biochemistry of the gene products from murine MHC mutants. *Ann. Rev. Genet.* **14**, 241-277.

Nizetic, D., Figueroa, F., and Klein, J. (1984). Evolutionary relationships between the t and H-2 haplotypes in the house mouse. *Immunogenetics* **19**, 311-320.

Pla, M., and Condamine, H. (1984). Recombination between two mouse t-haplotypes ($t^{w12}tf$ and t^{Lub1}): mapping of the H-2 complex relative to centromere and tufted (tf) locus. *Immunogenetics* **20**, 277-285.

Shin, H. S., Bennett, D., and Artzt, K. (1984). Gene mapping within the T/t complex of the mouse. IV) The inverted MHC is intermingled with several t-lethal genes. *Cell* **39**, 573-578.

Shin, H. S., Flaherty, L., Artzt, K., Bennett, D., and Ravetch, J. (1983). Inversion in the H-2 complex of t-haplotypes in mice. *Nature* **306**, 380-383.

Shin, H. S., Stavnezer, J., Artzt, K., and Bennett, D. (1982). Genetic structure and origin of t-haplotype of mice, analyzed with H-2 cDNA probes. *Cell* **29**, 961-968.

Silver, L. M. (1981). Genetic organization of the mouse t-complex. *Cell* **27**, 239-240.

Silver, L. M. (1982). Genomic analysis of the H-2 complex region associated with mouse t haplotypes. *Cell* **29**, 961-968.

Silver, L. M. (1983). Reevaluation of the evidence for the generation of mouse lethal t-haplotypes by mutation. *Immunogenetics* **18**, 91-96.

Silver, L. M., Uhan, J., Danska, J., and Garrels, J. J. (1983). A diversified set of testicular cell proteins specified by the mouse t-complex. *Cell* **35**, 35-45.

Silver, L. M., Lukrelle, D., Garrels, J. J., and Willison, K. R. (1984). Persistence of a lethal t-haplotype in a laboratory stock of outbred mice. *Genet. Res., Cambr.* **43**, 21-25.

Steinmetz, M., Moore, K. W., Frelinger, J. G., Sher, B. T., Shen, F. W., Boyse, E. A., and Hood, L. (1981). A pseudogene homologous to mouse transplantation antigens: transplantation antigens are encoded by eight exons that correlate with protein domains. *Cell* **25**, 683-692.

Sturm, S., Figueroa, F., and Klein, J. (1982). The relationship between t and H-2 complexes in wild mice. I) the H-2 haplotypes of 20 t-bearing strains. *Genet. Res. Comb.* **40**, 73-88.

Weiss, E. H., Golden, L., Fahrner, K., Mellor, A. L., Devlin, J. J., Bullman, H., Tiddens, H., Bud, H., and Flavell, R. A. (1984). Organization and evolution of the class I gene family in the major histocompatibility complex of the C57BL/10 mouse. *Nature* **310**, 650-655.

Weiss, E., Golden, L., Zakut, R., Mellor, A., Fahrner, K., Kvist, S., and Flavell, R. A. (1983). The DNA sequence of the H-2K^b gene: evidence for gene conversion as a mechanism for the generation of polymorphism in histocompatibility antigenes. *EMBO J.* **2**, 453-462.

ADAPTATION AND EVOLUTION IN THE IMMUNE SYSTEM

Jacques Ninio

Institut Jacques Monod
Tour 43
2 Place Jussieu
75251 Paris cedex 05, France

ABSTRACT

Evolution at the species level takes thousands of years to achieve only slight improvements in enzyme activity. On the other hand, it is often stated that the immune system of higher vertebrates is able to design, within a few days, specific antibodies against almost any novel foreign antigen. It is proposed here that in fact, antibodies are much less specific than usually stated, and that the immune system is designed to achieve globally specific responses with loosely specific antibodies.

Three designs of increasing complexity that permit, in theory, achievement of this goal are presented: i) Somatic variants of antibodies are generated and tested before birth; ii) somatic variants are generated at any time, but the late ones have limited viability; iii) somatic variants are generated at any time and have unlimited viability, but an untested B cell is allowed to proliferate only when a tested B cell is binding the same antigen as the tested one.

The true system, as it works today, with a circulation of information between three classes of cells (from B lymphocytes to APCs, from APCs to T cells and from T to B) is discussed and interpreted as a more elaborate version of the third design. Then the role of MHC molecules becomes crystal clear, and the necessity for their fast evolution is easily explained.

Possible steps in the evolution of immunological logic are presented. It is shown that the organization of the immune system may have followed successively each of the simple designs i), ii) and iii) before settling to the present design.

I. THE IMMUNE SYSTEM AS A PARADOX IN EVOLUTION

The evolutionary rate of amino acid changes in protein sequences is often expressed as the percentage of observed changes in a sequence in 100 million years. Typical rates would be 12, 3 or 0.06 of such units in the hemoglobin, the cytochrome c or the histone families, respectively. When evolution is directed by man, in the laboratory, much higher rates may be obtained, but the functional changes are less than moderate. In studies of acquisitive evolution (reviews in Lin *et al.,* 1976; Ninio, 1983) extensive selection can bring about a modification of the affinity of an enzyme for a new substrate by a factor of 2 or 3, but rarely by a factor of 10 (Hall, 1981). A few thousand different enzyme specificities is all that molecular evolution was able to achieve in four billion years.

Nevertheless, immunological orthodoxy holds that when a higher vertebrate is invaded by a foreign molecule, never encountered before by the species, the immune system is able to produce after selection, within a few days, antibodies with high binding affinities and specificities for the foreign antigen. The adaptive process is usually described as an almost trivial evolutionary one: clonal selection. Those immunological cells (B lymphocytes) that happen to produce antibodies capable of binding, albeit weakly, the antigen, grow faster than the others. They also have a high rate of mutation so that the B cell population is quickly enriched in cells having antibodies with a very high affinity for the antigen (review in Manser *et al.,* 1985). If one could make the growth of *E. coli* dependent upon a nutrient with mutagenic effects (like nitroso-guanidine) one would have a bacteriological implementation of the clonal selection idea.

My belief is that although some clonal selection does occur in the immune system, it cannot by itself explain the emergence of highly efficient and specific antibodies.   Specificity is not a property of individual antibodies, but a property of the global response of the system, achieved through an astute communication network between three classes of cells. We know, in computer science, that reliable calculations can be carried out with unreliable components (von Neumann, 1956), and in molecular biology, that there are ways of improving the specificity of an enzyme system beyond the specificities inherent to the enzyme-substrate interactions (Hopfield, 1974; Ninio, 1975). Such ways are effectively used by ribosomes in translating messenger RNA into protein (Ruusala *et al.*, 1982). Here, I will attempt to analyze the immune system as being a device for the production of well-adapted global responses with loosely specific antibodies. From this biased view point, the design of the immune system, complicated as it is, becomes almost obvious, and one can easily devise a scheme for its evolution from a system obeying much simpler logics.

II. MOLECULAR SPECIFICITY

Absolute specificity is unknown in molecular biology.   The informational enzymes of the cell can discriminate among the set of substrates normally present in the cell, and bind preferentially one or two "specific" substrates. Out of context, however, their resolving power is poor. Here are two examples illustrating how specificity is understood in molecular biology.

Repressor-operator interactions. The repressor must bind specifically to a small piece of DNA of about 20 base pairs, less than one part in 100,000 of the *E. coli* chromosome. Actually, the Lac repressor in *E. coli* binds to almost any stretch of natural or artificial DNA, including simple repetitive polymers. This is part of its strategy to reach the operator. It would take too long to find the operator through diffusion in three dimensions and random collisions. The repressor binds anywhere along the DNA molecule, then reaches the operator site by a random walk in one dimension along the DNA molecule (Winter *et al.*, 1981). Non-specific is a prerequisite to specific binding. If by mutation, the affinity of the repressor for its operator is increased, there is a correlative increase in affinity for many sequences resembling the operator sequence. Then, there are high chances that the repressor remains glued to one of these sequences, leaving the operator free as though no repressor at all was produced in the cell. This behavior was observed in mutant X86 of the Lac repressor (Pfahl, 1976).

Two features of this system are of a more general significance. First, a non-specific component is necessary to speed up specific recognition. A substrate will have little chance of fitting an enzyme's catalytic site by random collisions if this site matches the substrate's shape too tightly. Second, an increase in affinity usually implies a decrease in specificity. For a detailed discussion of this "kinetic modulation" effect, see Ninio (1985). The existence of an inverse relationship between antigen-antibody binding strength and specificity has also been noticed in immunology (Karush, 1978).

The tRNA-acylating enzymes system. The situation here is closer to immunology, since it involves interactions between two families of molecules, each comprising at least 20 different members. For each amino acid used in the genetic code, there is one acylating enzyme which links it to the tRNAs specific for the amino acid. Thus, the valine acylating enzyme of yeast will add valine to the valine specific yeast tRNAs and no other. What happens if the same acylating enzyme is given as substrates distantly related tRNAs, such as those extracted from *B. stearothermophilus* ?   The yeast valine enzyme acylates nearly all these tRNAs whether specific for valine or not!   (Giegé *et al.*, 1974). Some viral RNAs, hardly resembling a tRNA may also be acylated (review in Haenni *et al.*, 1982).

Within the cell, specific acylation is not obtained at once.   The acylating enzyme may bind a number of tRNAs of desired or undesired specificity, add the amino acid onto them, and remove it later through a proofreading step (Hopfield *et al.*, 1976).

Here, the message is that recognition is essentially negative.   The acylating enzymes have the capacity to bind a large class of RNA molecules, but have developed specific ways of rejecting the undesirable competing tRNAs of the same cell.   Foreign tRNAs are acylated because there are no specific measures directed against them.

The number of different antibodies that a mouse can manufacture is thought to be, at most of the order of 10^8 or 10^9.   This number is infinitely smaller than the number of potential antigens.   For example, with 20 amino acids, 10^{13} decapeptides can be constructed most of which are immunogenic since the antibody repertoire is practically complete (Coutinho *et al.*,1984).   Hence, antibody specificity must be extremely degenerate (Talmage, 1959; Inman, 1978).

Antibody "multispecificity" is often observed, and, in particular, it is widely found among monoclonal antibodies (Lane and Koprowski, 1982).   In a heterogeneous population of antibodies, raised against one antigen, the alternate specificities differ from one antibody species to another.   Thus, any alternate specificity that may be conspicuous in a monoclonal antibody preparation might be diluted to background level in the heterogeneous population.   The population as a whole is more specific than any of its components (Talmage, 1959; Inman, 1978).

In any event, the major specificity problem, as acknowledged by most immunologists, is the discrimination between self and foreign molecules. I do not consider the immune system capable of rapidly designing efficient antibodies against foreign antigens that would be at the same time totally unreactive towards the molecules of the organism. There is now a growing body of evidence indicating that normally circulating antibodies do indeed bind molecules of the self (Guilbert *et al.*, 1982) in the absence of auto-immune diseases. How then are the self-binding activities regulated to remain at a tolerable level?

III. IMAGINARY DESIGNS FOR A SPECIFIC IMMUNE SYSTEM OPERATING WITH LOOSELY SPECIFIC ANTIBODIES

Before explaining how , in my view, the immune system really works, I shall discuss simpler theoretical designs. These will clarify the nature of the problems the immune system has to solve, and point to the way the complex organization of today may have originated from a simpler system.

Design 1. Prenatal testing. If a large repertoire of antibodies is generated before birth, and all the cells that produce antibodies able to bind endogenous molecules are eliminated or suppressed, then we are left with antibodies potentially active towards many (but not all) foreign compounds, and inactive towards the self (Lederberg, 1959). The system is safe, provided no somatic mutation occurs afterwards. Hoffmann (1980) proposed an ingenious elimination stragegy: the binding of (self) antigens at the surface of B cells would induce, in the testing period, a cell division accompanied by somatic mutation. Thus, the antibody sequences would drift away from those sequences that correspond to self antigens. This plausible mechanism is, by the way, the opposite of what the clonal selection theory assumes (i.e., evolution towards higher antigen-antibody affinities). Perhaps both behaviors are predictable, depending upon the ratios of the rates of cell division to somatic mutations. This point of theory ought to be worked out carefully. The essential element in this design is the existence of two developmental stages, one in which antibodies are tested, but where foreign is confused with self, and one in which somatic diversification occurs. The expression "prenatal testing" is used here for convenience, but the passage between the first and the second stage need not be coincident with birth.

It is widely believed that antibody-producing cells go through a testing period at an early stage of development. But afterwards, somatic mutations occur. In paticular, the chicken has one gene for immunoglobulin light chains (Reynaud *et al.*, 1985). In this case, antibody light chain diversity is essentially due to late somatic mutations.

If antibody testing occurs, it must be of a rather complex kind. Any antibody produced by an organism acts as an antigen towards another antibody. There are a considerable number of interactions, between B cells, through pairs of "complementary" antibodies. This property of complementarity manifests itself quite early in development (Holmberg *et al.*, 1984). Inasmuch as antibodies are indeed self-antigens, this result leaves little room for extensive prenatal elimination of B cells capable of binding self-antigens. Furthermore, as Nossal (1983) puts it: "deletion of all cells with any reactivity to any self antigen cannot be absolutely correct, because there would be a real risk of deleting the total repertoire!".

Design 2. Control of somatic variants viability. A way of improving upon the first situation would be to allow B cells to mutate and produce new antibodies, but restrict the viability of the somatic variants generated after birth. For instance, somatic diversity would be generated through a recombination process that would result in the loss of a gene essential for the long-term survival of the clone derived from the mutant cell. Thus, the danger of anti-self activities would be real, but limited in time by the reduced viability of the late somatic variants. The (low) danger of self-destruction would be the normal price to pay for a well-diversified system of defense.

Bimodal viability is a known feature of B cell populations. According to Jerne (1984), half of the B lymphocytes in the blood of the mouse live just a few days while the others may live as long as the mouse. The origin of the differences in cell viability is unknown. Longevity may be acquired by cells that went successfully through a testing stage.

One important feature is lacking in this design: memory. We know in fact that the responses once developed by the immune system may be mobilized again many years later, and possibly be the source of further improvements.

Design 3. Conditional activation of somatic variants. In the previous design, tested B cells of the first stage, and somatically derived untested B cells of the second stage had unequal viabilities. Now, instead of restricting the viability of (untested) mutant B cells, we can make these cells as viable as the tested ones, but restrict in time the period during which they can excrete antibodies. Ideally, one would like to activate them only as long as the foreign antigen is present. Direct clonal selection is inapplicable according to our premises, since somatic variants often bind self-antigens. The problem can be solved in the abstract by using a new class of cells (homologous to T cells) that would activate the untested B cells only when an antigen is present at the surface of a tested B cell.

In this manner, the danger of self-destruction would exist only as long as the organism fights the foreign antigen. Once it is eliminated, the

T cell would no longer receive a signal from the tested B cells and thus would not stimulate the untested B cells. But if the foreign antigen cannot be eliminated, the organism is in permanent danger of self-destruction (Nossal, 1983). The danger would be reduced if the T cell were to send activating signals for a restricted period of time, or to switch after a while from an activating (helper) state to a deactivating (suppressor) state. Whether or not a same T cell, in the real immune system, alternates between the helper and the suppressor mode is presently unknown.

There is one flaw or difficulty in this design. We would like the T cell that has detected a foreign antigen at the surface of a tested B cell, to stimulate only those untested B cells that bear exactly the same antigen. But the antigen, being present as a complex with two different antibodies, may not look the same on the two cells. At this point, we are so close, in logic, to the real immune system, that it is more profitable to discuss the problem in its exact context.

IV. THE TRIANGULAR DESIGN OF THE IMMUNE SYSTEM

The immune system makes use of at least three broad classes of cells: B lymphocytes producing antibodies (Ab), T lymphocytes bearing receptors (R) at their surface, and antigen-presenting cells (APCs, usually macrophages) carrying proteins of the major histocompatibility complex (MHC). Activation of B cells that have bound antigen is carried out by T cells and is conditioned by the existence of the same antigen at the surface of an APC cell. Thus, we have a specificity of the kind described in the preceding section, with the exception that the role we assigned to tested B cells is now carried out by antigen-presenting cells. We need now to understand in more detail how specificity is achieved in the system.

The weapons factory. An analogy with visual perception will help us understand how a population of B cells might sharpen its response to a new antigen. We perceive objects delineated by sharp boundaries, but if we use a photometer we usually find smooth variations in the intensity of light across the edges. The sharp edges are illusory: They are the result of a processing of the visual input occurring just after the retina. Any nerve cell connected to a retinal receptor receiving light influences the neighboring nerve cells. The signal is an inhibitory one, and inhibition increases with the amount of light received by the first nerve cell and decreases with the distance between the first and second nerve cells (see Ratliff, 1972).

This scheme, which produces contour extraction can be transposed to B cell populations. A B cell that binds an antigen would be like a retinal receptor. It would influence other B cells according to two factors: the amount of bound antigen, and the relatedness of the interacting B cells. Physical proximity is replaced here by functional similarity in antigen combining site. How can a B cell appreciate its resemblance to another B cell? This is done in two steps. A B cell producing Ab1 stimulates a B cell producing Ab2 complementary to Ab1, and this cell in turn stimulates B cells producing Ab3's complementary to Ab2 and in small proportion, functionally similar to Ab1. This prediction of Jerne's network theory (Jerne, 1974) is well verified (Urbain *et al.*,1981). Instead of lateral inhibition, we have a kind of lateral facilitation, and instead of favoring the edges of a distribution, this method is likely to favor its center of gravity (Grossman, 1984).

Some authors argue very intelligently that the main purpose of the network may not be to fight pathogens and foreign molecules, but insure, by homeostasis, the permanence of the self (Coutinho *et al.*, 1984). In line with this view, it is often assumed that the network is closed and incapable of invention. Here, in fact, we are interpreting the network as an exploratory device working by lateral facilitation (Ninio, 1979, 1983).

The alarm system. Another perceptual analogy will help us to describe the relationship between APCs and T cells. When a familiar object is in the place where one would expect to find it, it is spotted at once even if only a very tiny portion of it is visible, the rest being concealed by other objects. The same object placed conspicuously in an unfamiliar place, may go completely unnoticed. Recognition is usually based, not upon an extensive description of the object to be recognized, but on a small set of attributes weighted against a background within which they constitute adequate signals.

In the immune system, the background, or better, the context, is set by the MHC molecules that we may view as sticky molecules, able to bind most antigens, whether foreign or belonging to the self. Now, one antigen will bind in a particular way to an MHC molecule, or perhaps in two or three different ways. In a sense, the MHC is selecting two or three "views" of the antigen. When the T cell receptor interacts with the MHC-antigen complex, it has a biased, restricted perception of the antigen. Actually, in the case of protein antigens, it would seem that the antigen is somehow processed by the APC before presentation. Processing takes almost one hour and most probably involves proteolysis. Thus, it is a peptide rather than the complete protein which is presented (Grey and Chesnut, 1985). Self-nonself discrimination is based upon the few attributes of the molecules that are visible when bound to the MHC. The T cell receptors constitute the eyes of the immune system. I assume that early in development there is an important diversification of T cell receptors, and an elimination (or a tagging) of those which interact with APCs having MHCs complexed with antigen.

Since self-nonself discrimination is based upon a very restricted view of molecules, many foreign compounds will be treated as self. There are holes in the immune system. To each definition of the self (through the MHC molecules) there corresponds a set of blind spots (Klein, 1984), of foreign antigens the organism cannot fight. The failure is in the alarm system (APC-T relationship) and not at the level of the weapons factory (B cells network). Foreign antigens that fail to activate the alarm system and therefore trigger the immune response, are nevertheless able to bind to normal B cells (Nagy *et al.*, 1981).

When a foreign cell is introduced in the organism, bearing a foreign MHC, then any self-antigen, being viewed under a new angle, will be treated as foreign and trigger a response. In my view, graft rejection is not due to a "recognition" of foreign MHC's, but to the resulting misperception of self-molceules.

Immunologists diverge considerably on basic facts and issues. Every statement made in this section must have been made, in one form or another, by one or a few immunologists. On the other hand, many immunologists would strongly reject a number of elements in my description. For instance, many immunologists still believe that the T cell receptor recognizes the antigen and the MHC molecule independently (review in Rock & Benacerraf, 1983). Many still question the existence of a wide diversification of T cell receptors. Many still look for a mechanism by which the immune system might learn to recognize foreign MHC molecules (Ritzel *et al.*, 1984). These are fundamental divergences, indicating that the role assigned here to MHC molecules as selecting a biased view of the self is far from being universally accepted.

How the alarm system and the weapons factory communicate. The MHC molecule presents the antigen to the T cell receptor under a certain angle. Antibodies provide a quite different view of the same antigen. How can the T cell receptor equate the two views? To be effective, T cells must stimulate only the B cells that bind the antigen presented to them by APCs. However, *in vitro* studies seem to indicate that the growth factors released by T cells upon APC stimulation are beneficial to all B cells whether they bind the same antigen as APCs or not (DeFranco *et al.*, 1984).

In my view, in the general case where B cells do not possess by themselves APC function, the communication between T and B cells might occur through a ternary cellular complex (Fig. 1). An APC cell would make a contact with a B cell and capture antigens from its surface. There is a technique for preparing "active" APCs called opsonization which consists in incubating the APCs with B cells having bound antigen. But immunologists do not yet seem to consider antigen transfer from B cells to APCs as a truly important physiological process. I believe that the circuits of antigen capture by MHC molecules are of fundamental importance. If, as I assume, the antigen is transferred from B cells to APCs, an antigen that binds too strongly to the antibodies at the surface of B cells will not be transferred to APCs, and will fail to be immunogenic. If the postulated B cell-APC complex lasts long enough, then upon binding of a T cell to the APC, and the subsequent release of growth factors of various kinds, the B cell will benefit from these factors. Thus a direct T Cell – B cell recognition is not needed, provided antigen presentation occurs in a ternary complex. This view is completely heterodox, but it does not disagree with experimental evidence, since there is no experimental evidence bearing on the question.

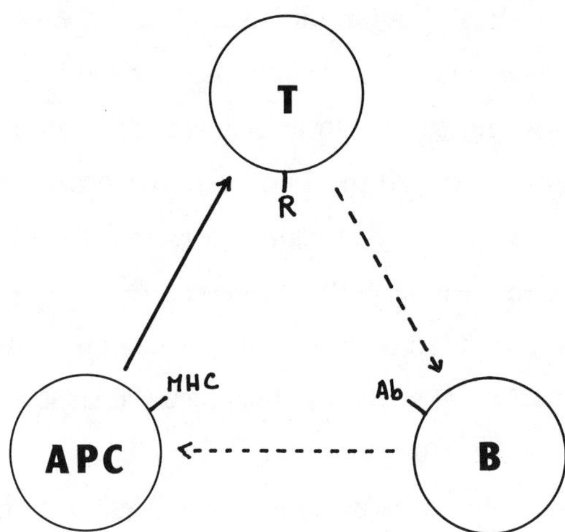

Figure 1. The triangular design of the immune system. When the receptor (R) of a T cell detects an antigen linked to an MHC molecule at the surface of an antigen presenting cell (APC), the T cell releases growth factors that stimulate B cell proliferation. Here, the hypothesis is made that in fact, there is a tertiary cellular complex between B and T cells and APCs. The APC would capture the antigen at the surface of a B cell. Therefore, the growth factors released by the T cell would preferentially stimulate the donor B cell.

Some B cells apart from lymphomas and other laboratory monsters have both the B cell and the APC characters (Ashwell *et al.*, 1984). In this case, a binary interaction with T cells is sufficient. Presumably-- using an anthropomorphic language--the T cell does not check that the antigen bound to the MHCs of these cells are the same as those that are bound to their antibodies, but just assumes this to be the case. It would seem that the normal sequence of events for a protein antigen includes its capture by the surface antibodies of the B cell, its processing inside the cell (via proteolysis?) and then its presentation outside on MHC molecules (Lanzavecchia, 1985). Possibly, the B cells with the dual function may be fully tested ones that have been promoted to a higher level of responsibility.

Cautionary comment. The above description is essentially based upon the genetic understanding of the immune system. In molecular biology, ideas on the mechanism of DNA replication or protein synthesis are commonly tested in cell-free systems. There is little equivalent work in immunology, and there is not even a hint of an answer to some of the most basic questions one would like to ask. For instance, excreted, cell-free antibodies in the blood of the mouse are 1,000 fold more numerous than antibodies at the surface of B cells (estimate by Jerne, 1984). A normal sequence of events, in such a situation, might be for an antigen to bind first to a circulating antibody, and this antigen-antibody complex could be a triggering signal for Jerne's network (Weill, J. C., personal communication). One can also note that Jerne's network is described in terms of interactions between B cell. We do not know whether this interaction is direct or indrect (i.e., through T cells for instance).

Molecular (co)evolution in the immune system. The viewpoint selected by any particular MHC molecule establishes a relationship between the molecules of the self and the blind spots (holes) in the immune system. Any novel molecule that becomes established in the species, keeping the same MHC, generates a new hole. Reciprocally, an environmental change may threaten the organism through one of its formerly safe holes. Thus, MHC molecules must evolve fast, for they need to integrate all modifications of the species with respect to the environment. Whereas fast molecular evolution is in general taken as a proof of neutrality (Kimura, 1983; Ohta, this volume), the MHC molecules would be under intense selective pressure to evolve rapidly. This postulated property of MHC molecules is in agreement with Hedrick *et* *al.'s* conclusions based upon population genetics (Hedrick *et al.,* this volume).

The antibodies are not submitted, in my view, to strong selective pressures. However, if Ab-Ab complementarity is fundamental, then there must be a good deal of co-evolution in the immunoglobulin families. As for the T cell receptor, its mechanical task is probably much more complex than that of MHC molecules. One expects a limited coevolution of T cell receptors along with MHC molecules, but with a good deal of conservatism, due to the constraints on the maintenance of function.

V. POSSIBLE STEPS IN THE EVOLUTION OF THE IMMUNE SYSTEM

It is conceivable that the immune system went through the various designs outlined in Section III.

At the begining, I imagine a class of "X" cells that express various products "K". The set of K molecules is assumed to exhibit the same kind of specificity and diversity as occurs in the set of an organism's proteases. Some proteases function as maintenance enzymes. They allow the cell to get rid of aged proteins with altered structures, or eliminate proteins containing too many translational errors (Murakami *et al.,* 1979). Proteases and antiproteases are weapons commonly used in warfare between phages and bacteria (Simon *et al.,* 1978; see also Levin, this volume). It does not seem (but this point has not been examined seriously) that proteases undergo a testing period in development. I thus take it for granted that higher organisms, despite the large amount of intra-species variation due to sexed reproduction, can afford to produce a set of proteases playing a role in the maintenance of the self and the protection against foreign proteins.

Whether the K's from which the antibodies (or MHC molecules) later evolved were proteases (Erhan and Greller, 1974), complement-like binding proteins, or proteins potentially toxic to cells (analogs of superoxide dismutase: see Richardson *et al.*, 1976) need not be determined here. We just emphasize the fact that, from the point of view of specificity and self non-self discrimination, our starting point is a valid one.

Prenatal testing, then prenatal diversification and testing may then come into the picture so that the system would work in accordance with our first design of Section III.

In a second stage, we imagine that when the K molecules at the surface of X cells are saturated with any compound (presumably, foreign), the X cells divide and mutate to cells Y of limited viability, producing variants "L" of the original K molecules. Our system obeys then the logic of the second design in Section III.

Actually, if we attempt to visualize the division of X cells and the appearance of Y cells in the progeny, we are led to imagine an intermediate stage where a daughter of the X cell has on its membrane both K molecules coming from the mother and newly synthesized L molecules (Fig. 2). The distribution of K and L patches would depend upon the detailed physics of cell division.

Assume now that the species has a permanent problem with a particular pathogen. The time is ripe for the appearance of a new cell (of the T type) that stabilizes the transient intermediate between the X and Y cells whenever the K molecules on the transient cell bind a pathogen's antigen. Binding of the antigen to the K molecules is taken as indirect evidence that it binds to the L molecules of this cell better than to L molecules of other cells (otherwise, the antigen would diffuse away and be picked up by the other cells). We are now in a situation somewhat similar to that of the third design of Section III, the main difference

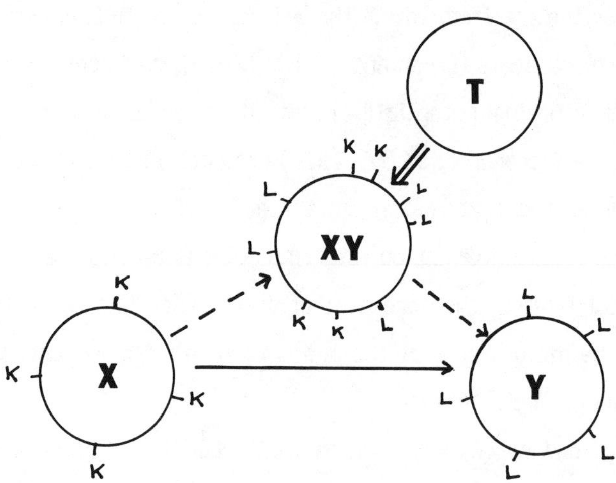

Figure 2. Possible steps in the evolution of the immune system. It is assumed that the defense of the organism was first taken up by a class of X cells expressing products K. Then, a mechanism of somatic diversification produced low-viability Y cells expressing products L. Actually, in the cellular division pathway leading from X to Y, the first cells that produced L molecules had patches of parental membranes with K molecules. The T cells arose to stabilize some of these XY intermediates, when antigen could be detected on K molecules. Later on, the K and L functions differentiated into MHC and antibody functions, respectively.

being that the tested and the untested antibodies, instead of being carried by different cells, are transiently carried on a same cell.

From now on, imagining further stages poses no difficulty. We may have a diversification of T cells, and the appearance of a sub-family of T cells that intervene to inhibit, at an early developmental stage, those X cells that bind self molecules. Note that the cytotoxic function of T cells involves two classes of molecules (MHC and R) while the helper function with which we have been mainly concerned so far involves antibodies as a third component. Furthermore, the most primitive forms of immunity (e.g., in sponges) are of the cytotoxic type. Nevertheless, I find it conceptually difficult to view humoral immunity as an outgrowth of cellular immunity, and prefer, as here, to describe the second as a degenerate form of the first.

Once this system operates, it is easy to imagine that at one point, in addition to the generation of L molecules by variation of K sequences, an autonomous production of L-like molecules is started. Thus, we had the first true immunoglobulins. Little by little, there was a takeover, by the immunoglobulin family, of the function previously exerted by K-derived L molecules. The notion of evolution by takeover is sufficiently sound (Cairns-Smith, 1982) to need no justification here.

As the repertoire of true antibodies diversified according to the needs of the system, the evolution of the K molecules followed a different route, leading to the MHC molecules of today. In this picture, the true ancestors of antibodies, from the functional point of view, were MHC-like molecules. Whether the antibodies evolved directly from these ancestors or had another origin and took over the function of pre-MHC somatic variants is not of capital importance here.

V. CONCLUDING REMARKS

The speculative model for the evolution of the immune system presented here derives from a speculative model of how the system works today. That important pieces of information are lacking should not deter us. On the contrary, studies in population biology too often take the form of heavy theoretical treatments which, in the end, help to justify some punctual observation. If all the tools that have been sharpened over half a century are really well constructed, they should help us understand situations where the basic facts are still poorly understood. The immune system, with its three interacting cell populations, its internal generation of diversity, its severe testing methods, its requirements for fast adaptation is, I belive, one of the most fascinating situations a population biologist could explore...if willing to take risks.

ACKNOWLEGMENTS

I am greatly indebted to Michel Seman for teaching me some of the intricacies of immunogenetics, to Donny Strosberg for constructive remarks on the manuscript and to Jean-Claude Weill for communicating his ideas on the openness of Jerne's network, and the concept of the antibody-antigen complex as a true antigen.

REFERENCES

Ashwell, J. D., DeFranco, A. L., Paul, W. E., and Schwartz, R. H. (1984). Antigen presentation by resting B cells. Radiosensitivity of the antigen-presentation function and two distinct pathways of T cell activation. *J. Exp. Med.* **159**, 881-905.

Cairns-Smith, A. G. (1982). "Genetic Takeover and the Mineral Origins of Life." Cambridge University Press, London.

Coutinho, A., Forni, L., Holmberg, D., Ivars, F., and Vaz, N. (1984). From an antigen-centered, clonal perspective of immune responses to an organism-centered, network perspective of autonomous activity in a self-referential immune system. *Immunol. Rev.* **79**, 151-168.

DeFranco, A. L., Ashwell, J. D., Schwartz, R. H., and Paul, W. E. (1984). Polyclonal stimulation of resting B lymphocytes by antigen-specific T lymphocytes. *J. Exp. Med.* **159**, 861-880.

Erhan, S., and Greller, L. D. (1974). Do immunoglobulins have proteolytic activity? *Nature* **251**, 353-355.

Giegé, R., Kern, D., Ebel, J.-P., Grosjean, H., de Hénau, S., and Chantrenne, H. (1974). Incorrect aminoacylations involving tRNAs or valyl-tRNA synthetase from Bacillus stearothermophilus. *Eur. J. Biochem.* **45**, 351-362.

Grey, H. M., an Chesnut, R. (1985). Antigen processing and presentation to T cells. *Immunol. Today* **6**, 101-106.

Grossman, Z. (1984). Recognition of self and regulation of specificity at the level of cell populations. *Immunol. Rev.* **79**, 119-138.

Guilbert, B., Dighiero, G., and Avrameas, S. (1982). Naturally occurring antibodies against nine common antigens in human sera. I. Detection, isolation and characterization. *J. Immunol.* **128**, 2779-2787.

Haenni, A.-L., Joshi, S., an Chapeville, F. (1982). tRNA-like structures in the genomes of RNA viruses. *Progress in Nucleic Acids Res. Mol. Biol.* **27**, 85-104.

Hall, B. G. (1981). Changes in the substrate specificities of an enzyme during directed evolution of new functions. *Biochemistry* **20**, 4042-4049.

Hoffmann, G. W. (1980). On network theory and H-2 restriction. *Contemporary Topics in Immunobiology* **11**, 185-226.

Hopfield, J. J. (1974). Kinetic proofreading: a new mechanism for reducing errors in biosynthetic processes requiring high specificity. *Proc. Nat. Acad. Sci. USA* **71**, 4135-4139.

Hopfield, J. J., Yamane, T., Yue, V., and Coutts, S. M. (1976). Direct experimental evidence for kinetic proofreading in amino acylation of tRNA^{Ile}. *Proc. Natl. Acad. Sci. USA* **73**, 1164-1168.

Holmberg, D., Forsgren, S., Ivars, F., and Coutinho, A. (1984). Reactions among IgM antibodies derived from normal, neonatal mice. *Eur. J. Immunol.* **14**, 435-441.

Inman, J. K. (1978). The antibody combining region: speculations on the hypothesis of general multispecificity. *In* "Theoretical Immunology" (G. I. Bell, A. S. Perelson and G. H. Pimbley Jr., eds), pp. 243-278. Marcel Dekker, New York.

Jerne, N. K. (1974). Towards a network theory of the immune system. *Ann. Immunol. (Inst. Pasteur)* **125c**. 373-389.

Jerne, N. K. (1984). Idiotypic networks and other preconceived ideas. *Immunol. Rev.* **79**, 5-24.

Karush, F. (1978). The affinity of antibody: range, variability and the role of multivalence. *In* "Comprehensive Immunology. Vol. V. Immunoglobulins" (G. W. Litman and R. A. Good, eds.), pp. 85-116. Plenum, New York.

Kimura, M. (1983). "The neutral theory of molecular evolution." Cambridge University Press, London.

Klein, J. (1984). What causes immunological unresponsiveness? *Immunol. Rev.* **81**, 177-202.

Lane, D., and Koprowski, H. (1982). Molecular recognition and the future of monoclonal antibodies. *Nature* **296**, 200-201.

Lanzavecchia, A. (1985). Antigen-specific interaction between T and B cells. *Nature* **314**, 537-539.

Lederberg, J. (1959). Genes and antibodies: do antigens bear instructions for antibody specificity or do they select cell lines that arise by mutations? *Science* **129**, 1649-1653.

Lin, E. C. C., Hacking, A. J., and Aguilar, J. (1976). Experimental models of acquisitive evolution. *Bioscience* **26**, 548-555.

Manser, T., Wysocki, L. J., Gridley, T., Near R. I., and Gefter, M. L. (1985). The molecular evolution of the immune response. *Immunology Today* **6**, 94-100.

Murakami, K., Voellmy, R., and Goldberg, A. L. (1979). Protein degradation is stimulated by ATP in extracts of *Escherichia coli. J. Biol. Chem.* **254**, 8194-8200.

Nagy, Z. A., Baxevanis, C. N., Ishii, N., and Klein, J. C. (1981). Ia antigens as restriction molecules in Ir-gene controlled T-cell proliferation. *Immunol. Rev.* **60**, 59-83.

Nino, J. (1975). Kinetic amplification of enzyme discrimination. *Biochimie* **57**, 587-595.

Ninio, J. (1979). "Approches Moléculaires de l'Evolution." Masson, Paris. English versions: "Molecular Approaches to Evolution (1982). Pitman, London, and (1983). Princeton University Press, Princeton.

Ninio, J. (1985). Kinetic and proabilistic thinking in accuracy. *In* "Accuracy in molecular biology" (D. J. Galas, T. B. L. Kirkwood, and R. Rosenberger, eds.). Chapman and Hall, London, in press.

Nossal, G. V. J. (1983). Cellular mechanisms of immunological tolerance. *Annu. Rev. Immunol.* **1**, 33-62.

Pfahl, M. (1976). Lac repressor-operator interaction. Analysis of the X86 repressor mutant. *J. Mol. Biol.* **106**, 857-869.

Ratliff, F. (1972). Contour and contrast. *Sci. Am.* **226**, 91-101.

Reynaud, A.-C., Anquez, V., Dahan, A., and Weill, J.-C. (1985). A single rearrangement event generates most of the chicken immunoglobulin light chain diversity. *Cell* **40**, 283-291.

Richardson, J. C., Richardson, D. C., Thomas, K. A., Silverton, E. W., and Davies, D. R. (1976). Similarity of three-dimensional structure between the immunoglobulin domain and the copper, zinc superoxide dismutase subunit. *J. Mol. Biol.* **102**, 221-235.

Ritzel, G., McCarthy, S. A., Fotedar, A., and Singh, B. (1984). Gene conversion may be responsible for the generation of the alloreactive repertoire. *Immunology Today* **5**, 343-345.

Rock, K. L., and Benacerraf, B. (1983). MHC-restricted T cell activation: analysis with T cell hybridomas. *Immunol. Rev.* **76**, 29-57.

Ruusala, T., Ehrenberg, M., and Kurland, C. G. (1982). Is there proofreading during polypeptide synthesis? *EMBO J.* **6**, 741-745.

Simon, L. D., Tomcozak, K., and St. John, A. C. (1978). Bacteriophages inhibit degradation of abnormal proteins in *E. coli*. *Nature* **275**, 424-428.

Talmage, D. (1959). Immunological specificity. *Science* **129**, 1643-1648.

Urbain, J., Wuilmart, C., and Cazenave, P. A. (1981). Idiotypic regulation in immune networks. *In* "Contemp. Topics in Molec. Immunol.," Vol. 8 (F. P. Inman and W. J. Mandy, eds.), pp. 113-148. Plenum Press, New York.

von Neumann, J. (1956). Probabilistic logics and the synthesis of reliable organisms from unreliable components. *In* "Automata studies" (C. E. Shannon and J. McCarthy, eds.), pp. 43-98. Princeton University Press, Princeton.

Winter, R. B., Berg, O. G., and von Hippel, P. H. (1981). Diffusion-driven mechanisms of protein translocation on nucleic acids. 3. The *Escherichia coli* lac repressor-operator interaction: kinetic measurements and conclusions. *Biochemistry* **20**, 6961-6977.

PART II. TEMPO AND MODE OF MOLECULAR EVOLUTION

MOLECULAR AND PHENOTYPIC ASPECTS OF THE EVOLUTION OF HYBRID DYSGENESIS SYSTEMS[1]

Margaret G. Kidwell[2]

Division of Biology and Medicine
Brown University
Providence, Rhode Island 02912

ABSTRACT

The molecular and phenotypic properties of the *P* and *I* mobile element systems that induce hybrid dysgenesis are briefly summarized. The distributions of the two element families in both natural and laboratory populations of *D. melanogaster* and in populations of other *Drosophila* species are described. Evidence is presented that *P* elements and active *I* elements are recent invaders of *D. melanogaster*. The properties of an element system that might be critical in producing hybrid dysgenic-like organismal effects are discussed. It is argued that hybrid dysgenesis might represent a transient early phase in the evolutionary history of certain types of mobile element systems.

[1] This work was supported in part by PHS Grant No. GM-25399.

[2] Present address: Department of Ecology and Evolutionary Biology, Biological Sciences West Building, University of Arizona, Tucson, AZ 85721.

I. INTRODUCTION

Despite growing evidence for the widespread distribution of mobile elements in both prokaryotes and eukaryotes, their evolutionary significance remains obscure. Since the idea of "selfish DNA" was put forward by Doolittle and Sapienza (1980) and Orgel and Crick (1980), intense speculation has continued as to whether these elements are maintained and spread by virtue of their competitive edge in replication at the DNA sequence level despite a possible selective disadvantage or whether they are maintained, at least in part, by selection at a higher level of organization. However, very little is actually known about possible selective advantages or disadvantages that mobile elements might confer on individual organisms or supraorganismal groups over either the short or long term.

A second, but possibly related, group of questions which have received rather less attention are concerned with the mode of origin and evolutionary history of mobile element systems. A mobile element family might originate in a species by either horizontal transmission from another, perhaps unrelated, species, or by the mutation or recombination of existing sequences. However, the detailed mechanisms and relative frequencies of these modes are completely unknown. If horizontal transmission, by interspecific gene transfer, should prove to be other than an isolated rare event, the implications could be revolutionary for phylogenetic reconstruction based on DNA sequence homology and for other aspects of evolutionary theory. A broad investigative approach which extends beyond the molecular to the organismal and population levels is required to attempt to answer these complex evolutionary questions.

Two mobile element families, the P elements and the I elements in *Drosophila,* have a number of properties which provide excellent opportunities for integrated molecular, chromosomal, population and evolutionary studies. These families are particularly noteworthy because, when they are mobilized, they induce striking arrays of phenotypic traits known collectively as "hybrid dysgenesis" (Kidwell *et al.*, 1977). This behavior is in contrast to that of many other known mobile element families which were first identified by their molecular properties and whose repertoire of phenotypic effects appears to be relatively limited. This situation raises the following question: are the striking phenotypic manifestations of hybrid dysgenesis explainable as a transient phase of a "young" or newly invasive element system, or is the high frequency of induction of phenotypic traits related to particular structural, functional or regulatory properties? Of course, these two explanations may not be mutually exclusive.

The purpose of this paper is briefly to compare and contrast the molecular and phenotypic properties of the *P* and *I* mobile element systems, to summarize present knowledge of their distributions and to discuss various hypotheses about their evolutionary history in *Drosophila* and whether this might be related to their seemingly unusual properties.

II. THE PHENOMENOLOGY OF HYBRID DYSGENESIS

About a decade ago observations in several laboratories revealed a surprising variety of unusual genetic phenomena when males from natural populations of *D. melanogaster* were crossed with females from long-established laboratory strains. These phenomena, collectively referred to as hybrid dysgenesis, include high frequencies of partial or complete sterility, male recombination, visible and lethal mutation, chromosomal rearrangement, transmission ratio distortion (including sex ratio distortion), chromosomal non-disjunction and modification of female recombination frequencies.

Two apparently independent systems of hybrid dysgenesis are known, each resulting from crosses between two mutually interacting categories of strains. These are called paternal (P) and maternal (M) in the P-M system and inducer (I) and reactive (R) in the I-R system. As shown in Figure 1, dysgenic traits usually appear in only one of the two possible reciprocal crosses between interacting strains and not in matings between the same type of strain. There are also "quasi neutral" strain categories called Q in the P-M system and N in the I-R system. Both phenotypic and molecular tests indicate that Q and N strains can be regarded as subsets of the P and R categories, respectively. Information about the strain types which are produced by all four possible combinations of the maternal and paternal components of the two systems is summarized in Figure 2.

There are a number of striking similarities in the phenomenological patterns produced by the P-M and I-R dysgenic systems. The most important of these are summarized in Table I. Differences between the two systems also exist (Table II) but appear to be of less significance than the similarities.

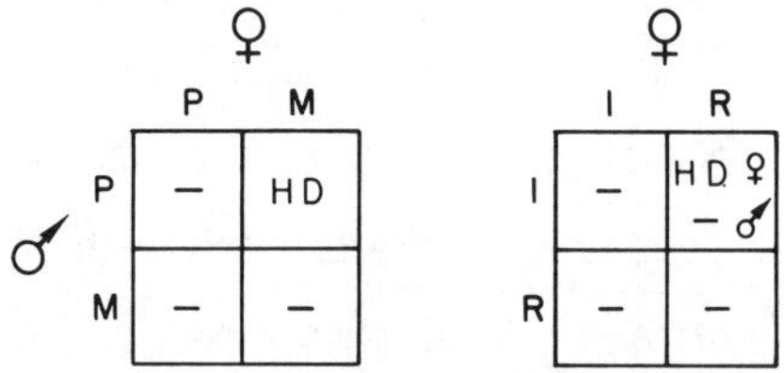

Figure 1. Expected outcomes of various intrastrain matings in the P-M and I-R systems. HD indicates the induction of hybrid dysgenesis in F_1 progeny, − indicates normal progeny.

P-M SYSTEM

<table>
<tr><td rowspan="2">I-R SYSTEM</td><td></td><td>Maternal (M)</td><td>Paternal (P)</td></tr>
<tr><td rowspan="2">Maternal (R)</td><td>RM
Typical of long-established laboratory strains, but absent in present day natural populations.</td><td>RP
Absent in natural and laboratory populations, but this combination was recently synthesized in the laboratory.</td></tr>
<tr><td rowspan="2">Paternal (I)</td><td>IM
Common in middle-aged laboratory populations.

IM'
Currently common in natural populations of certain regions such as Europe and Asia.</td><td rowspan="2">IP
Ubiquitous in natural & laboratory populations recently collected from certain geographical regions such as North and South America.</td></tr>
</table>

Figure 2. Strain types resulting from all four possible combinations of the maternal and paternal components of the two hybrid dysgenesis systems and their distributions in *D. melanogaster* populations.

Table I. Similarities between *P-M* and *I-R* systems of hybrid dysgenesis in *D. melanogaster*.

---

1. Hybrid manifestation.

2. Multiple, associated, dysgenic traits.

3. Reciprocal differences and other genetic evidence indicate both Mendelian and non-Mendelian mechanisms of transmission.

4. Germ line specificity.

5. Dependence on environmental factors for the manifestation of some dysgenic traits.

6. Patterns of regulation.

Table II. Differences between *P-M* and *I-R* systems of hybrid dysgenesis in *D. melanogaster*.

---

1. Restriction of *I-R* but not *P-M* dysgenesis to the female sex.

2. Dysgenic traits differ in type and frequency.

3. Environmental effects differ in stage-specificity, direction and reversibility.

4. Distribution patterns in natural and laboratory populations are overlapping but not identical.

III. THE MOLECULAR BASIS OF HYBRID DYSGENESIS

Two distinct families of mobile elements, the *P* elements and the *I* elements, have been demonstrated to be responsible for the two independent sets of dysgenic traits described earlier. *P-M* dysgenesis has been shown to be a consequence of the mobilization of members of the *P* family of elements (Rubin *et al.*, 1982; O'Hare and Rubin, 1983). The 2907 kbp *P* element is the largest of the family whose members are highly heterogeneous in size. A complete *P* element has 31 bp inverted terminal repeats and four open reading frames for translation. All four reading frames are apparently required to code for a transposase which catalyzes the transposition of the complete *P* element itself and that of other smaller *P* elements in the same genome (Karess and Rubin, 1984). Intact *P* elements that are autonomous with respect to transposition are referred to as *P* factors (Engels, 1984). Smaller, non-autonomous, *P* elements are deletion derivatives of the complete 2907 bp *P* factor. They share the same 31 bp inverted repeats which are thought to include recognition sites for the transposase enzyme but internal deletions evidently preclude the production of transposase. Insertion of a *P* element in a new genomic site generates an 8 bp duplication in the host DNA. Excision of *P* elements occurs frequently and can be precise or imprecise, leaving part of the element at the site of insertion (Daniels *et al.*, 1985a). Despite the broad similarities in phenotypic manifestations, the molecular structure of *I* elements is fundamentally different from that of *P* elements as illustrated in Table III.

Bucheton *et al.* (1984) have shown that the fully functional *I* factor is 5.4 kbp in length, almost twice that of the *P* factor. Unlike most known mobile elements, it has no repeated sequences at its termini. Similar to the *P* element family, there are deletion derivatives of the *I* factor which apparently lack the ability to code for a transposase. H. Sang (cited by Simmons and Karess, 1985) has shown that these deleted *I* elements are typically lacking a 2.3 internal sequence bounded by sites of the restriction enzymes Hind III and Pst I.

Table III. Comparison of the molecular properties of *P* and *I* elements.

	Element	
	P	I
Size of intact element	2.9 kbp	5.4 kbp
Terminal repeats	31 bp inverted	none
Open reading frames	3 large 1 small	At least 2 large
Size of host DNA duplication	8 bp	12 bp
Excision	Precise and imprecise	Not yet demonstrated

Table IV. Comparison of reported copy numbers per haploid genome and genomic distribution of intact and deleted elements in the *P-M* and *I-R* systems.

Element	Strain type	Element structure	
		Intact	Incomplete
P	P and Q	10–15	20–35
	Pseudo M (M')	Few or none	some
	True M	0	0
I	Inducer (I)	10–15	15–20
	Reactive (R)		
	or Neutral (N)	0–2	15–20

A. Genomic Distributions of *P* and *I* Elements in *D. melanogaster*

As shown in Table IV, both *P* and *I* strains carry a combination of both autonomous and nonautonomous elements belonging to their respective families. The two families are similar in that subsets of strains exist which carry a number of incomplete elements but few or no complete ones. These *M'* and *R* (or *N*) strains are apparently largely or completely deficient in transposition potential. The *P* family, however, differs from the *I* family in an important respect. Many *M* strains, usually those that have a long laboratory history, are completely lacking in *P* elements (Bingham *et al.,* 1982). However, all strains so far examined carry *I* elements (Bucheton *et al.,* 1984). An interesting difference also exists between the two families with respect to the structure and genomic distribution of the incomplete elements. Incomplete *I* elements are not simple deletion derivatives of complete elements (A. Bucheton, pers. comm.), and they tend to be restricted to the centric heterochromatin in both *I* and *R* strains. This is in contrast to the fully functional *I* factors which are usually found in the chromosomal arms. Both intact *P* elements and their deletion derivatives are found to be preferentially inserted in euchromatin; heterochromatic sites are found very infrequently. It should be stressed that the range of copy numbers indicated in Table IV is uncertain and based on a rather limited sample of strains. Detailed knowledge of the relationship between phenotype (strain type) and molecular characteristics (copy number, chromosomal insertion site and ratio of intact to incomplete elements) is currently unavailable.

The limitation of copy number per genome appears to be a characteristic common to both *P* and *I* elements families and also to many other *Drosophila* element families. However, there are no data yet available on the variance of copy number within a population. We do not know whether the variance is large in either newly invaded or equilibrium populations, or whether it is quite small as predicted by molecular drive theory (Dover, 1982).

B. Regulation of Transposition

Our knowlege of the phenomenology of hybrid dysgenesis indicates that the transposition of *P* and *I* elements is controlled by a number of genetic and environmental factors. The transposition of both element families is normally restricted to the germ line. *P* elements can transpose in both sexes but *I* element transposition is limited to females. In both systems, the frequency of transposition is strongly determined by the maternal cellular environment, referred to as cytotype in the *P-M* system (Engels, 1983) and degree of reactivity in the *I-R* system (Bregliano and Kidwell, 1983). Both chromosomal and extrachromosomal components are involved in the determination of cytotype and degree of reactivity. Despite this general knowledge of the existence of transpositional controls, very little is known about the detailed molecular and cellular mechanisms involved.

In the *P-M* system the cellular environment in which *P* factors are active has been defined as "M cytotype" and that in which they are quiescent has been defined as "P cytotype" (Engels, 1983). Two molecular models of the regulation of *P* element mobility have been developed. The two-component model of O'Hare and Rubin (1983) proposes that *P* element sequences code for both a transposase and a regulator molecule. Karess and Rubin (1984) hypothesized that such a regulator molecule and a transposase could both be generated from different portions of the same open reading frame by differential splicing of RNA transcripts. An alternative, one-component model, for *P* element regulation has been proposed by Simmons and Bucholz (1985). They hypothesize that transposase binds to sequences in the terminal repeats of both intact and internally deleted *P* elements. Competition for transposase, when this is limited in supply, may result in a reduction of transposition frequency. Such competition might be enhanced by the ability of the hypothesized extrachromosomal *P* elements to bind the transposase enzyme, in addition to binding by chromosomally integrated elements. The experimental results of both Kidwell (1985) and Simmons (cited by Simmons and Karess,

1985) are consistent with the one-component model of P element regulation.

C. Methods of Strain Characterization

Prior to the elucidation of the molecular basis of hybrid dysgenesis, *D. melanogaster* strains were characterized exclusively on the basis of phenotypic tests of the hybrid progeny of matings with standard reference stocks. These tests have been used to define the degrees of potential P or I factor activity (transposase activity) and regulation in the two systems. In the P-M system, female gonadal dysgenesis is the trait which has most frequently been employed (Engels and Preston, 1979; Kidwell and Novy, 1979). In the I-R system, SF sterility (the reduction of hatchability of eggs laid by dysgenic females) has been the trait of choice (Picard, 1976). Both P and I factor activity and regulation appear to have continuous distributions, but there are pronounced modes in the distributions of the main strain categories with respect to both their P factor activity and regulatory potentials as shown in Figure 3. Strains with P factor activity and cytotype values outside the indicated areas have been shown to exist transiently in the laboratory (M. Kidwell, unpublished results), but their condition is highly unstable. Such strains either become extinct or evolve to one of the stable conditions indicated.

The discovery of the molecular basis of the two dysgenic systems, which was described in an earlier section, has led to partial success in the development of molecular assays for strain classification. However, the detailed nature of the relationships between molecular and phenotypic characteristics of a strain have yet to be worked out. We do not yet know what are the critical molecular attributes which distinguish the various phenotypically defined types.

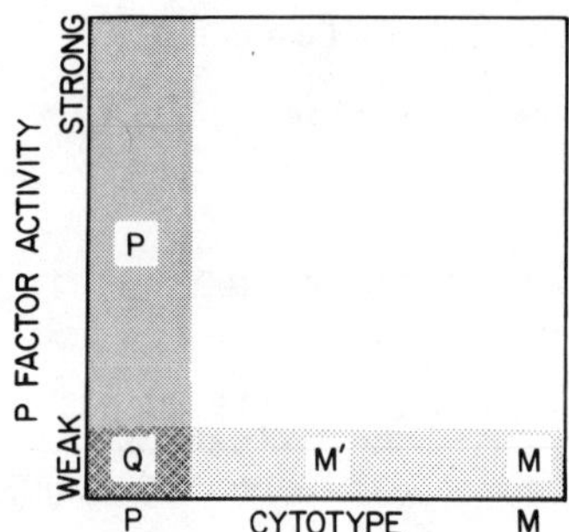

Figure 3. Representation of the distribution of *D. melanogaster* strain types according to their potential for *P* factor activity and regulatory characteristics (cytotype).

IV. CURRENT DISTRIBUTIONS OF HYBRID DYSGENESIS DETERMINANTS

A. Interspecific Distribution of *P* Elements

Species of the genus *Drosophila* have been assigned to a number of subgenera, the largest of which are the *Sophophora* and the *Drosophila*. These subgenera arose from successive radiations within the genus (Throckmorton, 1975). The subgenus *Sophophora* is comprised of a number of species groups. The *saltans-willistoni* lineage evolved in the New World tropics, and the *melanogaster* lineage in the Old World tropics. The *obscura* group which is most closely related to the *melanogaster* group is distributed throughout the Holarctic temperate zone.

P elements were first isolated from *D. melanogaster*, but they are not carried by all populations of this species. Consistent with the results of phenotypic tests (Kidwell *et al.*, 1977; Kidwell, 1983), *P* elements were typically completely absent from most long-established laboratory strains (Bingham *et al.*, 1982). *P* elements have not been found to occur naturally in any other species of the *melanogaster* group (Brookfield *et al.*, 1984; G. Dover, pers. comm.). Their absence in the

sibling species *D. simulans* and *D. mauritiana* is particularly noteworthy.

In contrast to the absence of *P* elements in those species most
closely related to *melanogaster*, these elements have recently been
found to have a widespread distribution in the more distantly related
willistoni and *saltans* groups which stem from a single sophophoran
lineage (Daniels *et al.*, 1984; Daniels and Strausbaugh, pers. comm.). All
of the species of the *willistoni* group so far examined, except
for *D. insularis*, carry sequences homologous to the *melanogaster*
P elements. Species of the *saltans* group make up five subgroups which
show an orderly progression from primitive to derivative. *P* elements
have so far not been found in species of *cordata* and *elliptica*, the two
most primitive of these subgroups, but they are found in *sturtevanti* and
austrosaltans, the more derivative species clusters (S. Daniels, pers.
comm.). The subgroup *parasaltans* has not yet been examined.

Comparison of the patterns of hybridization of *P* element sequences
in *D. melanogaster* with those of species in the *willistoni* and
saltans groups results in some striking contrasts (Daniels, pers. comm.).
In *D. melanogaster*, the pattern of hybridization in *P* strains is complex
and distinctive for each strain (Bingham *et al.*, 1982). Such a pattern is
consistent with the presence of active mobile elements. However, in the
willistoni and *saltans* groups, patterns of hybridization in different
isolates of the same species are often quite similar. These patterns are
also less complex than in *D. melanogaster*, the *P* element copy number
is lower and preliminary evidence indicates that the elements may be
localized predominantly in centric heterochromatin. Taken together, these
observations suggest that *P* element mobility in at least some of the
willistoni and *saltans* group species may be low or nonexistent. This

further suggests a relatively long-standing presence of *P* elements in these species compared to a recent invasion in *D. melanogaster.*

B. Interspecific Distribution of *I* Elements

Homology to the cloned *I* factor (Bucheton *et al.*, 1984) has been detected in all *D. melanogaster* populations examined and also by A. Bucheton (pers. comm.) in other species of the *melanogaster* group including *D. simulans, D. mauritiana, D. erecta, D. sechellia, D. yacuba,* and *D. tessieri.*   In contrast to the *I-R* dichotomy observed among strains of *D. melanogaster,* all of ten *D. simulans* strains examined from different locations and laboratory ages possessed *I* elements containing the 2.3 kbp Hind III-Pst I internal fragment (A. Bucheton, pers. comm.).   Despite this, no evidence has been found for hybrid dysgenic phenotypes in *D. simulans* which might be associated with the presence of the 2.3 kbp internal fragment. Perhaps this might be explained by the non-existence of the reactive category of strain in *D. simulans.*

C. Intraspecific Variability of *P* and *I* Characteristics in *D. Melanogaster*

P element homology has been found to exist in all natural populations of *D. melanogaster* which have been sampled thoughout the world during the last five years (Anxolabéhère *et al.*, 1984, 1985). This is in sharp contrast to their complete absence in most long-established laboratory strains. Although the distribution of *P* elements is currently widespread, the copy number, structure and functional potential of the genomic complements of these elements differs markedly according to geographic region.

A detailed survey was carried out of over two hundred isofemale lines collected in North America in the period 1977-80. Analysis of the functional potential of these lines with respect to the two components of *P-M* hybrid dysgenesis indicated that very little qualitative variability existed in North American populations (Kidwell and Novy, 1985). All populations proved to be polymorphic with respect to the *P* and *Q* types and, with very few exceptions, possessed a strong *P* cytotype. *P/Q* polymorphism implies a quantitative variability in *P* factor activity and the mean values of gonadal sterility potential did show some variability among locations. This and other data from these populations suggest a general tendency for high *P* factor activity in the southeast and midwest with lower frequencies in other areas. The Pacific northwest had the lowest frequencies and this area has provided the only evidence for the *M'* type in N. America (D. Holm, pers. comm.).

Populations in many other regions of the world do not show the same distributions as those in N. America (Fig. 4). The results of Anxolabéhère *et al.* (1984 and 1985) together with those of Kidwell *et al.* (1983) indicate that the *P* type exists in central Africa, but is essentially absent in North Africa, Europe and Asia. Anxolabéhère *et al.* (1985) present evidence for a *P* element cline extending from western Europe (*Q* type with relatively high *P* element copy number) to central Asian (*M'* type with low *P* element copy number). Strong evidence for a cline has also been found on the east coast of Australia by I. Boussy (pers. comm.). Both phenotypic and molecular data show high *P* element activity in tropical northeast Australia, through moderate activity at intermediate latitudes to low activity and *M'* cytotype in southern latitudes.

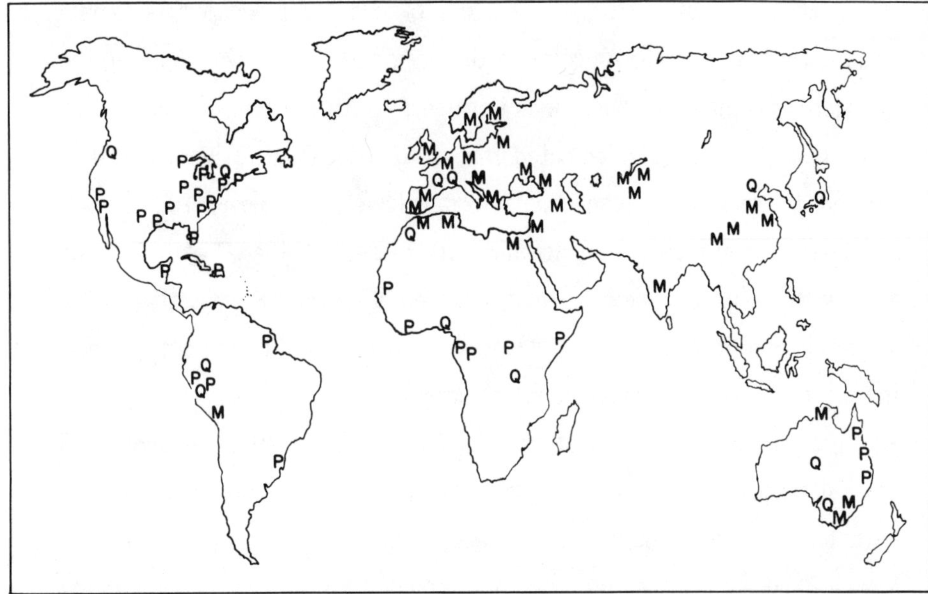

Figure 4. Worldwide distributions of *P*, *Q* and *M*'strains in present day populations of *D. melanogaster*. Only the predominant type of a region is shown. The sample sizes varied widely from one location to another.

In contrast to the marked differences in current geographical patterns which exist for the *P-M* system in *D. melanogaster*, the distribution of *I-R* characteristics in natural populations shows very little variability. In surveys of several hundred strains in two laboratories (Kidwell *et al.*, 1983), all those collected during the last decade were classified as inducer. However, there does appear to be considerable quantitative variability among different natural populations in inducer potential.

D. *P*-Element-Mediated Interspecific Transformation in *D. melanogaster*

It was first shown in *D. melanogaster* that, by microinjection of early embryos, it was possible to introduce *P* factors into the germ line of individuals previously lacking them. In the new cellular environment they were shown to transpose to new chromosomal sites (Rubin and Spradling, 1982; Spradling and Rubin, 1982). Subsequently, it has been demonstrated, using the same method, that functional *P*-elements can be introduced into the closely related but reproductively isolated sibling species, *D. simulans* (Scavarda and Hartl, 1984; Daniels *et al.*, 1985b) and the more distantly related Hawaiian species, *D. hawaiiensis* (Brennan *et al.*, 1985). In both species there was good evidence that the introduced *P* elements retained their functional ability to transpose in the cellular environment of a species in which they had presumably not existed previously. It is unclear whether dysgenic traits such as gonadal sterility are induced by the presence of the introduced elements. Transformed lines of *D. simulans* have produced gonadal sterility but the frequency is usually low and somewhat unpredictable (Daniels *et al.*, 1985b; M. Kidwell, unpublished results).

Thus, it is apparent that species barriers are no impediment to the ability of *P* elements to transpose and increase their copy number. However, a question of considerable interest has not yet been answered: whether the regulation of *P* element transposition will evolve in these transformed species, and if so, whether the mode of regulation will be found to be the same as it is in *D. melanogaster*.

V. HISTORICAL DISTRIBUTIONS OF HYBRID DYSGENESIS DETERMINANTS

When *D. melanogaster* strains having different laboratory ages were examined for their properties with respect to the two hybrid dysgenesis systems, marked temporal trends were observed (Kidwell, 1983). It was shown that the frequency of *P* and *M* strains was positively correlated with laboratory age. The oldest *I* strains had been cultured in the laboratory for about 50 years and the *I* type became increasingly frequent in strains with decreasing laboratory age. The *P* type was not found in strain samples collected before 1950. Again, there was a pattern of increasing frequency of the *P* type, the shorter the period of culture in the laboratory.

In contrast to the dichotomy which has been observed in *P*-element presence in *D. melanogaster* populations, with the exception of *D. insularis*, these elements are found to be present in all populations of species of the *willistoni* group so far examined, seemingly without regard for laboratory age (S. Daniels, pers. comm.). In fact, the first evidence for the presence of *P* elements in this group came from strains of *D. paulistorum* which had been maintained in the laboratory by Dr. Lee Ehrman for up to 25 years (Daniels *et al.*, 1984). These observations are consistent with the conclusion that the presence of *P* elements in most species of the *willistoni* group is evolutionarily of long standing.

A. Explanations for Historical Distributions

Two main types of explanation have been suggested to account for the observed temporal trends in the frequencies of *P* and *I* strains in *D. melanogaster*. The first, the recent-loss hypothesis, proposes that the negative correlations of *P* and *I* strain frequencies with laboratory age can be accounted for by loss of *P* and *I* elements during laboratory culture. According to this idea, the older the strain, the higher the probability of element loss. For example, Bucheton *et al.* (1976) suggested that reactive strains were derived from inducer strains as an artifact of laboratory culture methods. Engels (1981) extended the idea to the *P-M* system and the mechanism of "stochastic loss" of *P* and *I* elements. The second hypothesis, the recent invasion hypothesis (Kidwell,

1979, 1983), contends that P elements and active I elements have only very recently invaded natural populations of *D. melanogaster*. The two types of mechanisms are clearly not mutually exclusive; therefore, the relevant question is, which process was predominant in effecting the changes which have been recorded over a period of time that, in evolutionary terms, is practically instantaneous?

B. Recent Invasion Versus Stochastic Loss of P Elements

With respect to the $P-M$ system, stochastic loss as an explanation for the absence of P elements in laboratory populations, which are older than 10-20 years, seems unlikely for several reasons. First, many strains maintained in different laboratories, under different environmental conditions with different population sizes, have been found to be completely lacking P element homology. Even in small populations, the expected time to complete loss of as many as 30-50 P elements per haploid genome is very much longer (Charlesworth, 1985). Further, studies of the molecular properties of various types of strains suggest that deletion derivatives are abundant in all strains (P, Q or M') that carry P elements. It seems far more likely that loss of P factors would result in strains carrying only defective elements that no longer had the capacity to transpose than in strains with no elements at all. Preliminary observations of Engels and Preston (1980) are consistent with this prediction. They maintained a number of wild-caught P strains in the laboratory for 35 generations. Some lost their P cytotype and changed to the M' type which carry mainly defective P elements, rather than to the true M type in which P elements are completely absent (W. Engels, pers. comm.). Preliminary results in my laboratory indicated that not even one out of 20 P or Q population maintained under two different temperature regimes, evolved from a P to an M cytotype over a period of five years.

The recent invasion hypothesis implies a rapidity of spread throughout the cosmopolitan population of *D. melanogaster* that is unprecedented in terms of Mendelian genes. However, it can be argued that the amount of migration, both naturally occurring and that promoted by human activities, is sufficient to account for the rapid spread of a highly mobile element. On the other hand, a number of dysgenic traits, such as sterility, would be expected to act strongly against a rapid invasion by virtue of the low fitness of hybrids at the invasion front. However, it has been shown by Kidwell *et al.* (1981) and Kiyasu and Kidwell (1985) that small mixed *P* and *M* bottle populations will almost always rapidly evolve to the *P* type, even when the maintenance temperature allows maximum sterility to occur and when the initial frequency of *P*-bearing individuals in the original mixed population is as low as ten percent.

A second difficulty is that the recent invasion hypothesis implies too high a rate of origin of new mobile element families over evolutionary time. However, recent invasion does not necessarily imply recent origin. *P* and *I* elements could have existed in small geographical isolates for an unknown period of time and only recently rapidly invaded the species. Without knowing the time of origin, no judgment can be made about whether the rate is unreasonable. Also, the sample size of two element families is very small and the recent effects of man's activities on the biotic environment are likely to have been more disruptive and pervasive than in the past.

A third difficulty is that the presence of *M'* strains in many parts of the world is not explainable without also postulating loss of complete elements following the initial invasion. Again, this is not a serious difficulty, given the high frequency of internal mutability of the *P* element family to defective types. Indeed, in synthetic populations it has been observed many times that newly introduced P elements will initially increase in P factor activity, but that this activity is rapidly lost. As the P cytotype is never achieved in such populations, they appear to evolve to the M' type and have similar properties to many observed European populations (M. G. Kidwell, unpublished results).

The distribution patterns of P elements in current natural populations of both *D. melanogaster* and that of other *Drosophila* species is consistent with a recent invasion of *D. melanogaster* from a species of a taxonomically different group, such as *willistoni* or from some other source such as mites, yeast or other food. In addition to the fact that a temporal dichotomy with respect to the presence or absence of P elements exists within *D. melanogaster*, the complete absence of these elements in closely related species is consistent with a recent invasion of *melanogaster*. This is in sharp contrast to the widespread distribution of P elements in species of the *willistoni* group and some lineages of the closely related *saltans* group, which suggests that they have been present in these groups for a much longer period of time than in *D. melanogaster*.

Further indirect evidence that the P element is a recent invader of *D. melanogaster* comes from the fact that the AT:GC ratio of P elements is not typical of the average ratio for the *melanogaster* genome (Crow, 1983). The simulation studies of Uyenoyama (1985) are also consistent with the invasion hypothesis but do not address that of stochastic loss. She showed that an element can exist for a long time in a population at low frequency and thus may not be detectable by limited sampling. When the element eventually increases to a certain critical frequency, it can sweep to fixation in a very short period of time.

C. Recent Invasion Versus Stochastic Loss of *I* Elements

As pointed out by Bucheton *et al.* (1984), the status of the two hypotheses with respect to the *I-R* system is less clear than for the *P-M* system. Neither hypothesis in its simple form can adequately explain the data. The presence of *I* elements in all species of the *melanogaster* group suggests that their origin predates speciation events within the species group. The presence of mainly defective *I* elements in *R* strain *D. melanogaster* in fixed centromeric locations might be explained in two ways. Loss of active *I* factors by natural populations might have occurred over an extended period of time, leaving only defective elements in the genomes of most individuals. Reactivation of defective *I* elements, possibly by recombination, or by relocation, may have occurred very recently, followed by a new invasion of the species. Alternatively, active *I* factors may have persisted in natural populations and the observed reactivity of long established laboratory strains could be the result of "functional extinction" during the period of laboratory culture. Observations of at least thirty *I* strains provide no support for the rapid loss of *I* elements in laboratory culture. There was not a single case in which *I* strains had evolved to *R* strains over a period of 10 years (Bregliano and Kidwell, 1983).

VI. DISCUSSION AND CONCLUSIONS

The molecular characterization of the *P* and *I* element families has enabled an interesting comparison to be made. Despite the many close similarities in the phenotypic and genetic transmission patterns which provided the basis for including these two element systems in the single term "hybrid dysgenesis", the molecular structure of the two element families is quite different. Rubin (1983) classified known *Drosophila* transposable elements into broad types depending on their structure and properties. Neither the *P* nor *I* element families belong to the two most representative types, the *copia*-like and foldback (FB) elements; and thus, with the exception of the *hobo* element which has a similar structure to *P* (McGinnis *et al.*, 1983), neither family appears to be similar to any other major type of element in *Drosophila*. However, the *P-M* system has several important similarities with the *Ac* (Activator) and *Ds* (Dissociator) mobile element system in maize (Federoff, 1983). The *I* element is particularly unusual in not possessing terminal repeats, an attribute possessed by most other prokaryotic and eukaryotic elements.

Despite their structural differences, *P* and *I* elements are similar in exhibiting unusually high transposition rates in certain cellular environments, and there are some indications of strong similarities in regulatory mechanisms. It might be argued that these are essential determinants of a hybrid dysgenesis-like system. However, it is not known whether these characteristics are an unchanging functional property of the element families themselves or whether they are related to the evolutionary age of the system.

Kaplan *et al.* (1985) have discussed two broad classes of elements, those (like the *copia*-like elements) that transpose via an RNA intermediate, and those whose transposition depends on DNA replication. Transposition via an RNA intermediate imposes important constraints on the degree of mutability of the elements. Mutations resulting in loss of function are expected to be eliminated by natural selection because of the close relationship between sequence structure and phenotype. Thus the structure of this type of element tends to be highly conserved over evolutionary time. A second class of elements, such as *P*, apparently has a less constraining transposition mechanism, depending solely on DNA replication, and a propensity for frequent mutation which may be unrecognized by selection. Quite a different evolutionary history for this class is predicted, with fully functional elements being relatively abundant in the early stages. Mutation of element sequences leads to an increasing frequency of deletion or other derivatives of the original functional type. Eventually genomes are predicted to carry only defective elements and the element family may have a high rate of extinction. However, the time required to lose all the defective elements may be considerable because the final evolutionary phase is expected to be the loss of the ability for excision in the absence of fully functional copies. The rate of evolution of a highly mutable family might be increased by natural selection at the organismal level if fully functional copies induced high frequencies of traits which were detrimental to organismal fitness.

Indirect evidence for such an increase is provided by the mixed population studies of Kidwell *et al.* (1981) and Kiyasu and Kidwell (1985). The two studies were identical except for the maintenance temperature employed. At a temperature of 20º C, which was sufficiently low to avoid the induction of gonadal sterility, the equilibrium level of *P* factor activity achieved by the mixed populations was approximately equal to that of the *I* control population (Kidwell *et al.*, 1981). However, at a maintenance temperature of 27º, which was expected to induce maximum gonadal sterility in dysgenic hybrids, the equilibrium level of *P* factor activity was, with one exception, significantly lower than that of the *P* population control. These results may plausibly be explained by strong selection for *P* elements, with reduced capacity for the induction of gonadal sterility at the restrictive 27º temperature.

It is therefore argued that the ability to induce hybrid dysgenesis might depend, not only on the functional potential of the element system (transposition, regulation, etc.), but also on the stage of its evolutionary history. If there is strong natural selection against elements that induce low fitness traits at the level of the organismal phenotype, then an element family would be expected to exhibit an increasingly modified repertoire of such traits with time, provided that the genetic variability was present at the element sequence level.

One of the questions that arises from the possibility of recent invasion of a new species is to what extent preselection for some aspects of the regulation of these elements is implied. Regulation of transposition apparently occurs at several levels. The activity of both the P and I element families is limited to the germ line. If the invasion has been as recent as the evidence suggests, there will have been little or no time for the evolution of suppression of transposition in somatic cells. This suggests that a mechanism for suppression of transposition already existed at the time of invasion. Such a mechanism might have allowed the previous invasion of other element families because, without such a mechanism, a high rate of transposition in somatic cells, and its consequent disruptive effects, might have led to the death of the host. A similar argument holds for the observed suppression of I elements mobility in the male sex.

M. Simmons (pers. comm.) has suggested that, in the germ line, the titration mechanism may provide a rather primitive method for P element regulation. Such a regulatory mechanism is postulated not to be dependent on host regulatory functions, but to be a property of the element family itself and one which automatically accompanies the invasion of the element. It may be that this crude mechanism could be refined by interaction with host functions, but such refinement would be unnecessary for providing the degree of "self-restraint" in the increase of copy number which might allow successful invasion of a species. This mechanism alone could provide a means of restricting the copy number of an essentially parasitic element to the "carrying capacity" of its host.

In conclusion, it is proposed that the determinants of a hybrid dysgenesis-like mobile element system might include some or all of the following: 1) the potential for high transposition frequencies which would allow invasion of a species despite an initial reduction of fitness at the organismal level; 2) a non-specific mechanism for regulation of transposition that is largely independent of host functions and would allow a limited increase of copy number but prevent excessive deleterious effects of high copy number of host phenotype by "self-restraint"; 3) a replication mechanism that does not severely constrain the survival of mutated elements. These three general properties might allow the invasion of a host species by an element family and its subsequent evolution. If the requisite variability within the element family were present, then selection would be expected to favor those elements that increased organismal fitness but to act against those that reduced it. Because most effects, at the level of the organism, would be expected to be deleterious, at least initially, selection would tend to modify the production of large phenotypic effects. Thus, it is argued that hybrid dysgenesis might be a relatively transient early phase in the evolutionary history of certain element families that have the capacity for frequent transposition, self-regulation, and potential for internal mutability.

ACKNOWLEDGMENTS

The author thanks Stephen Daniels, Michael Simmons, Alain Bucheton, David Holm, David Finnegan and Brian Charlesworth for sharing with her their unpublished results. She is grateful to Michael Simmons and Stephen Daniels for comments on an earlier version of the manuscript.

REFERENCES

Anxolabéhère, D., Kai, H., Nouaud, D., Periquet, G., and Ronsseray, S. (1984). The geographical distribution of P-M hybrid dysgenesis in *Drosophila melanogaster*. *Génét. Sél. Evol.* **16**, 15-25.

Anxolabéhère, D., Nouaud, D., Periquet, G., and Tchen, P. (1985). P element distribution in Eurasian populations of *Drosophila melanogaster*: a genetic and molecular analysis. *Proc. Natl. Acad. Sci. USA* **82**, 5418-5422.

Bingham, P. M., Kidwell, M. G., and Rubin, G. M. (1982). The molecular basis of P-M hybrid dysgenesis: the role of the P element, a P-strain-specific transposon family. *Cell* **29**, 995-1004.

Bregliano, J. C., and Kidwell, M. G. (1983). Hybrid dysgenesis determinants. *In* "Mobile Genetic Elements," pp. 363-410. Academic Press, New York.

Brennan, M. D., Rowan, R. G., and Dickinson, W. J. (1984). Introduction of a functional P element in the germ-line of *Drosophila hawaiiensis*. *Cell* **38**, 147-151.

Brookfield, J. F. Y., Montgomery, E., and Langley, C. H. (1984). Apparent absence of transposable elements related to the P elements of *D. melanogaster* in other species of *Drosophila*. *Nature* **310**, 330-332.

Bucheton, A., Lavige, J. M., Picard, G., and L'Héritier, Ph. (1976). Non-Mendelian female sterility in *Drosophila melanogaster*: quantitative variations in the efficiency of inducer and reactive strains. *Heredity* **36**, 305-314.

Bucheton, A., Paro, R., Sang, H. M., Pelisson, A., and Finnegan, D. J. (1984). The molecular basis of *I-R* hybrid dysgenesis in *Drosophila melanogaster*: identification, cloning, and properties of the I factor. *Cell* **38**, 153-163.

Charlesworth, B. (1985). The population genetics of transposable elements. "Proceedings of Oji International Seminar."

Crow, J. F. (1983). Hybrid dysgenesis and the P factor in *Drosophila*. *Japan J. Genet.* **58**, 621-625.

Daniels, S. B., Strausbaugh, L. D., Ehrman, L., and Armstrong, R. (1984). Sequences homologous to *P* elements occur in *Drosophila paulistorum*. *Proc. Natl. Acad. Sci. USA* **81**, 6794-6799.

Daniels, S. B., McCarron, M., Love, C., and Chovnick, A. (1985a). Dysgenesis induced instability of rosy locus transformation in *Drosophila melanogaster*: analysis of excision events and the selective recovery of control element deletions. *Genetics* **109**, 95-117.

Daniels, S. B., Strausbaugh, L. D., and Armstrong, R. A. (1985b). Molecular analysis of the behavior of a P transposable element in *Drosophila simulans*. *Molec. Gen. Genet.* **200**, 258-265.

Doolittle, W. F., and Sapienza, C. (1980). Selfish genes, the phenotype paradigm and genome evolution. *Nature* **284**, 601-603.

Dover, G. (1982). Molecular drive: a cohesive mode of species evolution. *Nature* **299**, 111-117.

Engels, W. R. (1981). Hybrid dysgenesis in *Drosophila* and the stochastic loss hypothesis. *Cold Spring Harbor Symp. Quant. Biol.* **45**, 561-565.

Engels, W. R. (1983). The P family of transposable elements in *Drosophila*. *Ann. Rev. Genet.* **17**, 315-344.

Engels, W. R. (1984). A *trans*-acting product needed for P factor transposition in *Drosophila*. *Science* **226**, 1194-1196.

Engels, W. R., and Preston, C. R. (1979). Hybrid dysgenesis in *Drosophila melanogaster*: the biology of female and male sterility. *Genetics* **92**, 161-174.

Engels, W. R. , and Preston, C. R. (1980). Components of hybrid dysgenesis in a wild population of *Drosophila melanogaster*. *Genetics* **95**, 111-128.

Federoff, N. V. (1983). Controlling elements in maize. *In* "Mobile Genetic Elements," pp. 1-63. Academic Press, New York.

Kaplan, N., Darden, T., and Langley, C. H. (1985). Evolution of transposable elements in Mendelian populations. IV. Mutant elements and extinction. *Genetics* **109**, 459-480.

Karess, R. E., and Rubin, G. M. (1984). Analysis of P transposable element functions in *Drosophila*. *Cell* **38**, 135-146.

Kidwell, M. G. (1979). Hybrid dysgenesis in *Drosophila melanogaster*: the relationship between the *P-M* and *I-R* interaction systems. *Genet. Res.* **33**, 205-217.

Kidwell, M. G. (1983). Evolution of hybrid dysgenesis determinants in *Drosophila melanogaster*. *Proc. Natl. Acad. Sci. USA* **80**, 1655-1659.

Kidwell, M. G. (1985). Hybrid dysgenesis in *Drosophila melanogaster*: nature and inheritance of P element regulation. *Genetics* **111**, 337-350.

Kidwell, M. G., and Novy, J. B. (1979). Hybrid dysgenesis in *Drosophila melanogaster*: sterility resulting from gonadal dysgenesis in the *P-M* system. *Genetics* **92**, 1127-1140.

Kidwell, M. G., and Novy, J. B. (1985). The distribution of hybrid dysgenesis determinants in North American populations of *D. melanogaster*. *Drosophioa Info. Serv.* **61**, 97-100.

Kidwell, M. G., Frydryk, T., and Novy, J. B. (1983). The hybrid dysgenesis potential of *Drosophila melanogaster* strains of diverse temporal and geographical origin. *Drosophila Info. Serv.* **59**, 63-69.

Kidwell, M. G., Kidwell, J. F., and Sved, J. A. (1977). Hybrid dysgenesis in *Drosophila melanogaster*: a syndrome of aberrant traits including mutation, sterility and male recombination. *Genetics* **86**, 813-833.

Kidwell, M. G., Novy, J. B., and Feeley, S. M. (1981). Rapid unidirectional change of hybrid dysgenesis potential in *Drosophila. J. Hered.* **72**, 32-38.

Kiyasu, P. K., and Kidwell, M. G. (1985). Hybrid dysgenesis in *Drosophila melanogaster*: the evolution of mixed *P* and *M* populations maintained at high temperature. *Genet. Res.* **44**, 251-259.

McGinnis, W., Shermoen, A. W., and Beckendorf, S. K. (1983). A transposable element inserted just 5' to a Drosophila glue protein gene alters gene expression and chromatin structure. *Cell* **34**, 75-84.

Rubin, G. M. (1983). Dispersed repetitive DNAs in *Drosophila. In* "Mobile Genetic Elements," pp. 329-361. Academic Press, New York.

Rubin, G. M., Kidwell, M. G., and Bingham, P. M. (1982). The molecular basis of P-M hybrid dysgenesis: the nature of induced mutations. *Cell* **29**, 987-994.

O'Hare, K., and Rubin, G. M. (1983). Structures of P transposable elements of *Drosophila melanogaster* and their sites of insertion and excision. *Cell* **34**, 25-35.

Orgel, L. E., and Crick, F. H. C. (1980). Selfish DNA: the ultimate parasite. *Nature* **284**, 604-607.

Picard, G. (1976). Non-Mendelian female sterility in *Drosophila melanogaster*: hereditary transmission of I factor. *Genetics* **83**, 107-123.

Rubin, G. M., and Spradling, A. C. (1982). Genetic transformation of *Drosophila* with transposable element vectors. *Science* **218**, 348-353.

Scavarda, N. J., and Hartl, D. L. (1984). Interspecific DNA transformation in *Drosophila. Proc. Natl. Acad. Sci. USA* **81**, 7515-7519.

Simmons, J. J., and Bucholz, L. M. (1985). Transposase titration in *Drosophila melanogaster*: a model of cytotype in the P-M system of hybrid dysgenesis. *Proc. Natl. Acad. Sci. USA* (in press).

Simmons, M. J., and Karess, R. E. (1985). Special report: molecular and population biology of hybrid dysgenesis. *Drosophila Info. Serv.* **61**, 2-7.

Spradling, A. C., and Rubin, G. M. (1982). Transposition of cloned P elements into Drosophila germ line chromosomes. *Science* **218**, 341-347.

Throckmorton, L. H. (1975). The phylogeny, ecology and geography of *Drosophila*. *In* "Handbook of Genetics. Invertebrates of Genetic Interest," Vol. 3 (R. C. King, ed.), pp. 421-470. Plenum, New York.

Uyenoyama, M. K. (1985). Quantitative models of hybrid dysgenesis: rapid evolution under transposition, extrachromosomal inheritance and fertility selection. *Theor. Pop. Biol.* **27**, 176-201.

THE SPREAD AND SUCCESS OF NON-DARWINIAN NOVELTIES

Gabriel A. Dover

Department of Genetics
University of Cambridge
Downing Street
Cambridge CB2 3EH, England

ABSTRACT

Ever since Mendel there has been no alternative to Darwin. Mendel's laws of inheritance do not lead, in themselves, to evolutionary change and in the neo-Darwinian synthesis natural selection became the only means for achieving a highly improbable shift in the average useful phenotype of a population. Hence, all species-specific functions are defined as Darwinian adaptations: the current endpoints in a long series of 'solutions' to changing ecological 'problems'.

A decade of molecular studies on the non-Mendelian behavior of genes reveals that Mendelian populations in long-term genetic equilibria might not exist. Both 'single-copy' genes and multigene families are observed to be in a state of flux due to a variety of non-reciprocal mechanisms of DNA exchange which promote a gain or a loss of a genetic variant in an individual's lifetime. Detailed studies on diverse genes in a wide range of eukaryote organisms make it unlikely that genes exist which are refractory to such mechanisms. However, the generally low rates of non-reciprocal exchange ensure that a Mendelian analysis of small numbers of progeny over few generations would not reveal their presence.

Continual fluctuations in the copy-number of genetic variants can lead to a gradual and cohesive spread (molecular drive) of one or other variant throughout a sexual population. The slow rates of non-reciprocal exchange, in conjunction with the sexual process, ensure that at each

199

generation most individuals have a similar copy-number of the new variant gene, and a similarly changed phenotype.

A cohesively evolving population could gradually exploit a previously inaccessible component of its environment. In such circumstances, a novel function would have arisen as a consequence of a prior change in the average form, rather than as a consequence of a prior change in the environment. Simplistically, the organisms pose the 'problem' and the environment offers a 'solution'.

In some circumstances negative selection could eliminate a population as individual phenotypes reflect a threshold number of variants in a given gene family. Alternatively, normalizing selection might be eliminating the ill-formed progeny of parents that were maximally dissimilar in the copy-number of the variant, without however affecting the fluctuations in the mean copy-number between generations and the long-term population changes that might ensue. Positive selection could promote a molecular coevolution of other genes whose products interact with the gene family or its products, in order to maintain essential functions during the changes.

Evidence in support of molecular drive and molecular coevolution in the establishment of non-Darwinian novelties is discussed with reference to the eggshell chorion proteins of silkmoths, the ribosomal RNAs of diverse genera, and the mobile P elements inducing hybrid dysgenesis in *Drosophila*.

I. EVOLUTIONARY PROBLEMS AND SOLUTIONS: WHICH IS WHICH?

It is an axiom of classical evolutionary theory that all biological functions are Darwinian adaptations. They represent the current endpoints in a long series of past solutions to problems set by the environment. Function and natural selection have become inextricably linked. Species-specific functions and species-specific modes of development are perceived as the products of a process of continually refined responses to continually changing ecologies. Hence, all biological functions which are required for the general purposes of development, living and reproduction, are circumscribed as particular keys to particular environmental locks. In this way, the multitudinous and generally discontinuous ways of life of organisms are perceived as being isomorphic with a corresponding external set of environmental discontinuities. Natural selection and function have become synonymous to a degree that obscures the difference between process and product and which often makes of the latter the proof *sine qua non* of the former.

Defining form and function in terms of adaptations has been criticized by several investigators coming from a variety of disciplines (Gould and Lewontin, 1979; Lewontin, 1983; Gould and Vrba, 1982; Garcia-Bellido, 1983, 1985, 1986). There are several reasons for this unease with traditional formulations of the evolution of function. For example, Gould and Vrba have argued that some aspects of current form might have been functionless at first but coopted for function only later. Such "exaptations" cannot be regarded as adaptations in the traditional sense in that their initial establishment might not have been a direct consequence of selection. Furthermore, Lewontin has stressed that it is naive to produce models of evolution which regard organisms as passive functional solutions, produced by selection, to problems set by the environment, in that organisms and their niches can only be defined each in terms of the other. A change in one is compounded by a change in the other and cause and effect cannot be so easily dissociated.

The most cogent argument calling for a fresh look at our notions of function and adaptation is made by Garcia-Bellido: an argument based on the hard new facts of the genetic components responsible for the observed discontinuities in species morphologies. Large and non-serial discontinui-ties can be generated by different combinations of a relatively small and finite set of genetic operations, making evolution and development more a combinatorial than additive process. Hence, distinct species morphologies are not necessarily the result of a continual refinement by selection operating on a vast network of non-specific polygenes. The new data on the genetic autonomy of cells and lineages support the case for a limited number of independent, yet combinatorial, operations and against the notion of a cascade of polygenic effects modulating phenotype. Hence, the successful establishment of a novel genetic combination could depend largely on the need to maintain a number of specific molecular interactions during development rather than on the need to maintain a tight link with some component of the environment. Hence, the internal rules that govern ontogeny curtail the production of new forms and do not make available continuously variable phenotypes for selection to adapt effortlessly to convenient niches.

In the following pages I address this shift in emphasis from the external to the internal factors because the internal factors not only initiate new functions, but also promote their spread through a population. I develop the argument, based on our increased understanding of the unexpected, yet widespread, non-Mendelian behavior of the genetic material and the specific cohesive manner in which this causes long-term changes in the average phenotype of a population, that organisms and their niches do not reflect necessarily the supposedly successful adaptive solutions to problems of environment. Organisms do relate to their environment, but a prior change in form might have led to a subsequent exploitation of a new niche. The new niche is a solution to a problem posed by the organisms. Novel functions arising in this manner can be defined as non-Darwinian adaptations.

Before entering into specific case histories, it is necessary to back-track a little and to examine the reason why classical Darwinian theory equates function and adaptation. The heart of the problem is captured succinctly by R. A. Fisher's (1930) definition of selection as a means of achieving a highly improbable shift in the average genotype of a population. This is based on the realization that the Mendelian laws of inheritance, when transferred to whole populations of infinite size and idealized free mating, cannot in themselves lead to the spread of one allelic variant over another. Hence, selection becomes an absolute prerequisite for adaptive evolutionary change in a background of a continuous and stochastic recycling of chromosomes by the sexual process. Even though infinite, sexually unconstrained, populations do not exist and non-random sampling at the level of gametes and individuals (genetic drift) can increase the probability of gaining or losing an allele out of proportion to that expected in Mendelian populations, the classical view that mutations always follow the chromosomes on which they arose and are the passive providers of the grist for the evolutionary mill, has not been altered.

Today our knowledge of the behavior of genes and their evolution are radically different, for ample studies on the molecular behavior of eukaryote genomes over the past decade have revealed that, in addition to mutation, a vast proportion of the DNA is subject to a variety of non-reciprocal exchanges that can transfer mutational variants from one locus to another and from one chromosome to another. All such mechanisms of non-reciprocal exchange (gene conversion, unequal exchange, transposition, slippage replication and RNA-mediated transfers) induce rare but persistent non-Mendelian patterns of segregation which can induce the spread of mutations through a population over long periods of time. These mechanisms operate in both single-copy genes and multigene families, although their effects are more easily noticeable in multigene families which show relatively high species-specific patterns of homogeneity. It is conceivable that strict Mendelian genes and stable Mendelian populations in long-term Hardy-Weinberg equilibria do not exist, except when generally observed over short periods of time and in small numbers of progeny. It is of significance that the mechanisms of DNA turnover proceed generally at rates (10^{-2} – 10^{-5} per generation) which lie between the mutation rate and the rate at which chromosomes are continually randomized between generations. Hence, at the level of the chromosome, the Mendelian laws of segregation and stable Mendelian populations are relevant and accurate; at the level of DNA they are not, except on a short-term observational basis. The degree to which the behavior of the chromosomes and the behavior of DNA are out of synchrony could be of long-term evolutionary significance.

The process by which the genotypic composition of a population can be changed, as a consequence of the persistent non-Mendelian effects of DNA turnover, is called molecular drive (Dover, 1982; Dover *et al.*, 1982). It can promote, like selection, an improbable long-term shift in the mean genotype of a population. The shift takes place, however, not as a result of the selective sorting of allelic variants resulting from their effects on individual fitness, but as a consequence of the internal dynamics of DNA turnover.

The large disparity in rates between mutation, turnover and the randomization of chromosomes by sex leads to a specific prediction that at any given transition stage there will be a small population variance relative to the two extremes of no homogenization and full homogenization for a new variant (Dover, 1982; Ohta and Dover, 1984). This is to say, that with rates of turnover ranging from 10^{-2} to 10^{-5} per generation, it is improbable that one or a few individuals become homogenized in advance of the rest of the interbreeding population. Hence, molecular drive can effect a slow, cohesive shift in the mean composition of a gene family in a population, without the generation of large differences between individuals in the ratio of old to new variants, at any given generation.

Theoretical assessments (Ohta and Dover, 1984) and experimental data (Strachan *et al.*, 1985) on the variance are now available.

The maintenance of a relatively high genetic similarity between individuals throughout a period of transformation is of critical significance to our previous discussion on functions and adaptations, and to our understanding of the potential interactions with selection. In the following sections I illustrate, with reference to the evolution of eggshell proteins in silkmoths, the ribosomal RNAs in many species, and the mobile elements causing hybrid dysgenesis in Drosophila, how the specific cohesive mode of population change under molecular drive can (a) introduce a prior change in the average form leading to a concomitant new function and relationship with existing environments, and (b) promote the molecular coevolution of other internal components in order to maintain biological function. These facts require some re-evaluation of our concepts of function, adaptation and fitness, at least for all those many and varied aspects of phenotype that are under the control of multigene and non-coding DNA families. An examination of the contribution of the same internal forces to the complex evolution of the several gene families involved with mammalian immunity is made elsewhere (Dover and Strachan, 1986).

II. FIRST EXAMPLE: EVOLUTION OF SILKMOTH EGGSHELLS

In order to assess the relative contributions of selection and molecular drive to any one functional trait we would need to know i) the numbers and effects of single-copy genes, polygenes and multigene families contributing to the trait; ii) the modes, rates and biases in the molecular interactions between the relevant allelic and non-allelic genes; and iii) the heterogeneity of environments obtaining at each step.

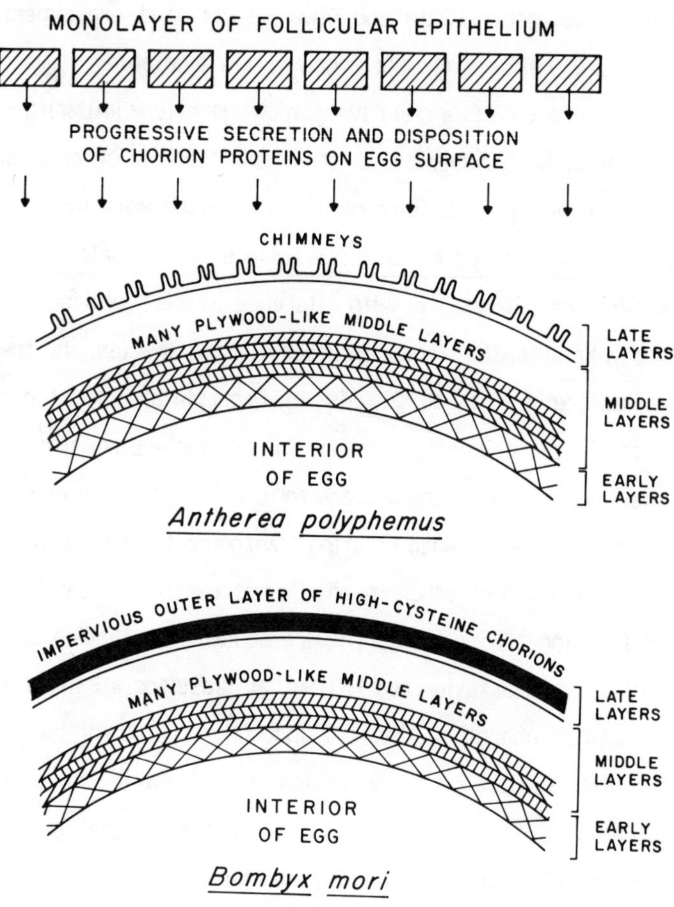

Figure 1. My 'artist's impression' of the differences between the two silkmoth species, *Antheraea polyphemus* and *Bombyx mori* regarding the outermost last-deposited chorion proteins. The 'chimneys' of *A. polyphemus* facilitate aeration, whilst the more impervious outer layer of *B. mori* (the unique products of the two high-cysteine subfamilies of genes Hc-A and Hc-B -- see Fig. 2) are thought to restrict gas exchange and relate to egg diapausy. It is not possible to indicate the tens of chorion layers and their intricate orientations. Nothing is to scale. (With apologies to Fotis Kafatos and his group).

There are some well-researched cases of morphology and function for which most of the requisite components for understanding their evolution are known. One striking example which introduces and illustrates most of what I wish to discuss, concerns the elaborate, highly-complex, species-specific architectures of eggshells in the family of Saturniidae silkmoths. These eggshells have been investigated extensively in a number of species by Fotis Kafatos and his associates, with respect to their biochemistry, ultrastructure, development, genetics, and genomic organization. The data when view *in toto* make it extremely difficult to explain eggshell evolution by natural selection alone, without an excessive use of *ad hoc* suppositions. The facts of the matter, drawn from the most recent review from which the references to the original papers can be explored (Goldsmith and Kafatos, 1984), are as follows.

The eggshell (chorion) of silkmoths serves as a protective layer against predation and pathogens, facilitates the exchange of gases, and prevents desiccation. It is composed of well over one hundred proteins which are secreted at precise developmental stages by the monolayer of follicular epithelial cells surrounding the developing oocyte (Fig. 1). The biophysical properties of the proteins are such that they are able to arrange themselves into fibrils with complex patterns of orientation and packing. These layers are topped at the egg surface by a final layer of chorion proteins which form elaborate species-specific surface structures. The chorion proteins, like many other structural proteins of animals and plants (for example, actins, tubulins, myosins, crystallins, collagens, histones, keratins, and other proteins of the cytoskeleton, etc.), as well as numerous non-structural proteins (for example, immunoglo-bulins, T-cell antigen receptors, globins, etc.) and functionally important RNAs, are the products of a multigene family. In the case of the chorion proteins the genes are organized into a large superfamily that has become structurally differentiated over time into families, subfamilies, and sub-subfamilies between which there are decreasing amounts of sequence identity. As with other complex families (for example, the mammalian immune genes; echinoderm histone genes; diverse globin genes, etc.), the member genes at the lowest level of the hierarchy show the highest level of genetic identity and are similarly expressed in time and space.

A biologically crucial difference between the two species *Antheraea polyphemus* (wild silkmoth) and *Bombyx mori* (cultivated silkmoth) revolves around the surface structures. In the former, these consist of chimney-like projections required for gas exchanges; whilst in the latter the surface is covered by a compact apparently impermeable layer in keeping with the diapausal behavior of the egg (Fig. 1). This compact layer consists of two cysteine-rich chorion proteins coded by two sub-families (Hc-A and Hc-B) that are limited to *B. mori* and are distinct from the subfamily genes controlling the chimneys of *A. polyphemus*. The outer layers of the eggshells, together with the other composite layers, are functional in that there is a relationship between structure and ecology for each species.

A close examination of the structure of the Hc-A and Hc-B sub-families poses problems for an argument based on natural selection alone for the origin of the new function. Each subfamily consists of approximately 15 near-identical genes arranged in pairs (one from each subfamily) in a head-to-head configuration with each pair transcrip-tionally regulated by similar 5' sequences in the spacer that separates them (Fig. 2). The sequences of the central domain of each gene show that the 15 genes of each subfamily are derived from two pre-existing subfamilies (A and B) in the superfamily, either by a process of *de novo* amplification from an A and B variant gene pair or by the

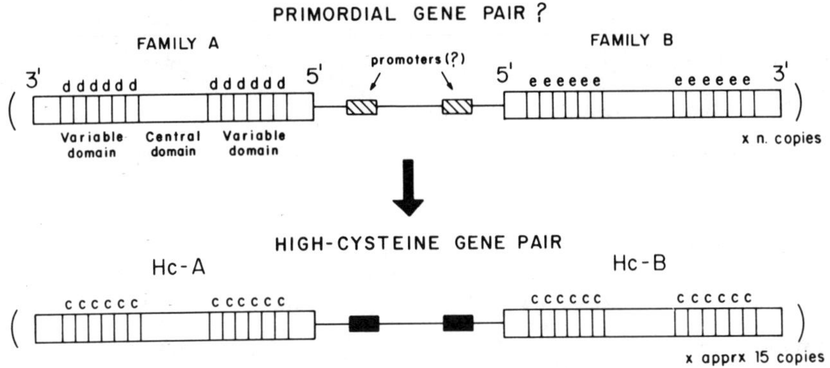

Figure 2. Schematic outline of a pair of genes of the two high-cysteine subfamilies, responsible for the outermost chorion layer of *B. mori* (see Fig. 1). They are orientated, like other gene pairs of other coordinately expressed subfamilies, in a head-to-head arrangement. The 'arms' contain variable numbers of a simple DNA motif (c) coding for cysteine/glycine amino-acids. There are approximately 15 gene-pairs derived from variant genes of two (A and B) of the five families of the superfamily. The primordial gene-pair would have consisted of other repetitive motifs (e.g., e and d) of variable number and sequence. The putative promoters and signals involved with developmental expression have also changed. (Drawings are not to scale.)

gradual replacement of a subfamily of 15 ancestral gene pairs by 15 derived gene pairs. Furthermore, the left and right hand domains surrounding the central domain are internally repetitive for tens of copies of a simple sequence motif coding for cysteine and glycine (Fig. 2). The genes of Hc-A and Hc-B contain a similar motif. Interspecific comparisons between genes of other subfamilies show that the right and left hand domains are highly variable due to the gain-and-loss of short tandem sequence motifs that are different for each subfamily type. The central portion is more conserved across all families although it too can be used to identify the subfamily and family origin of a given gene. The central portion gives rise to β-pleated sheets of protein important for the construction of fibrils, whilst the arms presumably contribute to the variable subfamily-specific functions.

How did the genes of Hc-A and Hc-B arise and why did they spread in the original progenitor population of *B. mori*? To answer these questions we need to explain the spread of tens of copies of a particular simple sequence motif in both the left and right hand arms of each of the approximately 15 genes of each subfamily, to all individuals of *B. mori*, bearing in mind that the pairwise orientation and synchronous developmental expression point to their close parallel evolution.

The argument for selection rests on there being an effect on phenotype of the first single gene with the first single cysteine/glycine motif in the left or right domain, in place of one copy of some other pre-existing repetitive motif in any one of a number of genes. Whilst there is no direct evidence that such an effect did not take place, it is difficult to conceive of an advantageous effect on survival dependent on what must be an extremely minor change in the surface sculpturing produced by a single partially variant gene in a background of several hundred other genes of the superfamily. Furthermore, even if we were to suppose that the effect on phenotype of the first partially mutant gene had been advantageous and fixed by selection, by what further mechanism could selection proceed to spread the cystein/glycine motif horizontally within the gene and to all other non-allelic loci of the relevant A and B subfamilies?

It seems more probable that the effect on phenotype would only become apparent once a considerable number of the cysteine/glycine motifs had accumulated by other means in each arm of each of the 30 genes in question. Mechanisms of *de novo* accumulation, or replacement of pre-existing genes with variant copies, are well characterized at the molecular level (for reviews see Ohta 1980, 1983; Arnheim, 1983;

Fedoroff, 1979; Smith, 1973; Davidson and Britten, 1973; Hood *et al.,* 1975; Tartof, 1974; Dover, 1982; Dover and Tautz, 1986; Flavell, 1982 1986). All give rise to stochastic gain-and-loss, or deterministic gain, of mutant genes, as a consequence of non-reciprocal exchanges between alleles and non-alleles of a gene family; and hence can affect the spread of a mutant gene through the family and through the population (see Section I).

Detailed examination of small gains and losses of sequences in the left and right hand domains of various chorion genes strongly suggest that the internally repetitive arms are subject to continuous fluctuations in the copy-number of the repeats by a mechanism such as slippage replication (Jones and Kafatos, 1982). Slippage replication, unequal exchange and other similar mechanisms operating over small stretches of DNA are well documented and are held responsible for the widespread occurrence and fluctuations of so-called simple repetitive sequences in eukaryote and some viral genomes (for references see Tautz and Renz, 1984; Dover and Tautz, 1985; Jeffreys *et al.,* 1985; Karlin, this volume). It is likely that the accidental accumulation of the cysteine/glycine motif in place of the preexisting motif took place by this mechanism. Furthermore, the close parallel evolution of the Hc-A and Hc-B gene pair suggests that a process such as gene conversion has continuously transferred such motifs between the arms of the two types of genes. Indeed there is evidence in the superfamily of chorion genes as a whole, in addition to the high cysteine subfamilies, that gene conversions have been directly involved in the evolution of the genes (Goldsmith and Kafatos, 1984). The synchronous coordinated expression of the Hc-A and Hc-B gene pairs, as with other subfamily pairs that are expressed at the same developmental period, might also be the result of the transfer of critical sequences around the 5' ends of the gene pairs that are involved with transcription promotion and regulation (Fig. 2). The existence of common sequences in all such key positions within a cluster of coordinately expressed gene-pairs of specific subfamilies and their absence from other clusters of other subfamilies, strongly supports the involvement of gene conversion in their generation.

Once a given gene-pair of the A and B families begins to evolve in parallel both with regards to the internal cysteine/glycine motifs and to the common regulation signals, the variant pair can accumulate as a unit by unequal crossing over. The clustered arrangement of the 15 Hc-A and Hc-B pairs suggests that a mechanism of this sort has been taking place. There is ample evidence for the simultaneous operation of several turnover mechanisms embracing different lengths of DNA and proceeding at different rates and biases, in the same multigene family (Coen *et al.*, 1982a,b; Coen and Dover, 1983; Smithies and Powers, 1986; Dover and Tautz, 1986; Flavell, 1982, 1986; Brown, 1983; Klein and Petes, 1981). The combined operations of slippage replication, gene conversion and unequal exchange, between allelic and non-allelic loci, could explain both the present day two-tiered repetitive organization of the chorion genes, their coordinate expression and their initial spread through the population. One biological consequence of these genetic processes would be the gradual and cohesive accumulation of a dense impervious late-forming outer layer of high cysteine proteins in *B. mori*. What role might selection have played in all this?

Before answering this important question (see Section V) it is necessary to examine two other examples of multigene families which illustrate additional features of evolution under the influence of the internal dynamics of the genes.

III. SECOND EXAMPLE: MOLECULAR COEVOLUTION IN THE RIBOSOMAL RNA GENES

In the case of the chorion proteins it is reasonable to suppose that the acquisition of the high-cysteine layer of *B. mori* has necessitated a concomitant evolution of other parts of the genetic system, attendant on the changed physiology. Studies on gene families in other species show that a coevolution has taken place of genes whose products functionally interact with the family in question.

Coevolution is receiving increased attention at the molecular level, as an explanation for the observed conservation of the three-dimensional structure and function of certain classes of proteins, despite the very high levels of divergence of the amino-acids (see Chothia, 1984 for review). In such cases coevolution is considered to be taking place in one part of a protein in compensation for changes which have occurred in another. It is clear that within all species, representing unique modes of relatively harmonious and integrated development, a widespread coevolution of interacting parts must be taking place. The phenomenon is the very essence of what we are trying to explain by evolutionary processes.

The combined studies from a number of laboratories investigating the molecular biology of the key ribosomal RNAs and their genes in diverse species of animals, plants and bacteria are providing data that indicate, for the first time, the evolutionary processes by which molecular coevolution is taking place. These studies are reviewed in a number of places, and I shall not enter into all details (see Reeder, 1984; Dover and Flavell, 1984; Arnheim, 1983, and Arnheim, this volume). The data, when viewed *in toto* indicate that in this particular gene family, and possibly in others for which data are available, molecular drive is not only affecting the biology of a given multigene family but is pressurizing other parts of the same family and other genes in the genome to follow suit.

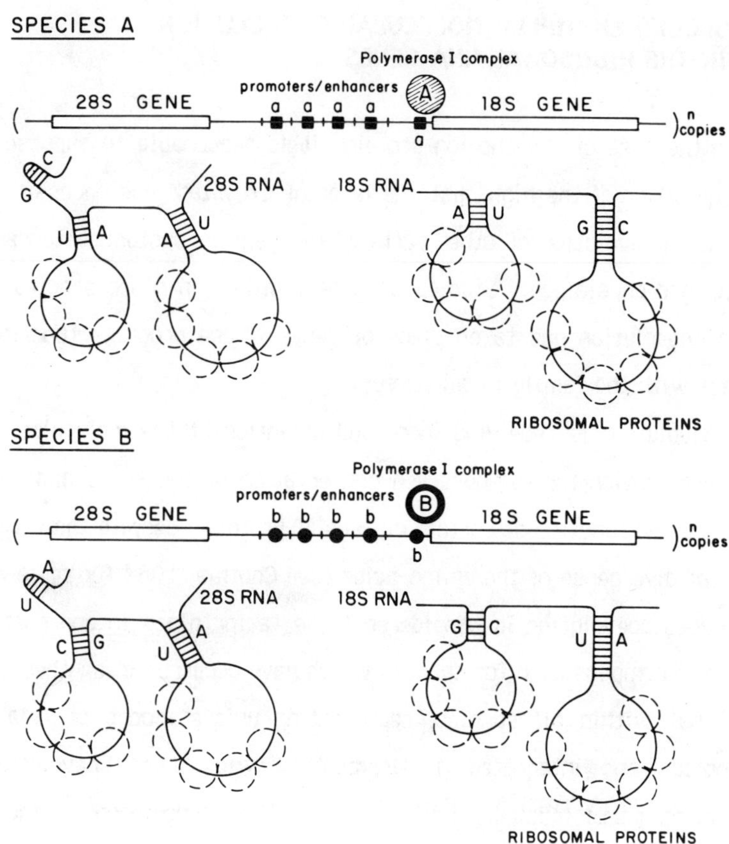

Figure 3. A typical eukaryote ribosomal RNA gene unit containing the 18S and 28S genes separated by a spacer, itself internally subdivided with multiple transcription promoters and enhancers. These differ between species (A and B) as do the polymerase/transcription complexes. The compound unit can be repeated up to several hundred fold according to species. The 18S and 28S RNA secondary structures are indicated crudely. For example, there can be up to 94 loops in the 28S RNA which are conserved across kingdoms. There is no conservation of the DNA or RNA sequences. See text for an explanation of the mechanisms of molecular coevolution involved with the maintenance of promoter/polymerase I complex compatibility and the base-base compensatory changes underlying RNA loop maintenance. (Drawings are not to scale.)

This can be envisaged in the first instance, as an interaction between molecular drive and natural selection with the former providing the internal driving force for the latter.

The two major rRNA genes (18S and 28S) occur in a compound unit together with the DNA spacer that separates them (Fig. 3). The unit is often repeated up to several hundred times (depending on species) in long tandem arrays that are located at the nucleolar organizers on homologous and sometimes non-homologous chromosomes. The great majority of units within and between individuals of a species can be identified by diagnostic mutations that are specific for the species. These also occur in each of the subrepeats within the spacer of some species. The observed pattern of high levels of species-specific homogeneity in the rDNA family, as in all other multigene and non-genic families, is called concerted evolution (for reviews, see Arnheim, 1983; Tartof, 1975; Dover, 1982; Fedoroff, 1979; Hood *et al.,* 1975; Coen *et al.,* 1982a,b). In the case of the rDNA, unequal exchange operating at the two levels of subrepeat and the complete unit is the genomic turnover mechanism responsible for the molecular drive process in the family (Coen *et al.,* 1982a,b; Dover, 1982): molecular drive being the population genetics *process* behind the final observed *pattern* of concerted evolution. Stages of transition during the linked processes of family homogenization and population fixation can be deduced by the existence of partially spread mutations in some non-genic families, irrespective of the locus, chromosome or individual in which the variant repeats occur (Strachan *et al.,* 1985).

The majority of homogenized rDNA mutations are found in the intergenic spacer. This region is of critical importance to the biology of the rDNA unit for recent evidence shows that it contains the signals required for the enhancement, promotion and termination of transcription;

for the initiation of replication; for the differential amplification of the genes in some species; and for rDNA specific modes of chromatin organization (reviewed by Reeder, 1984; Dover and Flavell, 1984; Grummt *et al.*, 1985). Some of these activities are controlled by sequences within the spacer subrepeats such that the efficiency of transcription not only reflects the sequence *per se* of the promoters/enhancers but also is affected by variations in their copy-number due to unequal exchanges at the level of the subrepeats. Interspecific comparisons of the sequences of the promoters/enhancers, and also those potentially involved with transcription termination in the genus Drosophila, reveal that these important regions are not refractory to mutation and to the spread of these through the family and population (Coen and Dover, 1982; Franz *et al.*, 1985; Dover and Tautz, 1985; Reeder, 1984). That these molecular driven changes have affected the biology of the rDNA, and in turn affected the evolution of other parts of the genome leading to species-specific phenotypes, has been revealed by two surprising observations. The first concerns interspecific incompatibilities, even between closely related species, of the RNA polymerase complex of one species and the rDNA promoters/enhancers of another; and the second concerns the remarkable conservation of rRNA secondary structure across biological kingdoms, despite DNA sequence divergence.

Enhancement, promotion and termination of transcription of all genes involves proteins, and possibly RNAs, transcribed and coded by other parts of the genome. rDNA is transcribed by RNA polymerase I and its cofactors, which, together with the proteins putatively involved with recognizing other DNA signals for regulation, enhancement, termination, replication and chromatin organization, would make up a substantial number of products of other genes that need to monitor the changes occurring in rDNA spacers.

Recent experiments involving related species of mammals, insects, frogs, protozoa and wheat have revealed an interspecific divergence in polyermase I and its cofactors (Fig. 3) seen as either an incompatibility between the rDNA of one species and the polymerase complex of another (when asayed in *in vitro* and *in vivo* heterologous transcription systems), or by the phenomenon of 'nucleolar dominance' in interspecific hybrids. Nucleolar dominance is revealed by the transcriptional activity of the rDNA of only one species in the hybrid. For details of these experiments see Arnheim (this volume) and Grummt *et al.*, 1982; Mishima *et al.*, 1982; Miesfeld and Arnheim, 1984; Skinner *et al.*, 1984; reviewed in Reeder, 1984; Dover and Flavell, 1984; Dover and Tautz, 1985).

The second surprising finding concerns the secondary structure of rRNA (that is the presumed folding patterns of the 28S and 18S RNA required to bind the over 50 ribosomal proteins that constitute the fully assembled ribosome necessary for mRNA to protein translation), which has been compared between bacteria, amphibia, fungi, slime-moulds, insects, mammals and plants (reviewed by Gerbi *et al.*,1982; Clark *et al.*, 1984; Noller, 1984; Brimacombe, 1984). The two RNAs are remarkably conserved in structure although they are very different at the DNA and RNA sequence level. For example, 94 stem-loop secondary structures can be formed by the 28S RNA sequence of *Xenopus laevis,* of which 74 can be formed by the corresponding 23S RNA sequence of *E. coli.* Surprisingly, of these 74 only 7 are conserved at the primary sequence level between *E. coli,* yeast, *Physarum* and *Xenopus.* The rest reveal primary sequence divergence and secondary structure conservation, in that a change in one part of the gene must have been followed by a coevolutionary compensatory change in another part. By this means 68 stem-loop structures have been maintained by ensuring that RNA base-base pairing, required in the formation of the stems (G with C; A with U) always obeys the usual rules (Fig. 3). Many of the loops of the 18S and 28S RNA are required for the proper functioning of the ribosome, with respect to protein binding sites; peptidyl transferase and GTPase centers; tRNA and 5S RNA binding and interaction sites; and RNA-RNA switching thought to be involved wiht ribosome sliding from one mRNA codon to another during protein translation.

What evolutionary processes have been reponsible for polymerase I complex divergence and RNA secondary structure conservation, both of which reflect the maintenance of function (albeit in a species-specific manner) despite the relatively rapid genetic divergence in the rDNA unit? An answer to this question is given in Section V.

IV. THIRD EXAMPLE: THE SUPPRESSION OF P ELEMENT INDUCED HYBRID DYSGENESIS IN DROSOPHILA

The presence of mobile elements in eukaryote genomes is not a satisfactory state of affairs either because they can jump into important functional regions of the DNA or because, as in Drosophila, some of them can lead to severe abnormalities and sterility unless suppressed. By phrasing the problem in this way I am adopting a classical view that organisms are strictly regulated and are in close harmony with their environments, such that any disturbances from the norm need to be eliminated. Elimination could be either by negative selection or by an increase in any other genetic component that confers resistance to the elements and stops them from moving. Many recent theoretical models of this assumed situation have arrived at solutions to the problems describing various equilibrium positions between two opposing forces: duplicative transposition increasing element copy-number and natural selection decreasing element copy-number. Such models also assume that there is an incremental drop in fitness with each additional element (for example, see Charlesworth and Charlesworth, 1983). Molecular investigations of dysgenesis suggest, however, an alternative approach to this problem.

The full biology of dysgenesis is described in the accompanying paper of Kidwell (this volume) and is reviewed by Engels (1983; 1986) and Bregliano and Kidwell (1983). Dysgenesis (a complex syndrome of maladaptive effects including rudimentary, sterile gonads) is usually seen in the progeny of fathers carrying fully intact P (or I) elements and mothers that do not carry the elements. The reciprocal cross and progeny from within both of the parental populations do not exhibit dysgenesis. The non-reciprocity of effects and the stability of wild populations has been interpreted as a result of the build-up of cytoplasmic suppressors of P element mobility. The genetics of the situation, when viewed *in toto*, suggest that such cytoplasmic suppressors have arisen, in part or in whole, from the P elements themselves. By analogy with prokaryote transposons, O'Hare and Rubin (1983) proposed a model how this might happen. Complete P elements would consist of two genes, one producing transposase (required for jumping: an activity necessary for dysgenesis induction) and the other producing repressor. The latter would positively feedback on itself and induce its own further production and negatively feedback on the transposase gene and prevent its production. I will call this model the two component model (Fig. 4A).

Several recent findings (Karess and Rubin, 1984; Kidwell, this volume) are not, however, in keeping with this model and generally support a one component model in which a complete P element is capable of producing transposase only (Fig. 4B). Evidence for the one component model is derived from molecular, genetic and population studies in a variety of laboratories which were collectively interpreted in support of

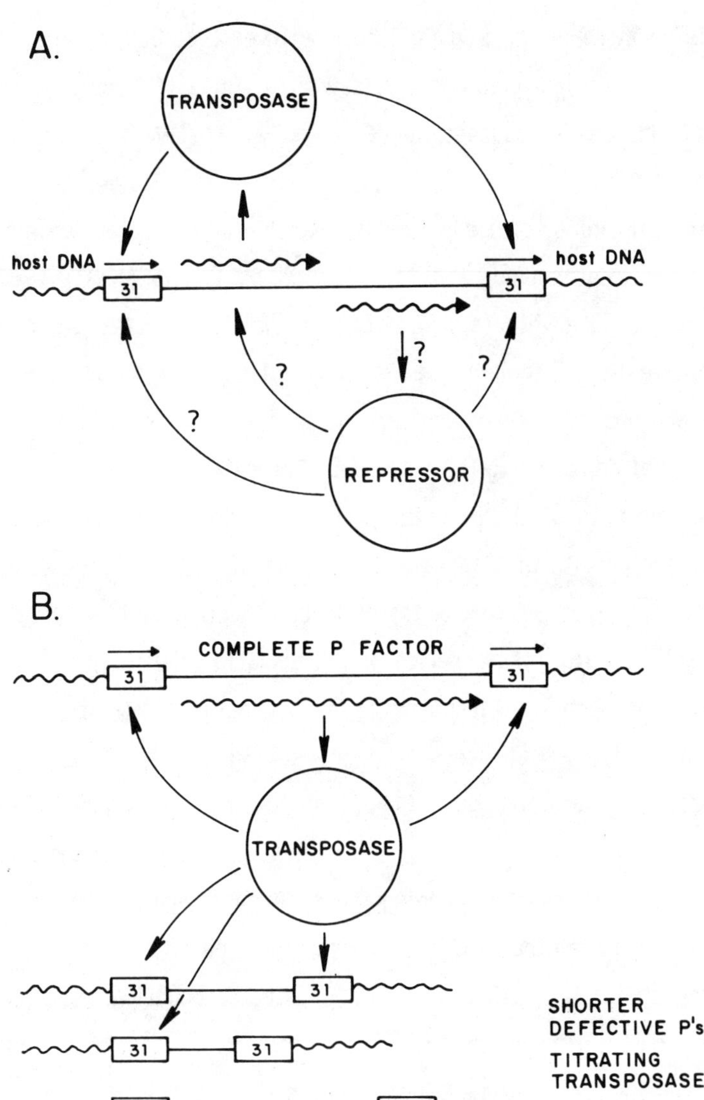

Figure 4. The two-component and one-component models of P-element mobility suppression in *D. melanogaster*. In the first the absence of dysgenesis in P populations is considered to be due to the cytoplasmic accumulation of suppressor protein coded from within the P-element and which interferes with transposase enzyme required for mobility. In the second, this is considered to be due to a progressive titration of transposase, by the ends of internally deleted elements (produced by the act of jumping) which are unable to produce transposase themselves (see text). (Not to scale.)

the model at a recent Cambridge Dysgenesis Workshop (see report: Simmons and Karess, 1985; and also Simmons, 1985, and Kidwell, this volume).

The one-component model supposes that suppression arises out of a situation in which the accumulation of defective P elements titrates transposase to the point where no complete P element has sufficient transposase for its mobility. Defective P elements, consisting of elements with deletions in their central portions, but with the requisite left- and right-hand ends capable of binding to any available transposase, are found in large numbers in many world-wide populations whether of the P, M or intermediate phenotype (Kidwell, this volume; Anxolabehere *et al.,* 1985; Black, Kidwell and Dover, in preparation). The one component model requires that shorter deletion-derivatives are incapable of producing transposase themselves but that their presence on the chromosomes and eventually as free elements in the cytoplasm absorb all available transposase produced by the complete elements. Suppression would then be an outcome of the ratio of complete to defective elements achieved in any one population. Hence both the accumulation of complete elements and suppression would be the result of the physical mechanics of the behavior of DNA. For reasons not understood, the act of transposition produces elements with internal deletions, a consequence of which is the probable eventual cessation of all transposition, as outlined. Although there are other complicating factors, the details of which can be found elsewhere, the simplest interpretation of the phenomenon is that both dysgenesis and its inevitable suppression might be a consequence of molecular drive acting alone. Selection might be involved only indirectly with the accumulation of suppression (see below).

V. MOLECULAR DRIVE AND NATURAL SELECTION: FUNCTION AND ADAPTATION

What role might selection play in the evolution of the high-cysteine chorion layers of Bombyx; in molecular coevolution in the rDNA family; and in the suppression of hybrid dysgenesis in Drosophila? Let us start with the last example in order to illustrate a general point.

A. Ontogeny and Fitness During Molecular Drive

Although it is reasonable to exclude selection from the one component model of suppression based on DNA mechanics, it is nevertheless necessary to examine the role of selection in terms of the dynamics of spread of P elements in a population.

Hybrid dysgenesis occurs between a male having complete P elements and a female without the means to suppress them: the two parents being derived from separate populations. A key question concerns the extent of dysgenesis within the population in which the elements initially accumulate. Selection would only be involved if a change of fitness is taking place. The degree of dysgenesis at any given generation might depend on the difference between any mating male and female in the copy-number of complete and defective P elements within them. The specific population dynamics under molecular drive (see above) predicts that at any given stage of transition the copy-number difference would be small relative to the initial condition of zero elements and the final condition of approximately 50 elements (complete and defective) (Fig. 5). As with other turnover mechanisms, the rate of duplicative transposition in wild populations may be as low as 10^{-2} to 10^{-4} per generation (Engels, 1983). At this rate there would be a gradual shift in the mean copy-number of

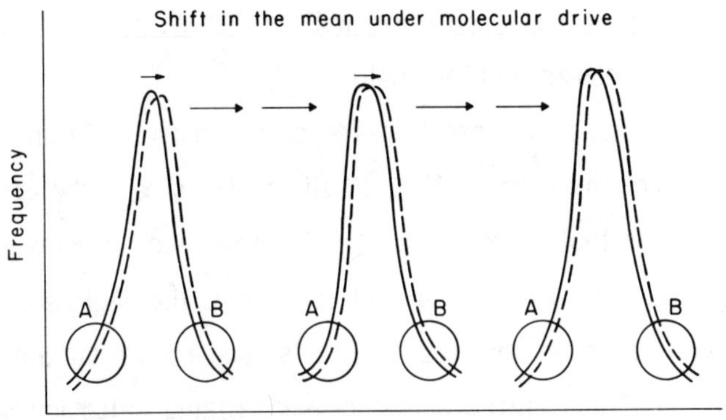

Figure 5. Schematic outline of a shift in the mean copy-number of complete and defective (see Fig. 4) P elements by molecular drive--in this case due to the mechanism of duplicative transposition. Normalizing selection could eliminate potentially dysgenic progeny of parents (from regions A and B of the distributions) which are maximally dissimilar in their P copy-number. This would slow down but not stop a longterm shift in the mean. At the end of the process full hybrid dysgenesis is observed between males of this population and females of other populations. During P accumulation, fitness might reflect the differences between developing zygotes in the extent to which their parental genomes were dissimilar and not necessarily to P copy-number *per se*. The applicability of this concept of fitness to a population experiencing a gradual and cohesive transformation with respect to a molecularly driven change of sequence in a multigene family, is described in the text.

elements, without the generation of a greater than Poisson variance at any given generation (see Ohta and Dover, 1984). There is no available evidence to suggest that the copy-number difference between males and females during the stages of transition in natural populations, is sufficient to induce dysgenesis. Furthermore, the concomitant build-up of defective elements and hence suppression, as a physical consequence of transposition, might also militate against the expression of dysgenesis. Only at the end of the process, with the occasional mating of a male fly from a fully suppressed population with a female from an unaffected population, might dysgenesis take hold.

We might wish to suppose that the copy-number difference between individuals at any given generation, albeit small, is nevertheless inducive to dysgenesis and that selection would eliminate such hybrids. Evidence that this is occurring during a burst of activity at the beginning of spread of P elements has been obtained by Engels (see report by Simmons and Karess, 1985). This limited action of normalizing selection would not necessarily stop the inexorable shift in the mean copy-number of complete and defective elements from one generation to the next, which is solely a consequence of duplicative transposition. In such circumstances selection might be simply eliminating the progeny of parents taken from the two extremes of the distribution (Fig. 5). Although the true situation is known to be more complex in that temperature, female age and possibly the genotypic background in which the elements find themselves, are also contributing to the final outcome, the P elements illustrate what could be an important general point about relative fitnesses at any given generation during molecular drive.

Individual fitnesses might be a reflection of the degree to which each individual's parental genomes were dissimilar. In the case of dysgenesis the maximum loss of fitness might occur in those progeny whose parents show the maximum difference in P element copy-number (Fig. 5). This type of situation can be envisaged for population genetic changes effected by other molecular mechanisms of turnover. For example, at any stage of transition in the transformation of a multigene family from an initial state of all *a* genes to a final state of all *A* genes, only those individuals with parents that maximally differ in their individual ratios of *a* : *A*, might suffer a loss of fitness relative to the rest of the population. As with dysgenesis, selection is a decision operating against individuals whose ontogeny and survival might have been adversely affected by the dissimilarity in the two parental sets of haploid genes. Hence, the action of normalizing selection, in eliminating the progeny of parents taken from the extremes of a distribution, is not necessarily a lack of fitness of parents and their progeny due to an incompatibility with the existing biotic and physical environment. The majority of the population, which by definition are the products of parents with the average ratio of $a : A$ (whatever this average might be at each point in a given time span of molecular drive), might survive simply as a consequence of the relatively high genetic compatibility between haploid sets of chromosomes. In other words, the success of ontogeny might depend simply on the degree to which molecular interactions can be maintained in an individual. A group of similarly developing individuals might be able to negotiate successfully a molecularly drive change in ontogeny because the maintenance of the molecular interactions is dependent on an ever-present compatibility between parental genomes in most individuals at any given generation. A convergent view of ontogenetic interactions being the internal reference for the success of new combinations of the genetic processes controlling development, has been expressed succinctly by Garcia-Bellido (1983, 1985, 1986).

There will always be small differences between individuals during molecular drive which could be of paramount importance in more traditional models of evolution that make selection the sole means for achieving a highly improbable shift in the average form and function of a population. However, there might not be *a priori* theoretical grounds for assuming that all individual differences, wherever they exist, are inevitably of survival value, given that there is an additional means for achieving the highly improbable. The relevance of individual differences and the action of selection in the evolution of a trait needs to be justified on ecological grounds and not subsumed in a general definition of all functions as continually refined adaptive responses to environmental changes.

Bearing this in mind, what might be the biological reasons for invoking the action of negative or positive selection in the evolution of the chorion multigene family and in the observed molecular coevolution involved with maintaining the functions of the rDNA family?

B. Molecular Drive and Eggshell Functions

It is clear that the accumulation of the outer layer had not been detrimental to the population of silkmoths that first underwent this change, for the physiological consequences of slow gas exchanges and prevention of dessication are functionally required for diapausal eggs. The involvement of negative selection in eliminating other less fortunate populations that might have accumulated a greater than tolerable layer of alternative variant proteins that might have blocked gas exchange altogether, is easy to envisage. In all such instances there would be some threshold level at which the molecularly driven accumulation of a variant gene would be of adverse effect. A cohesively evolving population of individuals approaching the threshold, would be at a disadvantage relative to other populations not being so transformed. Selection would always be at the individual level, nevertheless the cohesive mode in which an idealized population evolves (for details, see Ohta and Dover, 1983, 1984), would introduce an element of interpopulation selection, the dynamics of which needs further analysis. Notwithstanding such possibilities, our extant *B. mori* species is living proof that the accumulation of high cysteine genes within and between individuals--an event that could not have taken place in the absence of the internal genetic turnover

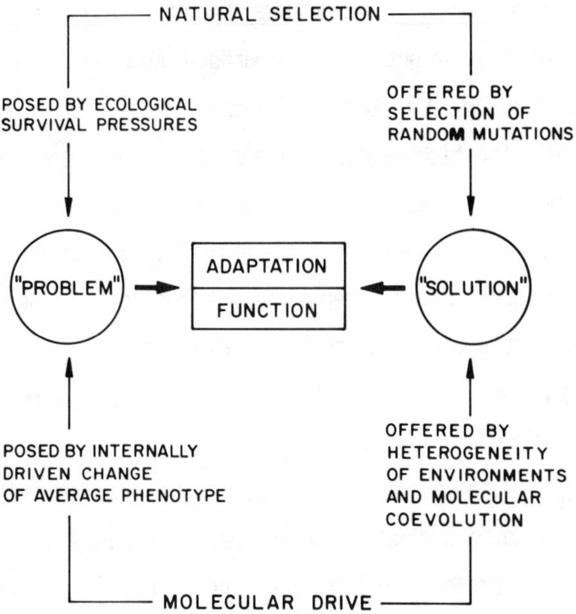

Figure 6. Illustration of the different modes of origin of 'functions' and 'adaptations' (see text). All biological functions as observed are not necessarily Darwinian adaptations, although they are essential for the unique ways of life that characterize each species.

mechanisms--has not been prevented by selection. To what extent, therefore, might selection have positively encouraged the internally driven events? The answer to this question brings us to the problem of function and adaptation, described above.

If our view of evolution is that the force which drives selection is the ever present pressure to survive arising out of ecological tensions, then the environment poses the 'problem' and selection perfects the 'solution' (Fig. 6). This view would imply that, as regards the outer layer of the Bombyx egg, although it is clear that selection could hardly have had a hand in the initial horizontal transfer of new information between loci, nevertheless it might have encouraged such activities either by accelerating directly the turnover mechanisms or by choosing those individuals with the highest copy-number of the new gene at any given generation. By these means we can retain, if we so wish, the Darwinian notion that the accumulation offers a 'solution' to a 'problem' of survival. Selection might indeed have taken some positive role in the rDNA family by these two routes, (see below). Whether or not positive selection contributed to the evolution of *Bombyx mori* depends on the hetero-geneity of existing environments at each stage during the accumulation of the outer layer. If the accumulation could have proceeded willy-nilly, and if the increasing diapausal consequences had been tolerated by an appropriate preexisting environment then, formally speaking, the environment has offered a 'solution' to the 'problem' set by the transformed population. In such circumstances the final observed function would be due to the population of organisms 'adopting' a new environmental component (Fig. 6). The new relationship between the organism and its environment could be described as a non-Darwinian adaptation.

This scenario and that offered by the selection process need to explain what happens to the organisms at intermediate stages of progressive diapausy, in that no egg wants to wake up in the middle of the winter. Under selection we would need to invoke our familiar vague notions of preadaption. Under molecular drive we would need to propose a variety of appropriate environments which are open to exploitation successively at each stage. It might be inappropriate to assume that diapausy has evolved through the direct tracking of a seasonal temperature gradient with increasing layers of chorion. As described above, in so long as most individuals at each generation have accumulated a new layer to similar extents, and that the parental genetic information entering each and every new zygote is similar--then the cohesively evolving population might be able to survive in several prevailing environments, notwithstanding the action of selection against the progeny of parents who were maximally dissimilar in the copy-number of the Hc-A and Hc-B gene pairs (see Fig. 5). The current endpoint in this process is observed as diapausy. However, this function, and all the intermediate steps in its evolution, might not have required necessarily a lock-and-key relationship with a single environmental gradient.

Finally, there might be some requirement for the coevolution of other egg functions attendant on the change in shell morphology in *B. mori*. Some solution to this problem could be at hand in terms of the mechanisms involved with molecular coevolution in other gene families.

C. Molecular Drive and Molecular Coevolution

An explanation for the observed molecular coevolution between the promoters/enhancers and the polymerase I complex, envisages this as arising from the pressures emanating primarily from the continual homogenization of new variant promoters/enhancers throughout the family and population and the effects this has on selection (Coen, 1983; Dover and Flavell, 1984). In species of *Drosophila* there are approximately 10 promoter repeats within each of the 500 rRNA units evenly divided between the X and Y sex chromosomes (Coen and Dover, 1982; Coen *et al.*, 1982; Kohorn and Rae, 1982; Glover, 1981). The slow and cohesive mode of population change would provide the time and relaxed conditions for selection to increase the frequency of available alleles of the polymerase I and other genes, which are more efficient at transcribing the slowly increasing sets of new promoters/enhancers. It is possible also that there exist a range of polymerases or cofactors, coded by independent genes in an individual, only one of which is optimally effective in recognizing a given set of promoters, whilst the others bind at suboptimal levels to give rise to metastable complexes. Under such circumstances, molecular drive might be promoting simply a switch from suboptimum to optimum efficiency of a given polymerase or cofactor as old and new promoters switch in frequency.

Extant species with their interspecific incompatibilities are clearly that subset of populations within which there were available the appropriate compensatory polymerase or cofactor alleles for molecular coevolution to take effect. Other, less well endowed, populations might have gone extinct.

It is unlikely that the pressure for coevolution emanates first from the selection of new alleles of polymerase I and its cofactors (in response to survival problems imposed by the environment), to be followed by a change in the whole spectrum of enhancers/promoters. An individual faced suddenly with such new alleles might be at a selective disadvantage, rather than the converse, with respect to the several thousand promoters/enhancers that are no longer efficiently recognized. Furthermore, as outlined above, it is difficult to envisage how all promoters/enhancers could achieve genetic identity via selection alone, unless we assume that the turnover mechanisms themselves are under selective control. Although one of the possible explanations for the coevolution of compensatory changes in the rRNA secondary structure might imply the interference of selection in the turnover mechanisms (see below), nevertheless, a large body of experimental evidence in rDNA and other multigene families indicates that the turnover mechanisms are continuous and proceed at their own prescribed rates, independently of selection (see Strachan *et al.*, 1985).

Although molecular coevolution arises out of the spreading effects of nonreciprocal exchanges in gene families, there would clearly be situations in which selection itself might reinforce the spread of a variant in order to maintain an adequate number of promoters/enhancers (or fully functional rRNA secondary structures). Hence the extent of the interplay between natural selection and molecular drive could vary at different stages during the change in sequence composition of a multigene family. Furthermore, if there are multiple genes for polymerase I and its cofactors, then there is the reciprocal opportunity emanating from these families to influence the rDNA family.

Molecular coevolution involved with the maintenance of rRNA secondary structure is more difficult to explain in that both the first mutation and the compensatory mutation are occurring within the same repeat unit and need to spread to all repeats in all individuals. We can envisage the gradual and cohesive spread of the first mutation in a region of the 18S or 28S gene that produces an RNA which fails to form a given rRNA stem structure; but by what means are the compensatory second mutations spread? It is clear that only genomic mechanisms of turnover can spread information between repeats. Hence, the observed spread of compensatory mutations has either been the result of selection directly affecting the rate of turnover or that the first individuals with the highest number of units with the double compensatory mutations have been favored by selection. Neither of these arguments is simple in that we need constantly to bear in mind that the effects on phenotype (including the important phenotype of protein synthesis) of the presence of initially few variant repeats (whether with the single or double mutant), might be minimal and inconsequential to survival in a background of several hundred repeat units. Whatever the precise details, which presumably differ from species to species, it is clear that molecular drive has been responsible for the change in the average genetic composition of the population with respect to the gene family and that seemingly this has been the driving force for a subsequent response by natural selection in these particular examples of molecular coevolution.

SUMMARY

It is possible that a slow, gradual and cohesively evolving population, with respect to its multigene families (and all other repetitive families potentially involved in the biology of gene regulation, RNA processing and chromosome behavior), might be cleansed continually of the progeny of genetically disharmonious parents by selection, but would nevertheless be capable of proceeding in its average change of form and function, by the stochastic and deterministic activities of the DNA turnover mechanisms. It is possible that long term improbable shifts in the average genotype of a population, the emergence of new functions and the progressive exploitation of new environments, might be as much a consequence of the internal pressures arising out of the genome, as they are a consequence of external pressures generated by the ecology. The success of this non-Darwinian and non-Mendelian mode of evolution would be reinforced by an internal coevolution of molecules which ensure the continual maintenance of the interactions required of ontogeny. The relative contributions of molecular drive and natural selection to the seemingly innumerable varieties of past and existing biological life styles have yet to be assessed (for a review of their potential contributions to the gene families of the immune system see Dover and Strachan, 1986). This assessment is a matter of experimentation and investigation of the molecular behavior of the genes, their effects on developing phenotypes and the heterogeneity of environments at each evolutionary step. For mathematicians seeking universal algorithms of evolutionary processes this picture may be disspiriting, for biologists learning to live with the unexpected complexities of genetic turnover and the genetics of development, it should be exhilarating.

ACKNOWLEDGMENTS

I am indebted to Antonio Garcia-Bellido for many thoughtful and constructive discussions on aspects of this paper.

REFERENCES

Anxolabehere, D., Nouaud, D., Periquet, G., and Tchen, P. (1985). P element distribution in Eurasian populations of *D. melanogaster*: a genetic and molecular analysis. *Proc. Natl Acad. Sci.* **82**, 5418-5422.

Arnheim, N. (1983). Concerted evolution of multigene families. *In* "Evolution of Genes and Proteins" (M. Nei and R. K. Koehn, eds.), pp. 38-61. Sinauer Assoc., Inc., Sunderland.

Bregliano, J. C., and Kidwell, M. G. (1983). Hybrid dysgenesis determinants. *In* "Mobile Genetic Elements" (J. A. Shapiro, ed.), pp. 363-404. Academic Press, New York.

Brimacombe, R. (1984). Conservation of structure in ribosomal RNA. *Trends. Biochem. Sci.* **9**, 273-277.

Brown, S. D. M. (1983). Concerted evolution and molecular drive: evolutionary consequences of repeat sequence change. *In* "Proc. XV Int. Congr. Genet." pp. 221-234. Oxford Publ.

Challoner, P. B., Amin, A. A., Pearlman, R. E., and Blackburn, E. H. (1985). Conserved arrangements of repeated DNA sequences in nontranscribed spacers of ciliate ribosomal RNA genes: evidence for molecular coevolution. *Nucl. Acid. Res.* **13**, 2661-2680.

Charlesworth, B., and Charlesworth, D. (1983). The population dynamics of transposable elements. *Genet. Res.* **42**, 1-28.

Chothia, C. (1984). Principles that determine the structure of proteins. *Ann. Rev. Biochem.* **53**, 537-572.

Clark, C. G., Tague, B. W., Ware, V. C., and Gerbi, S. A. (1984). *Xenopus laevis* 28 rRNA: a secondary structure model and its evolutionary implications. *Nucl. Acid. Res.* **12**, 6197-6220.

Coen, E. S. (1983). The dynamics of multigene family evolution in *Drosophila*, Ph.D. Thesis, University of Cambridge.

Coen, E. S., and Dover, G. A. (1982). Multiple promoter-like sequences in the nontranscribed rDNA spacers of *D. melanogaster. Nucl. Acid. Res.* **10**, 7017-7026.

Coen, E. S., and Dover, G. A. (1983). Unequal exchanges and the coevolution of X and Y rDNA arrays in *Drosophila melanogaster. Cell* **33**, 849-855.

Coen, E. S., Thoday, J. M., and Dover, G. A. (1982a). Rate of turnover of structural variants in the rDNA gene family of *Drosophila melanogaster. Nature* **295**, 564-568.

Coen, E. S., Strachan, T., and Dover, G. A. (1982b). Dynamics of concerted evolution of ribosomal DNA and histone gene families in the melanogaster species subgroup of *Drosophila. J. Mol. Biol.* **158**, 17-35.

Davidson, E. H., and Britten, R. J. (1973). Organisation, transcription and regulation in the animal genome. *Quart. Rev. Biol.* **48**, 565-613.

Dover, G. A. (1982). Molecular drive: a cohesive mode of species evolution. *Nature* **299**, 111-117.

Dover, G. A. (1986). "Evolution by Means of Molecular Drive." Cambridge University Press (in preparation).

Dover, G. A., Brown, S., Coen, E., Dallas, J., Strachan, T., and Trick, M. (1982). The dynamics of genome evolution and species differentiation. *In* "Genome Evolution" (G. A. Dover and R. B. Flavell, eds), pp. 343-372. Academic Press, London.

Dover, G. A., and Flavell, R. B. (1984). Molecular coevolution: DNA divergence and the maintenance of function. *Cell* **38**, 622-623.

Dover, G. A., and Strachan, T. (1986). Molecular drive in the evolution of the immune superfamily of genes: the initiation and spread of novelty. *In* "Evolution and Vertebrate Immunity" (G. Kelsoe and D. H. Shulze, eds.). Univ. Texas Press (in press).

Dover, G. A., and Tautz, D. (1986). Conservation and divergence in multigene families: alternatives to selection and drift. *In* "The Evolution of DNA Sequences" (B. C. Clarke, A. Robertson and A. J. Jeffreys, eds.). Phil. Trans. Roy. Soc. (in press).

Engels, W. R. (1983). The P family of transposable elements in *Drosophila. Ann. Rev. Genet.* **17**, 315-344.

Engels, W. R. (1986). The evolution and behaviour of transposable genetic elements. *In* "The Evolution of DNA sequences" (B. C. Clarke,

A. Robertson and A. J. Jeffreys, eds.). *Phil. Trans. Roy. Soc. B.* (in press).

Fedoroff, N. V. (1979). On spacers. *Cell* **16**, 697–710.

Fisher, R. A. (1930). "The Genetical Theory of Natural Selection." Clarendon Press, Oxford.

Flavell, R. B. (1982). Sequence amplification, deletion and arrangement: major sources of variation during species divergence. *In* "Genome Evolution" (G. A. Dover and R. B. Flavell, eds.), pp. 301–324. Academic Press, London.

Flavell, R. B. (1986). Repetitive DNA and the evolution of plant genomes. *In* "The Evolution of DNA Sequences" (B. C. Clarke, A. Robertson and A. J. Jeffreys, eds.). Phil. Trans. Roy. Soc. B. (in press).

Franz, G., Tautz, D., and Dover, G. A. (1985). Occurrence of major S1 sensitive sites in the non-conserved spacer region of rDNA in *Drosophila* species. *J. Mol. Biol.* **183**, 519–527.

Garcia-Bellido, A. (1983). Comparative anatomy of cuticular pattern in the genus *Drosophila*. *In* "Development and Evolution" (B. C. Goodwin, N. Holder and C. C. Wylie, eds.), pp. 227–255. Cambridge University Press.

Garcia-Bellido, A. (1985). Genetic analysis of morphogenesis XVI. *Stadler Symp.* (in press).

Garcia-Bellido, A. (1986). Cell lineages and genes. *In* "Single Cell Marking" (P. A. Lawrence and M. Graham, eds.). *Phil. Trans. Roy. Soc.* (in press).

Gerbi, S. A., Gourse, R. L., and Clark, C. K. (1982). *In* "The Cell Nucleus rDNA part A" (H. Busch and L. Rothblum, eds.), Vol. 10, pp. 351–386. Academic Press, New York.

Glover, D. M. (1981). The rDNA of *Drosophila melanogaster. Cell* **26**, 297–298.

Goldsmith, M. R., and Kafatos, F. C. (1984). Developmentally regulated genes in silkmoths. *Ann. Rev. Genet.* **18**, 443–487.

Gould, S. J., and Lewontin, R. C. (1979). The spandrels of San Marco and the Panglossian paradigm: a critique of the adaptionist program. *Proc. Roy. Soc. B.* **205**, 581–598.

Gould, S. J., and Vrba, S. E. (1982). Exaptations – a missing term in the science of form. *Paleobiol.* **8**, 4–10.

Grummt, I., Roth, E., and Paule, M. R. (1982). Ribosomal RNA transcription *in vitro* is species specific. *Nature* **296**, 173–176.

Grummt, I., Sorbaz, H., Hofmann, A., Roth, E. (1985). Spacer sequences downstream of the 28S RNA coding region are part of the mouse rDNA transcription unit. *Nucl. Acid. Res.* **13**, 2293–2304.

Hood, L., Campbell, J. H., and Elgin, S. C. R. (1975). Organization, expression and evolution of antibody genes and other multigene families. *Ann. Rev. Genet.* **9**, 305-349.

Jeffreys, A. J., Wilson, V., and Thein, S. L. (1985). Hypervariable 'mini satellite' regions in human DNA. *Nature*, 67-72.

Jones, W. C., and Kafatos, F. C. (1982). Accepted mutations in a gene family: evolutionary diversification of duplicated DNA. *J. Mol. Evol.* **19**, 87-103.

Karess, R. E., and Rubin, G. M. (1984). Analysis of P transposable element functions in *Drosophila. Cell* **38**, 135-146.

Klein, H. L., and Petes, T. D. (1981). Intrachromosomal gene conversion in yeast. *Nature* **289**, 144-148.

Kohorn, B. D., and Rae, P. M. M. (1982). Accurate transcription of truncated rDNA templates in a Drosophila cell-free system. *Proc. Natl. Acad. Sci. USA* **79**, 1501-1505.

Lewontin, R. C. (1983). Gene, organism and environment. *In* "Evolution from Molecules to Men" (D. S. Bendall, ed.), pp. 273-284. Cambridge University Press.

Miesfeld, R., and Arnheim, N. (1984). Species-specific rDNA transcription is due to promoter-specific binding factors. *Mol. Cell. Biol.* **4**, 222-227.

Mishima, Y., Financsek, I., Kominami, R., and Muramatsu, M. (1982). Factors required for accurate transcription of mammalian rDNA: identification of species-dependent initiation factor. *Nucl. Acid. Res.* **10**, 6659-6669.

Noller, H. F. (1984). Structure of ribosomal RNA. *Ann. Rev. Biochem.* **53**, 119-162.

O'Hare, K., and Rubin, G. M. (1983). Structures of P transposable elements and their sites of insertion and excision in the *D. melanogaster* genome. *Cell* **34**, 25-35.

Ohta, T. (1980). "Evolution and Variation of Multigene Families. Lecture Notes in Biomathematics, Vol. 37, Springer, New York

Ohta, T. (1983). On the evolution of multigene families. *Theor. Pop. Biol.* **23**, 216-240.

Ohta, T., and Dover, G. A. (1983). Population genetics of multigene families that are dispersed into two or more chromosomes. *Proc. Natl. Acad. Sci. USA* **80**, 4079-4083.

Ohta, T., and Dover, G. A. (1984). The cohesive population genetics of molecular drive. *Genetics* **108**, 501-521.

Reeder, R. H. (1984). Enhancers and ribosomal gene spacers. *Cell* **38**, 349-351.

Simmons, M. J., and Bucholz, L. M. (1985). Transposase titration in *D. melanogaster*: a model of cytotype in the P-M system of hybrid dysgenesis. *Proc. Natl. Acad. Sci.* (in press).

Simmons, M. J., and Karess, R. E. (1985). Molecular and population biology of hybrid dysgenesis. (Report of Workshop, organizers G. A. Dover and M. G. Kidwell, Cambridge, U.K. 1984). *Dros. Inf. Serv.* **61**, 2-7.

Skinner, J. A., Öhrlein, A., and Brummt, I. (1984). *In vitro* mutagenesis and transcriptional analysis of a mouse ribosomal promoter element. *Proc. Natl. Acad. Sci. USA* **81**, 2137-2141.

Smith, G. P. (1973). Unequal crossover and the evolution of multigene families. *Cold Spring Harbor Symp. Quant. Biol.* **38**, 507-513.

Smithies, O., and Powers, P. A. (1986). Gene conversions and their relation to homologous chromosome pairing. *In* "The Evolution of DNA sequences" (B. C. Clarke, A. Robertson and A. J. Jeffreys, eds.). Phil. Trans. Roy. Soc. B. (in press).

Strachan, T., Webb, D. A., and Dover, G. A. (1985). Transition stages of molecular drive in multiple-copy DNA families in Drosophila. *EMBO J.* **4**, 1701-1708.

Tartof, K. (1974). Unequal mitotic sister chromatid exchange and disproportionate replication as mechanisms regulating ribosomal RNA gene redundancies. *Cold Spring Harbor Symp. Quant. Biol.* **38**, 491-500.

Tartof, K. D. (1975). Redundant genes. *Ann. Rev. Gen.* **9**, 355-386.

Tautz, D., and Renz, M. (1984). Simple sequences are ubiquitous repetitive components of eukaryotic genomes. *Nucl. Acids. Res.* **12**, 4127-4138.

POPULATION GENETICS THEORY OF MULTIGENE FAMILIES WITH EMPHASIS ON GENETIC VARIATION CONTAINED IN THE FAMILY

Tomoko Ohta

National Institute of Genetics
Mishima, 411, Japan

ABSTRACT

The significance of repetitive DNA families in evolution has three aspects, i.e., to increase genetic information, to regulate expression of structural genes, and to increase themselves as selfish DNA. Various existing multigene families (or gene clusters, supergenes and gene complexes) are real examples of the first aspect. Their evolution is characterized by concerted change due to continued occurrence of unequal crossing-over, gene conversion and transposition. Population genetics theory of their evolution is based on identity coefficients. Transitional equations of allelism, an allelic identity coefficient, and two nonallelic coefficients are presented for a model of gene conversion and duplicative transposition under the assumption of constant copy number per genome. Results of numerical studies show that, by adjusting parameter values such as copy number, conversion rate and transposition rate, any desired gene family may be attained with respect to genetic variability contained in the family. Natural selection on these parameters is termed "indirect selection." Actual gene families range from those with highly uniform members to those made of quite diverse gene copies, and they are likely to be attained by indirect selection.

I. INTRODUCTION

There exist many kinds of repetitive DNA families in the genomes of higher organisms, and their implications for evolution are far-reaching. In my view, three aspects are of particular interest: 1) As exemplified in many multigene families, repetitive sequences provide efficient ways for accumulating the genetic diversity necessary for the evolution of complexity of higher organisms. 2) Because some repeating sequences contain promoter function and may transpose from one locus to another, they provide for the adjustment of expression of structural genes. 3) Some repeating sequences are selfish, in that they carry out no useful function for the organism. The second point has been recognized only very recently through rapid progress of molecular biology of transposons and other dispersed DNA families (see Shapiro, 1983, for a collection of reviews, and Mackay, 1984, and Mukai *et al.*, 1985, for quantitative genetic aspects). On the other hand, the first aspect has been considered to be important for more than ten years by molecular biologists (see Brown *et al.*, 1972; Tartof, 1975; Hood *et al.*, 1975).

As a population geneticist, I have been interested in the first aspect, and started to formulate the evolution of various multigene families by using the population genetics method about ten years ago. Present evidence suggests that not only the evolution of typical multigene families but also that of supergenes, gene clusters and gene complexes are governed by a common principle at least in the early period of their history (e.g., see Bodmer, 1981; Hood *et al.*, 1982). In other words, any repeating gene family is dynamically evolving through various interaction mechanisms among its members such as unequal crossing-over, gene conversion and transposition. These mechanisms may provide many opportunities to accumulate new genetic information. It has been customary to suppose that formation of new functions of genes or new genes has occurred only through gene duplication and subsequent differentiation of duplicated genes (Ohno, 1970). Dr. Ohno now argues that this process is too inefficient to cope with urgent need because the rate of nucleotide substitution is so low (Ohno, 1984). However, through various interaction mechanisms among the repetitive gene members, the efficiency should be greatly increased.

As a result of continued occurrence of unequal crossing-over, gene conversion and duplicative transposition, concerted evolution takes place in typical multigene families (Dover, 1982; Arnheim, 1983; Ohta, 1980a, 1983a). Here, evolution is much complicated by the fact that spreading of a gene copy or a DNA segment occurs at two different levels; within a genome by unequal crossing-over, conversion and transposition, and in a population by random genetic drift, natural selection, and biased conversion or transposition. Accordingly, the mechanisms for maintaining genetic variability in the population are quite different from the single locus case. I have worked out a population genetics theory of concerted evolution under simplifying assumptions (Ohta, 1980a, 1983a, for review). In the models studied, no selection and no bias of conversion or transposition are assumed, and also the number of repeating copies is set to be constant per genome. Nevertheless, the results are useful for understanding quantitatively the evolution and variation of important gene families such as those of immunoglobulins and histocompatibility antigens (Ohta, 1980a, 1983a, 1984a), and for estimating various parameters of unequal crossing-over, conversion and transposition. Of course, these models will need revision if more data become available that contradict the prediction. In the next section, the theory of concerted evolution is reviewed.

II. POPULATION GENETICS THEORY

The basic concept of the analysis is the identity coefficient that is defined as the probability of identity of genes or DNA segments belonging to the multigene family. In an ordinary single locus model, it is equivalent to virtual homozygosity which has been much used in statistical analyses on protein polymorphisms (see Kimura, 1983, for review). In the following, I shall start from a most simple model of intrachromosomal gene conversion.

Let us assume a random mating population of effective size, N. Each genome of the population contains a multigene family of n tandemly arranged copies. In one generation, the family undergoes gene conversion, recombination at meiosis, mutation and random sampling of gametes at reproduction. Let λ_c be the rate at which a gene is converted by the remaining (n-1) genes on the same chromosome (intrachromosomal conversion) in one generation (Ohta, 1982). Here symmetric conversion that results in reciprocal exchange of genes is not considered; for the analysis including such cases, see Nagylaki (1984a) and Ohta (1984a). Let v be the mutation rate per unit whose identity is compared per generation under the infinite allele model (Kimura and Crow, 1964). Recombination at meiosis is assumed always to be equal, and let β be the rate per generation between adjacent gene units. All evolutionary forces are assumed to be weak, i.e., v, λ_c, β and $1/N \ll 1$.

Three identity coefficients are defined as in Figure 1; f is the allelic identity probability, C_1 is the identity coefficient of genes at different loci on one chromosome, and C_2 is that of genes at different loci of two chromosomes chosen randomly from the population. Let the vector, $c_t = (f, C_1, C_2)$ at generation t. Then it can be shown that the

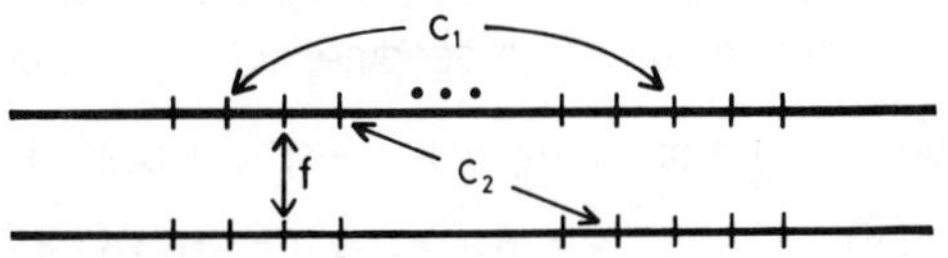

Figure 1.   Diagram showing the definition of three identity coefficients of the conversion model.

change of the three identity coefficients in one generation is expressed by the following transition equations (Ohta, 1982, 1983a,b; Nagylaki, 1984a).

$$c_t = Ac_{t-1} + b \tag{1}$$

where

$$A = (1-2v)\begin{bmatrix} 1 - \dfrac{1}{2N} - (n-1)\alpha_c & 0 & (n-1)\alpha_c \\[2ex] 0 & 1-\alpha_c - \dfrac{(n+1)}{3}\beta & \dfrac{(n+1)}{3}\beta \\[2ex] \alpha_c & \dfrac{1}{2N} & 1 - \alpha_c - \dfrac{1}{2N} \end{bmatrix} \tag{2}$$

and

$$b = (\dfrac{1}{2N}, \alpha_c, 0), \tag{3}$$

with $\alpha_c = 2\lambda_c/(n-1)$.

The above analysis may be extended into a more general model including duplicative transposition for dispersed repetitive gene families. Also included in the model are interchromosomal gene conversion and transposition. Let w be the proportion of conversion occurring within a genome, and $1-w$, between the genomes that takes place in diploid phase of germ cell lineages. If λ_c is the rate per generation at which a unit is converted, then $w\lambda_c$ is the rate at which a unit is converted by one of the remaining (n-1) units in the same genome, and $(1-w)\lambda_c$ is the rate of conversion by a unit in the other genome. Without bias, it is equivalent to the model that a unit converts another belonging to the same genome at the rate $w\lambda_c$, and another unit belonging to the other genome at the rate $(1-w)\lambda_c$. The model is shown in Figure 2, by depicting the genome with a line of chromosome.

Duplicative transposition is assumed to be accompanied by deletion of another copy, so that the copy number per genome does not change. Again, with the rate, w, transposition is assumed to occur within the genome, and with 1-w, between the genomes. Thus, in one generation, duplicative transposition of any one copy occurs at the rate $w\lambda_t$ in the same genome, and at $(1-w)\lambda_t$ to the other genome (see Fig. 2). Also, transposition is assumed to occur always to a new chromosomal site not previously occupied (see Langley *et al.*, 1983; Charlesworth and Charlesworth, 1983 on the adequacy of this assumption). Interchromosomal recombination is assumed to be equal, and let R be the average rate by which any two nonallelic units recombine. When conversion and transposition have no preference of chromosomal site, this treatment is accurate enough. It corresponds to the coefficient, $(n+1)\beta/3$, of the previous simple model. Note R may take any value between 0-0.5. The

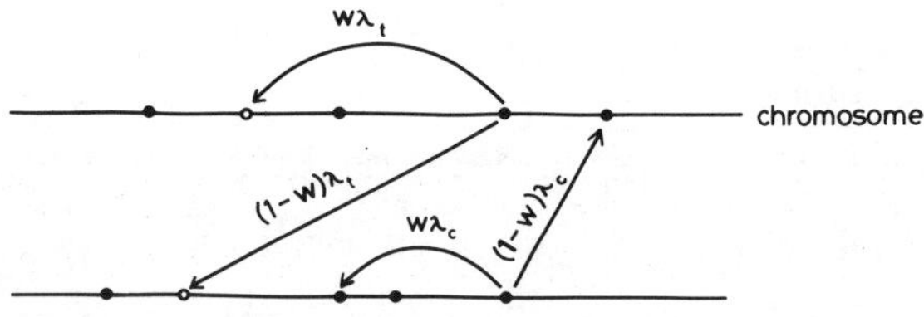

Figure 2. Diagram illustrating the model of duplicative transposition and gene conversion.

other parameters, v, λ_c, λ_t and $1/N$ are assumed to be much less than unity. In addition, n is assumed to be sufficiently large, and $1/n \ll 1$.

In the following formulation, interchromosomal recombination is assumed to take place after mutation, conversion and transposition have occurred in each generation. "Allelism" is defined as the probability that when two genomes are randomly chosen from the population and an element is randomly chosen from one of the genomes, a homologous copy exists at the same chromosomal site on the other genome. Let us denote it as F. It has been shown that the expected change of allelism in one generation becomes (Langley *et al.*, 1983; Charlesworth and Charlesworth, 1983; Ohta, 1985),

$$\Delta F = -2\lambda_t F + \frac{1}{2N}(1-F) . \qquad (4)$$

At equilibrium, the allelism becomes,

$$\hat{F} = \frac{1}{1 + 4N\lambda_t} , \qquad (5)$$

where $\hat{F}$ denotes an equilibrium value. Slatkin (1985) used a different measure of identity of site, and his measure is F/n.

By using the results of previous studies (Ohta, 1982, 1983a,b, 1984a,b; Nagylaki, 1984a,b), the expected changes of identity coefficients have been obtained (Ohta, 1985). As before, three identity coefficients are defined: allelic identity, f, identity coefficients of units within a genome, C_1, and that of units belonging to different genomes, C_2 (see Fig. 3). My previous formulation of the change of C_2 by duplicative transposition (Ohta, 1984b) contained an error and it has been corrected (Ohta, 1985).

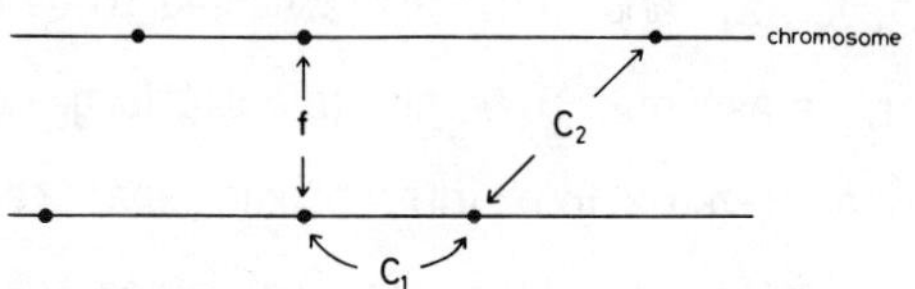

Figure 3. Definition of three identity coefficients for the model of duplicative transposition and gene conversion.

The vector c_t is transformed from one generation to the next according to the following equation, by letting $\alpha_c = 2\lambda_c/(n-1)$, $\alpha_t = 2\lambda_t/(n-1)$, and $F' = n(n-F)F/\{n(n-F) + 2n\lambda_t F\}$,

$$c_t = Ac_{t-1} + b , \qquad (6)$$

where the subscript, t, denotes the tth generation, and

$$A = \begin{bmatrix} a_{11} & 0 & 2\lambda_c \\[2ex] \begin{matrix} F(1-R)(1-w) \\ \times (\alpha_c+\alpha_t) \end{matrix} & a_{22} & \begin{matrix} R\{1-2v-(1-w)(\alpha_c+\alpha_t)\} \\ +2(1-R)(1-w)(\lambda_c+\lambda_t)(1- F/(n-1)) \end{matrix} \\[2ex] F\alpha_c + F'\alpha_t & {}^1/_{2N} & a_{33} \end{bmatrix}$$

$$(7)$$

with $a_{11} = 1 - 2\lambda_c - 2v - 1/(2NF)$, $a_{22} = (1-R)\{1 - 2v - w(\alpha_c + \alpha_t) - 2(1-w)(\lambda_c + \lambda_t)\}$ and $a_{33} = 1 - 2v - {}^1/_{2N} - F\alpha_c - F'\alpha_t$, and

$$b = \begin{bmatrix} 1/(2NF) \\[2ex] \{w(1-R) + (1-w)R\}(\alpha_c + \alpha_t) \\[2ex] 0 \end{bmatrix} \qquad (8)$$

By using equation (6), it is possible to examine various quantities of interest.

Table 1. Numerical examples of allelism (F) and identity coefficients (f, C_1, and C_2) at equilibrium.

Set	Parameters		F	f	C_1	C_2
I	λ_t	λ_c	($R = 0.0$, $w = 0.5$, $N = 2000$, $n = 20$, $v = 10^{-5}$)			
	0.05	0	0.0025	0.9998	0.5825	0.5607
	0.025	0.025	0.0050	0.7807	0.5824	0.5606
	0	0.05	1.00	0.5491	0.5705	0.5481
			($R=(1/3)\times10^{-4}$, $w=1.0$, $N=5\times10^4$, $n=100$, $v=10^{-7}$)			
	0	10^{-6}		0.8328	0.0800	0.0799
	0	10^{-5}	1.00	0.5478	0.3293	0.3272
	0	10^{-4}		0.7328	0.7320	0.7201
II	n	R	($\lambda_t=0.0$, $\lambda_c=10^{-5}$, $w=1.0$, $N=5\times10^4$, $v=10^{-7}$)			
	10	$1/3\times10^{-5}$		0.9212	0.9021	0.8910
	20	$2/3\times10^{-5}$	1.00	0.8510	0.7937	0.7850
	50	$5/3\times10^{-5}$		0.6877	0.5430	0.5384
	100	$1/3\times10^{-4}$		0.5478	0.3293	0.3272
			($\lambda_t=5\times10^{-5}$, $\lambda_c=0.0$, $w=0.5$, $N=5\times10^4$, $v=10^{-7}$)			
	10				0.8332	0.8332
	20				0.7039	0.7039
		0.5	0.0909	0.9982		
	50				0.4804	0.4804
	100				0.3141	0.3141

Table 1 gives some numerical examples of allelism and identity coefficients at equilibrium. Data of set I show that the values of the nonallelic identity coefficients (C_1 and C_2) are similar in the models of conversion and of transposition. In other words, exchange of λ_c and λ_t has a small effect on them. However, allelism and allelic identity are quite different in the two models. Also shown by the data is that the identity coefficients become larger as λ_c or λ_t increases. The last line of set I may be applicable to a uniform multigene family such as rRNA genes. Data of set II show that identity decreases with copy number (n). The upper half may be applicable to tightly linked families, and the lower half, dispersed families such as transposon families. Slatkin (1985) has shown that, for independently segregating transposon families, $C_1 \approx C_2 \approx 1/(4nNv+1)$, under ordinary rate of transposition. The data of set II approximately agree with the predictions of his formula.

The transition equation of identity coefficients is useful not only for calculating gene identity, but also for obtaining the rate of steady decay of genetic variability, from which the time until fixation in the whole population of a mutant belonging to a family may be predicted (Ohta, 1983b, Ohta and Dover, 1983; Nagylaki, 1984a,b). Also, by using the eigenvector of the transition matrix with no mutation ($v = 0$), the "cohesiveness" during spread of a mutant copy, which is defined as the variance of mutant copy number among individuals of the population, may be approximately obtained (Ohta and Dover, 1984). At any rate, the eigenvalue and eigenvector of the matrix give useful information on genetic structure of gene families, at the state of steady decay.

III. IMPLICATIONS OF THE RESULTS FOR MAINTENANCE OF GENETIC VARIABILITY

In assessing the role played by gene families in the maintenance of genetic variability, one problem needs attention. In the theory of the previous section, it is assumed that the unit whose identity is compared is not divisible by conversion, crossing-over or transposition during the course of evolution. Actually, however, any gene or DNA segment may be separated when conversion and other processes start or end in its region. If such "recombination" takes place, it may increase genetic variability of the gene family, whereas conversion and other processes are intrinsically factors which reduce variability. Indeed, Baltimore (1981) and others emphasized gene conversion as a mechanism to increase genetic diversity of immunoglobulin genes and others. If the unit is taken to be very small such as an amino acid site or a nucleotide site as in my study of the amino acid diversity of immunoglobulins (Ohta, 1980a), this problem does not arise. But, in general, this is an important problem not solved yet. One way of investigating it is to analyse linkage disequilibrium between amino acid (nucleotide) sites *within* the unit (Ohta, 1980b). "Recombination" would increase variability only when linkage disequilibrium exists.

Although the model studied is too simple and there are various unsolved problems such as the above, the results of the previous section indicate that, by adjusting the parameter values particularly those of λ_c , λ_t and n, it is possible to acquire any type of gene family with respect to gene diversity needed by the organism. Apparently, λ_c and λ_t are genetically determined and are therefore responsive to natural selection. On the other hand, natural selection may operate more directly on diversity itself. Let us call it direct selection against the former type of selection on λ_c and λ_t (indirect selection). The effect of the direct selection cannot be quantitatively predicted, but it can be shown that it may efficiently increase diversity or decrease identity (Ohta, 1980a). Direct selection works through changing genotype frequencies, and if it is relaxed, genetic structure tends to revert. The former type of selection on the rates of conversion, unequal crossing-over and so on (indirect selection) would be slow in response but efficient for maintaining diversity once attained. There exist many kinds of multigene families from those with uniform members like ribosomal RNA genes to those with highly variable members such as immunoglobulin genes. For the evidence that the ribosomal DNA enhances recombination in *S. cerevisiae,* see

Keil and Roeder (1984). Various multigene families seem to be quite well adapted with respect to the indirect selection. For example, it has been shown that amino acid diversity at the hypervariable region of immunoglobulins may well be understood by assuming that the rate of nucleotide substitution in the region is close to the maximum under the neutral theory of Kimura (1983) (Ohta, 1980a, 1984c; Gojobori and Nei, 1984). Direct selection for diversity would make the rate of nucleotide substitution higher than the neutral prediction, therefore the diverse members of immunoglobulin gene family have apparently been maintained by the indirect selection.

So far, the present discussion has been focused on the typical multigene families. The evolutionary significance of transposons and other dispersed repetitive families is more difficult to understand. As I pointed out in the Introduction, they are likely to be important for regulating expression of other genes. Indeed, Peterson and his associates have found that some transposon families increased their copy number in selected lines of maize, and the results indicate that their insertion contributes to selection response (for review, see Peterson, 1985). Anyway, the rise and fall of dispersed families seem to be very rapid and the copy number of each family is greatly changing in evolution (e.g., see Kidwell, 1983). Here the concept of molecular drive (Dover, 1982) might be worthy of examination, or most of these families may be merely selfish DNA. In any case, they provide one of the most stimulating subjects of research, both for theoretical and experimental evolutionary biologists.

ACKNOWLEDGMENTS

I thank Dr. Kenichi Aoki and two anonymous referees for their many useful suggestions to improve the presentation. This is contribution no. 1616 from the National Institute of Genetics, Mishima, Japan.

REFERENCES

Arnheim, N. (1983). Concerted evolution of multigene families. *In* "Evolution of Genes and Proteins" (M. Nei and R. K. Koehn, eds.), pp. 38-61. Sinauer, Sunderland, Mass.

Baltimore, D. (1981). Gene conversion: some implications for immunoglobulin genes. *Cell* **24**, 592-594.

Bodmer, W. F. (1981). The William Allan Memorial Award Address: Gene clusters, genome organization, and complex phenotypes. When the sequence is known, what will it mean? *Am. J. Hum. Genet.* **33**, 664-682.

Brown, D. D., Wensink, P. C., and Jordan, E. (1972). A comparison of the ribosomal DNAs of *Xenopus laevis* and *Xenopus mulleri*: the evolution of tandem genes. *J. Mol. Biol.* **63**, 57-73.

Charlesworth, B., and Charlesworth, D. (1983). The population dynamics of transposable elements. *Genet. Res* **42**, 1-28.

Dover, G. A. (1982). Molecular drive: a cohesive mode of species evolution. *Nature* **299**, 111-117.

Gojobori, T., and Nei, M. (1984). Concerted evolution of the immunoglobulin V_H gene family. *Mol. Biol. Evol.* **1**, 195-212.

Hood, L., Campbell, J. H., and Elgin, S. C. R. (1975). The organization, expression, and evolution of antibody genes and other multigene families. *Ann. Rev. Genet.* **9**, 305-353.

Hood, L., Steinmetz, M., and Goodenow, R. (1982). Genes of the major histocompatibility complex. *Cell* **28**, 685-687.

Keil, R. L., and Roeder, G. S. (1984). Cis-acting, recombination-stimulating activity in a fragment of the ribosomal DNA of *S. cerevisiae*. *Cell* **39**, 377-386.

Kidwell, M. G. (1983). Evolution of hybrid dysgenesis determinants in *Drosophila melanogaster*. *Proc. Natl. Acad. Sci. USA* **80**, 1655-1659

Kimura, M. (1983). "The Neutral Theory of Molecular Evolution." Cambridge University Press, London, New York.

Kimura, M., and Crow, J. F. (1964). The number of alleles that can be maintained in a finite population. *Genetics* **49**, 725-738.

Langley, C. H., Brookfield, J. F. Y., and Kaplan, N. (1983). Transposable elements in Mendelian populations. *Genetics* **104**, 457-471.

Mackay, T. F. C. (1984). Jumping genes meet abdominal bristles: hybrid dysgenesis-induced quantitative variation in *Drosophila melanogaster*. *Genet. Res.* **44**, 231-237.

Mukai, T., Baba, M., Akiyama, M., Uowaki, N., Kusakabe, S., and Tajima, F. (1985). Rapid change in mutation rate in a local population of *Drosophila melanogaster*. *Proc. Natl. Acad. Sci. USA* (in press).

Nagylaki, T. (1984a). The evolution of multigene families under intra-chromosomal gene conversion. *Genetics* **106**, 529-548.

Nagylaki, T. (1984b). Evolution of multigene families under inter-chromosomal gene conversion. *Proc. Natl. Acad. Sci. USA* **81**, 3796-3800.

Ohno, S. (1970). "Evolution by Gene Duplication." Springer-Verlag, Berlin, New York.

Ohno, S. (1984). Birth of a unique enzyme from an alternative reading frame of the preexisted, internally repetitious coding sequence. *Proc. Natl. Acad. Sci. USA* **81**, 2421-2425.

Ohta, T. (1980a). "Evolution and variation of multigene families." *Lecture Notes in Biomathematics*, Vol. 37. Springer-Verlag, Berlin, New York.

Ohta, T. (1980b). Linkage disequilibrium between amino acid sites in immunoglobulin genes and other multigene families. *Genet. Res. Camb.* **36**, 181-197.

Ohta, T. (1982). Allelic and non-allelic homology of a supergene family. *Proc. Natl. Acad. Sci. USA* **79**, 3251-3254.

Ohta, T. (1983a). On the evolution of multigene families. *Theor. Pop. Biol.* **23**, 216-240.

Ohta, T. (1983b). Time until fixation of a mutant belonging to a multigene family. *Genet. Res. Camb.* **41**, 47-55.

Ohta, T. (1984a). Some models of gene conversion for treating the evolution of multigene families. *Genetics* **106**, 517-528.

Ohta, T. (1984b). Population genetics of transposable elements. *IMA J. Math. Appl. Med. Biol.* **1**, 17-29.

Ohta, T. (1984c). Population genetics theory of concerted evolution and its application to the immunoglobulin V gene tree. *J. Mol. Evol.* **20**, 274-280.

Ohta, T. (1985). A model of duplicative transposition and gene conversion for repetitive DNA families. *Genetics* **110**, 513-524.

Ohta, T., and Dover, G. A. (1983). Population genetics of multigene families that are dispersed into two or more chromosomes. *Proc. Natl. Acad. Sci. USA* **80**, 4079-4083.

Ohta, T., and Dover, G. A. (1984). The cohesive population genetics of molecular drive. *Genetics* **108**, 501-521.

Peterson, P. A. (1985). Mobile elements in maize: A force in evolutionary and plant breeding processes. *In* "Proc. Stadler Genetics Symp." (in press).

Shapiro, J. A. (ed.) (1983). "Mobile Genetics Elements." Academic Press, New York, London.

Slatkin, M. (1985). Genetic differentiation of transposable elements under mutation and unbiased gene conversion. *Genetics* **110**, 461-468.

Tartof, K. D. (1975). Redundant genes. *Ann. Rev. Genetics* **9**, 355-385.

STATISTICAL ASPECTS OF THE MOLECULAR CLOCK

John H. Gillespie

Department of Genetics
University of California
Davis, CA 95616

ABSTRACT

Although it has been known for many years that the protein sequence data is not compatible with a molecular clock model based on the Poisson process, very little work has been aimed at finding a model for the clock that fits the data better than the Poisson process. Here we use a Cox process--a Poisson process with a randomly varying rate--to model the molecular clock. A major result is that the great time spans that separate typical species for which sequences are available tends to mask the variability in the rates of molecular evolution. When the data is adjusted for this masking effect, it appears that the variance in the rates is much larger than previously thought. So large, in fact, that molecular evolution appears to be an episodic process: bursts of substitutions are separated by long spells with no substitutions. It is argued that this property of molecular evolution is more easily accounted for by the action of natural selection than by the neutral allele model. A general discussion of the *ad hoc* assumptions required for modeling variable-rate evolution is included.

I. INTRODUCTION

If the primary sequence of a particular protein is examined in a number of related species, it is commonly inferred that the rate of evolution of the protein is nearly constant. The apparent generality of this inference has led to the concept of a "molecular clock." In discussing the molecular clock, most writers (e.g., Wilson *et al.*, 1977) are quick to point out that the ticking of the molecular clock is more similar to the ticks of a Geiger counter than to the ticks of an ordinary clock. In fact, a commonly employed model for both Gieger counters and the molecular clock is the Poisson process. This is so even though the Poisson process has repeatedly been shown to be incompatible with the protein sequence data (and, of course, with the Geiger counter). The reason for the incompatibility in the case of the sequence data is that the variance in the number of substitutions in a lineage is significantly higher than the mean, whereas for the Poisson process the variance should equal the mean (Ohta and Kimura, 1971; Langley and Fitch, 1974). Very little work has been reported that seeks to find a statistical model of the molecular clock that is compatible with the sequence data. Finding such a model should play a central role in our efforts to understand the mechanisms of molecular evolution.

This paper will begin with an effort to fit the sequence data to a model of the clock that is a natural generalization of the Poisson process. This will lead quite naturally to a view that the clock may be episodic, with periods of rapid evolution separated by periods of stasis. This description of the clock suggests a model of molecular evolution by natural selection that appears to be in better agreement with the statistical aspects of the sequence data than is the more population neutral allele theory.

II. THE COX PROCESS AS A MOLECULAR CLOCK

This section will attempt a straightforward fitting of the protein sequence data to a particular model of the molecular clock called a doubly stochastic Poisson process or, more concisely, a Cox process (Cox and Isham, 1980, contains all of the background material on point process that is needed to fully understand this paper). The Cox process is an example of a stationary point process. For such processes it is a common practice to make the primary object of study the total number of events that occurred during a span of time t. We will refer to the number of events, i.e., the number of substitutions, that occurred during such a span of time as $N(t)$. The data all come from "star" or "bush" phylogenies. That is, from phylogenies where all of the modern species arose in a short time relative to the time back to their radiation. Such phylogenies are ideal for statistical studies of the clock because they avoid biases introduced by the ancestral sequence inferences required of more complex phylogenies. We can estimate the mean and variance of the number of substitutions that have occurred on the lineages leading to the species under study in a bush phylogeny by an adaptation of a method introduced by Kimura (1983). The main extension that is required is a technique for correcting the data for multiple substitutions at a site without invoking a Poisson assumption. Such a correction may be found in Gillespie (1985). The resulting estimates of the mean and variance of $N(t)$ for six proteins may be found in Table I. Of particular interest in this table is the estimate of

$$R(t) = \mathrm{Var}\{N(t)\}/E\{N(t)\} \ .$$

$R(t)$ is called the index of dispersion in the point process literature and may be used as a measure of departure from the Poisson process. For a

Table I. Estimates of the Moments of $N(t)$ for Proteins.

Protein	Number of species	Time back to radiation[1]	Estimates of: $E\{N(t)\}$	$Var\{N(t)\}$	$R(t)$
β-hemoglobin	6	65 m.y.	14.79	44.93	3.04
α-hemoglobin	6	65 m.y.	12.59	14.74	1.17
myoglobin	6	63 m.y.	12.27	19.67	1.60
cytochrome-c	4	300 m.y.	8.16	26.26	3.22
ribonuclease	4	65 m.y.	20.39	43.81	2.15
α-crystalline	6	65 m.y.	4.19	11.36	2.71

[1] The proteins for which the time is 65 million years ago are from orders of mammals; the one case where the time is 300 million years ago is the reptile radiation that occurred in the Pennsylvanian.

Poisson process, $R(t)$ equals one. The protein sequence data exhibits $R(t)$ values in the range $1.1 < R(t) < 3.3$. Since the variance in the number of substitutions is larger than expected under the Poisson model, it is natural to inquire about the implications of the observed values of $R(t)$ on the variance in the rate of substitutions of a protein. A major goal of this section is to show that the observed values of $R(t)$ suggest that the variance in the rate of evolution can be very large relative to the mean.

For the Cox process the probability that a substitution occurs in the short time interval $(t, t+dt)$ is

$$E\{dN(t)\} = \lambda\theta(t)dt ,$$

where λ is the mean rate of substitution of the protein and $\theta(t) \geq 0$ is an integrable stochastic process with continuous sample paths and moment

$$E\{\theta(t)\} = 1 ,$$

$$r(x) = Cov\{\theta(t+x),\theta(t)\} ,$$

$$\sigma^2 = Var\{\theta(t)\} = r(0) ,$$

This model of the molecular clock is a natural extension of the Poisson process. It can be viewed as a molecular clock with a variable tick rate. The tick rate, $\lambda\theta(t)$, changes at random through time as dictated by the stochastic process, $\theta(t)$.

The final assumption that must be introduced concerns the relationship between the clocks on separate lineages of the star phylogeny. We will assume that at a branch point the clocks for all of the "daughter lineages" will initially have the same tick rates, but from that point on the clocks will change their tick rates independently. That is, if we call the tick rate for the ith lineage, $\lambda\theta_i(t)$, then we are assuming that $\lambda\theta_i(t_b) = \lambda\theta_j(t_b)$, where the time of the branch is called t_b. Any dependencies that exist between the clocks in different lineages will thus be due to correlations with their initial values rather than, say, similar environmental events. Our goal in this section is to use the data to characterize $\theta(t)$.

It is well known that for a Cox process the number of events during the interval $(0,t)$, given the trajectory of $\theta(t)$ on this same interval, is Poisson distributed with mean

$$E\{N(t)\,|\,\theta(s),\ 0 < s < t\} = \lambda \int_0^t \theta(s)ds \ . \qquad (1)$$

Of particular importance is that the number of substitutions depends on the integral of the tick rate. Since integration has a smoothing effect, it is reasonable to suppose that the variability in the rate of evolution will be masked if the lineages under study are long relative to the time scale of change of the process $\theta(t)$. To see that this is probably the case notice that in Table I most of the star phylogenies stem from the mammal radiation that occurred about 65 million years ago. If the variability in the tick rate of the molecular clock is due to environmental changes, then a time scale of thousands to tens of thousands of years would seem reasonable since this is the time scale of such major climatic changes as the ice ages. Thus the integral in (1) is of a process in which the autocovariance function, $r(x)$, will be nearly zero for values of x that differ from zero. As is well known, as the autocovariance function approaches zero, so does the variance of the integral of the process, unless the variance of the process itself, $r(0)$, grows appropriately.

The previous argument leads quite naturally to an investigation of a process,

$$\xi(t) = \lim \int_0^t \theta(s)ds \, ,$$

where the limit is taken as $r(x) \to 0$ for $x \neq 0$ and $r(0) \to \infty$ in such a way that the variance of the integral remains fixed. In Gillespie (1985) it is argued that $\xi(t)$ will be a process with independent increments that admits the Levy representation

$$E\{e^{-z\xi(t)}\} = e^{-t\psi(\lambda)} \, ,$$

$$\psi(z) = \int_0^\infty [1 - e^{-zx}]\mu(dx) \, ,$$

where $\mu(x)$ is the Levy measure that characterizes the process (see Breiman, 1968, Sec. 14.4 for a discussion of these processes). The value of the Levy formula is that it allows us to write down immediately the probability generating function (pgf) for $N(t)$ as

$$E\{e^{-\lambda\xi(t)(1-s)}\} = e^{-t\psi(\lambda(1-s))} \, .$$

The recognition that this is the pgf of a random variable that is a Poisson sum of positive random variables,

$$N(t) = Y_1 + Y_2 + \ldots + Y_{M(t)} , \qquad\qquad (2)$$

is our main result. Before discussing this result note that the Y_i have the pgf

$$\frac{1}{\psi(\lambda)} \int_0^\infty e^{-\lambda x} [e^{\lambda x s} - 1]\mu(dx) ,$$

and that the mean of the Poisson process $M(t)$ is

$$E\{M(t)\} = t\psi(\lambda) .$$

The representation of the substitution process, $N(t)$, as a Poisson sum of positive random variables suggests calling the molecular clock an "episodic clock" for which there are $M(t)$ episodes of evolution within a lineage and $Y_i > 0$ substitutions during the ith episode.

By specializing to particular versions of the clock, i.e., by specifying $\theta(t)$, it is possible to estimate the mean number of episodes of evolution within a lineage, $E\{M(t)\}$, and the mean of the number of substitutions per episode, $E\{Y_i\}$. Consider the following examples:

(1) *The two-state clock.* For this clock $\theta(t)$ is assumed to jump back and forth between two values, zero and β. The time spent in each state is assumed to be exponentially distributed. To meet our assumption that the autocovariance of the clock approaches zero we require that $\beta \rightarrow \infty$ while the mean time spent at β approaches zero. A straight-forward calculation (see Gillespie, 1985) shows that for this case the Y_i are geometrically distributed and $N(t)$ is Pòlya-Eeppli distributed. Table II gives the estimates for the moments of the process for this clock.

(2) *The gamma clock.* For this clock $\theta(t)$ is assumed to jump back and forth between zero and a gamma-distributed random variable with successive gamma variables being independent. To achieve the proper limit the time spent in any state must shrink to zero as the variance of

Table II. Inferred Episodic Structure of Evolution for Two Clocks.

Protein	mean number of substitutions per episode		mean number of episodes	
	Two-state	Gamma	Two-state	Gamma
β-hemoglobin	2.02	1.83	7.33	7.07
α-hemoglobin	1.09	1.08	11.60	11.62
myoglobin	1.30	1.28	9.42	9.60
cytochrome-c	2.11	1.90	3.87	4.30
ribonuclease	1.57	1.50	12.96	13.58
α-crystalline	1.86	1.71	2.26	2.44

the gamma random variables goes to infinity. At the limit the Y_i are distributed according to the logarithmic series distribution and $N(t)$ is negative binomially distributed. Table II also gives the estimates of the moments for this clock.

The striking feature of Table II is the similarity of the estimates for the two clocks despite the fact that the clocks are very different mathematically. For instance, the Levy measure for the two-state clock is finite while for the gamma clock it is infinite. The similarity must stem from the low values of $R(t)$ that dictate that the Y_i have a relatively small variance. Perhaps an asymptotic expansion of this model near $R(t) = 1$ would provide evidence for this interpretation.

It is important to realize that the episodic nature of the clock uncovered in this analysis is radically different from the behavior of the Poisson clock. The main difference lies in the occurrence of multiple substitutions within short periods of time. Such behavior suggests a model of evolution by natural selection that will be described in the next section.

IV. MOLECULAR EVOLUTION BY NATURAL SELECTION

A simple model of molecular evolution by natural selection that leads to an episodic clock involves three primary elements: a changing environment, a mutational landscape, and a simple form of epistasis. The environment will be viewed as a stationary point process, $P(t)$, where each event represents a change in the environment for which the locus under study can potentially respond with an allelic substitution. The changes in the environment will be referred to as environmental challenges. With each challenge the locus will respond with one or more substitutions. If we let X_i be the number of substitutions after the ith environmental challenge we can write the total number of substitutions that occurred in a time interval of length t as

$$S(t) = X_1 + X_2 + \dots X_{P(t)} . \qquad (3)$$

This looks exactly like our representation of $N(t)$ as a random sum of random variables. However, at this point in the development there is no reason to suppose that $P(t)$ is a Poisson process or that the random variables X_i have the properties suggested by the estimates in Table II. To achieve these properties we introduce a simple epistatic scheme.

With each environmental challenge we suppose that there are a number of roughly equivalent loci that can mollify the environmental challenge with one or more allelic substitutions. This is a form of epistasis since it implies that the substitution of an allele at one locus alters the fitness relationships between the alleles at other loci. Suppose there are L loci that can respond to a particular environmental challenge. The particular loci that respond will be chosen at random from among the L loci.

There are two extreme models that can be envisioned that characterize the way by which the loci are chosen. The first of these supposes that each of the L loci are equivalent in all regards, including the number of alleles available and the intensity of selection on the available alleles. Using the boundary-layer argument developed in Gillespie (1983), or common sense, it is clear that the probability of any particular locus experiencing a particular one of the substitutions following an environmental challenge is $1/L$. If there were k substitutions in response to a challenge, then the number of substitutions that will occur at a particular locus will be binomially distributed with parameters k and $1/L$. The distributions of X_i is just this binomial, conditioned on $X_i > 0$.

The second model assumes that the L loci are only equivalent in a long-term evolutionary sense, but with any particular environmental challenge one of the loci will have much stronger selective differences between its alleles than any of the others. Here also the boundary-layer/common-sense argument suggests that most, if not all, of the substitutions will occur at this one locus. In this model the probability of any particular locus being chosen is $1/L$ and upon being chosen, the locus will experience a random number of substitutions. For this model we can appeal to the results in Gillespie (1984b) on the nature of the mutational landscape to argue that for the chosen locus X_i will be likely to have a mean around 1.5 to 2.5.

Both of these models can be encompassed in a uniform framework by assuming that with each environmental challenge the probability of any particular locus being chosen is aL, and the number of substitutions that the locus experiences, given that it is chosen, is given by the probability generating function $g_L(s)$. If we focus our attention on a single locus, we note that it will no longer experience an allelic substitution with each environmental challenge.   Therefore the representation (3) should be modified to reflect this fact by defining $P^*(t)$ to be a "thinning" of $P(t)$. That is, the events of the process $P^*(t)$ will be those events of the process $P(t)$ for which the locus under consideration experiences at least one allelic substitution. The distribution of the X_i should be modified by conditioning on at least one substitution having occurred. That is, define X_i^* to have the distribution of X_i, given that $X_i > 0$. $S(t)$ may now be written as

$$S(t) = X_1^* + X_2^* + \ldots + X^*_{P^*(t)} .$$

The distribution of $S(t)$ has the pgf

$$E\{[1 + (a/L)(g_L(s) - 1)]^{P(t)}\}$$

where the expectation is taken with respect to the distribution of the process $P(t)$. If the number of environmental challenges and the number of loci grow such that $E\{P(t)\}/L \rightarrow \lambda t$ and $g_L(s) \rightarrow g(s)$, then this pgf becomes,

$$e^{\lambda a t (g(s)-1)}$$

which will be immediately recognized as the pfg of a Poisson sum of positive random variables. Thus our simple epistatic scheme has led to a representation of the substitution process that is identical to that obtained from the statistical analysis of the sequence data.

There is one detail of this argument that needs amplification. The final representation of the substitutions at a particular locus can now be written in the notation of the statistical analysis as

$$S(t) = N(t) = Y_1 + Y_2 + \ldots + Y_{M(t)} .$$

where $M(t)$ is a Poisson process. That is, we are saying that

$$X_i^* \rightarrow Y_i \, ,$$

$$P^*(t) \rightarrow M(t) \, .$$

The value of $R(t)$ for this representation is

$$R(t) = E\{Y_i\} + \text{Var}\{Y_i\}/E\{Y_i\} \, .$$

Perhaps the most important aspect of these results is to understand why this theoretical development implies that $R(t)$ will be restricted to a narrow range of values. Since $R(t)$ is independent of $M(t)$, the values of $R(t)$ depend only on the properties of the Y_i. In the two models that led to Y_i, the first yields, asymptotically, that Y_i is identically one and thus that $R(t) = 1$. The second model yields a value for $R(t)$ that is characterized by the mutational landscape as described in Gillespie (1984b). In this case $R(t)$ was shown to be in the range of around 2.5 to 3.5. Therefore this model predicts that the observed values of $R(t)$ should be in the range 1 to 3.5, exactly as observed in the data.

This, then, is perhaps the first model of molecular evolution by natural selection that seems to be in complete agreement with the statistical properties of the protein sequence data. While the development of the model might seem contrived at first, it is really quite biological in its motivation. Recall that its only assumptions are that the environment is continually changing, that a mutational landscape exists, and that there is an epistatic scheme in which more than one locus is capable of satisfying an environmental challenge.

V. PROBLEMS WITH THE NEUTRAL ALLELE THEORY

For many years it has been argued that the substitution process for the neutral allele model is a Poisson process. However, in 1979 Gillespie and Langley showed that samples from a phylogeny undergoing neutral evolution should exhibit a larger variance in the number of substitutions than predicted by the Poisson process. Their argument was based on an infinite-sites, no recombination model. The main result was that

$$R(t) = 1 + 4Nu/(1 + t/2N) ,$$

when a single sequence is sampled from each of two species. In this expression N is the population size, u is the neutral mutation rate for the locus, and t is the time back to the common ancestor of the two species under study. In order to account for the observed values of $R(t)$, $4Nu$ must be large, say in the range one to ten. This is higher than predicted by polymorphism data and suggests a fundamental problem with the neutral allele theory. Recently, Hudson (1983a) extended this analysis to more than pairs of species. His analysis, which is considerably more sophisticated than ours, also showed that $4Nu$ must be large.

It now appears that the problem of fitting the sequence data to the neutral model is even more severe than suggested by Hudson's analysis. This arises from the effects of intragenic recombination. As the amount of intragenic recombination increases, the value of $R(t)$ for the neutral model must obviously decrease. Following a suggestion by Richard Hudson (per. com.) to combine the arguments used in Gillespie and Langley (1979) and Hudson (1983b) we obtain

$$R_r(t) = 1 + 4NuV(4Nr)/(1 + t/2N) .$$

where $V(\ )$ is the function described in Hudson (1983b, formula 4) and r is the probability of a recombination within the locus under study. $V(\ )$ is a monotonically decreasing function of its argument that approaches one as its argument approaches zero, and approaches zero as its argument approaches infinity. Thus any intragenic recombination will make it even more difficult to account for the observed values of $R(t)$ with the neutral allele model. If r is of the order of 10^{-5}, as it is for the rosy locus in *Drosophila*, then $4Nr$ could easily be in the range 10 to 100 which would result in an inflation of the estimate of $4Nu$ by 2- to 4-fold compared to the no-recombination model. This magnifies the internal inconsistency of a purely neutral interpretation of both the polymorphism and molecular evolution data.

These observations about the neutral allele theory suggest that it does not fit the statistical properties of the molecular evolution data particularly well. There are, of course, other aspects of the neutral allele theory that provide potential difficulties in accounting for the data. A well known example is the clock-time rather than generation-time dependence of the molecular clock. This issue has been thoroughly examined by Kimura (1983) so will not be discussed further here. Here we will only note that all of the problems with the neutral allele theory may undoubtedly be corrected by introducing more elements into the theory. Among these might be variable population sizes, mutation rates, or generation times, asymmetric mutation processes, mildly deleterious alelles, or population subdivision. Doing this, however, will make the theory almost as parameter-rich as theories based on natural selection. The net effect will be to remove the neutral model from its special place as the most parsimonious explanation for the patterns of variation observed in natural populations.

VI. OTHER MODELS OF THE MOLECULAR CLOCK

The representation of the molecular clock as a Poisson sum of positive random variables is a direct consequence of five factors:

(i)   The representation of the molecular clock by a Cox process.

(ii) $\theta(t)$ is assumed to be stationary process.

(iii) $\theta(t)$ is assumed to change on a time scale that is much shorter than the length of the lineages.

(iv)   Clocks are assumed to be equal at branch points and to vary independently down lineages.

(v)   The observed values of $R(t)$ are significantly greater than one.

It is clearly of importance to assess the effects of relaxing the assumptions contained in (i) to (iv). This will be attempted in this section.

A. Processes Other Than the Cox Process

The Cox process was chosen as the natural extension to the Poisson process. The fact that it is parameterized by a rate, $\lambda\theta(t)$, makes it one of the few processes that fit our common verbal patterns used to describe the evolutionary process.   However, there are an infinity of other processes that would fit the data as well as the Cox process. Many of these would not exhibit the episodic structure. For example, a renewal process, for which the quotient of the variance of the time between events and the square of the mean time between events is in the range 1.1 to 3.3, fits the data quite well without the occurrence of episodes of multiple substitutions. There appears to be no *a priori* arguments for choosing one point process over another, and there are not enough data to distinguish between different processes. In this light, we must view our results as being totally model-dependent. There is no escape from this dilemma. The best that can be achieved is to arbitrarily choose a statistical model, estimate its parameters using the data, and try to attach biological significance to the results. If the final phase fails, then perhaps another model would be appropriate.

B. $\theta(t)$ Is Not a Stationary Process

There are many reasons why $\theta(t)$ may not be stationary. If the variations are due to environmental fluctuations, then it is quite possible that there have been trends throughout the history of the planet that cause trends in $\theta(t)$. Since most of the relevant environment is probably biological rather than physical, non-stationarity seems even more likely. There are not enough data available at the present time to reject the assumption of stationarity of $\theta(t)$. Anyone interested in molecular evolution should keep firmly in mind that molecular evolution could be a non-stationary process that appears stationary because of the current paucity of data.

C. The Time Scale of $\theta(t)$ Is Comparable to the Length of Lineages

If the time scale of $\theta(t)$ were similar to the length of the lineages, then a significant correlation in the tick rates would persist in separate lineages. This would lead to an underestimate in the variance in $\theta(t)$ because of the correlation in the number of substitutions in each lineage that would result (Gillespie, 1984a). Thus a longer time scale for $\theta(t)$ would lead to a lower estimate of the variance in the rate than suggested by our short time scale analysis, but would not lead to a variance in the rate that is as small as suggested by the naive interpretation that $R(t)$ is itself a variance to mean ratio of the rate. Another consequence of lengthening the time scale of $\theta(t)$ would be to prolong the episodes of high rates of evolution. There remains, of course, the problem of identifying an environmental factor that would effect the rate of evolution that changes on a time scale measured in millions of years.

D. The Relationship of The Clock to The Phylogeny Is More Complex

The most common interpretation of the fact that $R(t)$ exceeds one is that the rates of evolution differ in the lineages that are under study. Although a specific model is seldom mentioned, the implicit model appears to require that a rate of evolution be assigned independently to each of the lineages under study at the moment of the branching and to remain fixed from that point on. The evolution down each lineage is then supposed to follow a Poisson process. If, for example, we assign the ith lineage the rate, $\lambda\beta_i$, where the θ_i are now a collection of independent, identically distributed random variables with the moments as used for the process $\theta(t)$, then it is not difficult to show that

$$R(t) = 1 + \lambda t \sigma^2 .$$

This model obviously fits the data quite well without having to invoke an episodic clock.

The problem with this model is that it attaches special significance to the branching events that led to the species *under study* and ignores all other branches. For example, if we happen to be studying the rat and the cow, why should we assume that at the time that the lineages leading to the rat and to the cow first diverged, the rates of evolution leading to each of these species were fixed even though during the ensuing 65 million years innumerable other branches occurred in each of these lineages and none of these resulted in a different rate? Surely if we are to claim that the rates altered at one particular branch point, then they should be at least potentially alterable at all subsequent branch points. Furthermore, there are no obvious reasons why changes in the rates of evolution should only occur at branch points. Why not during the periods between branch points? It would seem to be very difficult to defend this model. Nonetheless, this is the model that seems to be the most prevalent one that is used to describe the variation in the rate of molecular evolution.

REFERENCES

Breiman, L. (1968). "Probability." Addison-Wesley, Reading.

Cox, D. R., and Isham, V. (1980). "Point Processes." Methuen, London, pp.1-188.

Gillespie, J. H. (1983). Some properties of finite populations experiencing strong selection and weak mutation. *Amer. Natur.* 121, 691-708.

Gillespie, J. H. (1984a). The molecular clock may be an episodic clock. *Proc. Natl. Acad. Sci. USA* 81, 8009-8013.

Gillespie, J. H. (1984b). Molecular evolution over the molecular landscape. *Evolution* 38, 1116-1129.

Gillespie, J. H. (1985). Natural selection and the molecular clock. *Molec. Biol. Evolut.* (submitted).

Gillespie, J. H., and Langley, C. H. (1979). Are evolutionary rates really variable? *J. Mol. Evol* 13, 27-34.

Hudson, R. R. (1983a). Testing the constant-rate neutral allele model with protein sequence data. *Evolution* 37, 203-217.

Hudson, R. R. (1983b). Properties of a neutral allele model with intragenic recombination. *Theor. Pop. Biol.* 23, 183-201.

Kimura, M. (1983). "The Neutral Theory of Molecular Evolution." Cambridge University Press, Cambridge, pp. 1-367.

Langley, C. H., and Fitch, W. M. (1974). An estimation of the constancy of the rate of molecular evolution. *J. Mol. Evol.* 3, 161-177.

Ohta, T., and Kimura, M. (1971). On the constancy of the evolutionary rate of cistrons. *J. Mol. Evol.* 1, 18-25.

Wilson, A. C., Carlson, S. S., and White, T. J. (1977). Biochemical Evolution. *Ann. Rev. Biochem.* 46, 573-639.

PART III.   COMPARATIVE ANALYSIS OF DNA AND PROTEIN SEQUENCES

PROCESSES OF CHLOROPLAST DNA EVOLUTION[1]

Michael T. Clegg
Kermit Ritland[2]

Department of Botany and Plant Science
University of California
Riverside, CA 92521

Gerard Zurawski

DNAX Research Institute
1450 Page Mill Road
Palo Alto, CA 94304

ABSTRACT

We briefly consider four aspects of chloroplast DNA evolution, based upon our own work with barley. First, we discuss the use of restriction fragment comparisons of cpDNA samples, to make inferences about the genetic consequences of plant domestication. While our studies of barley show that cpDNA diversity has been restricted during the domestication of barley, these studies also show rather low levels of cpDNA diversity at the intraspecific level. It seems likely that comparisons of cpDNA variation will have limited utility at the intraspecific level.

To investigate the elemental features of cpDNA evolution, it is necessary to compare complete DNA sequences among diverse plant

[1] Supported in part by National Science Foundation Grant BRS-8500206.

[2] Current address: Department of Botany, University of Toronto, Toronto, Ontario, Canada.

lineages. The remaining three issues considered in this article concern: (1) the evolution of protein-coding sequences; (2) the evolution of noncoding sequences that map between protein-coding genes; and (3) the evolution of chloroplast tRNA introns. The results of these comparisons show a relatively low rate of synonymous evolution in protein-coding genes, supporting the general impression that chloroplast genes evolve at a conservative rate. In addition, the distribution of missense substitutions is nonrandom in coding genes, probably as a result of functional constraints associated with the protein molecule. Evolutionary comparisons of noncoding regions reveal retarded rates of evolution, relative to the synonymous rate in coding genes, probably raising from functional constraints associated with promoter and ribosome-binding requirements.

These results suggest that evolutionary comparisons can be used as a kind of substitutional filter to identify sequence regions that code for important functions. To investigate this appliation of evolutionary data, we have compared the trnVI intron from barley, maize, pea, and tobacco. We present an algorithm for clustering the sequence into blocks characterized by different rates of evolutionary change. This analysis shows substantial variation in the accumulation of substitutional differences over blocks. Such analyses may prove useful in the search for intron-associated functions.

I. INTRODUCTION

The chloroplast organelle is the cellular site of photosynthesis in higher plants. This organelle contains its own DNA complement, and many photosynthetic functions are chloroplast encoded. The chloroplast genome ranges in size from about 120 to about 180 kilobase pairs (kbp) (Whitfeld and Bottomley, 1983) and, therefore, has the potential to code for more than 100 polypeptides. While the molecular characterization of the chloroplast genome is proceeding rapidly, fewer than 20 chloroplast-encoded genes have been chracterized to date (including ribosomal RNA cistrons). Among the chloroplast-encoded genes which have been characterized are *rbcL* (large subunit of ribulose-1, 5-bisphosphate carboxylase; McIntosh *et al.*, 1980; Zurawski *et al.*, 1981), the α, β, ϵ, and DCCD-binding proteolipid subunits of ATP synthase (*atpA*, *atpB*, *atpE*, and *atpH*; Alt *et al.*, 1983a; Deno *et al.*, 1983; Howe *et al.*, 1983a; Howe *et al.*, 1983b; Krebbers *et al.*, 1982; Zurawski *et al.*, 1982a), the 32 kilodalton thylakoid membrane protein (*psbA*; Zurawski *et al.*, 1982b), four ribosomal proteins (Subramanian *et al.*, 1983; Sugita and Sugiura, 1983; Zurawski *et al.*, 1983; Zurawski *et al.*, 1984), the p700 chlorophyll A apoprotein (Westhoff *et al.*, 1983), three polypeptides of the cytochrome b6-f complex (Alt *et al.*, 1983b; Wiley *et al.*, 1983), and numerous tRNAs (Alt *et al.*, 1983a; Deno *et al.*, 1983; Bergmann *et al.*, 1984; Holschuh *et al.*, 1984; Chu *et al.*, 1985). It is fair to conclude from this rather incomplete list of recent citations that the molecular characterization of the chloroplast genome is one of the most rapidly advancing areas of plant molecular biology.

The study of plant molecular evolution has also advanced most rapidly where the chloroplast genome is concerned. This rapid advance is a direct consequence of the acquisition of basic information on the molecular structure of chloroplast-encoded genes. A second important contributing factor is the relative ease with which pure chloroplast DNA preparations can be obtained. A general conclusion from comparative studies of chloroplast DNA (cpDNA) evolution is that this molecule evolves at a conservative rate (reviewed by Curtis and Clegg, 1984).

Our goal in this article will be to consider some of the elementary statistical features of cpDNA evolution. Our main emphasis will be on our work with barley, both at the level of restriction site variation and at the level of complete DNA sequence comparisons. For detailed discussions of chloroplast DNA evolution across the plant kingdom, see two excellent recent reviews by Palmer (1985a,b).

A. Chloroplast DNA Polymorphism in Wild and Cultivated Barley

Our initial goals were to assess the extent of cpDNA diversity within a major crop species (barley) and its wild progenitor, and to investigate the utility of molecular markers for the study of crop plant evolution. We began by obtaining a sample of 11 lines of *Hordeum spontaneum*, the wild progenitor of cultivated barley, and nine lines of cultivated barley (*H. vulgare*). The cultivated barleys included five primitive land race entries and four modern cultivars. Chloroplast DNAs were purified using the nonaqueous technique of Bowman and Dyer (1982). Genetic diversity among the cpDNAs was estimated by comparison of cpDNA fragmentation patterns produced by the digestion of each cpDNA sample with each of 10 hexanucleotide restriction endonucleases.

Even though the sample of materials is small (20 lines), these data represent one of the larger surveys of cpDNA diversity within a species. The results showed a low level of cpDNA diversity within the cultivated materials; one polymorphic fragment was detected in approximately 250 fragments. The wild materials exhibited five polymorphic fragments and, moreover, two lines from a single wild population differed by four fragments. The level of diversity in the cultivated materials was significantly below that of the wild materials indicating that the cultivated barleys probably trace back to a restricted base of cytoplasmic types. Finally, the cpDNA type shared by the modern cultivars is the most geographically widespread pattern among the wild barleys. We infer from these results that cultivated barleys contain a limited sample of the extant cpDNA variation (Clegg *et al.*, 1984).

Estimates of the proportion of nucleotide substitutions per nucleotide site (p) can be made using the methods of Nei and Li (1979). These estimates range from 0.0 to 0.0002 for the *H. vulgare* materials and from 0.0 to 0.001 for the *H. spontaneum* materials. These data are consistent with studies of other higher plant species which also show low levels of cpDNA diversity (Timothy *et al.*, 1979; Kung *et al.*, 1982; Bowman *et al.*, 1983; Palmer and Zamir, 1982; Terachi *et al.*, 1984).

While the comparison of restriction fragment patterns is useful for reconstructing plant genetic relationships above the species level, the low levels of diversity provide little information on networks of genetic similarity within species. This stands in marked contrast to mammalian mitochondrial DNA, which is especially well suited to intraspecific comparisons as a result of high levels of polymorphism (Avise *et al.*, 1983). In addition, the comparison of restriction fragment patterns provides little information on the kinds of mutational events which occur in the chloroplast genome. For these reasons, we regard the comparison of complete DNA sequence data as most informative about the processes of cpDNA evolution.

B. Comparison of Complete DNA Sequences for Chloroplast Protein-coding Genes

In order to determine the kinds of mutational events that occur in cpDNA evolution, and how these events are distributed with respect to function, we cloned and sequenced a portion of the barley chloroplast genome (Zurawski *et al.*, 1984; Zurawski and Clegg, 1984). The region sequenced contained the genes for *rbcL* (large subunit of ribulose 1, 5-bisphosphate carboxylase), *atpBE* (β and ϵ subunits of ATPase), *trnM2* (gene for tRNA2^{met}), *trnV1* (gene for tRNA1^{val}) and part of an open reading frame (*orfA*). The barley sequences for *rbcL* and *atpBE* can be compared to published data for the homologous region of the maize cpDNA (McIntosh *et al.*, 1980; Poulsen, 1981; Krebbers *et al.*, 1982). In comparing the barley and maize sequences for *rbcL*, *atpB*, and *atpE*, we seek answers to the following three questions: (1) What kinds of mutational events have occurred in coding sequences? (2) How do relative rates of mutation compare among functionally different genes? And, (3) do mutational events occur randomly along the sequence?

All of the mutational events observed in the three coding sequences are nucleotide substitution events. This observation is expected because addition or deletion events are likely to have a more drastic effect on coding function. Table I classifies the observed substitution events into six categories. Of the 16 possible substitution events among the four bases (A, G, T, C), exchanges among like nucleotides cannot be observed (e.g., A $\leftrightarrow$ A). In addition, the direction of substitution from the ancestral sequence is unknown (e.g., A $\rightarrow$ G or G $\rightarrow$ A), leaving just six classes of observed events. These classes can be further grouped according to whether they are purine-purine exchanges or pyrimidine-pyrimidine exchanges (transition events) or whether they are purine-pyrimidine exchanges (transversion events). Transversion events can be subdivided into those that preserve the number of hydrogen bonds (transversion I, e.g., A $\leftrightarrow$ T, G $\leftrightarrow$ C) or those that alter the number of hydrogen bonds (transversion II, e.g., A $\leftrightarrow$ C, T $\leftrightarrow$ G).

We first ask whether these six classes of events are occurring randomly. Denote the relative frequency of the four nucleotides in the sequence of interest by f_A, f_G, f_T, f_C, where $\Sigma f_i = 1$ (i = A, G, T, C). Because we cannot estimate the relative frequencies of the four nucleotides in the pool of introduced nucleotides, we assume that a nucleotide is substituted by each of the other three with equal probability. Thus, the conditional probability that an existing A is replaced by a G, given that A $\leftrightarrow$ A replacements cannot be observed, is $\frac{1}{3} f_A$. The total probability of A $\leftrightarrow$ G transitions is therefore $\frac{1}{3}(f_A + f_G)$. The expected

Table I. Transition bias among coding sequences for barley vs. maize

Type of substitution	*rbcL*		*atpB*		*atpE*	
	Exp.	Obs.	Exp.	Obs.	Exp.	Obs.
A $\leftrightarrow$ G	12.8	24	12.2	27	3.1	6
Transition						
T $\leftrightarrow$ C	11.9	23	10.8	20	2.5	5
A $\leftrightarrow$ T	13.9	8	13.3	2	3.2	1
Transversion I						
C $\leftrightarrow$ G	10.8	6	9.7	6	2.4	1
A $\leftrightarrow$ C	11.5	7	11.0	9	2.8	2
Transversion II						
T $\leftrightarrow$ G	13.2	6	12.0	5	2.8	1

1279 nucleotides were compared for *rbcL*, 1494 nucleotides were compared for *atpB*, and 411 nucleotides were compared for *atpE*.

Table II. Number of nucleotide substitutions per site for first, second and third codon positions for protein coding sequences in barley–maize contrast. Standard errors in parentheses.

Codon Position	rbcL	atpB	atpE	χ^2
1st codon position	0.036 (0.009)	0.022 (0.007)	0.015 (0.010)	2.4
2nd codon position	0.019 (0.007)	0.014 (0.005)	0.0	0.3
3rd codon position	0.135 (0.019)	0.112 (0.016)	0.121 (0.032)	0.9
3rd position synonymous	0.116 (0.018)	0.098 (0.010)	0.102 (0.015)	0.7
Total over all positions	0.060 (0.007)	0.048 (0.006)	0.041 (0.010)	2.8

Statistics and standard errors were calculated from equations 6, 11, 12, and 13 of Kimura (1981).

number of A $\leftrightarrow$ G transitions is $N/3(f_A + f_G) = E_{AG}$ where N is the total number of differences observed. These expected numbers are calculated for each of the three protein-coding genes in Table I. Comparison with the observed numbers shows about a two-fold excess of transition events, as compared to transversion events. This substitution bias is of the same

Table III. Tests for fit of the observed nucleotide runs to a geometric distribution for protein coding sequences in the barley-maize constrast. Degrees of freedom are given in parentheses.

Statistical test	*rbcL*			*atpB*		
	1st*	2nd	3rd	1st	2nd	3rd
Goodness of fit	1.6 (3)	0.0 (1)	8.4 (11)	0.4 (2)	2.3 (1)	7.2 (11)
Variance Ratio	17.5 (13)	15.3** (6)	93.2** (50)	16.2 (11)	17.2** (6)	54.6 (49)

* 1st, 2nd, and 3rd refer to the respective codon positions. ** $p < 0.05$.

magnitude for each of the three coding genes. It is note-worthy that this observed transition bias is small relative to the bias observed in mammalian mitochondrial sequence comparisons (Aquadro and Greenberg, 1983).

We estimated the number of nucleotide substitutions per nucleotide site using the estimator of Kimura (1981; eq. 6 and 11). The data are partitioned by codon position for each of the three coding genes. In addition, the number of third codon position synonymous substitutions are estimated separately (Table II). These data show a six-fold excess of synonymous substitutions over second position (missense) substitutions. Interestingly, the data are homogeneous over the three coding sequences for each codon position considered separately, as shown by the non-significant Chi-square heterogeneity tests in Table II.

A homogeneous rate of nucleotide substitution between *atpB* and *atpE* was unexpected because both genes have evolved at quite different rates when compared to homologous sequences from *Escherichia coli*. Specifically, derived amino acid sequence comparisons for *atpB* and *atpE* between spinash chloroplast sequences and *E. coli* show approximately

67% and 26% identity, respectively (Zurawski *et al.*, 1982b). Evidently, *atpE* evolved more rapidly prior to the monocot-dicot split.

Next, we consider the distribution of sites of substitution along the sequence. A simple model is to assume that each of the three codon positions have a constant probability $(\theta_1, \theta_2, \theta_3)$ of substitution. Substitution events are further assumed to be independent over codons. Subject to these assumptions, the distribution of the number of codons between substitution events (run length, x) is geometric with distribution function $g(x; \theta_i) = (1-\theta_i)^{x-1}\theta_i$ (i = 1,2,3).

We employ two different statistical tests to ask whether the observed distribution is geometric. First, we compare the variance of the geometric distributions, $(1-\hat{\theta}_i)/\hat{\theta}_i^2$, where the maximum likelihood estimate of θ_i is denoted $\hat{\theta}_i$, to the sample variance S_i^2 as the ratio, $(k_i-1)\hat{\theta}_i^2 S_i^2/(1-\hat{\theta}_i)$, and where the number of substitution events is k_i (for i = 1,2,3). This statistic is approximately distributed as Chi-square with k_i-1 degrees of freedom. This test is referred to as the variance ratio test. Second, we perform a Chi-square goodness-of-fit test by comparing expected and observed run lengths, subject to the constraint that the expected numbers in a class exceed three to satisfy normality assumptions. A consequence of pooling classes is that long runs with very low probability are consolidated with shorter runs. We, therefore, regard the variance ratio test as more powerful for the detection of an over-dispersed distribution.

Table III presents the results of both tests applied to the barley-maize comparison for *rbcL* and *atpB* (*atpE* is not analyzed because the number of events observed in this short sequence is small). Because of the effect on protein function, it is not surprising to find that the run lengths for second-position substitutions depart from a geometric distribution in *rbcL*. A number of missense events are clustered in one region of the sequence (Zurawski *et al.*, 1984). On the other hand, if is more surprising to find a significant departure for third-position events in *rbcL*. It is possible that there are some constraints on the occurrence of synonymous substitutions in particular regions of the sequence. Second-position substitutions depart from a geometric distribution in *atpB* by the variance ratio criterion. These events are more common in the N-terminal 4% and C-terminal 2% of the coding sequence (Zurawski and Clegg, 1984).

Barley and maize are at opposite ends of the grass family which began to appear in the fossil record about 50 to 65 million years ago (Stebbins, 1981). If we consider synonymous substitutions, then an estimate of the per site, per year rate of nucleotide substitution is approximately 1×10^{-9}. This compares with estimates of $2-3 \times 10^{-9}$ synonymous substitutions/site/year for animal nuclear genes (Kimura, 1983) and 5×10^{-9} substitutions/site/year for pseudogenes (Li, 1983). Thus, our data are consistent with a conservative rate of evolution for chloroplast-coding sequences. Sequences which do not code for proteins make up a considerable fraction of the chloroplast genome. It is therefore of interest to ask how noncoding sequences evolve relative to coding sequences.

C. Comparison of Noncoding Chloroplast DNA Sequences

The coding regions of *rbcL* and *atpBE* are separated by 784 bp of noncoding sequences in barley. Because *rbcL* and *atpBE* are transcribed divergently, the initiation of transcription for both genes maps into this noncoding region. We determined the sites of transcription initiation for both genes in barley using the primer extension method, as described by Zurawski *et al.* (1981). The 5' end of the *rbcL* mRNA maps to coordinates −320 (measured from the first AUG codon of *rbcL*), and the

5' end of *atpBE* (the β and ε genes are contained on a dicistronic message) maps between coordinates -296 and -309 (measured from the first codon of *atpBE*). Thus, there are 155 to 167 bp of untranscribed sequence separating these two genes (Zurawski *et al.*, 1984).

Comparison of the barley-maize sequence data reveals two major classes of mutational events: short addition/deletion events and nucleotide substitution events. Estimates of the number of nucleotide substitution events per site (K) for the noncoding region (excluding addition/deletion events) yield K = 0.076, as compared to a third codon position estimate for *rbcL* of K = 0.135. Evidently, the rate of nucleotide substitution is retarded in the noncoding region as compared to the third codon position rate. We can attempt to correct for addition/deletion events to obtain a total estimate of the evolutionary rate for the noncoding region by assuming that each addition/deletion event has occurred only once since the separation of barley and maize lineages. Our modified estimate then becomes K' = 0.093.

These calculations suggest that the *rbcL-atpBE* noncoding region is evolving at a reduced rate relative to the third position rate in coding regions. A more detailed analysis of the *rbcL-atpBE* noncoding region indicates that the retarded rate of evolution can, in large part, be accounted for by selective constraints associated with promoter and ribosome-binding functions (Zurawski *et al.*, 1984).

D. Evolution of a Chloroplast tRNA Intron

The gene for tRNA^{Val} (trnV1) is interrupted by a 597 base pair (bp) intervening sequence in barley. No known functions are associated with chloroplast tRNA introns, although models of secondary structure have been proposed (Michel and Dujon, 1983). In an initial attempt to identify conserved regions in the trnV1 intron, we have used evolutionary comparisons as a kind of mutational filter. Our goal is to highlight regions of sequence conservation that may have been constrained by functional requirements. Complete DNA sequence data are presently available from barley (Zurawski and Clegg, 1984), maize (Krebbers *et al.*, 1984), tobacco (Sugita and Suguira, 1983), and pea (Zurawski, unpublished data).

To ask whether the distribution of differences fits a geometric distribution, we define an event as a site with one or more nucleotide differences. Subject to this definition, the observed distribution of run lengths between events fits a geometric distribution by the goodness-of-fit criterion ($X^2_{10} = 10.74$), but is discrepant by the variance ratio criterion ($X^2_{143} = 260.46$). If the substitutional process is nonrandom along the line, as suggested by the variance ratio test, it would be desirable to find an algorithm that identifies contiguous blocks of nucleotides that differ in frequency of nucleotide substitution. One such algorithm is described below.

We begin by considering the number of nucleotide differences at the ith nucleotide site, where $i = 1,2,...,L$, and L is the length of the aligned sequence in nucleotides. At many sites, pairs of sequences share the same substitutional event due to co-ancestry (e.g., pea and tobacco versus barley and maize). A simple measure of substitution differences, not dependent upon phylogenetic history, is to classify each site as either (a) identical for all four nucleotides or (b) nonidentical for two or more nucleotides at a site. This classification discards information on multiple events. Our analysis also omitted regions of the aligned sequences where deletion/addition events occur, in order to compare the same number of nucleotides at each site.

Step one in the algorithm is to identify runs of consecutive identical or nonidentical sites. These runs constitute the initial clusters or "blocks" of substitutional events. Denote the number of sites in the kth block by N_k and the number of sites within this block that are non-identical as M_k. Define $U_k = M_k/N_k$ as the probability that a randomly chosen site within the kth block is nonidentical. Clearly, $U_k = 0$ or $U_k = 1$ following step one. Next, we seek a criterion for joining adjacent blocks into larger blocks.

One criterion is to calculate the likelihood that the number of differences in two adjacent blocks are sampled from populations with identical means, assuming that M_k is a binomial random variable with sample size N_k and parameter p_k. For two adjacent blocks $(k,k+1)$, we estimate the pooled value of p_k

$$\hat{p}_k = \frac{M_k + M_{k+1}}{N_k + N_{k+1}} \ .$$

For $M_k/N_k \le \hat{p}_k$, the probability of observing M_k or fewer differences in block k, given $\hat{p}_k$, is obtained from the binomial cumulative distribution function $B(M_k,N_k,p_k) = a$.

The probability of observing M_{k+1} or greater nonidentical sites in block $k+1$ is $1 - B(M_{k+1},N_{k+1},\hat{p}_k) = b$. The probability of either of these events is, therefore, $1 - (1-a)(1-b)$. A similar argument applies to the case $M_k/N_k \ge \hat{p}_k$. The pair of adjacent blocks with the maximum value of $P_{k,k+1}$ is then joined at each pass through the data, reducing the total number of blocks by one, and the algorithm is repeated. For example, if blocks k and $k+1$ are joined to create block j, then $N_j = N_k + N_{k+1}$ and $M_j = M_k + M_{k+1}$.

Figure 1 plots the behavior of the average value of $P_{k,k+1}(\overline{P})$ over all pairs of adjacent blocks and the maximum value of $P_{k,k+1}$ [i.e., the maximum value of $P_{k,k+1}$ defines the pair of blocks $(k,k+1)$, that are joined at a given iteration]. These values are plotted as a function of the

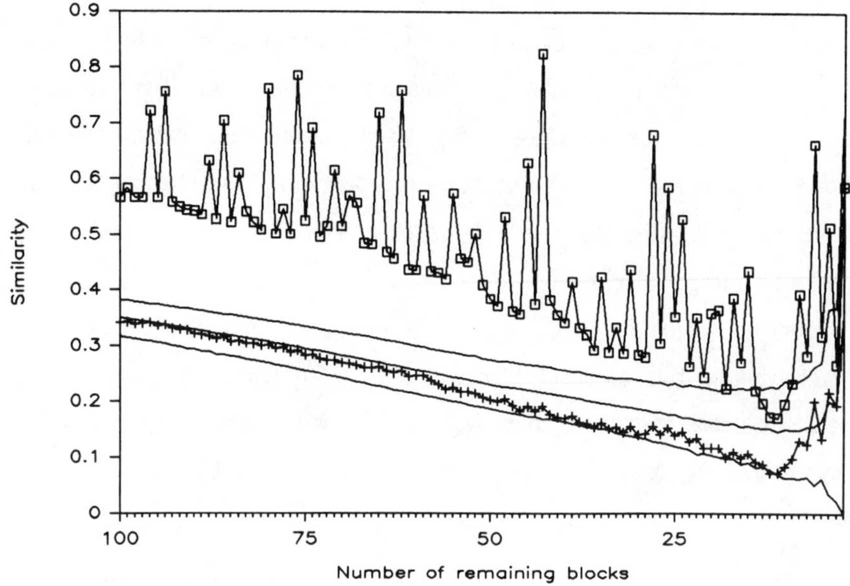

Figure 1. Graphs the average behavior of $P_{k,k+1}$ over blocks $(\overline{P})$ (++++), the maximum value of $P_{k,k+1}$ (□—□—□—□), and the mean of $\overline{P}$ $(\overline{\overline{P}})$ for 100 Monte Carlo simulations of the clustering algorithm with 95% confidence bands (——). The Monte Carlo trials were performed on random sequences of 0's and 1's, where each sequence had a length of 513. The plots are functions of the number of blocks remaining in the clustering process.

number of blocks (or the number of iterations remaining before all blocks are joined). Because the number of blocks is reduced by one at each iteration, the plots show how the measure of similarity among adjacent blocks behaves during the course of the iterations. Both the average and maximum values decrease to a minimum at 12 blocks and then rapidly increase. We take the minimum value of 12 blocks as a reasonable stopping point for the clustering algorithm.

To investigate whether the clusters formed depart from those expected with a uniform distribution of sequence differences, we performed Monte Carlo trials of the algorithm applied to 100 random sequences of zeros and ones. These sequences were generated by

Figure 2. trnV1 intron sequences of tobacco, barley, maize, and pea aligned to minimize the number of mutational differences. Regions in which one or more of the sequences had an addition or deletion event are omitted. The length, in nucleotides, of omitted regions is indicated by a number in parentheses below the sequence where the addition/deletion maps. The final 12 blocks resulting from the clustering algorithm are identified by a line above the sequences, together with the value of U_k.

assuming a probability, 0.28 for a difference at each site and independence over sites. The value of 0.28 is the observed fraction of nonidentical sites in the original set of four sequences. The average value of $\overline{P}$ ($\overline{\overline{P}}$) over the 100 Monte Carlo trials, together with 95% confidence bands, are also plotted in Figure 1.

Note that the minimum value of $\overline{P}$ for the observed sequence is below the minimum value of $\overline{\overline{P}}$ (0.07 versus 0.14) and just outside the 95% confidence bands of $\overline{\overline{P}}$ at the stopping point of 12 blocks. We conclude that there are probably regions of significant substitutional differences among adjacent blocks. Figure 2 gives the nucleotide sequence comparisons and identifies the 12 blocks together with the fraction of nonidentical sites for each block (regions with additions or deletions have been omitted). Figure 3 plots the values of U_k as a histogram for the final 12 blocks. An interesting feature of the result of the clustering process is the distribution of blocks along with intron length. The left half of the sequence is characterized by alternating high-low blocks of substitutional change, while the right is characterized by a uniformal rate of substitutional change. Moreover, addition/deletion events appear to occur less frequently in the last 120 nucleotides of the sequence (indicated by numbers in parentheses in Figure 2).

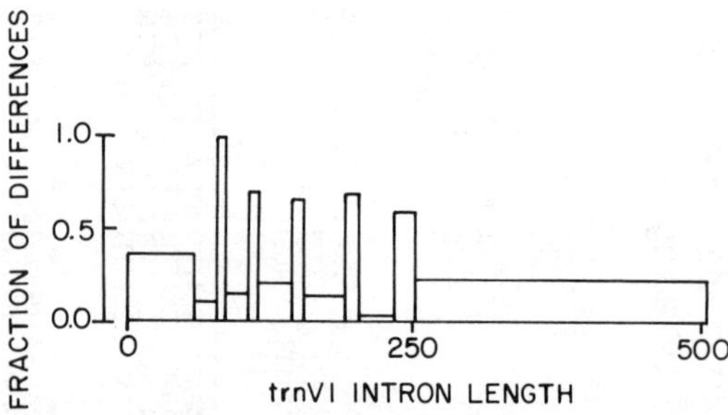

Figure 3. The inferred 12 blocks of mutational difference, plotted as a histogram with heights equal to U_k (for the kth block) and horizontal segments equal to the number of nucleotides in each region.

Evidently, different regions of the trnV1 intron evolve at different rates. This result supports the earlier analysis of Zurawski and Clegg (1984) for barley and maize. The causes of the differential rates of evolution are unknown, but two possibilities suggest themselves. First, it may be that certain regions of the intron are most liable to mutation, perhaps for structural reasons. The second possibility is that mutations in certain regions are selected against because of functional constraints. This latter possibility is particularly interesting because no known function has been associated with chloroplast tRNA introns. It is possible that evolutionary comparisons of this kind will aid the molecular biologist in the search for functional regions.

REFERENCES

Alt, J., Sebald, W., Moser, J. G., Schedel, R., Westhoff, P., and Herrmann, R. G. (1983). Localization and nucleotide sequence of the gene for the ATP synthase proteolipid subunit on the spinach plastid chromosome. *Current Genet.* **7**, 129-139.

Alt, J., Westhoff, P., Sears, B. B., Nelson, N., Hurt, E., Hauska, G., and Herrmann, R. G. (1983b). Genes and transcripts for the polypeptides of the cytochrome b6/f complex from spinach thylakoid membranes. *EMBO J.* **2**, 979-986.

Aquadro, C. F., and Greenberg, B. D. (1983). Human mitochondrial DNA variation and evolution; analysis of nucleotide sequences from seven individuals. *Genetics* **103**, 287-312.

Avise, J. C., Shapira, J. F., Daniel, S. W., Aquadro, C. F., and Lansman, R. A. (1983). Mitochondrial DNA differentiation during the speciation process in *Peromyscus. Molec. Biol. Evol.* **1**, 38-56.

Bergmann, P., Seyer, P., Burkard, G., and Weil, J.-H. (1984). Mapping of transfer RNA genes on tobacco chloroplast DNA. *Plant Molec. Biol.* **3**, 29-36.

Bowman, C. M., and Dyer, T. A. (1982). Purification and analysis of DNA from wheat chloroplasts isolated in nonaqueous media. *Analytical Biochem.* **12**, 108-118.

Bowman, C. M., Bonnard, G., and Dyer, T. A. (1983). Chloroplast DNA variation between species of *Triticum* and *Aegilops.* Location of the variation of the chloroplast genome and its relevance to the inheritance and classification of the cytoplasm. *Theoret. Appl. Genet.* **65**, 247-262.

Chu, N. M., Shapiro, D. R., Aishi, K. K., and Tewari, K. K. (1985). Distribution of transfer RNA genes in the *Pisum sativum* chloroplast DNA. *Plant Molec. Biol.* **4**, 65-80.

Clegg, M. T., Brown, A. H. D., and Whitfeld, P. R. (1984). Chloroplast DNA diversity in wild and cultivated barley: implications for genetic conservation. *Genet. Res.* **43**, 339-343.

Curtis, S. E., and Clegg, M. T. (1984). Molecular evolution of chloroplast DNA sequences. *Molec. Biol. Evol.* **1**, 291-301.

Deno, H., Shinozaki, K., and Sugiura, M. (1983). Nucleotide sequence of tobacco chloroplast gene for the subunit of proton-translocating ATPase. *Nucleic Acids Res.* **11**, 2185-2191.

Holschuh, K., Bottomley, W., and Whitfeld, P. R. (1984). Organization and nucleotide sequence of the genes for spinach chloroplast tRNA^{Glu} and tRNA^{Tyr}. *Plant Molec. Biol.* **3**, 313-318.

Howe, C. J., Bowman, C. M., Dyer, T. A., and Gray, J. C. (1983). The genes for the alpha and proton-translocating subunits of wheat chloroplast ATP synthase are close together on the same strand of chloroplast DNA. *Mol. Gen. Genet.* **190**, 51-55.

Howe, C. J., Suffret, A. D., Doherty, A., Bowman, C. M., Dyer, T. A., and Gray, J. C. (1982). Location and nucleotide sequence of the gene for the proton-translocating subunit of wheat chloroplast ATP synthase. *Proc. Natl. Acad. Sci. USA* **79**, 6903-6907.

Kimura, M. (1981). Estimation of evolutionary distances between homologous nucleotide sequences. *Proc. Natl. Acad. Sci. USA* **78**, 454-458.

Kimura, M. (1983). The neutral theory of molecular evolution. *In* "Evolution of Genes and Proteins" (M. Nei and R. K. Koehn, eds.), Sinauer Associates, Sunderland, Mass.

Krebbers, E. T., Larrinua, I. M., McIntosh, L., and Bogorad, L. (1982). The maize chloroplast genes for the β and ϵ subunits of the photosynthetic coupling factor CF_1 are fused. *Nucleic Acids Res.* **10**, 4985-5002.

Krebbers, E. T., Steinmetz, A., and Bogorad, L. (1984). DNA sequences for the *Zea mays* tRNA genes tV-UAC and tS-UGA: tV-UAC contains a large intron. *Plant Molec. Biol.* **3**, 13-20.

Kung, S. D., Zhu, Y. S., and Shen, G. F. (1982). *Nicotiana* chloroplast genome. III. Chloroplast DNA evolution. *Theoret. Appl. Genet.* **61**, 73-79.

Li, W.-H. (1983). Evolution of duplicate genes and pseudogenes. *In* "Evolution of Genes and Proteins" (M. Nei and R. K. Koehn, eds.), Sinauer Associates, Sunderland, Mass.

McIntosh, L., Poulsen, C., and Bogorad, L. (1980). Chloroplast gene sequence for the large subunit of ribulose bisphosphate carboxylase of maize. *Nature* **288**, 556-560.

Michel, F., and Dujon, B. (1983). Conservation of RNA secondary structures in two intron families including mitochondrial-, chloroplast- and nuclear-encoded members. *EMBO J.* **2**, 33-38.

Nei, M., and Li, W.-H. (1979). Mathematical model for studying genetic variation in terms of restriction endonucleases. *Proc. Natl. Acad. Sci. USA* **76**, 5269-5273.

Palmer, J. D. (1985a). Evolution of chloroplast and mitochondrial DNA in plants and algae. *In* "Monographs in Evolutionary Biology: Molecular Evolutionary Genetics" (R. J. MacIntyre, ed.), Plenum Publishing Co., New York.

Palmer, J. D. (1985b). Phylogenetic analysis of chloroplast DNA variation. *Ann. Missouri Bot. Gard.* (in press).

Palmer, J. D., and Zamir, D. (1982). Chloroplast DNA evolution and phylogenetic relationships in *Lycopersicon. Proc. Natl. Acad. Sci. USA* **79**, 5006-5010.

Poulsen, C. (1981). Comments on the structure and function of the large subunit of the enzyme ribulose bisphosphate carboxylase-oxygenase. *Carlsberg Res. Commun.* **46**, 259-273.

Subramanian, A.R., Steinmetz, A., and Bogorad, L. (1983). Maize chloroplast DNA encodes a protein sequence homologous to the bacterial ribosomal protein S4. *Nucleic Acids Res.* **11**, 5277-5286.

Sugita, M., and Sugiura, M. (1983). A putative gene of tobacco chloroplast coding for ribosomal protein similar to *E. coli* ribosomal protein S19. *Nucleic Acids Res.* **11**, 1913-1918.

Stebbins, G. L. (1981). Coevolution of grasses and herbivores. *Ann. Missouri Bot. Gard.* **68**, 75-86.

Terachi, T., Ogihara, Y., and Tsunewaki, K. (1984). The molecular basis of genetic diversity among cytoplasms of *Triticum* and *Aegilops*. III. Chloroplast genomes of the M and modified M genome-carrying species. *Genetics* **108**, 681-695.

Timothy, D. H., Levings, C. S., III, Pring, D. R., Conde, M. F., and Kernicke, J.L. (1979). Organelle DNA variation and systematic relationships in the genus *Zea*: teosinte. *Proc. Natl. Acad. Sci. USA* **76**, 4220-4224.

Westhoff, P., Alt, J., Nelson, N., Bottomley, W., Bunemann, H., and Herrmann, R. G. (1983). Genes and transcrips for the P_{700} chlorophyll a apoprotein and subunit 2 of the photosystem I reaction center complex from spinach thylakoid membranes. *Plant Molec. Biol.* **2**, 95-107.

Whitfeld, P., and Bottomley, W. (1983). Organization and structure of chloroplast genes. *Annu. Rev. Plant Physiol.* **34**, 279-326.

Willey, D. L., Huttley, A. K., Phillips, A. L., and Gray, J. C. (1983). Localization of the genes for cytochrome f in pea chloroplast DNA. *Mol. Gen. Genet.* **183**, 85-89.

Zurawski, G., Bohnert, H. J., Whitfeld, P., and Bottomley, W. (1982a). Nucleotide sequence of the gene for the M_r 32,000 thylakoid membrane protein from *Spinacia oleracea* and *Nicotiana debneyi* predicts a totally conserved primary translation product of M_r 38,950. *Proc. Natl. Sci. USA* **79**, 6799-7703.

Zurawski, G., Bottomley, W., and Whitfeld, P. R. (1982b). Structures of the genes for the β and ε subunits of spinach chloroplast ATPase indicate a dicistronic mRNA and an overlapping translation stop/start signal. *Proc. Natl. Acad. Sci. USA* **79**, 6260-6264.

Zurawski, G., and Clegg, M. T. (1984). The barley chloroplast DNA *atpBE*, *trnM2*, and *trnV1* loci. *Nucleic Acids Res.* **12**, 2549-2559.

Zurawski, G., Clegg, M. T., and Brown, A. H. D. (1984). The nature of nucleotide sequence divergence between barley and maize chloroplast DNA. *Genetics* **106**, 735-749.

Zurawski, G., Perrot, B., Bottomley, W., and Whitfeld, P. R. (1981). The structure of the gene for the large subunit of ribulose 1, 6-bisphosphate carboxylase from spinach chloroplast DNA. *Nucleic Acids. Res.* **9**, 3251-3169.

ESTIMATION OF THE NUMBERS OF SYNONYMOUS AND NONSYNONYMOUS SUBSTITUTIONS BETWEEN PROTEIN CODING GENES[1]

WEN-HSIUNG LI

Center for Demographic and Population Genetics
University of Texas
Houston, Texas 77030

ABSTRACT

A comparison of three current methods for estimating the numbers of synonymous and nonsynonymous substitutions between two protein coding genes is made. The three methods are briefly described and the advantages and disadvantages of each method are discussed. Difficulties in estimating the above two numbers between two distantly related genes are also discussed.

I. INTRODUCTION

A very basic quantity in the comparative study of DNA sequences is the number of nucleotide substitutions that have occurred since the separation of two sequences. Thus, how to obtain an accurate estimate of this number is a fundamental problem in molecular evolution and to date many models have been proposed for this purpose (e.g., Jukes and Cantor,

[1] This work was supported by NIH grant GM30998.

1969; Holmquist, 1972; Kimura, 1980; see Li *et al.*, 1985a for a review). In this article, I am concerned only with protein coding genes. In such genes it is desirable to consider separately the number of synonymous substitutions (causing no amino acid changes) and the number of nonsynonymous substitutions. Therefore, I shall focus on the methodological aspects of estimating these two numbers.

Currently, the most frequently used methods for estimating the above two numbers are the method by Perler *et al.* (1980) and that by Miyata and Yasunaga (1980). Recently, we (Li *et al.*, 1985b) have proposed a new method. I shall compare these three methods, pointing out the advantages and disadvantages in each method. For this purpose, it is necessary to review the three methods. I shall also discuss difficulties in estimating the above two numbers between distantly related genes.

II. METHODS

A. The Percentage Corrected Divergence (PCD) method

The PCD method was proposed by Perler *et al.* (1980). In this method nucleotide sites are classified as follows. A site is called a potential silent (synonymous) site if a change at that site can be synonymous and a potential replacement (nonsynonymous) site if a change at that site can be nonsynonymous (or nonsense). Potential silent (replacement) sites are further classified into three categories, according to whether a site can afford one, two, or three synonymous (nonsynonymous) changes. Table I shows three examples. For the codon TGG (tryptophan), all three possible changes at each of the three positions are nonsynonymous (or nonsense) so that all three positions are potential replacement sites of category 3 and none of them is a potential silent

Table I. Classification of nucleotide sites in a codon into potential silent and replacement sites and further into categories 1, 2 and 3.

Codon	Silent			Replacement		
	1	2	3	1	2	3
0 0 0 T G G (Trp) 3 3 3	0	0	0	0	0	3
0 0 1 T T T (Phe) 3 3 2	1	0	0	0	1	2
0 0 3 G T C (Val) 3 3 0	0	0	1	0	0	2

Legend. The number of potential silent or replacement changes is shown above or below each site in a codon, respectively.

site. For the codon TTT (phenylalanine), the third position is a potential silent site of category 1 and a potential replacement site of category 2 because one of the three possible changes at the third position is synonymous and the other two are nonsynonymous; the first two positions are potential replacement sites of category 3 because all three possible changes at each of the two positions are nonsynonymous. For the codon GTC (valine), the third position is a potential silent site of category 3 but is not a potential replacement site because all three possible changes at this site are synonymous. When comparing two sequences, one counts the number of potential silent sites and the number of potential replacement sites in each category in each sequence and then computes their averages between the two sequences.

The second step is to compare the two sequences codon by codon, score one point for each synonymous or nonsynonymous difference, and categorize it according to the type of site at which it has occurred. Table II shows three examples. In the first example, the difference between the

Table II. Classification of nucleotide differences between codons into silent and replacement substitutions and further into categories 1,2,and 3.

	Silent			Replacement		
Codon pair	1	2	3	1	2	3
CCC (Pro) – CCT (Pro)	0	0	1	0	0	0
GTC (Val) – GCC (Ala)	0	0	0	0	0	1
AAT (Asn) – ACG (Thr)	0	0	1/2	0	1/2	1

Legend. The number of potential silent or replacement changes is shown above or below each site in a codon, respectively.

two codons CCC and CCT occurs at a silent site of category 3; therefore, the substitution is a silent substitution of category 3. In the second example, the difference between the two codons GTC and GCC occurs at a replacement site of category 3; therefore, the substitution is a replacement substitution of category 3. In the third example, the two differences between the two codons AAT and ACG may arise by either of the following two pathways: AAT (Asn) ↔ ACT (Thr) ↔ ACG (Thr) and AAT (Asn) ↔ AAG (Lys) ↔ ACG (Thr). Both paths are assumed to be equally probable, so that the difference at the third position between AAT and ACG is half silent (category 3) and half replacement (category 2); in either path, the difference at the second position is nonsynonymous (category 3).

In the third step one divides the sum of silent or replacement substitutions in each category by the corresponding number of potential sites. Each percentage change (p) is then corrected for multiple events by one of the following formulas:

$$\text{Category 1: } -3\left[\frac{1}{2}\ell n(1-2p)\right], \tag{1}$$

$$\text{Category 2:} \quad -\frac{3}{2}\left[\frac{2}{3}\,\ell n(1-\frac{3}{2}\,p)\right]\,, \qquad\qquad (2)$$

$$\text{Category 3:} \quad -\left[\frac{3}{4}\,\ell n(1-\frac{4}{3}\,p)\right]\,. \qquad\qquad (3)$$

The brackets are correction formulas relating the observed frequency to the Poisson-distributed frequency of mutational hits for mutations restricted to one, two or three substitutions, respectively. The factors outside the brackets correct those restricted values up to a total mutation rate, assuming that substitution occurs randomly among the four types of nucleotides.

Finally, for each pair of sequences the overall synonymous or nonsynonymous divergence is the average of two weight averages of the corrected percentages, using separately as weighting factors the number of sites or the number of changes in each category.

B. Miyata and Yasunaga's Method

In this method nucleotide sites are classified as follows. Consider a particular position in a codon. Let i be the number of possible synonymous changes at this site. Then this site is counted as i/3 synonymous and (3-i)/3 nonsynonymous. For example, in the codon TTT (phenylalanine) the first two positions are nonsynonymous because no synonymous change can occur at these two positions and the third position is counted as one-third synonymous and two-thirds nonsynonymous because one of the three possible changes at this position is synonymous. As another example, the codon ACT (threonine) has two nonsynonymous sites (the first two positions) and one synonymous site (the third position) because all possible changes at the first two positions are nonsynonymous while all possible changes at the third position are synonymous. When comparing two sequences, one first counts the number of synonymous sites and that of nonsynonymous sites in each sequence and then computes the averages between the two sequences. (The averaging procedure proposed by Miyata and Yasunaga (1980) is more complicated than just started, but the two procedures would give essentially the same results.)

Nucleotide differences are classified into synonymous and nonsynonymous differences. For two codons that differ by only one nucleotide, the difference is easily inferred. For example, the difference between the two codons CCC (Pro) and CCT (Pro) is synonymous while the difference between the two codons GTC (Val) and GCC (Ala) is nonsynonymous. For two codons that differ by more than one nucleotide, all possible pathways with the minimum number of substitutions are considered but different weights may be assigned to different possible pathways. For example, for the two codons AAT (Asn) and ACG (Thr), the pathway AAT (Asn) $\leftrightarrow$ ACT (Thr) $\leftrightarrow$ ACG (Thr) is assumed to be more likely than the other pathway AAT (Asn) $\leftrightarrow$ AAG (Lys) $\leftrightarrow$ ACG (Thr), because the former requires one nonsynonymous change and one synonymous change, but the latter requires two nonsynonymous changes. The weights for different pathways are based on the relative frequencies of exchanges between amino acids observed in protein sequences (see Miyata and Yasunaga, 1980).

Correction for multiple hits at the same site is done as follows (Miyata *et al.,* 1982). Let N_S (N_A) be the average number of synonymous (nonsynonymous) sites between the two sequences compared and M_S (M_A) be the number of synonymous (nonsynonymous) differences between the two sequences. Further, let K_S (K_A) be the number of synonymous (non-synonymous) substitutions per synonymous (nonsynonymous) site. Then

$$K_S = -\frac{3}{4}\,\ell n(1 - \frac{4}{3}\frac{M_S}{N_S}),\qquad(4)$$

$$K_A = -\frac{3}{4}\,\ell n(1 - \frac{4}{3}\frac{M_A}{N_A}),\qquad(5)$$

These formulas are obtained by assuming that Jukes and Cantor's (1969) formula applies both to synonymous sites and to nonsynonymous sites.

C. Li, Wu and Luo's Method

We (Li *et al.*, 1985b) have recently developed a new method. The description here is based on the genetic code for nuclear genes; the method actually applies better to mammalian mitochondrial genes. We classify nucleotide sites into nondegenerate, two-fold degenerate, and four-fold degenerate sites. A site is four-fold degenerate, if all possible changes at the site are synonymous. The third positions of 32 of the 61 sense codons, e.g., GTT (Val), are of this type. A site is two-fold degenerate if one of the three possible changes at the site is synonymous. The third positions of 24 of the 61 sense codons, e.g., CAT (His), and the first positions of four leucine codons (TTA, TTG, CTA, CTG) and four arginine codons (CGA, CGG, AGA, AGG) are of this type. We also include the third positions of the three isoleucine codons in this class, although they are actually three-fold degenerate sites. A site is nondegenerate if all possible changes at that site are nonsynonymous or nonsense. The second positions of all sense codons and the first positions of most codons belong to this class and so do the third positions of ATG (Met) and TGG (Trp). We count the numbers of sites in the three classes in each of the two sequences compared and then compute the average numbers, denoting them by L_0 (nondegenerate), L_2 (two-fold), and L_4 (four-fold), respectively.

Next, we compare the two sequences codon by codon and infer the nucleotide differences between each pair of codons. We classify each difference according to the type of site at which it has occurred. This can be done by considering all possible paths between the two codons compared according to the weights obtained from the relative frequencies of codon changes in mammalian genes (for details, see Li *et al.*, 1985b). The nucleotide differences in each class are further classified into transitional (P_i) and transversional (Q_i) differences, $i=0,2,4$. We note that in the class of two-fold degenerate sites transitions are synonymous and transversions are nonsynonymous. The only exceptions are the third positions of the three isoleucine codons and the first positions of the four arginine codons mentioned above (no exceptions occur in the genetic code for mammalian mitochondrial genes). In all these cases, all synonymous changes are included in P_2 and all nonsynonymous changes are included in Q_2.

Correction for muliple hits at the same site is made by considering the three classes of sites separately. The means and approximate sampling variances of the numbers of transitional (A_i) and transversional (B_i) substitutions per i^{th} type site are given by

$$A_i = \frac{1}{2} \ell n(a_i) - \frac{1}{4} \ell n(b_i) , \tag{6}$$

$$V(A_i) = [a_i^2 P_i + c_i^2 Q_i - (a_i P_i + c_i Q_i)^2]/L_i , \tag{7}$$

$$B_i = \frac{1}{2} \ell n(b_i) , \tag{8}$$

$$V(B_i) = b_i^2 Q_i (1 - Q_i) , \tag{9}$$

where $a_i = 1/(1 - 2P_i - Q_i)$, $b_i = 1/(1 - 2Q_i)$, and $c_i = (a_i - b_i)/2$. These results are based on Kimura's (1980) two-parameter model. The total number of substitutions per i-fold degenerate site is given by

$$K_i = A_i + B_i , \tag{10}$$

with an approximate sampling variance given by

$$V(K_i) = [a_i^2 P_i + d_i^2 Q_i - (a_i P_i + d_i Q_i)^2]/L_i , \tag{11}$$

where $d_i = b_i + c_i$. We note that A_2 (B_2) denotes the number of synonymous (nonsynonymous) substitutions per two-fold degenerate site, $K_4 = A_4 + B_4$ the number of synonymous substitutions per four-fold degenerate site, and $K_0 = A_0 + B_0$ the number of nonsynonymous substitutions per nondegenerate site.

From the above results we can easily obtain K_S, the number of (synonymous) substitutions per synonymous site, and K_A, the number of (nonsynonymous) substitutions per nonsynonymous site. Following the convention, we count each four-fold degenerate site as a synonymous site, each two-fold degenerate site as one-third synonymous and two-thirds nonsynonymous, and each nondegenerate site as a nonsynonymous site. Then $K_S = (L_2 A_2 + L_4 A_4)/(L_2/3 + L_4)$ and $K_4 = (L_2 B_2 + L_0 K_0)/(2L_2/3 + L_0)$. Approximate formulas for the sampling variances of K_S and K_A can be readily obtained (Li *et al.*, 1985b).

One can also compute the K_S and K_A values using the procedure of Miyata and Yasunaga. This is readily done by putting $M_S = P_2 + P_4 + Q_4$, $M_A = Q_2 + P_0 + Q_0$, $N_S = L_2/3 + L_4$ and $N_A = 2L_2/3 + L_0$ into formulas (4) and (5). This computational procedure has also been included in our computer program.

III. COMPARISON OF METHODS

There are several aspects to be considered. First we examine the assumptions involved in making corrections for multiple substitutions at the same sites. The PCD method assumes that substitution occurs randomly among the four types of nucleotides. This assumption is unrealistic (Gojobori *et al.*, 1982a; Li *et al.*, 1984) and tends to underestimate the number of substitutions, particularly if the degree of sequence divergence is large (Takahata and Kimura, 1981; Gojobori *et al.*, 1982b;   Gojobori, 1983).   The PCD method further assumes that substitutions can occur between only two states at each category 1 site and between only three states at each category 2 site [see formulas (1) and (2) above]. For example, the third position of TTT (Phe) is a potential silent site of category 1, so that TTT is allowed to change to TTC, but not to TTA or TTG, in the estimation of the number of synonymous substitutions. This assumption tends to give underestimates, particularly when the degree of sequence divergence is large. In Miyata and Yasunaga's method, synonymous differences at two-fold and four-fold degenerate sites are considered together and so are nonsynonymous differences at two-fold degenerate and nondegenerate sites. They then use Jukes and Cantor's (1969) method to make corrections for multiple substitutions [see formulas (4) and (5)]. This approach is rather *ad hoc*. Further, it implicitly assumes that nucleotide substitution occurs randomly both at synonymous sites and at nonsynonymous sites. In our model, two-fold degenerate sites, four-fold degenerate sites, and nondegenerate sites are considered separately. Under this classification, corrections for multiple substitutions can be done more rigorously and the error variances can be derived in a rigorous manner. Formuals (6) to (9) are based on Kimura's (1980) two-parameter model. This model allows for the difference between transitional and transversional rates, which is usually the largest deviation from random substitution (Li *et al.*, 1984). Actually, one can use a more elaborate model, e.g., the six-parameter model of Kimura (1981) and Gojobori *et al.* (1982b), instead of the two-parameter model, but the computational procedure becomes complicated and the results may not be easy to interpret (Li *et al.*, 1985b).

Next, we consider the weighting of alternative paths between two codons with more than one nucleotide difference. The PCD method assumes equal probabilities for alternative paths. This assumption tends to give an underestimate for the rate of synonymous substitution and an overestimate for the rate of nonsynonymous substitution because, in reality, synonymous substitution occurs considerably more often than nonsynonymous substitution. Miyata and Yasunaga do assign different weights for different paths. However, their weights are based on extrapolations from the relative frequencies of amino acid changes in protein evolution and tend to be more in favor of nonsynonymous than synonymous substitutions (Li *et al.*, 1985b). By contrast, our weights are based on direct comparisons of DNA sequences and should therefore be more reasonable than those of Miyata and Yasunaga.

I shall use an example to demonstrate the importance of having appropriate weights for alternative paths when considering highly conserved genes. It is known that the protein sequence of glucagon is conserved among all mammalian species sequenced to data (Dayhoff, 1978; Lopez *et al.*, 1984). However, the arginine at the 17th residue is encoded by the codon CGC in hamster (Bell *et al.*, 1983) but by the codon AGG in cow (Lopez *et al.*, 1983). There are two possible paths between these two codons: CGC (Arg) ↔ CGG (Arg) ↔ AGG (Arg) and CGC (Arg) ↔ AGC (Ser) ↔ AGG (Arg). The former path requires two synonymous substitutions while the latter, two nonsynonymous substitutions. Under the PCD method, both paths are equally probable, and we would infer that one nonsynonymous substitution and one synonymous substitution have occurred at this residue position since the divergence between hamster and cow; note that glucagon consists of only 29 amino acids. This inference is unreasonable because in reality no nonsynonymous substitution seems to have occurred. Such unreasonable inferences may also ocur when comparing leucine codons, e.g., TTA and CTT (Li *et al.*, 1985b). In our method we assume that no nonsynonymous substitution has occurred between the synonymous codons for leucine or those for arginine, when comparing highly conserved genes such as histone genes and actin genes (Li *et al.*, 1985b). For histone, actin and glucagon genes, the rate of nonsynonymous substitution estimated by Miyata and Yasunaga's method is more than two times higher than that estimated by our method and the rate estimated by the PCD method is one order of magnitude higher than that estimated by our method.

Third, we consider the information provided by a method. Miyata and Yasunaga's method provides only the number of synonymous substitutions and the number of nonsynonymous substitutions. In the PCD method, nucleotide sites are classified into potential synonymous and nonsynonymous sites and then further into three categories. Thus, it can provide detailed information on the relative rates of substitution at different types of nucleotide sites. In our method, nucleotide sites are classified into four-fold degenerate, two-fold degenerate and nondegenerate sites and substitutions are classified into transitional and transversional. Thus, it can provide information about the rates of substitution at the three different types of sites. Application of this method to 35 mammalian genes has revealed several interesting properties (Li *et al.*, 1985b). One is that at all three types of sites the rate of transitional substitution tends to be higher than the rate of transversional substitution, though at each nucleotide site two types of transversional change and only one type of transitional change can occur. Another interesting observation is that K_4, the number of nucleotide substitutions per four-fold degenerate site, is generally substantially lower than K_S, the number of substitutions per synonymous site, although all changes at four-fold degenerate sites are synonymous. This difference occurs because in computing K_S the transitional changes at two-fold degenerate sites are also included. As mentioned above, only one-third of a two-fold degenerate site is counted as synonymous, but transitional (synonymous) substitution at a two-fold degenerate site occurs at a rate higher than one-third of the total rate of a four-fold degenerate site. For this reason, the synonymous rate obtained by the conventional definition tends to be higher than the rate of nucleotide substitution at four-fold degenerate sites. In inferring the stringency of functional constraint on synonymous mutations, it has been customary to compare the synonymous rate with the substitution rate in pseudogenes, which has been taken as the neutrality standard. Our results indicate that for this purpose one should compare instead the substitution rate at four-fold degenerate sites with the substitution rate in pseudogenes. Thus, our method can provide additional information that cannot be obtained from Miyata and Yasunaga's method.

In estimating the number of nucleotide substitutions, it is desirable to have both the mean and the standard error. The PCD method provides no standard error. In Miyata and Yasunaga's method, the error variances have been obtained in a rather *ad hoc* manner (Miyata *et al.*, 1982). In contrast, in our method, the error variances have been derived in a fairly rigorous manner.

Another aspect to be considered is the range of applicability of a method. We note that all correction formulas involve at least one logarithmic function. When the degree of sequence divergence becomes large, the argument of a logarithmic function may become negative because of sampling errors. If this occurs, no result can be obtained. In the PCD method, nucleotide sites are classified into six classes, so that the number of sites in a class may become small and the percentage change (p) may be subject to large sampling errors. For this reason, the chance of being inapplicable is expected to be higher for the PCD method than for the other two methods. In Miyata and Yasunaga's method nucleotide sites are classified into only two classes, while in our method they are classified into three classes. Therefore, the chance of being inapplicable would be smaller for their method than for ours. However, as mentioned above, our computer program also includes a computational procedure in which nucleotide sites are classified into only two classes.

In summary, from the theoretical point of view our method appears to have several advantages over the other two. In addition, it also includes a computational procedure similar to that of Miyata and Yasunaga so that it can provide two different kinds of results for comparison. The PCD method has some undesirable features and seems to be inferior to the other two. It should, however, be pointed out that when the degree of sequence divergence is relatively small, say K_S and K_A both smaller than 0.5, the three methods are expected to give very similar results (Li *et al.*, 1985a,b). This is because at this stage of divergence corrections for multiple substitutions are small. Therefore, when the degree of sequence divergence is small, any of the three methods can be applied to estimate K_S and K_A.

Gojobori (1985) has conducted a simulation study to compare the performance of the PCD method and that of Miyata and Yasunaga's method. He considered four mutational schemes. In each scheme he assumed that the probability of fixation is 1 for a synonymous mutation but only 0.2 for a nonsynonymous mutation. Figure 1 shows his simulation results. It is seen that in all four schemes both methods give good estimates of the expected K_A values. This is not surprising because the K_A values are smaller than 1 and all nonsynonymous mutations have the same probability of fixation (see below). In Figure 1A,. mutation is random. The PCD method gives a good estimate of the expected K_S value when K_S is smaller than 1.5 but an underestimate when K_S is larger than 1.5. Miyata and Yasunaga's method tends to overestimate the expected K_S value, particularly when K_S is large. It is not clear how this has happened because he did not present the actual numbers of substitutions. In Figure 1B, mutation follows the four-parameter model of Takahata and Kimura (1981), and in Figure 1C, mutation follows the six-parameter model of Kimura (1981). Both the PCD method and Miyata and Yasunaga's method, particularly the former, tend to give severe underestimate of the expected K_S value when K_S is large. In Figure 1D, mutation follows the pattern of nucleotide substitutions observed in pseudogenes (Gojobori *et al.*, 1982a). Miyata and Yasunaga's method appear to perform well under this mutational scheme but the PCD method tends to give a severe underestimate of the expected K_S value when K_S is larger than 1.

Since the substitution pattern observed in pseudogenes would be close to the true pattern of mutation, Gojobori's simulation results suggest that in practice Miyata and Yasunaga's method would perform better than the PCD method. However, it should be pointed out that his simulation study is very limited in scope. First, in his simulation, all nonsynonymous mutations have the same probability of fixation. Obviously, this is not true in reality. Actually, this selection scheme does not follow the assumption of Miyata and Yasunaga. In a recent simulation study in which different evolutionary paths are assigned the weights given by Miyata and Yasunaga, Gojobori (unpublished) found that

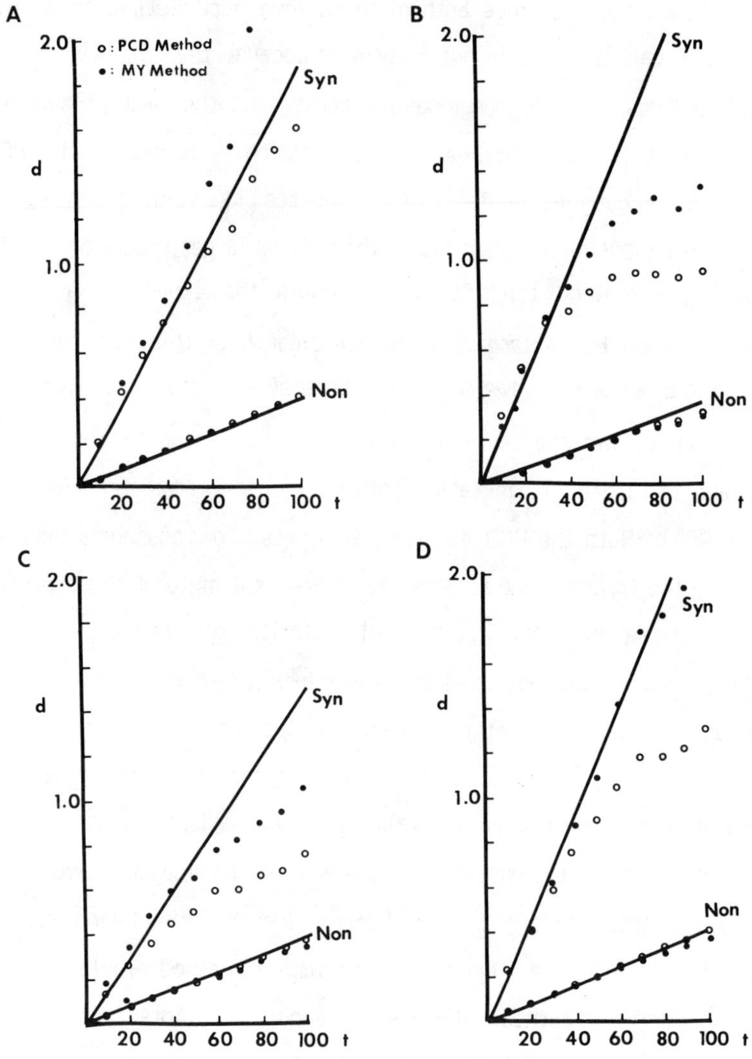

Figure 1. Comparison of the PCD method and Miyata and Yasunaga's method using computer simulation (modified from Gojobori, 1985). The straight lines indicated by Syn and Non represent, respectively, the expected number of substitutions per synonymous site and the expected number of substitutions per nonsynonymous site. The estimates by the PCD method are denoted by circles and those by Miyata and Yasunaga's (MY) method by black dots. The mutation schemes used are (A) random

both Miyata and Yasunaga's method and the PCD method give underestimates of K_A, though the former gives fairly good estimates of the expected K_S values. Second, for each mutation scheme, Gojobori simulated only one initial sequence. Although the sequence is fairly long, 1000 codons, it is not clear how much effect the initial condition may have on the performance of a method. Third, he did not compare the estimated value with the actual number of substitutions. We note that in practice estimating the actual number of substitutions may be more meaningful than estimating the expected number of substitutions, because the latter is unknown.

IV. DISCUSSION

The problem of how to obtain a reliable estimate of the evolutionary distance between sequences is quite simple when the distance is small but becomes increasingly difficult as the distance increases. In fact, the problem becomes very challenging when the number of nucleotide substitutions per site is larger than, say, 1.5. This can be seen from Gojobori's (1985) simulation results cited in Figure 1. It should be emphasized that his simulation was conducted under oversimplified conditions: all synonymous mutations have the same probability of fixation and so do all nonsynonymous mutations. Even under these simple conditions, both the PCD method and Miyata and Yasunaga's method often give biased estimates of K_S when K_S is 1.5 or larger. In the case where the mutation scheme follows the substitution pattern in pseudogenes, Miyata and Yasunaga's method gives good estimates for both K_S and K_A. We note, however, that both the mutation scheme and the selection scheme in this case do not follow Miyata and Yasunaga's assumptions. As mentioned above, when the selection scheme follows their assumption, their method gives underestimates of K_A. Our method is also expected to have similar difficulties as do the other two.

There are at least two difficulties in obtaining a reliable estimate of the evolutionary distance between two distantly related genes. The first one is how to take into account various possibilities of nonrandom substitution among nucleotides. The classification of nucleotide sites into four-fold degenerate, two-fold degenerate and nondegenerate sites allows the incorporation of many possibilities of nonrandom substitution. But Gojobori *et al*.'s (1982b) simulation study has shown that when a more elaborate model is used to correct multiple substitutions at the same site the chance of inapplicability increases. The second difficulty is how to take into account variation in substitution rate over different regions of the gene. We noted above that when the nonsynonymous rate is uniform over codons both the PCD method and Miyata and Yasunaga's method give good estimate of K_A for $K_A \leq 1$ (Figure 1), whereas both methods give underestimates of K_A when different nonsynonymous mutations are subject to different selective constraints (Gojobori, unpublished). All current methods assume that the substitution rate for each type of codon is the same over the whole coding region. This assumption is obviously unrealistic and tends to underestimate the substitution rate. One way to improve the estimation is to partition the coding sequence into different regions according to their degree of sequence conservation. There are, however, two problems. First, we often do not know which parts of the gene are conservative and which parts are not. Second, partitioning reduces the number of sites in a class and thus increases the chance of inapplicability of a formula.

In summary, there remain difficult problems in the estimation of evolutionary distance between genes. Fortunately, data on DNA sequences are accumulating rapidly. This will increase our knowledge of the pattern of evolutionary changes in genes and enable us to construct better models.

REFERENCES

Bell, G. I., Santerre, R. F., and Mullenbach, G. T. (1983). Hamster preproglucagon contains the sequence of glucagon and two related peptides. *Nature* **302**, 716-718.

Dayhoff, M. O. (1978). "Atlas of Protein Sequence and Structure," Vol. 5, supp. 3. National Biomed. Res. Found., Silver Spring, Maryland.

Gojobori, T. (1983). Codon substitution in evolution and the "saturation" of synonymous changes. *Genetics* **105**, 1011-1027.

Gojobori, T. (1985). A mathematical model of codon substitution and the constancy of evolutionary rate. *Proc. XII Internatl. Biometric Conf.* (to appear).

Gojobori, T., Li, W.-H., and Grauer, D. (1982a). Patterns of nucleotide substitution in pseudogenes and functional genes. *J. Mol. Evol.* **18**, 360-369.

Gojobori, T., Ishii, K., and Nei, M. (1982b). Estimation of average number of nucleotide substitutions when the rate of substitution varies with nucleotide. *J. Mol. Evol.* **18**, 414-423.

Holmquist, R. (1972). Theoretical foundations for a quantitative approach to paleogenetics, part 1: DNA. *J. Mol. Evol.* **1**, 115-133.

Jukes, T. H., and Cantor, C. R. (1969). Evolution of protein molecules. *In* "Mammalian Protein Metabolism" (H. N. Munro, ed.), pp. 21-123. Academic Press, New York.

Kimura, M. (1980). A simple method for estimating evolutionary rates of base substitutions through comparative studies of nucleotide sequences. *J. Mol. Evol.* **16**, 111-120.

Kimura, M. (1981). Estimation of evolutionary distances between homologous nucleotide sequences. *Proc. Natl. Acad. Sci. USA* **78**, 454-458.

Li, W.-H., Wu, C.-I., and Luo, C.-C. (1984). Nonrandomness of point mutations as reflected in nucleotide substitutions in pseudogenes and its evolutionary implications. *J. Mol. Evol.* **21**, 58-71.

Li, W.-H., Luo, C.-C., and Wu, C.-I., (1985a). Evolution of DNA sequences. *In* "Molecular Evolutionary Genetics" (R. J. MacIntyre, ed.). Plenum, New York. (in press)

Li, W.-H., Wu, C.-I., and Luo, C.-C. (1985b). A new method for estimating synonymous and nonsynonymous rates of nucleotide substitution considering the relative likelihood of nucleotide and codon changes. *Mol. Biol. Evol.* **2**, 150-174.

Lopez, L. C., Frazier, M. L., Su, C.-J., Kumar, A., and Saunders, G. F. (1983). Mammalian pancreatic preproglucagon contains three-glucagon-related peptides. *Proc. Natl. Acad. Sci. USA* **80**, 5485-5489.

Lopez, L. C., Li, W.-H., Frazier, M. L., Luo, C.-C., and Saunders, G. F. (1984). Evolution of glucagon genes. *Mol. Biol. Evol.* **1**, 335-344.

Miyata, T., and Yasunaga, T. (1980). Molecular evolution of mRNA: a method for estimating evolutionary rates of synonymous and amino acid substitution from homologous nucleotide sequences and its application. *J. Mol. Evol.* **16**, 23-36.

Miyata, T., Hayashida, H., Kikuno, R., Hasegawa, M., Kobayashi, M., and Koike, K. (1982). Molecular clock of silent substitution: At least six-fold preponderance of silent changes in mitochondrial genes over those in nuclear genes. *J. Mol. Evol.* **19**, 28-35.

Perler, R., Efstratiadis, A., Lomedico, P., Gilbert, W., Kolodner, R., and Dodgson, J. (1980). The evolution of genes: the chicken preproinsulin gene. *Cell* **20**, 555-566.

Takahata, N., and Kimura, M. (1981). A model of evolutionary base substitutions and its application with special reference to rapid change of pseudogenes. *Genetics* **98**, 641-657.

A HIDDEN BIAS IN THE ESTIMATE OF TOTAL NUCLEOTIDE SUBSTITUTIONS FROM PAIRWISE DIFFERENCES[1]

Walter M. Fitch

Physiological Chemistry
University of Wisconson - Madison
Madison, Wisconsin 53706

ABSTRACT

A nomographic method is presented that estimates the number of nucleotide substitutions since the common ancestor of two nucleotide sequences with no assumption about the proportion of transition and transversion substitutions except that it is constant over time. Of two previous methods of estimating this number, that of Kimura (1981) obtains the same result, and is thus confirmed by this work, while that of Brown *et al.* (1982) does not get the same result. The method presented here also obtains the fraction of all substitutions that are transitions. If one has three or more homologous sequences to compare, one can test the validity of the model by examining the constancy of the estimated proportion of substitutions that are transitions across the various pairs of sequences in a simple visual way. The method is general for any pair of mutually exclusive nucleotide substitutional categories, not just transitions and transversions. Mitochondrial data provide evidence that, for this and probably other current models correcting for superimposd substitutions, one or more of the underlying assumptions is incorrect. This is because there is some unknown systematic bias affecting this evolutionary process. It is suggested that at least part of the bias arises from incorrectly assuming that all sites are variable. In absence of

[1] This work was supported in part by National Science Foundation grant BSR-8400682.

evidence that this bias is not present in other data, all estimates of the number of substitutions based upon pairs of sequences and current methods of estimating superimposed substitutions at a single site should be viewed as uncertain.

I. INTRODUCTION

It is a well recognized phenomenon that in certain circumstances, some classes of nucleotide substitutions are more frequent than others. For example, in the third coding position, transitions are more frequent than transversions even in codons that are four-fold degenerate (Fitch, 1980). This inequality is expected for the totality of third positions because only transitions are silent (i.e., don't change the encoded amino acid) for the two-fold degenerate codons. This inequality is not necessarily expected for the four-fold degenerate codons unless one invokes a specific mechanisms of mispairing during replication, such as tautomerization, that specifically favors transitions (Topal and Fresco, 1976). What is perhaps less generally appreciated, as Brown *et al.* (1982) have astutely pointed out, is that as two sequences that are not otherwise constrained diverge, their observed *differences* in homologous positions will ultimately tend to show an excess of transversion differences over transition differences, even though most of the *substitutions* (mutations fixed) were transitions. This property of having the two classes of events appear, as time passes, as if all types of substitutions were more equally probable compared to opportunity, even though the underlying substitutional rates are not equal, will apply to any two mutually exclusive types of substitutional events, not just transitions and transversions. Thus most estimates of their underlying relative frequency of occurrence from raw observational counts of the types of differences is systematically biased to underestimate that frequency as well as to underestimate the total number of substitutions. It is the purpose of this paper to provide a simple nomographic procedure to correct this bias. In this respect it accords with the purposes of work by Kimura (1981) and Brown *et al.* (1982). This work gives results in agreement with those of Kimura's formula but not those of Brown *et al.*

In addition, however, the present work estimates the fraction of substitutions that are transitions. This fraction, in contrast to the reasonable expectation that it has some average rate over evolutionary time, is shown to have smaller and smaller estimates as evolutionary divergence increases, at least in mitochondrial DNA. This shows that a systematic bias remains in the data that is unaccounted for by any commonly used procedure that is intended to estimate rates of nucleotide substitutions on the basis of examining extant nucleotide sequences.

II. METHODS

One wishes an estimate of the total number of nucleotide substitutions and the fraction of them that are transitions from an observation of the number of transition and transversion differences between two sequences. The method used here is different in its derivation than any previously given, but the result gives values identical to those of Kimura (1981), thus confirming both methods given the assumptions.

The following analysis assumes that one has compared two homologous sequences and observed d differences per site, of which d_v are transversions/site and d_s are transitions/site. Hence $d = d_v + d_s$. The task is to obtain r, the total rate of substitution per site and v and

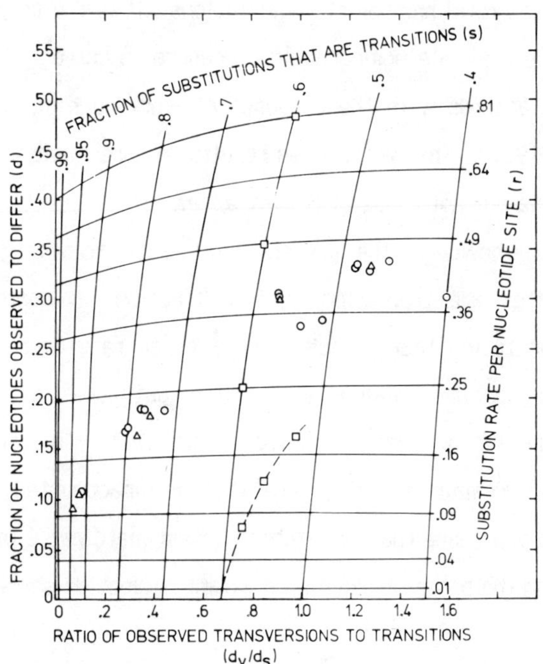

Figure 1. Real and hypothetical data plotted on the nomogram. The ordinate and abscissa are the variable d and d_S/d_S, directly observable in the data. These two values determine a point on the graph that relates to the lines that represent a second set of variables, the rate of substitution per nucleotide site (r), and the fraction of those substitutions that are transitions (s), and whose values may be estimated by interpolation. The squares represent points showing the effect of some sites being invariable (see text). The other 21 points are for pairs of taxa based on 896 nuclotides in mitochondrial DNA as provided by E. Prager. They appear from the lower left to the upper right in clusters of increasing size from 1 to 6. Each group includes one point, represented by a Δ, that is for the human and the taxon which is common to comparisons in that group. These taxa are, respectively, chimpanzee, gorilla, orangutan, gibbon, mouse and cow. A group of n points is a comparison of the nth taxon in this sequence with all the taxa to its left plus humans.

s, the fraction of r that are transversions and transitions respectively. Hence $v+s = 1$. The computer program was run using values of $r = x^2$ where x ranges from 0.1 to 2.0 in increments of 0.1 (x is a dummy variable to get r to increase exponentially rather than linearly). Values of s ranged from 0.20 to 1.00 in increments of 0.05 plus values of $0.\overline{3}$ and 0.99. The two pieces of information that one would normally possess that seem most useful in examining data nomographically are the fraction of sites, d, that are different, and the ratio of observed transversions to transitions, d_v/d_s. These constitute the axes of Figure 1. Values of d and d_v/d_s were computed for the values of r and s noted, plotted on the figure, and lines drawn to connect points of constant r (horizontal curves) and constant s (vertical curves).

Given any pair of observed values of d and d_v/d_s, one finds the point on the figure and interpolates to get the values of r and s that would best account for the data. If, for example, $d = 0.10$ and $d_v/d_s = 1.09$, then the best estimates of the underlying process are that there have been 0.11 substitutions per nucleotide (r) and 0.47 of all substitutions were transition(s).

III. RESULTS

The method was applied to mitochondrial DNA data kindly supplied by Dr. Ellen Prager, and the results are shown in Fig. 1 for all 21 pairwise comparisons.

The use of Figure 1, or any of the other methods currently in use, could lead to errors if not all of the sites were free to change. If, for example, only one out of three sites were free to vary according to the model, the observed ratio, d_v/d_s, would remain unchanged but d would be estimated at only one-third its appropriate value. An example is shown in Figure 1 for a series of three distances for which the fraction of all substitutions in the variable sites is a constant 0.6 (the three squares on that line). If, however, there were twice as many invariable sites as variable sites, d would be proportionately reduced to the set of three squares shown on the dashed line. If one were to assume that the fraction of substitutions that are transitions is relatively constant over time, then the failure of the points all to lie on a single line of that s value is evidence for the existence of invariable sites and the factor by which the d values would need to be multiplied to make them fall on that line, could be used to estimate the fraction of invariable nucleotides.

IV. DISCUSSION

A. General Properties

All curves of r and s are concave upward and to the left except for the strictly vertical line at $d_s = 0.\overline{3}$ ($d_v/d_s = 2.0$) and the x-axis at $d = 0$. Hence all values of d are underestimates of r except at $d = 0$ (at time 0!) and all observed fractions of differences that are transitions underestimate s except when $d_v/d_s \geq 2.0$. The latter tendency was noted by Brown *et al.* (1982). Moreover, all curves converge to $d = .75$, $d_v/d_s = 2$ as noted by Kimura (1981). That is, with the occurrence of a sufficient number of substitutions per site, two sequences will come to differ in three quarters of their sites and the ratio of transversion to transition differences will approach 2.00 (i.e., they look like otherwise random sequences) regardless of the actual rate of transition substitutions relative to transversion substitutions. More complicated base substitutional schemes need not lead to a ratio of two however (Holmquist, 1982).

B. Comparisons with Other Methods

Table I compares the results of this method with those of Kimura (1981) and of Brown *et al.* (1982). The method given here agrees exactly with those given by Kimura except for ±1 in the second significant digit which represents the limits of interpolation on a larger, complete version of Figure 1. Since these two methods were derived independently and by rather dissimilar approaches (continuous vs. discrete), their agreement verifies the mathematical correctness of each.

The method of Brown *et al.* (1982) gives results that do not agree with those of Kimura nor with mine. I can see no difference in our stated assumptions that would account for this discrepancy, but, on the contrary, believe all three parties are attempting to make the same correction.

Table I. Substitutions/Nucleotide by Three methods

	Brown *et al.*	Kimura	Fitch	s
Chimp (*Pan trogladytes*)	.09	.10	.10	.94
Gorilla (*Gorilla gorilla*)	.11	.11	.11	.92
Orang (*Pongo pygameus*)	.25	.18	.19	.78
Gibbon (*Hylobates lar*)	.31	.21	.21	.75
Cow (not given)	.88	.38	.39	.57
Mouse (not given)	1.24	.44	.44	.44

Legend. Substitutions per nucleotide are for each of the taxa on the left compared to human. Sequences are the 896 nucleotides studied by Brown *et al.* (1982) and include the C-terminus of URF4, three t-RNA's, and the N-terminus of URF5, all from mitochondrial DNA. The last column is the fraction of the substitutions calculated to be transitions. The values relating to cow and mouse were based on information kindly supplied by E. Prager (personal communication), and who calculated the values according to the method of Brown *et al.* (1982). The column labeled Kimura contains values obtained by his method (1981). The orangutan, cow and mouse have, respectively, 1 less, 2 less and 3 extra nucleotides, compared to the others. These data appear in Figure 1 as the points marked by a Δ .

What I believe to be the difference is in the attempt to estimate transition rates on the basis of transition differences independently of the transversions. The Jukes and Cantor formulation (1969), $\Lambda = -(3/4)\ln(1-4\lambda/3)$, can easily be seen to be divisible into two parts where λ_i are for different sites an the weighted average for the different Λ_i are used to give an overall Λ for all sites combined. It is not at all clear that it is legitimate to partition the transition and transversion processes at a single site (Λ_i and Λ_v), as Brown *et al.* do, to get an overall Λ for both processes combined. This doubt is applicable as well to the formulations of Perler *et al.* (1980) although the magnitude of the error may be less in their case to the extent that four-fold and two-fold degenerate codon positions persist in the degree of their degeneracy.

As the taxa become more distantly related, the disparity between Brown *et al.*'s Λ_i and Λ_v becomes greater, and so the method of averaging these two values could also contribute to the difference, especially since they seem to bracket the values obtained by Kimura and by this method. Their averaging method is shown by their equation (6) which is more complicated than necessary since it readily simplifies to

$$= [(1 + 2f_i)\Lambda_i + (1 + 2f_v)\Lambda_v]/4$$

where f_i and f_v are the fraction of the differences that are transitions and transversions, respectively (or d_s and d_v in this work). This is a peculiar weighting because in the case of recently diverged species such as human-chimpanzee with f_i close to one, it weights Λ_i nearly three times as much as Λ_v, while in the limit where $f_v/f_i = 2$, it would weight Λ_i at only 5/7 of Λ_v. Moreover, if one computes Λ_i and Λ_v for the human-cow sequences (0.21 and 1.50, respectively), one would believe, in complete contrariety to the human-chimp data, that the rate of transversion substitutions was more than 7.5 times greater than that for transitions. If this were in fact true, why should one observe more transition than transversion differences? Under the circumstances, it seems that Λ_i and Λ_v are not properly conceived and hence there is no weighting scheme for Λ_i and Λ_4 that in general is likely to yield the correct Λ.

C. Invariable Nucleotide Positions

Fitch and Margoliash (1967) demonstrated that nucleotide substitutions producing amino acid replacements in structural genes distribute themselves as if a portion of the codons were invariable with respect to such changes. But if that is true for codons, it must be true for nucleotides as well. How would one observe such an effect?

As shown in the results with the squares in Figure 1, the plot of the data do not lie on a line for a single value of s but give successively smaller values of s as d increases. Knowing the problem, the correction is easy. One simply determines by what number the d values of the data need be multiplied to cause them to lie on the same s line. This does, of course, presuppose that the fraction of all substitutions that are transitions, s, be relatively constant over time.

What do real data show? The data for the 896 mitochondrial nucleotides studied by Brown *et al.* are plotted in Figure 1 along with values for the cow and mouse comparisons to the primates. Clearly, they do not fall on a single s value, for which there are two readily available explanations: (1) the fraction of substitutions that are transitions is quite variable; (2) the assumptions employed in the model are not sufficiently comprehensive to reflect accurately the evolutionary processes in this part of the mitochondrial genome.

With respect to explanation 1, the proposed variablity may well be true but is not an adequate explanation because the data do not simply fall randomly about some median s value. Rather, they fall in a very well-defined band that says there is some undefined underlying systematic bias that causes more distantly related sequences to appear to have had a lower fraction of transition substitutions in their past. This is true even after deliberately correcting for the problem recognized by Brown *et al.* (1982). It is difficult to believe that the bias is a change from predominantly transversion substitutions 100 million years ago to an overwhelming dominance of transitions in recent primate lineages. Explanation 2 therefore appears the more reasonable.

The implications of these results are considerable as they imply the models upon which substitution rates have been estimated from pairwise sequence differences may be inadequate for that purpose. Of course, these results may be peculiar to this segment of mitochondrial DNA, but there is as yet no evidence that the bias present here does not occur generally and it certainly has not been ruled out where these methods have been applied. Given that caveat with respect to generality, we may make several inferences.

The first inference is that, since none of the data points fall on the line $d_v/d_s = 2.0$, no method that ignores differences in transition and transversion rates, i.e., that assumes the ratio is indeed 2.0, can be expected to estimate accurately the number of nucleotide substitutions separating two genes. This includes the methods of Jukes and Cantor (1979), Kimura and Ohta (1972) and Perler *et al.* (1980), among others.

The second inference is that, since none of the data points fall on any one line, even approximately, no method that ignores the problem of invariable sites (or, more generally, of sites with different rates) can be expected to estimate accurately the number of nucleotide substitutions separating two genes. This includes the three methods of the previous paragraph plus the methods of Kimura (1981), Brown *et al.* (1982) and this paper, among others.

But we have shown at the beginning of this section how to allow for a fixed fraction of the sites being invariable. An important aspect of the Brown *et al.* (1982) data plus the additional data supplied by Prager is that no adjustment based upon d being underestimated will bring the mitochondrial data onto, or respectably near, a single s value line. Thus the third inference is that since no adjustment of d can cause the data points to fall on any one line, even approximately, no method currently available can be expected to estimate accurately the number of nucleotide substitutions separating two genes. These include all of the preceding methods plus the methods of Holmquist and Pearl (1980) and this paper.

D. Covariotides

After Fitch and Margoliash (1967) discovered that amino acid replacements in cytochrome c distributed themselves as if some positions might be invariable, Fitch and Markowitz (1970) then demonstrated that the estimates of the number of invariable positions increased as the range of taxa narrowed, suggesting that a considerable number of the positions may be invariable in any one taxon but that as replacements are fixed, those positions change. This leads naturally to the concept of <u>co</u>comitantly <u>vari</u>able cod<u>ons</u>, or covarions for short. Fitch (1971) showed that only about two-thirds of the 60 plus invariable amino acid positions of cytochrome c estimated for fungi and metazoans could be common to both groups. This lent further support to the covarion concept. The interpretation of those results in terms of numbers of invariable positions may be overly narrow in that all that is really necessary is that there be significantly different rates of replacements at different sites. Nevertheless, the concept that where the next nucleotide substitution or the next amino acid replacement may be fixed must be a function of what changes have preceded it can hardly be denied in an evolutionary framework and its applicability to nucleic acid data is surely as reasonable. Hence, covariotides or concomitantly variable nucleotides may be considered as a component to the perplexing problem of the location of the mitochondrial DNA points in Fig. 1.

The relatively horizontal march of the mitochondrial data across Figure 1 would be expected under the following somewhat forced model. Let some fraction of the variable sites permit only transition substitutions and be able to evolve rapidly and thus be sooner saturated with changes. Let another fraction of the variable sites, considerably larger than the former constitute the remaining variable sites which evolve much more slowly but will permit both transitions and transversions. These two fractions might be exemplified by the third codon position of two-fold degenerate amino acids and the first and second codon position of all amino acids, respectively. Since about two-thirds of the mitochondrial data examined here are coding sequences, this model is at least possible and may account for the large proportion of transition differences in recently diverged sequences. It can certainly be tested and suggests that the cautionary remarks made about all current methods may prove to be unnecessary for non-coding sequences. Still, non-coding sequences may possess other problems to plague us and caution about the robustness of our methods is seldom out of place for, while inconsistencies may demonstrate the invalidity of our assumptions, consistency cannot validate them because you cannot prove a null hypothesis, only reject it.

ACKNOWLEGMENT

Thanks are expressed to Ron Niece, Ellen Prager and Allan Wilson for their assistance.

REFERENCES

Brown, W. M., Prager, E. M., Wang, A., and Wilson, A. C. (1982). Mitochondrial DNA sequences of primates: Tempo and mode of evolution. *J. Mol. Evol.* **18**, 225-239.

Cantor, C. R., and Jukes, T. H. (1969). Evolution of protein molecules. *In* "Mammalian Protein Metabolism," Vol. III (H. N. Munro, ed.), pp. 21-132. Academic Press.

Fitch, W. M. (1971). The non-identity of invariant positions in the cytochrome c of different species. *Biochem. Gen.* **5**, 231-241.

Fitch, W. M. (1980). Estimating the total number of nucleotide substitutions since the common ancestor of a pair of homologous genes: Comparison of several methods and three beta hemoglobin messenger RNA's. *J. Mol. Evol.* **16**, 153-209.

Fitch, W. M. (1981). The Old REH theory remains unsatisfactory and the new REH theory is problematical. *J. Mol. Evol.* **18**, 60-67.

Fitch, W. M., and Margoliash, E. (1967). A method for estimating the number of invariant amino acid coding positions in a gene using cytochrome c as a model case. *Biochem. Gen.* **1**, 65-71.

Fitch, W. M., and Markowitz, E. (1970). An improved method for determining codon variability in a gene and its application to the rate of fixations of mutations in evolution. *Biochem. Gen.* **4**, 579-593.

Holmquist, R. (1982). Transitions and transversions in evolutionary descent. *J. Mol. Evol.* (in press).

Holmquist, R., and Pearl, D. (1980). Theoretical foundations for quantitative paleogenetics. Part III: The molecular divergence of nucleic acids and proteins for the case of genetic events of unequal probability. *J. Mol. Evol.* **16**, 211-267.

Kimura, M. (1981). Estimation of evolutionary distances between homologous nucleotide sequences. *Proc. Natl. Acad. Sci. USA* **78**, 454-458.

Kimura, M., and Ohta, T. (1972). On the stochastic model for estimation of mutational distance between homologous proteins. *J. Mol. Evol.* **2**, 87-90.

Perler, F., Efstratiadis, A., Lomedico, P., Gilbert, W., Kolodner, R., and Dodgson, J. (1980). The evolution of genes: The chicken preproinsulin gene. *Cell* **20**, 555-566.
Topal, M. D., and Fresco, J. R. (1976). Complementary base pairing and the origin of substitution mutations. *Nature* **263**, 285-289.

COMPARATIVE ANALYSIS OF STRUCTURAL RELATIONSHIPS IN DNA AND PROTEIN SEQUENCES[1]

Samuel Karlin

Department of Mathematics
Stanford University
Stanford, California 94305

ABSTRACT

Various new analytic and statistical concepts and methods are employed to help assess and interpret DNA sequence structural repeats. Applications are given to mitochondrial genomes, globin and immunoglobulin genes and various virus sequences. In particular, we found that T7 and λ-phage select against palindromes of all lengths whereas various mammalian viruses (e.g., Papilloma, Epstein Barr (EBV)) exhibit significantly long palindromes. A virtually unprecedented potential hairpin loop formation is identified in the 3 Kb repeat of EBV that can pair for stem length 208 base pairs of predominantly G+C stacking. The strongest similarities in the globin and immunoglobulin genes emphasize splice junctions. The most conserved mitochondrial gene segments reside in the 16S and 12S rRNA and tRNA clusters. The Xenopus and mouse compared to human and bovine mitochondrial genome exhibit significantly more highly repeated oligonucleotides of 6-8 bp length and only the Xenopus genome contains significant long repeats. The high frequency repeats in the immunoglobulins highlight the consensus nonamer flanking the V and J gene segments. A number of comparisons across genomic sequences using multiple alphabets (alternative DNA and amino acid classifications) is also given.

[1] Supported in part by NIH Grants GM10452-22 and HL30856-02, and NSF Grant MCS 82-15131.

I. INTRODUCTION

The problem of identifying and classifying patterns and relationships within and between nucleic acid and protein sequences is of importance. Several categories of statistics and data representations are proposed to help ascertain and interpret DNA sequence similarities and structures. These include: (i) occurrence distributions of structural repeats; (ii) high frequency repeats, repeat clusters, iterations of various letter types; (iii) special functionals (e.g., the longest repeat, dyad symmetries of specified minimal stem length and constrained loop length). The concepts and statistics are defined and applied in Sections II-V. Means for assessing statistical significance based on theoretical models and permutation procedures are described in Section VI. The examples considered, unless stated explicitly otherwise, involve DNA sequences from the Genbank data base. The interpretations of the analysis relate to issues of evolutionary processes and to problems intrinsic to comparing DNA sequences (see also Karlin and Ghandour, 1985a,b).

II. SEQUENCE CONCEPTS AND REPRESENTATIONS

Consider a sequence S of *letters* drawn from the alphabet $A = (A_1, A_2, \ldots, A_m)$ (e.g., m = 4 for DNA; m = 20 in the amino acid alphabet). A k-*word* is a set of k consecutive letters of the sequence.

A. Repeat-Count Distribution

Relative to S for each given word length k we define $f_k(\nu) = n_\nu$ to be the number of k words repeated ν times, $\nu = 0,1,2,\ldots$. The number of distinct k-words is $n^* = \Sigma_{\nu \geq 1} f_k(\nu)$. It is clear that $\Sigma_{\nu \geq 1} \nu f_k(\nu)$ is precisely the number of k-words in the sequence.

In Table I we present the count occurrence distributions of 6- and 8-words for the three papovavirus genomes, SV40 (N = 5243 base pairs [bp]), Polyoma (N = 5293), and BKV (N = 5153, Dunlop strain).

Polyoma is distinguished from both SV40 and BKV by the lower number of highly repeated 6-words for every repeat level (see also Fig. 1). These observations suggest a greater degree of duplication or homogenization events or a lesser mutation rate during the evolution of SV40 (and BKV) as compared to that of the Polyoma virus.

Table I. Repeat-occurrence distribution of oligonucleotides (words) of length 6 and 8.

ν	$f_6(\nu)$														
	1	2	3	4	5	6	7	8	9	10	11	12	13	14	18
SV40	898	627	361	175	87	57	30	14	7	1	0	0	2	0	1
Polyoma	1174	740	421	198	78	22	5	2	1	0	0	0	0	0	0
BKV-Dun	943	627	363	183	90	47	29	11	7	5	1	3	1	2	0

ν	$f_8(\nu)$					
	1	2	3	4	5	6
SV40	4198	434	46	7	1	0
Polyoma	4710	269	11	3	0	0
BKV-Dun	4158	383	57	12	2	1

Legend. The entries indicate the number of distinct k-words that occurred exactly ν-times for the specified sequence.

The papovaviruses' repeat occurrence distributions show an excess of repeated words when compared to the distributions of the same sequences with the letters randomly permuted (e.g., see Figure 2). In contrast, the amino acid sequences of the VP1 and VP2 capsid proteins of polyoma (data not shown) have 2-word (diamino acids) counts with significantly *fewer* repeated words than for corresponding random sequences. Thus, in the VP1 gene more unique diamino acids are selected for, possibly suggesting that

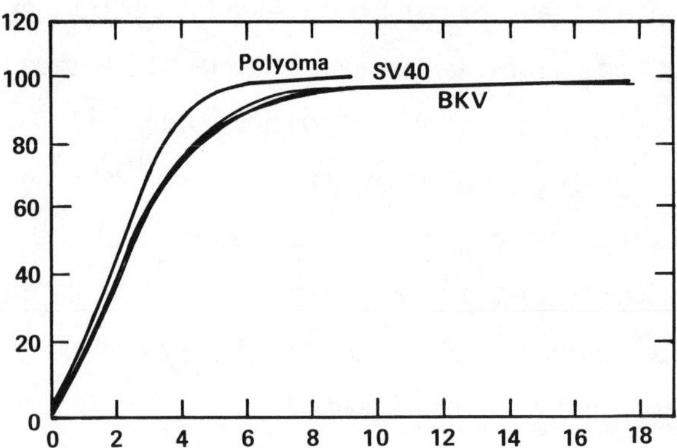

Figure 1. Cumulative count repeat distribution of words of length 6.

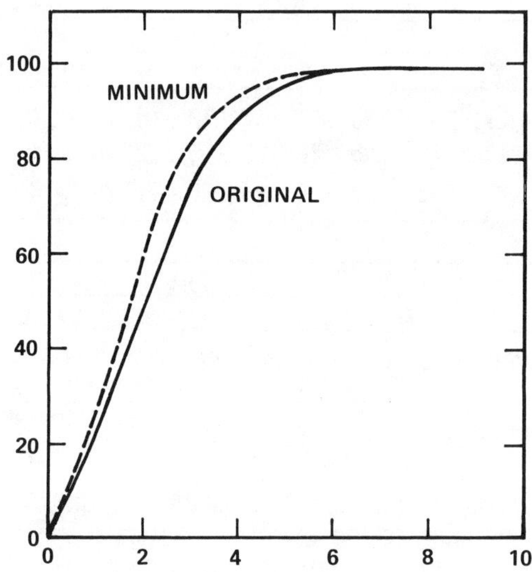

Figure 2. Polyoma count repeat distribution of words of length 6.

Frequency	1	2	3	4	5	6	7	8	9
orig.	.22	.50	.74	.89	.96	.98	.99	.99	1.00
min. of all permutation plots	.26	.60	.84	.94	.98	.992	.996	.998	1.00

Legend for Figures 1 and 2. The ordinate x is the percentage of words of the prescribed length that occur in $\leq x$ repeats. For each permutation there is a corresponding plot $F_1(x), F_2(x), \ldots, F_{10}(x)$. The dotted plot is $F^*(x) = \min_{1 \leq i \leq 10} F_i(x)$.

assembly of the icosohedral nucleocapsid is facilitated by providing more distinct components for each face.

The greater number and extent of long repeats for SV-40 and BKV compared to Polyoma is underscored by the tabulation (Table II) of non-embeddable repeats of length k (a repeat is said to be non-embeddable if the corresponding matching oligonucleotides cannot be extended to a longer matching length).

Can this be a reflection of a greater mutation rate, that disrupts longer repeats, in the mouse species and its viral inhabitants than in the primates (e.g., Fitch and Atchley, 1985; Wu and Li, 1985)? Or are SV40 and BKV of more recent origin still preserving more nascent repeat elements? Or have they had more short duplications?

Table II. Counts of non-embeddable repeats $\geq$ 12 bp length and $\geq$ 10 bp length for the papovaviruses.

Sequence	Number of nonembeddable repeats of length	
	$\geq$ 12	$\geq$ 10
SV40	6	38
BKV-Dun	8	30
Polyoma	1	13

B. High Frequency (hf) Repeats

A k-word in S is identified as high frequency if it shows exceptionally high occurrence among all k-words. We construct a "random" sequence corresponding to S and determine $\nu_0(k)$ such that a k-word repeated at least ν_0 times occurs by chance (for the random sequence) with probability $\leq$ .01. The hf words are defined to be those repeated at least ν_0 times.

Table III. Oligonucleotides with high representation* in human, mouse and rabbit Ig-kappa gene (J-C region) sequences.

Length 6: TTTTGT (9,11,6); AAATAA (11,8,9); TTTTTC (8,7,8); AATAAA (9,9,6); TTATTT (6,6,8)
Length 7: GTTTTTG (4,4,3); TTTTTGT (5,5,3)
Length 8: GTTTTTGT (4,4,3)
Length 9: AGGTTTTG (2,2,2); GGTTTTTGT (2,3,2)

*Criteria are ≥ 6, ≥ 3, ≥ 3, ≥ 2 in each sequence for length 6, 7, 8 and 9, respectively. Probability of hf in a corresponding random sequence is < .01, see text.

Legend. The array (m_1, m_2, m_3) indicates the number of occurrences (m_α) in the α-th sequence of the specified oligonucleotide. $(1,2,3)$ = (human, mouse, rabbit)

Table IV. High frequency oligonucleotides with skewed representations relative to the human, mouse and rabbit Ig-kappa gene sequences.

Length 5:   AAAAA (44,9,24); CAGGG (12,5,18); CAGTT (6,18,10); TATAA (5,10,19)
Length 6:   TTTAAA (12,3,10); TAAAAAT (2,9,9); AAAAAA (25,0,8); TAAAAA (9,2,6); ATTTTA (11,3,11); AGGGAG (6,1,10); GGGAAA (11,2,6); AAGATT (10,3,5); TAATTT (3,7,10); TATTTT (8,3,9)
Length 7:   AAAAAAA (15,0,3); AAATAAA (5,6,2); TATTTTA (4,1,7); AATAAAA (2,7,2); TTTTAAA (6,0,5)
Length 8:   AACGTAAG (4,4,1)
Length 9:   GTTTTGTA (2,4,1); ACCAAGCTG (1,3,1); CCAAGCTGG (1,3,1); CAAGCTGGA (1,3,1); TCAAACGTA (3,1,1); CAAACGTAA (3,1,1); AAACGTAAG (4,4,1); AACGTAAGT (4,4,1); ACGTAAGTA (2,4,1); GAAATAAAA (1,4,0); AGAGCTTCA (1,1,3)
Length 10: GGTTTTTGTA (2,3,1); GGAAATAAAA (1,3,0)
Length 11: AAACGTAAGTA (2,4,1); ACCAAGCTGGA (1,3,1)

Legend. The array (m_1, m_3, m_3) indicates the number of occurrences $(m_\alpha$ in the α-th sequence) of the specified oligonucleotide in the three sequences. An oligonucleotide of length k is labeled of high frequency (hf) provided $m_1 + m_2 + m_3$ exceeds an appropriate level L_k . In the case at hand, the requisite levels are $L_5 = 20$, $L_6 = 15$, $L_7 = 10$, $L_8 = 9$, $L_9 = 7$, $L_{10} = 6$, $L_{11} = 5$. A hf oligonucleotide is described as unevenly represented if the ratio of occurrences is at least 3:1 for some of the three sequences.

1. Examples of hf Repeats

(i) Ig-kappa genes: High frequency repeats of length 6, 7, 8 and 9 bp in the human (N = 5062 bp), mouse (N = 5495) and rabbit (N = 5235) immunoglobulin kappa gene (J-C region) with high representation in all three sequences are presented in Table III. It is perhaps telling that the common form of the polyadenalyation signal (AATAAA) occurs in aggregate 24 times suggesting that this 6-mer alone cannot be a decisive signal for transcription termination. However, in the Ig-kappa gene about 170 bp 3' to the constant C-domain the 10 bp oligonucleotide AATAAAGTGA is conserved (this is statistically significant) which extends to a 20 bp identity in all three Ig-sequences with a single mismatch in the middle of the mouse sequence (Karlin *et al.*, 1985). Parenthetically in the α-globin gene comparisons for human, mouse, goat, and rabbit the polyadenylation signal is embedded in a 13 bp oligonucleotide common to the mammalian species in approximate alignment 3' to the gene. Duck and chicken agree on a stretch of 28 bp containing the hexamer AATAAA. These facts suggest that there is much more to the polyadenylation signal than the canonical 6-mer.

All hf repeats of length ≥ 7 relate to the "consensus" nonamer GGTTTTTGT suggested as a control site in V-J rearrangement (e.g., Hieter *et al.*, 1982; Max *et al.*, 1979).

Table IV lists the oligonucleotides whose cumulative occurrence is significantly high in the aggregate 16,000 bp Ig-κ-gene DNA sequences, but is unequally represented (favoring by a factor of 3 one or two of the sequences compared to another) in the Ig-kappa gene human, mouse and rabbit sequences. Observe that the 10-mer GGTTTTTGTA (embracing the concensus nonamer) is most prominent in mouse. The hf oligonucleotides of lengths ≥ 9 bp over the aggregate human, mouse and rabbit sequences mostly occur in the mouse sequence and generally overlap the J-gene segments. By contrast, for oligonucleotides of lengths 5-7 bp the human and rabbit sequences show the most cases among the hf repeats. Most hf oligonucleotides of lengths 5-7 bp are A+T-rich and can be produced by iterations of a particular nucleotide involving a single point mutation with subsequent tandem duplications.

(ii) Hf Repeats in Mitochondrial Genomes: Table V records the hf repeats in the four complete mitochondrial genomes (human 16569 bp, bovine 16338, mouse 16295, Xenopus 17550). Generally, mouse and Xenopus tend to show more hf repeats compared to human and bovine.

Short and long multiple repeats tend to differ in both their content and distribution. Short ($k = 4,5,6$) hf repeats often reflect iterations of one or two base elements coupled to at most a single point mutation. Longer repeats ($k \geq 9$) tend to be G+C rich possibly providing some regulatory purposes (data not shown). Both short and long repeats tend to occur in noncoding regions.

Mechanisms for generating short versus long word repeats are probably many and varied. Short clustered repeats often result from the actions of DNA repair enzymes and processes of local gene conversion which may be associated with mutational biases. Unequal crossing-over

Table V. High Frequency Repeats in Human, Mouse, Bovine and Xenopus Mitochondrial Genomes.

		hf repeat number			
k	ν_0	Human	Mouse	Bovine	Xenopus
	20	42	63	31	59
6	30	3	8	1	10
	40	0	1	0	0
	5	65	88	48	93
8	6	18	32	13	48
	7	2	11	2	20
10	3	10	12	12	25
	4	2	1	0	1

Legend. The entries give the number of k-word repeats $\geq \nu_0$ times.

may require alignment of similar segments of sufficient length (a short stretch of homology would usually not suffice to establish a stable alignment to mediate recombination). Longer repeats often accompany transposition operations. Transposon termini and sites of genomic insertions are associated with repeat structures (direct and inverted) of longer lengths. Other events leading to longer repeats may relate to RNA reverse transcription, multiple re-replications coupled to recombinational events, and gene amplifications which often arise during the replication cycle under stress conditions. Clustered long ($\geq$ 9 bp length) word repeats may serve as positive or negative regulatory binding sites that convert on/off (binary) activation of a single copy into a progressive control.

III. HOMOLOGIES AND ASSOCIATED FUNCTIONALS FOR MULTIPLE SEQUENCES

For s sequences, $\{S_1, S_2, ..., S_s\}$, of lengths $\{N_1, N_2, ..., N_s\}$, respectively all from the alphabet A, an r-of-s homology (common word = block identity) of length k is a k-word appearing, at least once, in each of at least r out of the s sequences. The number of representatives of a specific k-word across the various sequences is described by the vector $v = \{v_1, v_2, ..., v_s\}$, where v_i is the number of times the k-word appears in sequences S_i.

A. Common Word Occurrence Distribution

We define $f_k(v_1, v_2, ..., v_s) = f_k(v) = n_v$ as the number of k-words that are represented across the sequences according to the vector v. We illustrate with an example of the bivariate distribution of common 8-words for human and mouse β-globin DNA sequences (Table VI).

Table VI. Bivariate distribution of DNA 8-word occurrences in human and mouse β-globin genes.

no. of occurrences	human β-globin (N = 2165)				
	0	1	2	3	4
mouse 0	--	1998	37	5	2
β-globin 1	1464	36	8	0	0
(N=1567) 2	13	9	0	0	0
3	2	0	0	0	0

Legend. The entry n_{ij} in the (i,j) cell is the number of 8-words that appear exactly i times in the mouse β-globin sequence **and** j times in the human β-globin sequence.

A permutation analysis (see Section VI) establishes that the unique common 8-word (36 occurrences) are statistically significant. 80% of the common words overlap the exons. The asymmetry in favor of multiple repeat words in the human β-globin sequence can be explained by extra iterations on T and many short tandem duplications in the second intron.

The utility of the concept of the count distribution is enhanced by its application to other alphabets. We present in Tables VII and VIII the pairwise and three-dimensional distribution, respectively corresponding to all common 2-words in the *amino acid* alphabet for the constant (C) domain of the human (106 residues), mouse (106), and rabbit b4-allotype (104) Ig-kappa genes.

Observe from Table VIII that the number of shared unique words exclusive to human and mouse (17) is substantially more than shared by the other species pairings. The number of shared diamino acids (11) common to all three species is significantly large. The rabbit is more heterogeneous (has fewer repeats) over the Ig-C-domain compared to human and mouse (Karlin *et al.*, 1985). This contrasts sharply with the case of the globin genes where human and rabbit are more similar than either is to mouse. The observation that, in both the Ig and globin flanking region (for the globin genes also in the coding domains), the human and rabbit sequences have greater similarity to one another than either has to the mouse sequence, may indicate that the mouse lineage separated earlier from the human and rabbit branch.

Examination of Tables VII and VIII reveal that human and mouse share significantly more amino acid doublets than human with rabbit and mouse with rabbit. This fact is underscored in comparing the number of common 4-word amino acid quadruplets between the species in the chemical alphabet (Table IX)*. In this alphabet there are 33 singly (occur once in each) common 4-words between human and mouse compared to 19 for human and rabbit, and only 12 for rabbit and mouse.

Table VII. Common word count distribution in the amino acid alphabet of 2-words (diamino acids) of the human, mouse and rabbit Ig-kappa C-domain.

Human Ig-κ-C

		0	1	2	3	> 4
	0	268	40	2	3	0
	1	40	31	5	1	0
Mouse	2	3	1	3	0	0
Ig-κ-C	3	1	2	1	0	0
	> 4	0	0	0	0	0

Human Ig-κ-C

		0	1	2	3	> 4
	0	267	50	5	0	0
	1	38	18	2	3	0
Rabbit	2	5	3	2	0	0
Ig-κ-C	3	2	4	0	0	0
	> 4	0	0	0	1	0

Mouse Ig-κ-C

		0	1	2	3	> 4
	0	266	50	5	1	0
	1	36	24	1	0	0
Rabbit	2	4	2	1	3	0
Ig-κ-C	3	4	1	1	0	0
	> 4	1	0	0	0	0

* An eight-letter chemical alphabet, Mahler and Chordes (1980) entails the following classification of the 20 standard amino acids: Acidic (Asp, Glu); Aliphatic (Ala, Gly, Ile, Leu, Val); Amide (Asn, Gln); Aromatic (Phe, Trp, Tyr); Basic (Arg, His, Lys); Hydroxyl (Ser, Thr); Imino (Pro); Sulfur (Cys, Met).

The three way count distribution of 2-words (amino acid doublets) of the Ig-kappa C-domain is displayed next.

Table VIII. Ig-kappa-C amino acid 2-words.

Count	235	27	29	31	4	3	0	11	7	17	2	0	0	0	0	0	0	0	4	11	
Hu	0	0	0	1	0	0	2	0	1	1	0	0	3	0	0	1	2	1	2	1	i*
Ms	0	0	1	0	0	2	0	1	0	1	0	3	0	1	2	0	0	2	1	1	j*
Rb	0	1	0	0	2	0	0	1	1	0	3	0	0	2	1	2	1	0	0	1	k*

$$i* + j* + k* \geq 4$$

Legend to Table VIII. The count of the (i,j,k) index gives the number of common 2-words occurring i times in the human C-gene, j times in the mouse C-gene, and k times in the rabbit C-gene. Thus there are 11 common diamino acids common to only mouse and rabbit. There are 4 diamino acids represented at least four times ($i* + j* + k* \geq 4$) over the three sequences.

Table IX. Common 4-word count distribution in the chemical classification amino acid (alphabet) with respect to the human, mouse and rabbit Ig-kappa C-domain.

		Human Ig-κ-C			
		0	1	2	> 3
	0	3931	65	0	0
Mouse	1	63	33	1	0
Ig-κ-C	2	2	0	0	1
	> 3	0	0	0	0

		Human Ig-κ-C			
		0	1	2	> 3
	0	3929	77	0	0
Rabbit	1	64	19	0	0
Ig-κ-C	2	3	1	0	0
	> 3	0	1	1	1

		Mouse Ig-κ-C			
		0	1	2	> 3
	0	3923	81	2	0
Rabbit	1	71	12	0	0
Ig-κ-C	2	1	3	0	0
	> 3	1	1	1	1

IV. STATISTICALLY SIGNIFICANT LONG BLOCK IDENTITIES (COMMON OLIGONUCLEOTIDES) BETWEEN SEQUENCES

Highly conserved DNA segments over multiple species may indicate regions important in biological functions that either are retained by virtue of selective forces, functional constraints, or are a consequence of convergent evolution. Alternatively, similarities may be nonfunctional evolutionary remnants of ancestral DNA segments. Within a single sequence, the presence of repeated DNA segments can result from duplications, rearrangements, and amplifications. These events may modulate gene expression and provide opportunities for partially differentiated gene function.

For s sequences we define the quantity $K_{r,s}$ as the maximal length of a word occurring in at least r out of the s sequences. The asymptotic distributional properties of $K_{r,s}$ are derived for a collection of random sequence models (Theorem 1, Section VI). From these properties the statistical significance of a long r-of-s shared word can be assessed.

We present a variety of examples of the several distinct longest r-of-s common words.

A. Papovaviruses (SV40 – N = 5243, Polyoma – N = 5293, BKV (Dun) – N = 5153)

The longest block identity is 32 bp between SV40 and BKV. This corresponds to the last 11 residues in a 23-residue identity of the VP_2 gene. There is a common 20 bp oligonucleotide after one mismatch with the 32 bp identity and therefore in essence the similarity extends to 52 bp. The second longest block identity is 23 bp relating SV40 to BKV in the VP_1 gene and this corresponds to the first eight residues in an 18-residue amino acid homology. There also occur two 22 bp block identities in the second exon of the large T gene.

There are only two distinct statistically significant oligonucleotides common to the three papovaviruses: one connecting the initial part of VP_1 (a 12 bp part of an 18 bp SV40-BKV identity); the other (11 bp) relates VP_2 in all three (it is part of the 32 bp SV40-BKV identity).

B. Papilloma Viruses (HPV – N = 7811, BPV – N = 7945)

There are three significant block identities, 17 bp, 15 bp, 14 bp, all clustered in the 3' half of the E_1 gene in perfect alignment (gaps in bp counts between them up to three nucleotides is the same for each sequence). Also there are other block identities (lengths 13, 12 and 11 bp) in the same region in alignment with the significant ones. Note that the most conserved DNA region in the papovaviruses is in the capsid (late) genes but occurs in a transforming (early) gene of the papilloma viruses (cf. Karlin *et al.*, 1984).

C. Mitochondrial Genomes (Human – N = 16569, Mouse – N = 16295, Bovine – N = 16338, Xenopus Laevis – N = 17550).

There are 35 statistically significant distinct identity blocks with respect to the four genomes. Fourteen (total 250 bp) are in the 16S rRNA gene. These are concentrated in the 3' half of the gene and divide into two clusters. The 5' half has diverged more than the remainder of the 16S rRNA gene. The longest oligonucleotide common to all four genomes is 30 bp in the 16S rRNA gene; the next longest is 28 bp found in the 12S rRNA gene.

Seven statistically significant block identities are in the 12S rRNA (total 117 bp). These are distributed evenly all in alignment over the 12S gene adjusting for a 136 bp displacement (insertion or deletion event) in the 5' end of the gene for the Xenopus versus the three mammalian genomes. The high conservation of the rRNA's undoubtedly reflects the importance in the translation of the mitochondrial genes.

The four mitochondrial genomes share significant block identities in the genes COI, COIII, URF2, and URF5. All of these genes are close to one another and straddle the origin of replication of the light strand (O_L). The other genes (e.g., Cytochrome b, COIII genes, and the other URF's) do not exhibit any strong similarity across all these species.

Each of the tRNA genes Asn, Ile, Met, Trp, Ser-1, Ser-2 show significantly long common oligonucleotides covering the four genomes. These tRNA are found except for Ser-2 close to the COI, URF2 and URF5 genes the most conserved genes of the four genomes. By contrast, the tRNA genes positioned proximal to the D-loop show no significant conservation even assessed in terms of pairwise comparisons.

The mammalian species possess significant block identities in 13 tRNA genes of which 12 show strong anticodon arm homology, four have additional long common oligonucleotides in one or both strands of the amino acyl stem.

One might anticipate strong similarity at the junction between genes and flanking sequences because initial concatenated transcripts might presumably be cleaved here to produce mature tRNA's. The tRNA genes (Met, Tyr, Leu), indeed show conserved segments which extend beyond the gene boundaries. Asn tRNA involves a highly conserved oligonucleotide which extends into a noncoding region containing O_L. However, tRNA gene boundaries are not generally part of significant block identities. It appears that tRNA genes which are part of the most clustered tRNA genes are also the most conserved. This may be in part due to the close spacing of processing sites.

We describe next briefly the nature of the significant block identities for all three-species comparisons.

Human-bovine-mouse: These three genomes share 53 distinct statistically significant block identities the highest number among all four three way comparisons, including the longest for three out of four sequence comparisons (lengths 52 and 44 both in the 16S rRNA).

Human-bovine-Xenopus: There are 11 statistically significant distinct block identities. The longest two (38 bp, 27 bp), both occur toward the 3' end of 16S rRNA gene, the gap distance between these two is 1 bp on each sequence, thus totaling a stretch of 66 bp with a single mismatch.

Human-mouse-Xenopus: This three way comparison shows the least number (five) of statistically significant block identities.

Bovine-mouse-Xenopus: This combination has the second largest number of statistically significant block identities (22) of the three way comparisons. These three species share three significant block identities in the Cytochrome-B gene not present in the human corresponding gene.

With respect to the three mammals (human, bovine, mouse), Xenopus is decisively the most similar to mouse in cumulative length of conserved block identities (data not shown). It is the least similar to human. An interesting distinction between Xenopus and the three mammals is in the D-loop region where only the three mammalian mitochondria display any significant shared oligonucleotides.

Among the three mammals, human and bovine appear to be the closest in terms of number and extent of their shared oligonucleotides. Second are bovine and mouse, followed by human and mouse. This is consistent with the impression emerging from immunoglobulin-kappa gene sequence and globin gene comparisons suggesting again that mouse separated very early during mammalian radiation or mutated more rapidly.

D. Hepatitis B-Genomes (Human – N = 3182, Duck – N = 3021, Ground Squirrel – N = 3311).

The human and ground squirrel share a 37 bp oligonucleotide aligned in the Core gene. The next longest is a 21 bp block identity that occurs late in the B-gene. There are lengths 20 bp, two 18 bp, and four 17 bp block identities clustered in the central part of the polymerase (Pol) gene. The only significant block identity between human and duck is 18 bp in the Pol gene. Duck versus ground squirrel show no significant pairwise block identities. There are two significant block identities for the three species of lengths 12 bp and 10 bp both in alignment centered in the Polymerase gene corresponding to four residues identities. We compared this with a number of retroviruses but found no significantly shared similarity.

E. β-Globin (Human – N = 2165, Mouse – N = 1567, Rabbit – N = 1827, Chicken – N = 2157)

The most conserved block identities among the β-globin DNA sequences emphasize exon-intron splice junctions. These encompass conservation of considerable stretches of DNA on both sides of the splice boundaries. Actually the longest block identity of β-globin sequences shared by the four species is of 17 bp length and overlaps the 3' splice junction of the second exon. This similarity extends to length 35 bp between human, mouse and rabbit where this segment involves only two mismatches with chicken. The second longest block identity (14 bp) involving all four sequences covers the 3' splice junction of Exon 1 which extends to 27 bp if chicken is excluded.

The β-globin processed mRNA was available for duck and Xenopus. The concluding 8-mer of exon 2 is identical over all six species and probably extends into the intron. Similar strong identities prevail for all six species at the 5' boundaries of both exons. A length 16 bp block identity contained in human, mouse and chicken starting 8 bp 5' to exon 3 and a length 15 bp identity between human, mouse and rabbit overlaps the same splice junction.

The tendency for conserving DNA sequences traversing splice junctions is also evident from the Ig-kappa gene comparisons. The longest shared oligonucleotides across the J-gene segments of human, mouse and rabbit singles out the 3' splice junction; Karlin et al. (1985).

The Xenopus β-globin DNA sequence (N = 1989) was compared with the corresponding sequences of human, mouse, rabbit and chicken. The longest common block identities are two 8 bp oligonucleotides both aligned in all five sequences, one upstream from the end of the second exon (E_2) 10 bp from the splice junction, and the second at 20 bp after the start of E_2. These segments relate to heme positions, α-β globin and inter α-helix contacts. The conservation of the neighborhood of the splice junction with the Xenopus is not significant. Thus, the degree of similarity of the β-globin sequence of Xenopus and chicken with the mammalian species contrast sharply. Chicken and Xenopus are considered to have diverged from the mammalian line, respectively, about 250 and 300 million years back (e.g., Dickerson and Geis, 1983) so that the time duration does not appear to be decisive. In what respect is the fact that chicken is warm blooded while Xenopus is cold blooded a discriminating factor pertinent to hemoglobin function? Is the fact that α and β are located on the same chromosome in Xenopus but exist on separate chromosomes in mammals and avian species a determining factor?

V. WORD RELATIONSHIPS AND ASSOCIATED FUNCTIONALS

The well-known example of dyad symmetries (inverted complements) suggests a further generalization of the repeat occurrence distribution to allow for related words. Let Δ be a one-to-one mapping of the alphabet A to itself. For instance, DNA base pairings possess a complementarity association on the four letter alphabet {A,C,G,T} leading to the special mapping {Δ^c: A → T, T→ A, G → C, C → G}. Let $\pi = \{\pi_k\}$ be a family of permutations such that π_k rearranges the set of letters in a k-word. For example, for each given k, we designate $\pi_k^{(I)}$ as the inversion permutation, moving the ith letter to the $k+1-i^{th}$ position, e.g., π_8^I(ATGCCGCT) = TCGCCGTA. A relationship of words R is defined by combining a mapping Δ and a set of permutations $\{\pi_k\}$ operations. In particular, the composed mapping of Δ^c and $\pi^{(I)}$, defines the relationship D, independent of the order of application of Δ^c and $\pi^{(I)}$. The two 5-words ATTCG and CGAAT, D(ATTCG) = CGAAT, are said to be in *dyad symmetry* (inverted repeat) relation. A word is called a self-dyad (abbreviated self-D = palindrome) if it is invariant under the mapping D.

Other possible relationships among DNA words can involve simple inversion (i.e., applying only $\pi_k^{(I)}$) or block inversions. A speculative role of the existence of close inverted words is that they can furnish recognition sites for a DNA-interacting dimer (or multimer) protein whose individual peptides are oppositely oriented such that the separate units most expeditiously bind to inverted words.

For amino acid sequences represented in the charge alphabet (grouping amino acids by their charge character yielding the three letters, + (Arg, Lys, His), - (Asp, Glu), and 0 — the other amino acids) we define the mapping Δ^*: {+ → - , - → + , 0 ↔ 0}. The composed mapping of Δ^* and $\pi^{(I)}$, called the *charge complement relationship,* inverts the order of the residues with subsequent complementation of their charge. Thus, the 7-word "0 + 0 0 + - 0" is in charge complement relationship to the word "0 + - 0 0 - 0". Identifying such charge complementary words may help discern specific regions in a polypeptide where paired β-sheets are likely to form. The charge complementarity relationship specifically identifies anti-parallel segments that exhibit aligned electrically attractive side chains. Alternatively, the complementary mapping Δ^* alone identified β-sheets in parallel pairings.

A. Count Distribution of Occurrences of R-Related Pairs

The occurrence distribution of a word relationship within a single sequence is encompassed in the two-dimensional distribution $g_{R,k}(\nu,\mu)$, which enumerates for each length k, the numbers of k-word pairs (w,w*),

Table X. Count occurrence distribution of $g_{D,8}$ for SV40, Polyoma, HPV and BPV compared with $g_{D,8}$ for corresponding permuted sequences.

Original sequences:

SV40

	1	2	3	4	5	6
1	276	68	9	3	0	1
2	---	15	9	1	0	0
3	---	--	0	0	0	0

Polyoma

	1	2
1	248	36
2	---	2

HPV

	1	2	3	4
1	478	114	11	1
2	---	12	1	0
3	---	--	2	0

BPV

	1	2	3	4	11
1	486	115	10	0	0
2	---	17	2	0	1
3	---	--	3	0	0

Permutation ranges (see Section V for a discussion of permutation constructions):

SV40

	1	2	3
1	(195,260)	(0,25)	(0,3)
2	---	(0,4)	(0,1)
3	---	---	(0,0)

Polyoma

	1	2	3
1	(192,229	(0,16)	(0,3)
2	---	(0,2)	(0,0)
3	---	---	(0,0)

HPV

	1	2	3	4
1	(341,407)	(40,92)	(0,7)	(0,2)
2		(0,6)	(0,1)	(0,0)
3			(0,0)	(0,0)
4				(0,0)

BPV

	1	2	3	4
1	(327,418)	(43,101)	(0,8)	(0,1)
2		(0,8)	(0,1)	(0,0)
3			(0,0)	(0,0)

in relation R where w occurs ν times and its R-word, w*, occurs μ times in S. Table X presents as an example the bivariate distribution $g_{D,8}(\cdot,\cdot)$ of dyad symmetry pairs of length 8 for SV40, polyoma, HPV and BPV.

The count distributions $g_{D,8}$ for polyoma and SV40 display the same distinction observed for repeat occurrence distributions of these two viruses: Polyoma shows a reduced number of D-related 8-words compared to SV40. Furthermore, comparison to the $g_{D,8}$ permutation ranges indicates that for both viruses $g_{D,8}(1,1)$ and $g_{D,8}(1,2)$ of the original sequences exceeds the maximum from the shuffled sets. This excess is evident for all $g_{D,8}(i,j)$, $i = 1,...,3$, $j = 1,...,6$ in SV40.

The HPV and BPV genome adjusted for genome size is comparable to SV-40 in counts of dyad symmetry combinations.

B. Close Dyad Symmetries (DS)

The importance of identifying DS pairs for their potential function in secondary structures and in transcription and replication control is well recognized. Dyad symmetries (inverted repeats) are considered, inter alia, to be associated with promoter sites, transcription termination locations, enhancer elements, DNA rearrangement control, and recombinational hot spots. In particular they can provide multiple sources for origin of replication and/or target sites for protein binding leading to transcription initiation. Strong DS at distances can link DNA segments mediated by protein interactions. Properly distributed close dyads can provide stability to RNA transcripts during the transition from the nucleus to the ribosomes.

Table XI reports counts of all exact close dyad symmetries for a variety of sequences. These tabulations show that short close DS (particularly stem length 5) tend to occur in excess compared to corresponding random sequences; this is not the case for close DS stem

Table XI. Number of exact dyad symmetries (the count in parentheses is self-dyads) with stem length ≥ 5 and loop length ≤ 50.

Stem length	5	6	7	8	9	10	11	12	13	14
		(SD)		(SD)		(SD)		(SD)		(SD)
λ	1797+	500+	113-	40+	11	4	4+	0	0	0
	(48502)	(332)-		(102)-		(20)-		(4)-		(0)
T7	1024+	218	49	6	0-	0	0-	0	0	0
	(39936)	(204)-		(47)-		(9)-		(3)-		(1)
SV40	176+	56+	14	6	1	0	0	0	1	0
	(5243)	(65)		(17)		(5)		(1)		(0)
PLYMA	158+	35	11	6	0	0	0	0	0	0
	(5293)	(49)-		(11)		(4)		(1)		(0)
BKV	202+	59+	13	4	0	0	0	0	0	0
	(5153)	(78)		(16)		(7)		(1)		(0)
HPV [a]	247	65	19	3	0	1	1	0	0	0
	(7814)	(87)		(20)		(6)		(1)		(1)
BPV [b]	279+	81+	30+	8	0	0	0	0	0	0
	(7945)	(90)		(19)		(6)		(1)		(1)
EBV [c]	7347	1918	574	204	68	7	13	4	2	1
	(172282)	(1973)		(485)		(154)		(60)		(3)
Hu-Mt [d]	337	92	19	6	1	2	1	0	0	0
	(16569)	(155)		(40)		(11)		(5)		(0)
Bv-Mt	469+	112	38+	10+	0	0	1	0	0	0
	(16338)	(190)		(48)		(9)		(2)		(0)
Ms-Mt [e]	501+	109	49+	10+	4+	3+	0	1+	0	0
	(16295)	(218)		(61)		(13)		(2)		(0)
Xe-Mt	580+	156+	41+	9	3	1	0	0	0	0
	(17550)	(222)		(64)		(17)		(4)		(2)
Hu-Ig	208+	47	15+	1	1	0	0	0	0	0
	(5062)	(64)		(19)		(4)		(1)		(0)
Ms-Ig	192+	55+	19+	8	1	0	0	0	0	0
	(5495)	(60)		(9)-		(1)		(0)		(0)
Rb-Ig	248+	67+	26+	5	4+	0	0	0	0	0
	(5235)	(57)-		(27)		(8)		(5)+		(2)+
E-74	307+	72+	18+	6	4+	1	0	0	0	0
	(7169)	(89)		(45)		(8)		(5)		(1)

Legend. Statistical significance is assessed by comparison with the range of the number of close DS based on 40 permutations of the observed sequence. The superscript + − indicates that the original sequence shows more (fewer) close DS than the maximum (minimum) of the permutation range. SD = self-dyad is an oligonucleotide identical to its inverted complement. [a] HPV has one 16 bp SD. [b] BPV has one 18 bp SD. [c] EBV has two close 15 bp DS, four SD of length 16 bp, three SD of length 18 bp, one 20 bp SD, and one 22 bp SD. [d] Hu-Mt has one 16 bp SD. [e] Ms-Mt has one 16 bp SD.

length $\geq$ 8 bp.   A paucity of self-dyads of length 6 compared to chance is observed.   Thus short self-dyads appear to be selected against; such hairpin formations may encumber transcription (create polymerase pause sites) or present unstable secondary structures.   It is documented in laboratory experiments that when short self-dyads are coerced into a stem loop they generally snap back to a linear configuration (Buchman and Berg, 1984).

It is interesting that both E-coli phage, λ and T-7, exhibit a significant deficiency of self-dyads of all lengths. Selection against self-dyads in T7 and λ-phage is consistent with the results of Wyman *et al.*, (1985) (see also Leach and Stahl, 1983) who found it difficult to clone self-dyads of any length (k $\geq$ 6) in E-coli . The inability to maintain long palindromic sequences in a variety of cloning vectors in E-coli has been reported by several laboratories.   This could be realized by the formation of deletions, possibly through the recognition of cruciform DNA con-figurations that are cleaved by endonucleases.   Along these lines, Leach and Stahl (1983) have indicated that simultaneous absence of two specific recombination nucleases confers viability to a derivative of λ-phage carrying a palindrome.

One might consider that since virtually all hexamer restriction sites are self-dyads that a broad range phage evolving in a variety of entero-bacteria might tend to eliminate 6-word and longer palindromes from its genome.   However, as the λ-phage and T7 host ranges are considered to be limited it seems more likely that E-coli embodies suitable enzymatic machinery that rapidly degrades palindromic DNA molecules.   Counts of palindromes in E-coli and other bacterial and associated viral DNA sequences could help clarify the degree to which negative selection of palindromes is in force.   There are documented cases of the existence of inverted repeats which are unstable in E-coli but which are stable upon cloning in yeast cells (Szostak and Blackburn, 1982).

There is a striking disparity in the number of close DS of λ versus T7 of lengths 9 and 11. T7 is significantly deficient in DS of lengths 9 and 11 whereas λ shows an excess of close DS of length 11. The marked deficiency in self dyads and excess of close short dyads for λ and T7 raises interesting questions.

Counts of close DS of stem length 8–16 bp of mammalian viruses do not differ from random sequences. However, BPV and EBV possess significantly long self-dyads of stem length ≥ 18 bp located in noncoding regions. We did not find any statistically significantly long self-dyads in the globin, immunoglobulin, and Drosophila (E-74, AdH) genes examined.

The Ms-Mt and Xe-Mt mitochondrial genomes show significantly more close dyad symmetries than the human mitochondrium sequence although the genome size and sets of nucleotide frequencies are similar. The same ordering applies with respect to high frequency repeats (data not shown).

C. Long Dyad Symmetries (DS)

We illustrate the extent of significant long dyad symmetries on some selected DNA sequences. The assessment of statistical significance is based on theoretical analysis parallel to Theorem 1 of Section VII (see Karlin and Ost, 1985, 1986).

Papovaviruses

SV40:   The longest observed exact dyad pair has 13 bp stem length with 1 bp loop located at the origin of replication. There is an 11 bp multiple DS having AATTAGTCAGC and its two dyad words after a gap of 85 bp and 119 bp in the enhancer region. An enhancer element's capacity to operate in either orientation could imply one or several DNA features that is also maintained in its inverted complementary sequence. A series of close dyad-symmetry pairs qualify in this respect.

Polyoma: Longest exact dyad pair is 12 bp with a long loop length (not statistically significant).

Papillomaviruses

HPV: Three long self-dyads of lengths 16, 14 and 12 bp (none statistically significant). There is a 13 bp DS with the stem pieces at opposite ends of the L_1 gene.

BPV: An 18 bp statistically significant self-D TCAGGTTTATAAACCTGA is in the L_2 gene. There exists a 13 bp dyad symmetry (loop length 223 bp) with starting location 149 bp 3' to the E_2 gene.

Epstein Barr Virus (EBV)

See text, part D.

Bacteriophages

λ-phage: The longest DS is 16 bp (first element AGAAAGGAAACGACAG) with 26 bp loop located 82 bp from the NuI gene. This is a substantial close DS pairing which may confer biological function.

Mitochondrial genomes

Human, mouse, bovine, Xenopus:

None contain statistically significant dyad symmetry pairings.

Ig-kappa gene

Mouse: There is a single significant 14 bp exact DS pair linking two "control" sites one 5' to J_1 and the other in the J/C intron part of an enhancer element; cf. Karlin and Ghandour (1985c).

Hepatitis Viruses

Duck: Longest is 11 bp DS (first element CTATATTTCTC) with 26 bp gap to its dyad word. This is at the 5' end of the polymerase gene.

Ground squirrel: There exists one self-dyad of length 12 and a 12 bp DS with 1462 bp loop.

Human: Two self-dyads of length 12.

Globin gene-family

Human α: 13 bp DS pair with first element CAGAGAGAACCCA located 15 bp 5' to E_1 (exon 1) with the dyad word 124 bp 3' to E_3 thus connecting flanks of the gene. There exists a 14 bp self-dyad exclusively G+C (CGCCCGGCCGGGCG) located 97 bp 5' to exon 1.

Human ζ: Near the terminus of intron 2 there exists an 11 bp segment (GGCCCCGCCCC) starting 35 bp 5' to E_3 with its inverted complement upstream after a gap of 7 bp. The 9 bp element CCCCGCCCC can dyad pair with 16 elements distributed over the first 180 bp of the second intron. There are also multiple repeat elements and associated close dyad symmetries in the first intron.

Human β: Longest DS pair is 13 bp (first element is TTTCTGATAGGCA) located 54 bp 5' to exon 2 with dyad word at 73 bp into exon 3. There exists a 20 bp statistically significant self-dyad T.AATATGTGTACACATATT.A (a dot indicates a mismatch position) located central to the second intron. There is a preponderance of close dyad symmetries in this neighborhood.

Human δ: There is a close (55 bp loop) 13 bp DS in the second intron starting 313 bp 5' to exon 3. The 16 bp self-D ATATATGTACATATAT occurs about central to the second intron. Note the resemblance of this SD with the 20 bp SD of the human β sequence.

D. Dyad Symmetries of the EBV Genome

We use the coordinate numbering of the EBV genomic sequence recorded in the Genbank data base.

There are two prolific regions of long and close DS pairings in EBV genomes, namely the ori-P region (Reisman *et al.*, 1985) coordinates 7315-9312 (known to be essential for EBV replication in its latent infectious stage) and the 5' third of the 3 kb repeat (Karlin and Ghandour, 1986). The ori-P region contains an abundance of statistically significant DS combinations which strongly correlate with the "21 × 30 bp" tandem repeat units and the truncated copies 1 Kb downstream of this region. Each of the units contains a central self-dyad (with few mismatches) of length 18 bp and these can strongly pair with each other. Multiple close dyads can provide alternative starting points for replication. There are two long close dyad pairings overlapping the downstream repeat region of ori-P, the first a 31 bp stem dyad pair (including three mismatch positions) with a 3 bp loop (Reisman *et al.*, 1985); a second 25 bp stem pairing with 19 bp loop presents a secondary structure that can attain a striking cruciform shape. The role of strong stem-loop structures is unknown but they exist at the origin of DNA synthesis in bacterial strains, mitochondrial genomes and in the papovaviruses. It is noted in Riesman *et al.* (1985) that ori-P of EBV is structurally similar with the γ origin of E-coli plasmid R6K which contains "7 × 22" tandem repeats augmented with a substantial A + T-rich stretch.

A striking number of strong close dyad symmetries are associated with the 5' end of the 3 Kb repeat unit of EBV (coordinates 12001–47643, which contains in the case at hand 11.6 tandem copies of a 3.072 kbp segment). This 3 Kb unit also relates to the only significant long range dyad pairings. They are listed next.

The three most significant global dyad symmetries are:

20 bp TAACGAGCAGAGAAGAAGTA at 12266 (and repeated in all the subsequent 3 Kb repeats) with dyad at 88913
18 bp CAGGCAGGGCCCCCCGGC at 12393 with dyad at 60804
20 bp GGCGCCTCCTCGGGGCCAGC at 12412 with dyad at 12779.

Notice that *all* these significant dyad pairs involve one component in the 200 bp stretch from 12260 to 12460.

The 600 bp stretch from about 12329 to 12950 in the 3 Kb unit can establish an extended hairpin structure of great energy $((G+C)/(A+T) \geq 5)$ involving possible stem length of 270 bp with hydrogen bonding potential in excess of 208 bp. In this stem formation there are 9 segments ranging from 9 to 20 bp length that can hydrogen bond perfectly.

This region also contains a preponderance of close DS and long palindromes. This virtually unprecedented stretch of the 3 Kb repeats entails multiple significant dyad symmetry combinations which allow for a myriad of strongly stable secondary structures; see Karlin and Ghandour (1986) for details. Is it possible that the lytic origin of replication is contained therein? Does this element of the 3 Kb repeat play a role in transactivation during lytic infection? Is it possible that high repeat occurrences (11.6 copies) of the 3 Kb element coupled to its versality in distinctive secondary formations (of great potential stability owing to the preponderance of $G+C$ stacking) present a series of targets for an array of enzymes that may propitiously interact during lytic infection?

The longest self-dyad of EBV is of 22 bp located at coordinate 153637. This is almost exactly at a place of a 15 kbp deletion in the B95-8 strain and could have been formed in sealing the genome in reaction to this deletion.

The long self-dyad (of length 18 bp (AATAACCTATAGGTTATT) located at coordinate 53262 is followed by a long 20 bp self-dyad (TATATACCTATAGGTATATA) 23 bp downstream. This is unusual.

E. Comments on Long and Close DS for Several Globin Genomic Regions

There is a striking dyad symmetry pairing of 30 bp stem length (with three mismatches connecting the 5' flank (of the human zeta globin gene) coordinates 721-690 5' to E_1) with the 3' flank (located at coordinates 300-330 bp 3' to E_3). This dyad pairing can conceivably establish long DNA loops amenable for easy transcription of the gene. These distant dyad pairings connecting the 5' flank to the 3' flank of the human α-globin genes tend to be pyrimidine rich in the 3' flank.

The phenomenon of substantial dyad symmetry combination between the 3' and 5' flanks of the α-globin genes also exists in the corresponding human, mouse, goat and rabbit sequences and possibly in most mammal species, e.g. the human α sequence contains the oligonucleotide TGGGTTCTCTCTG 124-3' to E_3 with its dyad at 15-5' to E_1; mouse α exhibits CTGAGTCTG at 237-3' to E_3 with its inverted complement at

18-5' to E_1. Goat α_1 shows TTCTCTCTGAG at 256-3' to E_3 and its dyad at 17-5' to E_1. Note the similar locations of the element in the 5' flank (around 17-5' to E_1). Substantial self-dyads (length $\geq$ 12bp) tend to be G+C rich and generally are concentrated in 5' or 3' flanks, e.g., human α contains the 14bp SD CGCCCGGCCGGGCG at 97-5' to E_1, and human-ζ exhibits the self-dyad GCTGGGCCCAGC at 103-5' to E_1 at about the same distance to the gene start.)

The 10bp self-dyad CCACATGGAG right over the beginning of the gene 4-5' to E_1 is conserved in the human, mouse and goat α, and in the human β and δ, rabbit β-1, and also in chicken ϵ and ρ. Why is CCAC abutting the initiation codon ATG maintained in these animal species?

There is an abundance of close strong dyad pairings in the 5' flank of the α-globin genes. We list representative examples:

Human α: 14bp self dyad CGCCCGGCCGGGCG 97-5' to E_1

Human α: GCCGGGC at 91-5' to E_1 with dyad at 44-5' to E_1
(the first element is part of the previous 14bp self-dyad)

Human ζ: 12bp SD GCTGGGCCCAGC 103-5' to E_1

Mouse α: CTCCAAGGGC at 109-5' to E_1 with dyad location 76-5' to E_1

Goat α_1: GCCC.GCGCCAG at 123-5' to E_1 with dyad location 81-5' to E_1 (the dot represents a mismatch position)

Duck α: CAGGGCGGGGC at 152-5' to E_1 with dyad location 133-5' to E_1

Chicken α: GGCTGGGGGG at 170 5' to E_1 and dyad location 95 5' to E_1

The human α-globin genes tend to show multiple close strong DS pairings in the 3' flank about 30-45 3' to E_3, e.g., human α has GGCCG.TGGG at 28 3' to E_3 and its dyad at 41 3' to E_3, human zeta contains the self-D (GCCCGGGC) at 46 3' to E_3.

The β-globin second intron is abundant in long self-dyads or close inverted complementary pairs. These tend to be composed of either predominantly weak bases (especially for SD) or dyads with one element primarily purine and the other primarily pyrimidine.

VI. ASSESSING STATISTICAL SIGNIFICANCE

Repeats and dyad symmetry relations *can occur by chance*. For example, a randomly generated 5,000 nucleotide sequence with each base equally likely, is expected on the average to have one repeated 12-word and one 9bp-word occurring three times. We employ two methods to help distinguish nonrandom sequence features in the data. (i) Comparisons of the original sequences to corresponding random sequence models, and (ii) data shuffling methods.

A. "Random" Sequence Models

We use a "random" model appropriate to the data as a standard in order to ascertain the distributional properties of the various data statistics. The following *independence random model* often serves well as a standard by which to assess significance for a data example of s sequences. For this model, the s sequences of lengths $N_1,...,N_s$, respectively, are generated such that the successive letters of the νth sequences are sampled independently having letter ℓ_i occur with probability $p_i^{(\nu)}$. (In the case of DNA, the $p_i^{(\nu)}$ are usually specified as the actual A, T, C and G frequencies of the observed sequences.)

Theoretical Results on the Length of the Longest Shared Word in r out of s Sequences

Theorem I. Karlin and Ost (1985). The expected length of the longest common word present in at least r out of s sequences, $K_{r,s}$, when $N_\nu \to \infty$ has order growth

$$[\log n(N_1,\ldots,N_s) + \log \lambda(1-\lambda) + 0.577] / (-\log \lambda) + \delta(\lambda)$$

where $n(N_1,\ldots,N_s) = \sum \prod_{\nu=1}^{r} N_{1_\nu}$, *the sum extending over all r-tuples of index integers from* $(1,2,\ldots,s)$ *and* $|\delta(\lambda)| \leq 1/2$. *The parameter* λ *for the independence models is* $\lambda = \max_{1 \leq \nu_1 < \nu_2 < \ldots < \nu_r \leq s} \left(\sum_{i=1}^{m} \prod_{j=1}^{r} p_1^{(\nu_1)} \right)$ *and* $\gamma = \lambda$. *The asymptotic standard deviation is approximately* $(1.283) / |\log \lambda|$ *independent of the sequence lengths.*

For the corresponding Markov dependent model and $r = s$ the characteristic value λ is the largest eigenvalue of the matrix $\|P_{ij}^{(1)} P_{i2}^{(2)} \ldots P_{ij}^{(r)}\|$ in which the elements consist of the corresponding products of the entries in the matrices $\|P_{ij}^{(\nu)}\|$.

A more elaborate model that allows for first order neighbor dependencies is constructed as follows. In the νth sequence, let $p_{ij}^{(\nu)}$ be the conditional probability of sampling letter ℓ_j following letter ℓ_i. The generation of successive letters of the νth sequence is now governed by the transition probability matrix $\|p_{ij}^{(\nu)}\|$. (In the case of DNA, $p_{ij}^{(\nu)}$ would correspond to the dinucleotide frequencies.) We refer to this sequence model as the *first order Markov dependent random models.* Other versions would allow for higher order Markov dependence or more complex long-range neighbor dependencies; see Karlin and Ost (1985).

A common word length is designated statistically significant relative to the corresponding random sequence model provided its length exceeds the expected length of the longest common word of the random model by at least two standard deviations. A statistically significant common word has probability less than .01 of occurring by chance in the corresponding random model.

B. Data Shuffling Methods

These methods are widely practiced in many areas of data analysis (e.g., Edgington, 1980; Pratt, 1981, Karlin and Williams, 1984). The idea is to compare the observed sequences against a collection of independent reconstructed sequences each obtained by repeatedly randomizing or shuffling the data. Usually a data statistic evaluated on the original sequence is compared with its values determined from the randomly reconstructed data sets. (In the context of protein sequences permutation methods have been used by Dayhoff (1978), and reviewed in Doolittle (1981), among others.)

We use a spectrum of shuffling (permutations) procedures that selectively shuffle the sequence letters. This process alters certain alphabet relationships while keeping others intact. We identify each permutation model by the class of letters to be permuted; for example, purine-only shufflings permute randomly all the A+G positions while all C and T letters are left in their original positions.

Sequence statistics are computed for 20 to 100 permutations of the data to assess the variability of the statistics under each permutation class. The value of the original statistic relative to the collection of values of the statistic for the permutations of the data provides an opportunity to assess the special (nonrandom) characteristics of the original data. If the original statistic is more extreme than most of the values of this statistic when evaluated over the permuted data sets, then the original statistic is considered to be significant (e.g., if less than 5% of the permutations yields a value for the statistic that is more extreme than the original value).

Insights into the data can also be gained by comparisons between different classes of permutations. In order to highlight nonrandom features depending say on the purine composition of the sequence we could implement a class of permutations which shuffle only the A+G bases and compare the repeat structure for these permuted sequences with the original C+T stretches which would be retained in the permuted sequence set. Comparing between the permutation models of shuffled purines only and that of shuffled pyrimidines only affords a way to assess the relative preservation of purines versus pyrimidines.

ACKNOWLEDGMENT

I am grateful to E. Blaisdell for valuable suggestions on the manuscript.

REFERENCES

Buchman, A. R., and Berg, P. (1984). Unusual regulation of SV40 early region transcription in genomes containing two origins of DNA replication. *Mol. Cell. Biol.* **4**, 1915-1928.

Dayhoff, M. O. (1978). "Atlas of Protein Sequences and Structure," Vol. 5, Suppl. 3. National Biomedical Research Foundation, Washington, D.C.

Dickerson, R. E., and Geis, I. (1983). "Hemoglobin." Benjamin Cummings Publ. Co., Menlo Park, Calf.

Doolittle, R. F. (1981). Similar amino acid sequences: chance or common ancestry. *Science* **214**, 149-159.

Edgington, E. S. (1980). "Randomization Tests." Marcel Dekker, New York.

Fitch, W. M., and Atchley, W. R. (1985). Evolution in inbred strains of mice appears rapid. *Science* **228**, 1169-1175.

Hieter, P. A., Maizel, J. V., and Leder, P. (1982). Evolution of human immunoglobulin κ-J region genes. *J. Biol. Chem.* **257**, 1516-1522.

Karlin, S., and Ghandour, G. (1986). An unusual potential secondary structure of the Epstein Barr virus. *Proc. Natl. Acad. Sci. USA* (in press).

Karlin, S., and Ghandour, G. (1985a). Comparative statistics for DNA and protein sequences: Single sequence analysis. *Proc. Natl. Acad. Sci. USA* **82**, 5800-5804.

Karlin, S., and Ghandour, G. (1985b). Comparative statistics for DNA and protein sequences: Multiple sequence analysis. *Proc. Natl. Acad. Sci. USA* **82**, 6186-6190.

Karlin, S., and Ghandour, G.(1985c). Alignment maps and homology analysis of the J-C intron in human, mouse, and rabbit immunoglobulin kappa gene. *Mol. Biol. Evol.* **2**, 53-65.

Karlin, S., and Ghandour, G. (1985d). The use of multiple alphabets in kappa-gene immunoglobulin DNA sequence comparisons. *EMBO Jr.* **4**, 1217-1223.

Karlin, S., and Ghandour, G., and Foulser, D. E. (1985). DNA sequence comparisons of the human, mouse and rabbit immunoglobulin kappa-gene. *Mol. Biol. Evol.* **2**, 35-52.

Karlin, S., Ghandour, G., Foulser, D., and Korn, L. (1984). Comparative analysis of human and bovine papillomaviruses. *Mol. Biol. Evol.* **1**, 357-370.

Karlin, S., and Ost, F. (1985). Maximal segmental match length among random sequences from a finite alphabet. *In* "Proc. of the Berkeley Conf. in Honor of Jerzey Neyman and Jack Kiefer," Vol. I (L. Lecam and R. A. Olshen, eds.). Wadsworth Inc., Belmont, Calif.

Karlin, S., and Ost, F. (1986a). Comparisons of random letter sequences: Length and counts of long aligned matching blocks. *Adv. in Appl. Prob.* (in press).

Karlin, S., and Ost, F. (1986a). Comparisons of random letter sequences: Maximal length of unrestricted matching blocks between random letter sequences (submitted).

Karlin, S., and Williams, P. T. (1984). Permutation methods for the structured exploratory data analysis (SEDA) of familial trait values. *Amer. J. Hum. Gen.* **36**, 873-898.

Leach, D. R. F., and Stahl, F. (1983). Viability of λ-phages carrying a palindrome in the absence of recombination nucleases. *Nature* **305**, 448-451.

Mahler, H. R., and Chordes, E. H. (1966). "Biological Chemistry." Harper & Row, New York.

Max, E. E., Seidman, J. G., and Leder, P. (1979). Sequences of five potential recombination sites encoded close to an immunoglobulin κ-constant region gene. *Proc. Natl. Acad. Sci.* **76**, 3450-3454.

Pratt, J. W. (1981). "Concepts of Nonparametric Theory." Springer Verlag, New York.

Reisman, D., Yates, J., and Sugden, B. (1985). A putative origin of replication of plasmids derived from Epstein Barr virus is composed of two cis acting components. *Mol. & Cell. Biol.* **5**, 1822-1832.

Szostak, J. W., and Blackburn, E. H. (1982). Cloning yeast telomeres on linear plasmid vectors. *Cell* **29**, 245-255.

Wu, C.-I., and Li, W.-H. (1985). Evidence for higher rates of nucleotide substitution in rodents than in man. *Proc. Natl. Acad. Sci. USA* **82**, 1741-1745.

Wyman, A. R., Wolfe, L. B., and Botstein, D. (1985). Propagation of some human DNA sequences in bacteriophage λ vectors requires mutant *Escherichia coli* hosts. *Proc. Natl. Acad. Sci.* **82**, 2880-2884.

RELATION OF HUMANS TO AFRICAN APES:
A STATISTICAL APPRAISAL OF
DIVERSE TYPES OF DATA[1]

Alan Templeton

Department of Biology
Washington University
St. Louis, MO 63130

ABSTRACT

The molecular evidence unambiguously indicates that the African Apes are the closest evolutionary relatives of humans. Unfortunately, the molecular similarity between these primates is so extreme that it has proven to be difficult to determine the exact evolutionary relationship between humans, chimpanzees, and gorillas. The recent literature has focused primarily upon two alternatives: one, an evolutionary tree that regards the African apes as a clade with humans as a separate evolutionary lineage, versus two, an evolutionary tree that regards chimpanzees and humans as a clade with gorillas as a separate evolutionary lineage. Various authors have estimated one or the other of these phylogenies in the recent literature from diverse types of data, but the statistical problem is not one of estimation but rather of hypothesis testing; namely, phylogeny 1 versus 2. A variety of statistical tests are applied to diverse types of data, including mitochondrial and nuclear DNA restriction maps and sequences, isozyme data, immunological distances, DNA-DNA hybridization data, and karyotypic data. Although some of these data sets yield phylogeny 2 as an estimate, none of them allow the rejection of phylogeny 1 at the 5% level of significance. However, the restriction maps of the mitochondrial DNA allow the rejection of phylogeny 2 in favor of phylogeny 1 at the 5% level. Hence, all of the data is statistically compatible with phylogeny 1, but the same cannot be said for phylogeny 2.

---

[1] Supported by NIH Grant RO1 GM31571

I. INTRODUCTION

Charles Darwin long ago predicted that the closest relatives to humans were the African apes. Modern studies have certainly confirmed this prediction, showing that at the molecular level humans and African apes are remarkably similar. Indeed, the similarity is so extreme that it has proven difficult to resolve the exact phylogenetic relationships among humans, chimpanzees and gorillas. However, recent studies have focused upon only two alternatives, illustrated in Figure 1 (Sibley and Ahlquist, 1984; Templeton, 1983a). Under phylogeny 1 in Figure 1, gorillas and chimpanzees form a clade that had a common ancestor after the human lineage had split off. Under phylogeny 2, humans and chimpanzees had a common ancestor after the gorilla lineage had split off. In this paper I will examine statistically several data sets in an attempt to resolve between phylogenies 1 and 2.

My statistical approach throughout is a nonparametric one based upon scores or ranks that make as few assumptions about the evolutionary process as possible. Many of the parametric approaches to analyzing molecular data assume an underlying Poisson process and rate homogeneity across lineages, across time and across the molecule. Sometimes variation in one or more of these parameters is introduced, but this requires the addition of even more parameters. Unfortunately, it is difficult to know *a priori* which, if any, of these assumptions are valid for a particular data set. Consequently, a non-parametric approach is taken that sacrifices some statistical power for greater robustness concerning these types of assumptions.

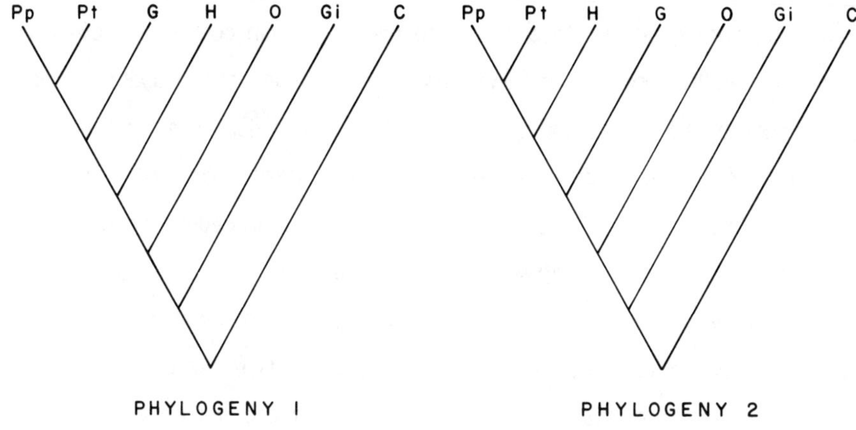

Figure 1. Two alternative branching orders for the hominoid primates. The following abbreviations are used: Pp is the pygmy chimpanzee (*Pan paniscus*), Pt the common chimpanzee (*Pan troglodytes*), G the gorilla (*Gorilla gorilla*), H humans (*Homo sapiens*), O the orangutan (*Pongo pygmaeus*), Gi the gibbon (*Hylobates lar*), and C cercopithecidae monkeys.

The ranking procedures used in this non-parametric approach are also chosen for their robustness. For example, in analyzing restriction endonuclease site maps, a pattern that can be explained by only a single substitution is ranked as more likely than a pattern that requires multiple substitutions. This ranking assumption is a valid one when comparing recently evolved groups, but it can be violated (Templeton, 1983a). Fortunately, criteria exist for evaluating the validity of this assumption (Templeton, 1983a). Given that single events are more likely than multiple events, certain multiple events can also be reliably ranked. First, creating a convergent event by adding on a substitution resulting in a loss of a restriction site is much more probable than one resulting in a gain of a restriction site (Templeton, 1983b). Second, convergent events created by adding on one loss are more probable than convergent events created by adding on two or more losses. However, the probability of convergence by adding on one gain is of the same order of magnitude as the probability of convergence by adding on two losses under the relevant conditions (Templeton, 1985a). Hence, the ranking procedure should not and does not discriminate between these two types of events.

Finally, when given an option, it is preferable to analyze character states as opposed to distance matrices. There are two reasons for this. First, the character states inherently contain more information than distances. Distances can always be calculated from the character states, but the reverse is not true. Obviously, information is being lost in going from character states to distances. For example, the ranking procedure for restriction enzymes takes into account not only how many mutations were fixed, but also the qualitative type (losses versus gains). As shown by the analysis of the hominoid primate data (Templeton, 1983a), much phylogenetic information is contained in the qualitative types of mutational events, yet this information is ignored in distance analyses.

Second, genetic distance often gives considerable weight to substitutions that character state analysis indicates have no branching order information. Many substitution are unique to a single taxon, and thus contain no information concerning branching order relationships to other taxa unless one is willing to make additional assumptions about the evolutionary process, such as a molecular clock. Nevertheless, such substitutions are sometimes major contributors to the genetic distances between taxa.

In some circumstances, the nature of the data is inherently of a distance type. In that case, the difficulties mentioned above cannot be avoided, and additional assumptions are needed in order to make phylogenetic inference. In these cases, the impact of these additional assumptions can be minimized by using the ranks of the distances rather than magnitudes of the distances. By using only ranks of distances, one can make phylogenetic inference by assuming that more recently diverged taxa have smaller distances than more anciently diverged taxa. This assumption is far weaker than the assumption of rate constancy over both time and lineages, and hence the conclusions drawn will not be so dependent upon making *a priori* assumptions about the process of molecular evolution. Once again, this robustness is purchased at the expense of statistical power.

II. RESTRICTION SITE DATA

Ferris *et al.* (1981) constructed restriction endonuclease cleavage site maps for the mitochondrial DNA (mtDNA) of hominoid primates. To see if phylogenies 1 and 2 (and other alternatives as well) could be resolved with these data, phylogeny 1 was assumed to be true, and scores were assigned to the set of substitutions detected by a particular restriction enzyme using the ranking procedures mentioned above. Next, the scoring procedure was repeated for each restriction enzyme system under the assumption of phylogeny 2. A matched pair score was then obtained by subtracting the raw score under phylogeny 2 from the raw score under phylogeny 1 for a particular restriction enzyme system. The set of matched-pair scores over all restriction enzymes was then converted into a statistical statement via the non-parametric Wilcoxon matched-pairs signed rank test. From this analysis, it was concluded that phylogeny 2 could be rejected in favor of phylogeny 1 at the 5% level.

Recently DeBry and Slade (1985) have reexamined this analysis. Although they concluded that the basic approach was sound, they criticized the use of sets of sites defined by a particular restriction enzyme as the unit of analysis. The purpose of using these sets in a matched-pair analysis is to control for potential heterogeneity in rates of evolution between different types of recognition sequences (Templeton, 1983a). However, DeBry and Slade (1985) point out that it is more likely that there will be heterogeneity between different segments of the piece of DNA being examined than between different recognition sequences. Hence, they argue that the individual restriction sites should be the unit of analysis. This argument has considerable merit, and single sites are undoubtedly a better unit of analysis than sets of sites defined by a single restriction enzyme. Nevertheless, there are still some difficulties with treating each individual restriction site as the unit of analysis.

The purpose of both the matched-pair analysis and the pooling is to control for heterogneity within the piece of DNA being examined (Templeton, 1983a). The trouble with following DeBry and Slade's (1985) suggestion of treating each restriction site as a unit is that it is frequently found that the segments of DNA being examined in such studies show one or more substitutional "hot spots." As discussed in detail in Templeton (1983a), the sites that have a higher rate of substitution are the ones most likely to be misinformative because of convergent evolution. If individual restriction sites are used as the unit of analysis, equal statistical weight is given to each site regardless of whether or not it came from a "hot" region. Moreover, a disporportionate number of the sites are expected to come from the "hot" regions. Hence, using individual sites as the unit of analysis effectively gives much weight to "hot" regions; yet, such areas are precisely the regions most likely to give erroneous phylogenetic information (Templeton, 1983a). One way of avoiding this difficulty is to pool the sites found in a segment that can be identified as a "hot" region, but otherwise treat individual sites as the unit of analysis as recommended by DeBry and Slade (1985).

The importance of correcting for "hot" regions can be illustrated by a reanalysis of the mtDNA data of Ferris *et al.* (1981). Table 2 of Ferris *et al.* (1981) gives the number of site changes within five equally sized genomic regions of the mtDNA. The chi-square test of the hypothesis of uniformity of site changes over these five genomic segments is 10.11 with four degrees of freedom, a value significant at the 5% level. Hence, the hypothesis of uniformity over the mtDNA molecule for the number of site changes can be rejected. Inspection of the numbers given in Table 2 of Ferris *et al.* (1981) clearly indicates that the genomic region between 40 -59.9 map units has experienced far more site changes than the other regions. Excluding this region from the analysis, the chi-square of uniformity is 3.76 with 3 degrees of freedom, indicating that the remainder of the mtDNA does not have a statistically significant non-uniform distribution of site changes. The Wilcoxon matched-pair signed rank test was then applied to the mtDNA data, first using the DeBry and Slade (1985) recommendation of treating each site as a unit of analysis, and then pooling the sites from the "hot" region into a single unit by taking the sum of the individual site scores. The results are shown in Table I.

For the individual site analysis, a total of 16 sites contain information that discriminates between phylogenies 1 and 2, with 11 sites favoring 1, and 5 favoring 2. Although the data clearly favor 1 over 2, this result is not significant, in either a one-tailed or two-tailed Wilcoxon test. However, when one corrects for the heterogeneity in the distribution of site changes by pooling all "hot" region sites together, the resulting test is significant at the 5% level (both one- and two-tailed). The reason for this is obvious upon inspection fo Table I. The distribution

Table I.   Analysis of the mitochondrial restriction site data using individual restriction sites as the unit of analysis and by pooling sites from a "hot" region.   Sits are designated as in Table 1 of Ferris *et al.* (1981). The matched-pair score for each site is determined as discussed in the text.

Site	Score		Signed Rank (Individual Sites)	Signed Rank (pooled)
101	1		8.5	5.5
17x	1		8.5	5.5
21o	1		8.5	5.5
23m	1		8.5	5.5
32g	1		8.5	5.5
41z	-1	"HOT"	-8.5	
45z	-1	REGION	-8.5	
47w	-1	POOLED	-8.5	-11
50x	-1	SCORE:	-8.5	
51j	1		8.5	
52i	-1	-4	-8.5	
60h	1		8.5	5.5
61w	1		8.5	5.5
86c	1		8.5	5.5
95o	1		8.5	5.5
95h	1		8.5	5.5
	Sum of negative ranks:		-42.5	-11*
	No. of non-zero ranks:		16	11

* Significant at the 5% level

of the sites discordant with phylogeny 1 is tightly clustered, and all of them fall into the "hot" region.   In contrast, the sites concordant with phylogeny 1 are found in all genomic regions.   The probability of finding all discordant sites in the "hot" region is only .022 even after taking into account the disporportionate number of site changes that occur in this region (42 of the 147 site changes).   This tight clustering of discordant sites into the "hot" region is exactly what is predicted by Templeton (1983a), and the clustering also means that the pooling procedure used above is an effective means of correcting for rate heterogeneity within the DNA molecule. Therefore, the mtDNA restriction site data favor phylogeny 1 over 2 at the 5% level of significance when rate heterogeneity within the mitochondrial genome is taken into account.

Besides mtDNA, some additional restriction site data are available from nuclear genes. There is one informative site in the globin gene maps of Zimmer (1981) that favors phylogeny 1 over 2 (Templeton, 1983a), and there are two informative sites in primate ribosomal DNA (PVU II a and b in the maps given by Wilson *et al.*, 1984), one of which favors 1 over 2 and the other which favors 2 over 1. Besides restriction sites, restriction site mapping also reveals some structural changes in the DNA such as deletions and insertions. No phylogenetically informative structural changes are present in the mtDNA, but one exists in the globin DNA (Zimmer, 1981) and one exists in the ribosomal DNA (Wilson *et al.*, 1984). Both favor phylogeny 2 over 1 (under the assumption that single events are more likely than convergent events). Pooling all the restriction mapping data from nuclear DNA, a total of 2 changes favor 1 and 3 changes favor 2. Hence, the nuclear restriction maps weakly favor 2 over 1, but the result is obviously without statistical significance.

III. DNA SEQUENCE DATA

Brown *et al.* (1982) sequenced 896 base pairs of mtDNA of humans, the African apes, orangutans and gibbons. This data set was analyzed using individual nucleotide sites as the statistical units and with a scoring procedure that ranked single events as more likely than convergent events, and transitions as more likely than transversions because of an observed strong transitional bias (Brown *et al.*, 1982). The results of this analysis are given in Templeton (1983b), with the test of phylogeny 1 versus 2 giving a sum of negative ranks of -133 with 22 non-zero scores. This result weakly favors phylogeny 2 over 1, but the level of significance (one-tailed) is only between .5 and .45. Hence, the mtDNA sequence data are virtually neutral as to which of these two phylogenies is the better.

More recently, Goodman *et al.* (1984) summarized the DNA sequences of the $\psi\eta$ globin pseudogene in humans, chimpanzees, gorillas and owl monkeys. Using the same ranking criteria as for the mtDNA, one obtains the results given in Table II. Using the Wilcoxon test, phylogeny 2 is favored over 1 with a one-tailed probability of .055 or a two-tailed probability of .11. Although the results are not statistically significant, they clearly favor phylogeny 2 over 1.

Table II. Wilcoxon matched-pairs signed ranks test of phylogeny 2 versus 1 and 2 versus 3 based on the DNA sequence of the primate $\psi\eta$ blogin pseudogene. Phylogeny 3 is given in Templeton (1983a) and groups humans and gorillas together as a clade.

Site	2 vs. 1		2 vs. 3	
	Score	Rank	Score	Rank
1013	1	+4	1	+4
1108	1	+4	1	+4
1363	0	--	-1	-4
1548	1	+4	1	+4
1811	1	+4	1	+4
1905	1	+4	1	+4
1988	1	+4	1	+4
2285	-1	-4	0	--
Sum of negative ranks:	-4		-4	
No. of non-zero ranks:	7		7	

As will soon be discussed, the nuclear DNA sequence data and the mtDNA restriction site data give the strongest statistical discriminations between phylogenies 1 and 2. Unfortunately, their respective discriminations are discordant. Although the nuclear DNA results are not significant at the 5% level, it is possible that such a discrepancy may be biologically real. The seemingly discordant patterns can arise because mtDNA is maternally inherited in many animals and because there may be selective differences betwene mitochondrial and nuclear genomes in interspecific hybrids.

If phylogeny 2 is true, the mtDNA results could be explained by female gorillas mating with male chimpanzees after the chimpanzee-human split. The female gorillas that engaged in such interspecific matings and their female progeny would then have to associate with chimpanzees coupled with selection against the gorilla nuclear genome but not the gorilla mitochondrial genome. However, this pattern of gene flow is incompatible with what is known of the social structure of these species. Gorilla females travel and feed in groups of other gorilla females (Wrangham, 1979), thereby making it unlikely for females to associate with the wrong species. Male mediated introgression is far more likely for these primates since it would require only a mating but no social integration. Such male mediated gene flow would cause nuclear introgression but no mitochondrial introgression with no need to assume any sort of selection. These arguments imply that if, indeed, a real discrepancy exists between the nuclear and mitochondrial based phylogenies, it is more likely that the mitochondrial phylogeny has been less influenced by past episodes of interspecific hybridization.

At this point it is still premature to invoke interspecific hybridization because the nuclear/mitochondrial DNA discrepany remains to be statistically substantiated. Moreover, the inferences about which phylogeny is favored for the pseudogene data are strongly determined by the sequence of the outgroup, which in this case is the owl monkey. Unfortunately, the owl monkey is not the most appropriate outgroup for this type of analysis, as it is much less closely related to humans or African apes than orangutans or gibbons, the outgroups available for the mitochondrial DNA analysis. Consequently, the conclusions based on the results shown in Table II must be regarded as tentative until sequence data becomes available for orangutans or gibbons.

IV. AMINO ACID REPLACEMENTS

At the protein level, humans and the African apes are extremely similar. Goodman *et al.* (1983) report that one amino acid replacement in the hemoglobins favors phylogeny 2 over 1. Unfortunately, monkeys are once again the outgroup, so the authors cautioned that a cleaner resolution would have to await data from orangutans or gibbons. In any event, this result is without statistical significance.

V. KARYOTYPIC EVOLUTION

Three recent karyotypic studies have been performed upon these primate species, with one concluding that the karyotypic data favor 2 over 1 (Yunis and Prakash, 1982) and the other two concluding that 1 is favored over 2 (Marks, 1982; Stanyon and Chiarelli, 1982). Pooling the data from all of these studies (which are overlapping in part), one obtains the results given in Table III. Here, the scoring procedure is simply whether the karyo- typic change evolved only once under a given phylogeny versus more than once. Six karyotypic changes favor phylogeny 1 over 2,

TABLE III. The informative karyotypic changes that distinguish between phylogenies 1 and 2.

Type of Change	Chromosome No.	Phylogeny Favored
C terminal bands	Several	1
Pericentric inversion	16	1
Pericentric inversion	12	1
Pericentric inversion	2q	1
Pericentric inversion	2p	2
Pericentric inversion	9	2
Pericentric inversion	4	1
Deletion	15	2
Paracentric inversion	7	2
NOR	Several	1

and four favor 2 over 1. Consequently, the karyotypic data weakly favor phylogeny 1 over 2.

VI. DNA-DNA HYBRIDIZATION

Sibly and Ahlquist (1984) recently compared the single-copy nuclear DNA sequences of the hominoid primates using DNA-DNA hybridization. This type of data is inherently a distance measure and cannot be analyzed with statistical procedures assuming discrete character states. Sibley and Ahlquist (1984) argued for phylogeny 2 on the basis of a t-test that shows that humans and chimpanzees have a significantly smaller distance than either species does to gorillas. This implies that phylogeny 2 is true. However, the results of this test procedure are not so clean-cut as presented in Sibley and Ahlquist (1984). Using the data given in Table I of Sibley and Ahlquist (1984) on the *Gorilla-Pan troglodytes* and *Gorilla-Homo* comparisons, the resulting t-value is -2.78 with 18 degrees of freedom.   This implies that the gorilla-chimpanzee distance is significantly smaller (at the 1% level) than the gorilla-human distance. This significant t-test implies that phylogeny 1 is corect. Obviously, the data set of Sibley and Ahlquist (1984) contains internal inconsistencies with regard to the correct phylogeny. Moreover, the t-tests are not appropriate for testing phylogeny in this case because several informative t-tests are possible, but they are not independent even though internally inconsistent.

An alternative analysis (Templeton, 1985b) is based upon a modification of the Q statistic of Pielou (1979, 1983) in which distance ranks are compared across columns as opposed to diagonals. The modified Q statistic measures how well an hypothesized phylogeny orders the ranks of the genetic distances with larger Q values indicating better fits. The difference between the Q value for phylogeny 1 and that of phylogeny 2 is -4, implying that the DNA-DNA hybridization data of Sibley and Ahlquist (1984) do indeed favor phylogeny 2 over 1. The significance of this difference in Q values can be evaluated exactly by generating all 120 equi-probable (under the null hypothesis of no phylogenetic discrimination) permutations of five informative distances (Templeton, 1985b). This distribution is given in Table IV. From Table IV, the probability of obtaining a delta Q value of -4 or smaller is 0.167. Hence, the data of Sibley and Ahlquist (1984) do not support the rejection of phylogeny 1 in favor of phylogeny 2 even with a one-tailed test. Table IV also shows that if a delta Q value of -6 had been obtained, phylogeny 1 could have been rejected at the 5% level in favor of phylogeny 2. The lack of resolution of Sibley and Ahlquist's data set is therefore attributable to the data set's internal inconsistencies that were noted earlier.

Table IV. The probability distribution of the delta-Q statistic when the third and fourth species in the order are interchanged.

Delta-Q value	Probability
-6	.0333
-5	.0667
-4	.0667
-3	.0667
-2	.1000
-1	.0667
0	.2000
1	.0667
2	.1000
3	.0667
4	.0667
5	.0667
6	.0333

Another reason for being cautious about the conclusion of Sibley and Ahlquist (1984) is that the recent study by O'Brien *et al.* (1985), using the same technique, did not replicate the Sibley and Ahlquist results. O'Brien *et al.* (1985) obtained the smallest distance between humans and gorillas rather than between humans and chimpanzees. It is not clear if this discrepancy is due to different laboratory procedures or if it is due to the intraspecific variation being of a comparable order of magnitude to the interspecific variation. Obviously, conclusions based upon the hybridization data must await a more detailed assessment of the magnitude of intraspecific variation in the relevant hominoid species. Hence, the DNA-DNA hybridization data have not resolved between phylogenies 1 and 2, and it is not even clear if this type of data even weakly favors 1 over 2 or vice versa given the discrepancies between laboratories.

VII. ISOZYME GENETIC DISTANCES

The delta Q test can be used to evaluate the statistical significance of genetic distance matrices based on isozymes. Using the isozyme data of Bruce and Ayala (1979), the delta Q value is -1.5, indicating a very weak and statistically non-significant favoring of 2 over 1. Moreover, Nozawa *et al.* (1982) looked at nine additional loci, and when their data are combined with that of Bruce and Ayala (1979), the resulting delta Q is -0.5. Lucotte and Lefebvre (1981) also performed an isozyme analysis, and the delta Q value stemming from their data is 0. The most extensive isozyme survey to date is that of O'Brien *et al.* (1985) who examined some 46 loci in these primate species. Because O'Brien *et al.* (1985) only examined one species of chimpanzee, the range of delta Q values for their data is -2 to 2. The observed delta Q was 2, indicating that this data set favored phylogeny 1 over 2 as strongly as possible and with no internal inconsistencies. Unfortunately, because there are so few informative distance contrasts in this data set, it is impossible for the delta-Q value to be significant at the 5% level.

It is also possible to analyze isozyme data in terms of character states rather than distances. This type of analysis should prove to be more powerful than the delta-Q test, but unfortunately O'Brien *et al.* (1985) did not include an outgroup species, which is necessary to perform such an analysis. Thus, all one can conclude at present is that the isozyme data supports phylogeny 1 over 2, but how strong this support is has yet to be determined.

VIII. MORPHOLOGICAL DATA

Phylogeny 1 is the pattern most often suggested by morphological studies (e.g., Oxnard, 1981; Martin, 1985). Moreover, recent studies using comparative electromyography on muscle usage during locomotion (Stern and Susman, 1981; Vangor and Wells; 1983), comparative joint motion studies (Prost; 1980), allometric studies on both recent and fossil hominoids (Aiello, 1981), the fossil finds that bipedality is very old in the hominids with no vestiges of knuckle-walking adaptations (Johanson and White, 1979; Lovejoy, 1981; Charteris *et al.* (1982), and an analysis of the type of vertebral pathologies found in the oldest hominid fossils (Cook *et al.*, 1983) all suggest that knuckle-walking was never a primitive condition in the hominids. It is very difficult to maintain that knuckle-walking is not primitive under phylogeny 2 because to do so implies that knuckle-walking, which represents a complex suite of traits, must have evolved independently in gorillas and chimpanzees. However, this interpretation fits well with phylogeny 1 since knuckle-walking could have evolved in the common ancestor of chimpanzee and gorillas after the hominid lineage had split off (Templeton, 1983a). Hence, the morphological data is far more easily explained under phylogeny 1 than 2. However, there is no obvious way to place a probabilistic confidence limit upon just how strong this support is.

IX. SUMMARY CONCERNING BRANCHING ORDER INFERENCE

Of all the data considered, only the mtDNA restriction sites could discriminate between phylogenies 1 and 2 at the 5% level. The phylogeny favored by the mtDNA analysis is phylogeny 1, and this phylogeny is also favored by the karyotypic data, the isozyme genetic distances, and morphology. In contrast, the DNA sequence data, nuclear restriction maps, amino-acid replacements, and possibly the DNA-DNA hybridization data favor 2 over 1, but in all cases the discrimination is not significant at the 5% level. Consequently, at present, all data can be viewed as statistically consistent with phylogeny 1, but phylogeny 2 is significantly inconsistent with the mtDNA restriction maps.

X. THE MOLECULAR CLOCK

Besides discriminating between branching orders, the molecular data can also provide statistically powerful inferences concerning the molecular clock hypothesis. This hypothesis was previously tested with the mtDNA restriction site data using a non-parametric test, and it was rejected in favor of the hypothesis that humans are evolving more slowly than the African apes (Templeton, 1983a). This conclusion is also robust to the different mutational allocation schemes one obtains by changing convergent events from gains into multiple-losses (Templeton, 1985a).

The data summarized by Goodman *et al.* (1984) on globin pseudo-genes can also be used to test the validity of the molecular clock hypothesis for nuclear DNA. Starting with the common ancestor of two lineages to be compared, the number of mutations allocated to the branch leading to one lineage is determined, and likewise for the second lineage. By tracing back to the common ancestor of the lineages being compared, the test is sensitive only to the relative rates of evolution and not the absolute rates. A major advantage of a relative rate test is that no fossil calibrations are required (Langley and Fitch, 1974).

There is no difficulty in allocating unique substitutions, as all estimation algorithms place such unique events in the terminal branches. However, parsimony will allocate the phylogenetically informative sites for a particular species pair differently under different phylogenies. Because phylogenies 1 and 2 cannot be unambiguously resolved with these data, the test is repeated under both allocations. Once the substitutions have been identified, a score of 1 is assigned to each site if it became fixed during the relevant time period and 0 otherwise. Rates in different lineages are obtained simply by calculating the difference between their scores at each relevant site. The resulting signed scores are then ranked, and a Wilcoxon matched-pairs signed rank test is used to convert the results into a statistical statement.

Table V gives the results of such tests for chimpanzee-human and gorilla-human and gorilla-human comparisons under both phylogenies 1 and 2. The clock hypothesis is rejected for the gorilla-human contrast only under phylogeny 1, but the clock is rejected for the chimpanzee-human contrast regardless of which phylogeny is true. Moreover, there are only two mutations that have ambiguous allocations in the chimpanzee-human contrast. Excluding those two sites, humans and chimpanzees have some 26 species-unique substitutional differences in their ψn globin pseudogene,

Table V. Wilcoxon matched-pairs signed ranks test of the hypothesis of equal rates of evolution in humans versus African apes, based on the DNA sequence of the primate $\psi\eta$ globin pseudogene.

Phylogeny	Chimps vs. humans		Gorillas vs. human	
	1	2	1	2
No. of non-zero ranks	28	28	33	33
sum of negative ranks	-101.5*	-116*	-119*	-238

   * Significant at the 5% level in a two-tailed test.
  ** Significant at the 1% level in a two-tailed test.

with 24 of them occurring in chimpanzees and only 2 in humans. Obviously, the rate of molecular evolution for this nuclear gene is much slower in humans than in chimpanzees.

This result is consistent with tests of the molecular clock performed with the mtDNA restriction site data (Templeton, 1983a). In addition, Marks (1982) concluded that humans have a slower rate of karyotypic evolution than the African apes, the Stanyon and Chiarelli (1983) concluded that humans have a slower rate of karyotypic evolution than gibbons. This pattern raises an interesting question: what factor or factors could cause such a slowdown in the human lineage that would simultaneously affect both nuclear and mitochondrial genome evolution?

The neutral theories of Ohta (1976) and Kimura (1979) provide a straightforward answer to this question. Ohta (1976) and Kimura (1979) have shown that under a model of neutral and nearly neutral mutations, the rate of evolution is a decreasing function of population size and of generation time. One of the important features of human evolution has been a dramatic slowdown in rate of development and an attendant increase in generation time over the past 3 million years (Cutler, 1976; Lovejoy, 1981). In addition, the human lineage greatly expanded in size, starting at the very least with *Homo erectus* which had a fossil distribution covering the entire Old World and not just Africa (Lovejoy, 1981). The simultaneous increase in both generation time and population size would result in a dramatic slowdown in human evolutionary rates under Ohta's (1976) and Kimura's (1979) theory.

XI. CONCLUSION

Only the mtDNA restriction data provide a statistically significant (5% level) resolution of the human/African ape phylogenetic relationships.

This resolution favors phylogeny 1 of Figure 1. The strongest support for phylogeny 2 comes from the DNA sequence data on globin pseudogenes, but the results are not significant at the 5% level. Although the sequence data at present are statistically compatible with phylogeny 1, the trend observed in the nuclear sequence data raises the possibility of a discrepany between the nuclear and mitochondrial-based phylogenies. This potential discordance may be real if past interspecific hybridization occurred. If so, the social behavior of these primates is such that it is more likely that the mitochondrial phylogeny is the true one. This phylogeny, which groups the African apes into a clade, also provides a simpler explanation of the non-molecular data.

Regardless of which phylogeny is true, the molecular data indicate that the human lineage is evolving more slowly than the African ape lineages. This result can be explained by the increased generation length and population size of humans.

ACKNOWLEDGMENTS

I wish to thank Dr. Walter Fitch for his constructive criticisms of an earlier draft.

REFERENCES

Aliello, L. C. (1981). The allometry of primate body proportions. *Symp. Zool. Soc. Lond.* **6**, 331-358.

Brown, W. M., Prager, E. M., Wang, A., and Wilson, A. C. (1982). Mitochondrial DNA sequences of primates: tempo and mode of evolution. *J. Mol. Evol.* **18**, 225-239.

Bruce, E. J., and Ayala, F. J. (1979). Phylogenetic relationships between man and the apes: electrophoretic evidence. *Evolution* **33**, 1040-1056.

Charteris, J., Wall, J. C., and Nottrodt, J. W. (1982). Pliocene hominid gait: new interpretations based on available footprint data from Laetoli. *Am. J. Phys. Anthrop.* **58**, 133-144.

Cook, D. C., Buikstra, J. E., DeRousseau, C. J., and Johanson, D. C. (1983). Vertebral pathology in the Afar Australopithecines. *Am. J. Phys. Anthrop.* **60**, 83-101.

Cutler, R. G. (1979). Evolution of human longevity and genetic complexity governing aging rate. *Proc. Natl. Acad. Sci. USA* **72**, 4664-4668.

DeBry, R. W., and Slade, N. A. (1985). Cladistic analysis of restriction endonuclease cleavage maps within a maximum likelihood framework. *Syst. Zool.*

Ferris, S. D., Wilson, A. C., and Brown, W. M. (1981). Evolutionary tree for apes and humans based on cleavage maps of mitochondrial DNA. *Proc. Natl. Acad. Sci. USA* **78**, 2432-2436.

Goodman, M., Braunitzer, G., Staugl, A., and Schrank, B. (1983). Evidence on human origins from haemoglobins of African apes. *Nature* **303**, 546-548.

Goodman, M., Koop, B. F., Czelusniak, J., Weiss, M. L., and Slightom, J. L. (1984). The η-globin gene. Its long evolutionary history in the β-globin gene family of mammal. *J. Mol. Biol.* **180**, 803-823.

Johanson, D. C., and White, T. D. (1979). A systematic assessment of early African hominids. *Science* **203**, 321-330.

Kimura, M. (1979). Model of effectively neutral mutations in which selective constraint is incorporated. *Proc. Natl. Acad. Sci. USA* **76**, 3440-3444.

Langley, C. H., and Fitch, W. M. (1974). An examination of the constancy of the rate of molecular evolution. *J. Mol. Evol.* **3**, 161-177.

Lovejoy, O. (1981). The origin of man. *Science* **211**, 341-350.

Lucotte, G., and Lefebvre, J. (1981). Distances electrophoretiques entre l'Homme, le Chimpanze (*Pantroglodytes*) et le Gorille (*Gorilla gorilla*) basees sur la mobilite des enzymes erythrocytaires. *Human Genetics* **57**, 180-184.

Marks, J. (1982). Evolutionary tempo and phylogenetic inference based on primate karyotypes. *Cytogenet. Cell Genet.* **34**, 261-264.

Martin, L. (1985). Significance of enamel thickness in hominoid evolution. *Nature* **314**, 260-263.

Nozawa, K., Shotake, T., Kawamoto, Y., and Tanabe, Y. (1982). Electrophoretically estimated genetic distance and divergence time between chimpanzee and man. *Primates* **23**, 432-443.

O'Brien, S. J., Nash, W. G., Wildt, D. E., Bush, M. E., and Benveniste, R. E. (1985). Riddle of the giant panda's phylogeny: a molecular solution, manuscript.

Ohta, T. (1976). Rate of very slightly deleterious mutations in molecular evolution and polymorphism. *Theor. Pop. Biol.* **10**, 254-275.

Oxnard, C. E. (1981). The place of man among the primates: anatomical, molecular and morphometric evidence. *Homo* **32**, 149-176.

Pielou, E. C. (1979). Interpretation of paleoecological similarity matrices. *Paleobiol.* **5**, 435-443.

Pielou, E. C. (1983). Spatial and temporal change in biogeography: Gradual or abrupt? *In* "Evolution, Time and Space: The Emergence of the Biosphere" (R. W. Sims, J. W. Price and P. E. S. Whalley, eds.), pp. 29-56. Academic Press, New York.

Prost, J. H. (1980). Origin of bipedalism. *Am. J. Phys. Anthrop.* **52**, 175-189.

Sibley, C. G., and Ahlquist, J. E. (1984). The phylogeny of the hominoid primates, as indicated by DNA-DNA hybridization. *J. Mol. Evol.* **20**, 2-15.

Stanyon, R., and Chiarelli, B. (1982). Phylogeny of the *Hominoidea*: the chromosome evidence. *J. Hum. Evol.* **11**, 493-504.

Stanyon, R., and Chiarelli, B. (1983). Mode and tempo in primate chromosome evolution: implications for hylobatid phylogeny. *J. Hum. Evol.* **12**, 305-315.

Stern, J. T., Jr., and Susman, R. L. (1981). Electromyography of the gluteal muscles in Hylobates, Pongo, and Pan: Implications for the evolution of hominid bipedality. *Am. J. Phys. Anthrop.* **55**, 153-166.

Templeton A. R. (1983a). Phylogenetic inference from restriction endonuclease cleavage site maps with particular reference to the evolution of humans and the apes. *Evolution* **37**, 221-244.

Templeton A. R. (1983b). Convergent evolution and nonparametric inferences from restriction data and DNA sequences. *In* "Statistical Analysis of DNA Sequence Data" (B. S. Weir, ed.), pp. 151-179. Marcel Dekker, New York.

Templeton A. R. (1985a). Non-parametric phylogenetic inference from restriction cleavage site. Submitted to *Mol. Biol. Evol.*

Templeton A. R. (1985b). The phylogeny of the hominoid primates: A statistical analysis of the DNA-DNA hybridization data. *Mol. Biol. Evol.* **2**, 420-433.

Vangor, A. K., and Wells, J. P. (1983). Muscle recruitment and the evolution of bipedality: evidence from telemetered electromyography of spider, wooly and patas monkey. *Annales des Sciences Naturelles, Zoologie* **5**, 125-135.

Wilson, G. N., Knoller, M., Szura, L. L., and Schmickel, R. D. (1984). Individual and evolutionary variation of primate ribosomal DNA transcription initiation regions. *Mol. Biol. Evol.* 1, 221-237.

Wrangham, R. W. (1979). On the evolution of ape social system. *Social Sci. Info.* **18**, 335-368.

Yunis, J. J., and Prakash, O. (1982). The origin of man: a chromosomal pictorial legacy. *Science* **215**, 1525-1530.

Zimmer, E. A. (1981). Evolution of globin genes. Ph.D. Dissertation, University of California, Berkeley.

PART IV. MODELS AND EVIDENCE OF SPECIATION

SEXUAL SELECTION AND SPECIATION[1]

Hampton L. Carson

Department of Genetics
University of Hawaii
Honolulu, HI 96822

ABSTRACT

In "The Descent of Man," Darwin proposed that sexual selection had "...acted powerfully on man, as on many other animals." No exposition of the all-pervasive quality of this kind of selection is more perceptive than the 350 text pages that follow the above statement. Starting with the Crustacea, Darwin surveys sexual phenomena through the arthropod classes and orders to the fish, reptiles, amphibians, birds and mammals. Remarkably, geneticists oriented towards populations have given little attention to the genetic properties of sexual selection. An exception was R. A. Fisher whose brilliant chapter in "Genetical Theory" stimulated the modern theoretical work of Peter O'Donald and indeed, of the co-chairman of this conference, Samuel Karlin (1978). Experimental studies, however, have been few. During a visit to Hawaii the late Th. Dobzhansky recognized the possibilities of an experimental approach to sexual selection, labeling the endemic drosophilas "The Birds of Paradise of the Drosophila World." Over the past 10 years our field and laboratory investigations have centered on population genetics and variability of a geologically very recently-evolved species, *Drosophila silvestris*, endemic to the island of Hawaii. Like many species in Hawaii, it differs from its close relatives principally in secondary sexual characters of males; electrophoretic, chromosomal and ambient ecological differences

[1] The experimental and geographical work of the author has been supported by grant DEB 792-6692 of the National Science Foundation.

are minor although genetic variance remains high. Two races based on a large secondary sexual difference are recognized; many natural local populations and some laboratory studies show differences in sexual behavior. Laboratory study of wild-caught specimens and their progeny reveals that the heritability of one of these characters, male tibial foreleg cilia, is high. An interracial cross shows this cilliary difference to have a genetic basis consisting of genes on at least two autosomes in addition to the X chromosome. These cilia, which are brought into tactile contact with the female only at a final crucial stage of courtship, respond to artificial selection and appear to be maintained in natural populations by some sort of stabilizing selection. Discriminatory behavior of females also responds to artificial selection. Census population sizes are ostensibly low in nature; epigamic sexual selection results in effective sizes that are even smaller. Thus, among laboratory tests of males emerging in a single stock, about one-third accomplish two-thirds of the observed copulations. It is proposed that the genetic basis of adaptations to the sexual environment are easily perturbed by population events such as hybridization, population crashes or founder effects. If such a genetic disorganization occurs, selection must rebuild the sexual adaptations of the daughter population, causing a shift to a novel balance. Such selective alterations in the pattern of sexual reproduction may be pivotal in the process of speciation in forms, from Crustacea to man, that display sexual selection.

I. INTRODUCTION

Contemporary formulations of the sexual selection hypothesis have generally retained the original scheme proposed by Darwin (1871) in his book, "The Descent of Man and Selection in Relation to Sex." Two facets are recognized: intrasexual and epigamic selection. Intrasexual selection is most characteristic of the male sex. Thus, in male-male interactions, behavior and its morphological and physiological embellishments, serves inter-male competition providing for position relative to access to females. A second and apparently equally important aspect concerns an active choice of sexual partner that may be exercised by the female. Certain male characters appear to relate exclusively to this facet. In both male competition and female choice the individuals participating appear to be engaging in a process of maximizing the Darwinian fitness of the participants in the basic prezygotic reproductive unit, with the energy investment of the female greater than that of the male.

As Darwin argued, these accompaniments of the sexual process are widespread in animals, although they are probably absent in most of the invertebrate classes below the arthropods. He recounts the apparent action of sexual selection from the various arthropod orders to the fish, amphibians, reptiles, birds and mammals.

II. SCARCITY OF EMPIRICAL DATA ON SEXUAL SELECTION

For various curious reasons, this view of the sexual process has not been closely followed by evolutionists in the modern neoDarwinian era. Refocused attention, however, has followed the strong interest in the evolutionary aspects of ethology following Wilson's (1975) treatise on sociobiology. Neglect of the theory of sexual selection was not confined to the ecological viewpoint. Most genetical theorists, with the exception of Fisher (1930), Karlin and O'Donald (1978), and O'Donald (1980), have not considered it as an important ingredient of the genetic system, related to the evolutionary process.

Throughout the extensive work on the genetics of natural populations between 1940 and 1970, little attention was given to intraspecific natural variability in characters relating to sexual behavior and courtship. Although interspecific comparative ethology appeared to be sometimes useful in phylogeny (e.g., Spieth, 1968), the contribution of secondary sexual characters to the variability system has been only very recently approached from the empirical point of view, especially that variability found in unhybridized single natural populations. The papers of Cade

(1981), Carson *et al.* (1982) and Carson (1985a) are examples of attempts to obtain data on the subject of natural genetic variance in secondary sexual characters.

III. SEXUAL SELECTION AND SEXUAL ISOLATION

I believe one of the reasons for a failure to appreciate the importance of sexual selection is due to a very strong emphasis on the role of reproductive isolation in the thinking of evolutionary biologists in this century. Viewed narrowly, sexual selection is an intrapopulational process and does not appear to contribute directly to the population splitting process characteristic of species formation. Such important thinkers as Mayr (1963) and Dobzhansky (1970) have been strong advocates of the "biological" species concept which, as Paterson (1982) has argued, really should be called the "isolation" concept of the species. In this view, the new or incipient species is best characterized in terms of the origin of those biological conditions that serve to isolate its gene pool from all others.

IV. DO ISOLATING "MECHANISMS" ARISE THROUGH SELECTION?

Two further important implications arise from a consideration of the biological species concept. First, natural selection is considered to be the agent responsible for the acquiring of "mechanisms" that serve to isolate a particular gene pool from other similar pools. At various stages during the extended process of species formation, it is proposed that natural selection serves to "reinforce" or deepen any tendencies to isolation. A second implication from this is that any hybridization is considered to be a serious threat to the genetic integrity of the incipient species. Thus selection favoring isolation is considered an essential ingredient of the speciation process in its initial phases.

V. ISOLATION AND ALLOPATRY

The above theory has been a highly persuasive one for a generation of evolutionists, myself included. Empirical evidence for it, however, is not strong.   Indeed, it is possible that the scheme is seriously flawed theoretically as well (Paterson, 1982). This author points out that most credible evidence argues that speciation modes are predominantly allopatric, a view also strongly held by Mayr. If this is true, it means that selection for isolation has few opportunities for refinement in natural populations since the entities concerned are physically separated from one another at the crucial stages. It follows that the genetic basis of the widely observed reproductive isolation between species may simply be acquired as a pleiotropic effect of some other selective influence, and is not the outcome of selection for isolation *per se*. This "incidental" theory of the origin of reproductive isolation has been strongly championed by Muller (1942). I adhere to this view but wish to point out here that both pre- and post-mating isolation need not characterize the crucial early stages of speciation. To use the term "mechanism" to refer to such conditions is prejudicial, as Paterson has argued.

In this connection, it should not escape us that sexual selection is basically an intrapopulational process for maximizing fitness. Under sexual selection, the details of the sexual processes and behavior of a species can be profoundly altered genetically. Such events might incidentally confer on individuals from the population characters that may reduce the ease of interbreeding between members of this population and other similar ones, if and when they happen to regain sympatry. When two populations have been subjected to different modes of sexual selection, they acquire different pre-mating characters. Some facet or facets of these pre-mating characters may be responsible for mating incompatibilities.

As two populations become genetically more distinct and acquire different adaptations, post-mating developmental disturbances may be observed if hybrids are formed. As in pre-mating incompatibilities, these deleterious effects appear to be the by-products of certain kinds of genetic divergence. In my view, these effects do not serve as selective forces that elicit and strengthen pre-mating incompatibilities, as is often assumed.

VI. SPECIATION IN PLANTS

In plant population biology, reproductive isolation between newly-formed species of flowering plants is not a general characteristic. For this reason, the plant biologist interested in speciation is less concerned with the origin of reproductive isolation than is the animal biologist. Indeed, this large difference has been responsible for the disparate views held by plant and animal biologists on the subject of the biological species. It has been my view that such a lack of unification of speciation theory suggests theoretical weaknesses in theories that depend on concepts not applicable to both kingdoms. I feel that this weakness stems mostly from adherence by the zoologist to the concept that equates speciation to a selective process involving the buildup of "isolating mechanisms."

VII. THE DISORGANIZATION-REORGANIZATION THEORY
OF SPECIATION

In 1982, I proposed the following unifying theory of species formation. The intention was not to add just another mode of species formation. Rather, it was an attempt to identify basic processes that appear to be common to speciation in plants and animals generally. The theory rests on a large data base that suggests that the genetic structure of a species, both animal and plant, is basically balanced and multigenic. This is organized into a complex genetic system. The theory calls for a disorganization of this system by a basically stochastic populational event such as a founder effect, a population crash, an interspecific hybridization or a sudden release from the constraints of normal natural selection, as in a population flush.

The disorganization phase is considered to be accomplished over a small number of generations and is primarily the outcome of the process of genetic recombination within the genome. It is noteworthy that the key events following the founder effect, for example, have to do with the disturbance of gametic equilibrium, effectively causing a new genetic milieu in the small regenerating population.

The active phase of speciation following the disorganization phase consists of the rebuilding of a new genetic organization in the daughter population. Accordingly, this idea does not emphasize simple random drift of gene frequencies as the driving force for the appearance of differences. Rather, the restructuring of the gene pool by natural or sexual selection is required to take place under altered genetic, sexual or ambient environments.

Essentially, then, the theory calls for the disorganization phase to be brief and to be followed by a quantum phase involving reorganization of the genome by selection around a particular selective mode. This mode may not necessarily take a strong new evolutionary direction. For example, a small population may undergo a disorganizing event that effectively destroys a well-established balanced system of sexual selection. As reorganization begins, the building of a new sexual selection system would not, due to the loss or disruption of key blocks of genes, be able to build back the same detailed system that existed previously. The changed system that results would have the potential of forming the nucleus of a novel behavioral syndrome in the population which might at some later time be recognized as a new species.

Alternatively, but in like manner, the processes of genic disorganization might specifically affect adaptive complexes of genes relating the organism to the ambient rather than the sexual environment (see Carson, 1985b). Under such circumstances certain derivatives of the disorganized gene pool may be more easily guided by natural selection into new adaptive relationships. As the precision of the adaptations becomes intensified by selection, this population might also come to be recognized as a new or incipient species. In certain flowering plants, such an event could give rise to a new species that is allopatrically separated from the parent species that is, at the same time, differentially adapted. Such entities may show little or no intrinsic reproductive isolation yet may serve as good examples of incipient species, as long as spatial isolation is retained.

VIII. AN EVOLUTIONARY PARADIGM: THE PICTURE-WINGED DROSOPHILIDS OF HAWAII

The basic facts on the diversification of this group of insects are to be found in several reviews (e.g., Carson *et al.*, 1970; Carson and Kaneshiro, 1976). Recent emphasis has centered on the detailed examination of the population biology of several selected species, particularly *Drosophila silvestris* (Perkins) (see Carson, 1982a). In this manner, we have sought clues to the mode and process of evolution within the group.

There are somewhat over 100 species of the large picture-winged flies, distributed over the six main islands of the archipelago. With only a few exceptions, each species is endemic to a single island or to several islands that are closely linked geographically, such as the Maui-complex (Maui, Molokai and Lanai). Of the 25 species found on the island of Hawaii, the newest and largest island, all but one are clearly endemic to the island. Potassium-argon and magnetic declination dating methods indicate that the island is less than a half-million years old. To the northwestward, each island is successively older, with Kauai, the oldest high island, showing an age of approximately 5.6 million years

When studied morphologically and chromosomally, these 100 species appear to represent a monophyletic group. In fact, the non-picture winged species, numbering perhaps 700, also show adherence to a basic and quite conservative body plan and no one has seriously challenged Throckmorton's (1975) bold suggestion that not only the picture wings but all the rest of the endemic species in the family *Drosophilidae* could have been derived from a single ancestral introduction into the island chain. Little data exist as to the time before the present that this introduction occurred. The fact that the islands northwest of Kauai have been eroded to near sea level means that ancestors of these characteristically high-altitude flies would not be expected on these islands at the present day, so direct evidence of their existence there appears to be lacking, although few collecting attempts have been made. On the basis of micro-complement fixation studies of a small number of species using a single larval protein, Beverley (1979) concludes that considerable evolution must have occurred on ancient islands now submerged. These studies show, however, that with the exception of the *adiastola* subgroup (about 15 species) all of the remaining picture-winged species (a total of about 85 species) are so close together phylogenetically as to be indistinguishable immunologically on the basis of this one protein.

Accordingly, the bulk of the picture-winged species are not only monophyletic but are also probably less than 5 million years of age. Despite this, the species are divisible into four or five subgroups based on morphology and similarity of polytene chromosome sequences. Even within each subgroup, there is great behavioral, morphological and ecological diversity. Ecological diversity embraces not only a variety of host plants but some species have become adapted to the extremely xeric forests of the leeward slopes whereas others are confined to the windward rainforests. The species range in altitude from 50 to 2500 meters above sea level.

IX. SEXUAL DIMORPHISM IN THE PICTURE-WINGED DROSOPHILIDS

Mention was just made of various adaptations relating to the ambient environment. Much more remarkable, however, is the extreme sexual dimorphism displayed in many of these species. This involves structural embellishments of the male antennae, aristae, maxillary palps, forelegs, labellum, face color, wing shape and wing maculations. These are spoken of as "embellishments" since they appear to have evolved as elaborations of the behavioral use of these body parts in male intrasexual contests and in courtship of females. Thus the most striking species-to-species variations involve characters that suggest an overriding intraspecific role for sexual selection.

X. SEXUAL SELECTION IN *DROSOPHILA SILVESTRIS* FROM HAWAII ISLAND (see review in Carson, 1982a)

This species is endemic to the island and is quite widespread, existing in small populations on all five volcanoes. There are two allopatric morphotypes or races based on the number of long recurved cilia on the tibia of the male foreleg. This character is evidently related to epigamic sexual selection as they are put to use only for a few seconds in the lifetime of the male. In his final approach, after a long series of courtship movements directed at the female, the dorsal hairy surface of the male tibia is vibrated against the dorsal surface of the abdomen of the female. These cilia are not used in male-to-male encounters.

Populations on the south and west side of Hawaii island (Kona-side) have about 40 of these cilia, deployed in two marginal rows on the tibia. This condition resembles that found in four closely related species, one on Hawaii (*D. heteroneura*) existing in partial sympatry with *silvestris*, and the others endemic to the older islands of Maui, Molokai and Oahu to the northwest.

Unlike Kona-side males, those from the north and east side of the island (Hilo-side) have a mean of 30-40 additional tibial cilia, a clearly apomorphic embellishment of the character. The facts indicate that these populations are phylogenetically newer and, conversely, specifying Kona *silvestris* as the ancestral type. Kona and Hilo races of *silvestris* are wholly allopatric, and no intermediate populations appear to exist.

In *D. silvestris,* as in a number of other Hawaiian species, a curious kind of one-sided mating preference exists such that females of certain populations strongly discriminate against males from various allopatric populations of the species. Kaneshiro (see review, 1983) has ascribed this to the tendency for a buildup of the strength of sexual selection in phylogenetically older populations. Thus, females from such ancestral populations are highly discriminatory in their choice of male, even among males of their own population. Contrariwise, females from the newer populations are less discriminatory against both their own and foreign males. Kaneshiro's experiments suggest that the population of Hualalai Volcano (having an ancestral cilia condition) is the oldest extant *silvestris* population found on Hawaii island. His results suggest that the Hilo-side apomorphic population from Kohala is derived and is in turn ancestral to the populations in the Kilauea area that are existing near the present area of vulcanism.

XI. CROSSES BETWEEN THE RACES *D. SILVESTRIS*

Since the publication of the 1982a review, several new studies of *silvestris* has been carried out. Although the one-sided mating preference situation makes certain hybrids difficult to obtain, it is usually possible to obtain crosses between any two *silvestris* populations. Hybrids are universally fertile in F_2's and backcrosses. Indeed, crosses between *silvestris* and its partially sympatric relative *D. heteroneura* are not difficult to obtain and also show fertility and high viability to the F_2. By analyzing the F_2's from these interspecific crosses, Val (1977) has shown that the morphological differences between the species depend on 15-19 segregating genetic differences. This substantial number does not include the many behavioral differences associated with lek and mating behavior. A small number of natural hybrids between these species have been observed at two localities, but breakdown if neither species has occurred.

Interfertility between th subspecies of *silvestris* has made possible a genetical analysis of the secondary sexual difference in cilia number between the races. Carson and Lande (1984) showed that there is an important segregating genetic unit on the X chromosome and others on at least two autosomes that contribute to the difference between the strains. The data provide evidence for the existence of genetic variance in one parent strain (Kilauea). Thus, the character may be spoken of as polygenic in nature. The existence of a fairly strong sex-linked factor underlying a secondary sexual character that is important in sexual selection is a very interesting finding. Despite these differences the strains involved are very similar biochemically (Sene and Carson, 1977).

XII. GENETIC VARIANCE FOR THE CILIA CHARACTER AT A SINGLE LOCALITY

A series of six isofemale lines of *D. silvestris* were established from wild female flies collected at the Olaa tract, Hawaii Volcanoes National Park, in 1978 (see Carson, 1985a). Cilia counts were made on about 20 F_1 males from each of these isolines. These were reared under uniform conditions in the laboratory and a sample of wild males captured at the same time as the isofemales was also studied. Highly significant differences in cilia number were found among these isolines; the heritability was calculated at .74 ± .33. A second study provides further evidence for the presence of genetic variance in this Olaa population (Carson and Teramoto, 1984). In 1979, one high and one low isofemale line was selected and, in succeeding generations, selection for cilia number was continued in the same direction as the F_1, using a cyclical family mode of selection. A substantial and highly significant difference in cilia number (a mean of about 25 cilia) was established between the two lines. Both lines, however, died out, due either to the deleterious effects of inbreeding or to disturbance by directional selection of a balanced polygenic system responsible for the determination of cilia number. Both influences may be present.

That some sort of stabilizing selective system may be operating in the natural population is indicated by the above data on artificial selection and by the data from a long-range study of cilia in the wild-caught males from this same population. The mean cilia number was stable from 1976 through 1978. In 1979 it rose significantly but fell back again to the previous level in 1980 (Carson, 1985a).

XIII. SEXUAL BEHAVIOR OF MALES FROM LABORATORY STOCKS OF *D. SILVESTRIS*

A series of experiments were begun (see Carson and Teramoto, 1980, and Spiess and Carson, 1981), and are continuing, in which attempts are made to document the operation of sexual selection in isofemale laboratory strains of *D. silvestris*. The following account represents a progress report on these experiments. In a variant of the experiment most commonly employed at present, the following procedure is being followed. An aliquot of 25 males of very similar age is taken from a section of the stock that was very well-nourished as larvae. These males are aged for four weeks in small groups (4 or 5) in regular food vials.

Following the maturation period, those flies still alive at four weeks of age (80-90%) were marked individually with small droplets of paint on the mesonotum. The flies were divided into two lots and placed into plexiglass cages (20 × 20 × 10 cm), containing moistened cellulose sponges and food cups. Various behavioral data are then obtained, including the scoring of male-to-male interactions, courtship movements, etc.

Table I. Performance of *D. silvestris* males in laboratory mating tests.

No. of times mated	No. of Males			No. of Matings	Percent males	Percent matings
	Kohala stock	Kilauea stock	Total			
0	34	42	76	0	30.4	0.0
1	24	33	57	57	36.4	30.3
2	15	19	34	68		
3	16	39	55	165	33.2	69.7
4	7	12	19	76		
5	4	3	7	35		
6	0	2	2	12		
Total	100	150	250	413	100.0	100.0

Following the male-male observations, two virgin females were introduced into each cage. As soon as a copulation occurred, the pair was covered with a shell-vial. After separation, the female was replaced with another virgin. Females not mating in approximately one-half hour were replaced with new virgins. This procedure was followed for about two hours each day for six consecutive days, after which the experiment was terminated. Males from two different stocks were tested in this manner with the results being very similar. Table I gives the basic result. Approximately one-third of the males, despite engaging in active male-male jousts and courtship movements, never succeed in copulating; apparently they are rejected by the females. Conversely, another one-third of the males engage in 3-6 copulations over the same time period. This one-third accomplishes about two-thirds of the matings. The final third of the matings are engaged in by males copulating once or twice during the six days. Far from mating at random, it seems that mating success is concentrated in males having certain characteristics. The nature of these characteristics and their possible genetic basis is currently being investigated.

Evidence that sexual selection is important in laboratory stocks as well as in nature provides an explanatory framework for a puzzling observation of long standing in the rearing of many Hawaiian picture-winged drosophilids. A wild female will very often produce a vigorous F_1 generation. When these F_1's are mated in small numbers in vials, very poor reproduction is observed; many of the vials produce no offspring at all. This F_1 "bottleneck" may often lead to a failure of the stock to become established or at best to a small F_2 generation. After some generations of doing poorly the stock often improves, although this is not invariably the case. During such a period, the stock often passes through a very small number of individuals, with quite drastic effects on behavior (see Ahearn, 1980). It is also noteworthy that pair matings are very difficult in these species.

Although much of the above is anecdotal, the following hypothesis may be advanced. Successful reproduction of these flies entails mate choice involving both intrasexual and epigamic factors. Successful operation of the system entails reasonably large aggregations of males in order to maximize participation in reproduction by high-fitness males. Two interesting corollaries are apparent. The effective population size is rendered smaller since few males participate and there is a tendency for the overall simplification of the sexual selection system. The latter effect has been observed in lines of *D. silvestris* selected for low discriminatory behavior in females and in turn forcing them to mate with males having low copulatory success in prior tests (Kaneshiro and Carson, 1982).

XIV. TENTATIVE CONCLUSIONS ABOUT SEXUAL SELECTION AND SPECIATION

The upshot of this presentation is that population biologists have tended to overlook a major type of genetic adaptation, namely that involving the sexual environment. Coadaptation of male and female behavior appears to assure that those that reproduce have maximal fitness among members of the deme. Such adaptations, based on complex balances, may be easily perturbed by population events such as hybridization, population crashes or founder effects. If such a genetic disorganization occurs, selection must rebuild the sexual adaptations of the isolated daughter population. This may cause a shift to a novel balanced system. Such selective alterations in the pattern of sexual reproduction may be pivotal in the process of speciation, especially as they may alter pre-mating characters in such a way as to render the new population clearly distinguishable as a novel adapted biological system. Sexual selection may be pivotal as a process contributing to speciation in many forms of animals, from Crustacea to man.

REFERENCES

Ahearn, J. N. (1980). Evolution of behavioral isolation in a laboratory stock of *Drosophila silvestris*. *Experientia* **36**, 63-64.

Beverley, S. M. (1979). Molecular evolution in *Drosophila*. Ph.D. Thesis, University of California, Berkeley.

Cade, W. H. (1981). Alternative male strategies: genetic differences in crickets. *Science* **212**, 563-564.

Carson, H. L. (1982a). Evolution of *Drosophila* on the newer Hawaiian volcanoes. *Heredity* **48**, 3-25.

Carson, H. L. (1982b). Speciation as a major reorganization of polygenic balances. *In* "Mechanisms of Speciation" (C. Barigozzi, ed.), pp. 411-433. Alan R. Liss, New York.

Carson, H. L. (1985a). Genetic variation in a courtship-related male character in *Drosophila silvestris* from a single Hawaiian locality. *Evolution* **39**, 678-686.

Carson, H. L. (1985b). Unification of speciation theory in plants and animals. *Systematic Botany* **10** (in press).

Carson, H. L., Hardy, D. E., Spieth, H. T., and Stone, W. S. (1970). The evolutionary biology of the Hawaiian Drosophilidae. *In* "Essays in Evolution and Genetics in Honor of Theodosius Dobzhansky" (M. K. Hecht and W. C. Steere, eds.), pp. 437-543. Appleton-Century-Crofts, New York.

Carson, H. L., and Kaneshiro, K. Y. (1976). *Drosophila* of Hawaii: systematics and ecological genetics. *Ann. Rev. Ecol. Syst.* **7**, 311-346.

Carson, H. L., and Lande, R. (1984). Inheritance of a secondary sexual character in *Drosophila silvestris. Proc. Natl. Acad. Sci. USA* **81**, 6904-6907.

Carson, H. L., and Teramoto, L. T. (1980). Differences in copulatory success among laboratory males of *Drosophila silvestris. Genetics* **94**, s14.

Carson, H. L., and Teramoto, L. T. (1984). Artificial selection for a secondary sexual character in males in *Drosophila silvestris* from Hawaii. *Proc. Natl. Acad. Sci. USA* **81**, 3915-3917.

Carson, H. L., Val, F. C., Simon, C. W., and Archie, J. W. (1982). Morphometric evidence for incipient speciation in *Drosophila silvestris* from the island of Hawaii. *Evolution* **36**, 132-140.

Darwin, C. (1871). "The Descent of Man and Selection in Relation to Sex." John Murray, London.

Dobzhansky, Th. (1970). "Genetics of the Evolutionary Process." Columbia University Press, New York, pp. 505.

Fisher, R. A. (1930). "The Genetical Theory of Natural Selection." Clarendon Press, Oxford, pp. 291.

Kaneshiro, K. Y. (1983). Sexual selection and direction of evolution in the biosystematics of the Hawaiian Drosophilidae. *Ann. Rev. Entomol.* **28**, 161-178.

Kaneshiro, K. Y., and Carson, H. L. (1982). Selection experiments on mating behavior in *Drosophila silvestris. Genetics* **100**, s34.

Karlin, S., and O'Donald, P. (1978). Some population genetic models combining sexual selection with assortative mating. *Heredity* **41**, 165-174.

Mayr, E. (1963). "Animal Species and Evolution." Harvard University Press, Cambridge, Mass., pp. 797.

Muller, H. J. (1942). Isolating mechanisms, evolution and temperature. *Biol. Symp.* **6**, 71-125.

O'Donald, P. (1980). "Genetic Models of Sexual Selection." Cambridge University Press, London, pp. 250.

Sene, F. M., and Carson, H. L. (1977). Genetic variation in Hawaiian Drosophila IV. Allozymic similarity between *D. silvestris* and *D. heteroneura* from the island of Hawaii. *Genetics* **86**, 187-198.

Spiess, E. B., and Carson, H. L. (1981). Evidence for sexual selection in *Drosophila silvestris* of Hawaii. *Proc. Natl. Acad. Sci. USA* **78**, 3088-3092.

Spieth, H. T. (1968). Evolutionary implications of sexual behavior in *Drosophila*. *Evol. Biol.* **2**, 157-193.

Throckmorton, L. H. (1975). The phylogeny, ecology and geography of *Drosophila*. *In* "Handbook of Genetics" (R. C. King, ed.), pp. 421-469. Plenum Press, New York.

Val, F. C. (1977). Genetic analysis of the morphological differences between two interfertile species of Hawaiian *Drosophila*. *Evolution* **31**, 611-629.

Wilson, E. O. (1970). "Sociobiology." Harvard University Press, Cambridge, Mass., pp. 697.

ALLOPATRIC AND NON-ALLOPATRIC SPECIATION;
ASSUMPTIONS AND EVIDENCE

Guy L. Bush

Department of Zoology
Michigan State University
East Lansing, Michigan 48824

Daniel J. Howard

Biology Department
Museum of Northern Arizona
Route 4, Box 720
Flagstaff, Arizona 86001

ABSTRACT

We examine the assumptions and evidence used to support and reject non-allopatric modes of speciation in the light of recent evidence emanating from the fields of molecular biology, theoretical population genetics, and evolutionary biology. Such an undertaking appears to be especially timely because the major works refuting non-allopatric speciation are now more than 20 years old, and their supporting evidence and premises are in need of reappraisal. We particularly focus on the traditional view that in the absence of an extrinsic barrier such factors as the swamping effect of gene flow, the coadaptation of the genome, the polygenic nature of reproductive isolation, and the inability of selection to mold isolating mechanisms make non-allopatric speciation unlikely. We conclude that new evidence forces a rejection of this traditional view and removes it as an obstruction to the acceptance of the view that non-

411

allopatric speciation may occur. On the other hand, increasing evidence supports assumptions invoked in models of non-allopatric speciation. We also contrast and evaluate the evidential foundation of allopatric speciation with that of non-allopatric speciation. We find that proponents of allopatric speciation have rarely documented their examples with the rigor demanded for the documentation of a non-allopatric speciation event. The relative frequency of allopatric versus non-allopatric speciation as well as the underlying processes will only be understood by combining analyses of geographic distribution patterns with studies of the genetics, ecology, and developmental biology of specific traits responsible for reproductive isolation.

I. INTRODUCTION

In 1963, Ernst Mayr wrote that all known phenomena in the areas of species distribution patterns, geographic variation of species, and the ranking of lower taxonomic categories are consistent with the theory of geographic speciation. Despite this confident outlook, papers and books advocating non-allopatric modes of speciation proliferated in the 1960's, 70's and 80's (e.g., White, 1968, 1978; Bush, 1969a, 1969b, 1974, 1975; Tauber and Tauber, 1977a, 1977b; Caisse and Antonovics, 1978; Wood and Guttman, 1982; Rice, 1984). This proliferation of papers has given rise in turn to several recent papers again rejecting the idea of non-allopatric speciation (Futuyma and Mayer, 1980; Paterson, 1981, 1982).

Of course these arguments did not begin only 20 years ago. In fact, the debate over the exclusiveness of geographic speciation is such a time honored ritual in evolutionary biology that the positions taken sometimes seem to be more a matter of faith and tradition than of critical logical thought based on an evaluation of the assumptions and evidence. We therefore wish to sharpen the focus of the debate by reviewing the assumptions and evidence used to support allopatric and non-allopatric modes of speciation and by contrasting the two concepts in the light of recent developments in molecular, theoretical, and evolutionary biology.

The theory of allopatric speciation which was first proposed by Wagner in 1868, emerged from the observation that closely related species are often distributed in a disjunct fashion. Jordan (1905) regarded the non-overlapping distributions of closely related species as not only suggestive of the conditions necessary for species formation, but as the most important prediction generated by the theory of allopatric speciation. He did not claim that such distributions are occasional, but that they are virtually universal, and he raised this observation to a general law of distribution.

> "Given any species in any region, the nearest related species is not likely to be found in the same region, but in a neighboring district separated from the first by a barrier of some sort." (Jordan, 1905, p. 557)

Mayr's (1942, 1947, 1963) major contributions to the theory of allopatric speciation were the provision of numerous examples of distribution patterns in accord with "Jordan's Law," the clear description of more predictions made by the theory, and the furnishing of evidence that these additional predictions are fulfilled in nature. Mayr recognized that if allopatric speciation actually occurred, and if it was a gradual and continuous process as he envisioned it, then it should be possible to find intermediate stages in nature. These intermediate stages were: (1) geographic variation in characters separating closely related species, (2) reduced fertility between geographic races of one species, and (3) subspecies that are on the verge of becoming species. Mayr provided numerous examples of the first stage and modest documentation of the second two stages. Regardless of the numbers, the existence of these intermediate stages "proved" that allopatric speciation occurred, according to Mayr (1942, 1963). Thus the evidence for allopatric speciation is that examples of situations predicted by the theory can be found in nature.

But what of non-allopatric speciation? Have alternative theories of speciation been rejected because they generate no testable predictions or because the predictions are not matched in nature? Quite the contrary. Theories of non-allopatric speciation provide a series of predictions fairly discrete from those of allopatric speciation. For example, sympatric speciation, a mode of non-allopatric speciation, predicts the occurrence of: (1) sympatric host or habitat races, (2) sympatric sister species, and (3) endemic speciation in isolated areas that are not broken up by barriers to dispersal, such as oceanic islands and freshwater lakes. Convincing examples of prediction (1) have been found among insects and plants (e.g., Bradshaw, 1952; Brown, 1959; Clarke *et al.*, 1963; Bradshaw *et al.*, 1965; Bush, 1969a; Antonovics, 1971; Antonovics *et al.*, 1971; Khasimuddin and de Bach, 1976; Muller, 1976; Price and Willson, 1976; Miller and Kosztarab, 1979; Claridge and Nixon, 1981; Katakura *et al.*, 1981; Kreslavsky *et al.*, 1981; Menken, 1981; MacNair and Christie, 1983; Diehl, 1984). Illustrations of predictions 2 and 3 are so numerous that we will not even begin to list them here. Many are admirably summarized by White (1978).

Jordan and Mayr acknowledged the above predictions, and they were not oblivious to the fact that their fulfillment represented a challenge to the exclusivity of allopatric speciation. During Jordan's time the frequent occurrence of sympatric sister species and of endemic speciation in isolated areas had not yet been recognized, so it was possible for him to regard as mere curiosities those rare examples that were known. Such an attitude was no longer possible by the 1940's, so to defend the universality of allopatric speciation Mayr was forced to take a more assertive stance. He denied the possibility of non-allopatric speciation on the grounds that it was inconsistent with the known facts of genetics

(Mayr, 1942, 1947, 1963), and with concepts regarding gene pool coadaptation (Mayr, 1954). He also attempted to accommodate the biological examples consistent with non-allopatric speciation within the framework of allopatric speciation. Host races were simply inadequately understood or arose through "microallopatric" processes. Sympatric sister species diverged in allopatry and invaded each other's territory only after attaining reproductive isolation. Some freshwater lakes became ancient repositories of old species which had long since become extinct in surrounding bodies of water. Endemic speciation on oceanic islands or isolated terrestrial habitats was attributed to multiple invasions from neighboring land masses or appropriate terrestrial habitats (Mayr, 1942, 1947, 1963). Although there was no direct evidence for the elaborate scenarios of multiple invasions invented by Mayr and others to explain endemic speciation and the occurrence of sympatric sister species, they were accepted as true because non-allopatric speciation was considered so highly improbable. When these scenarios are subjected to careful scrutiny, we suspect that many will prove wrong, at least if recent studies are any indication. For example, Barendse (1984) and Maxson and Roberts (1984) using modern biochemical systematic tools have refuted a well known multiple invasion scenario proposed by Main *et al.* (1958) to explain the occurrence of endemic Western Australian Leptodactylid frogs.

Thus the model of sympatric speciation was not rejected on the basis of insufficient evidence, but on the basis of theoretical implausibility. Even today, whether one accepts or rejects the evidence for non-allopatric speciation is heavily dependent on the genetic and ecological assumptions one holds as true. Many evolutionary biologists continue to cling to the same assumptions that motivated Mayr to reject the evidence for non-allopatric speciation. By an application of questionable assumptions to selected examples they dismiss all evidence for non-allopatric modes of speciation. Consequently, there are compelling reasons to review the assumptions made by the advocates of allopatric and non-allopatric modes of speciation and to establish which assumptions are supported by current evidence, and which are irrelevant or misleading.

II. DEFINITION OF TERMS

Because so many disagreements in evolutionary biology hinge on definitions of terms, and because we wish to avoid misunderstanding and to concentrate on the biologically meaningful issues, we accept the definitions of *sympatric, parapatric, allopatric, speciation* and *species* adopted by Futuyma and Mayer (1980, pp. 254-256). Briefly "two populations are *sympatric* if individuals of each are physically capable of encountering one another ... with moderately high frequency." "*Parapatric* populations occupy separate but adjoining areas, such that only a small fraction of individuals in each encounters the other." "*Allopatric* populations are separated by uninhabited space (even if it is only very short distance), across which migration (movement) occurs at very low frequency." "... *speciation* is a process whereby one species splits into two or more species." "A *species* is a group of populations whose evolutionary pathway is distinct and independent from that of other groups; a distinct and independent evolutionary path is achieved by the group's reproductive isolation from other groups." This last definition is very similar to the evolutionary species concept of Wiley (1978, 1981) and incorporates attributes of both the biological (Mayr, 1942) and evolutionary species concept (Simpson, 1961). We further accept the idea that groups have achieved full species status "if they are or could be (given the opportunity) truly sympatric without losing their separate identities through interbreeding" (Futuyma and Mayer, 1980, p. 256).

Although speciation theory can be recast in non-geographic terms, such as in the framework of population genetics (Templeton, 1981), geographical distribution remains an important factor affecting the selective regime of a population and the pattern and level of gene flow between populations. Moreover, within the context of a geographic outlook many controversial issues remain unresolved, and are worthy of continued attention from evolutionary biologists. The terminology of traditional speciation theory is also familiar to most biologists. Therefore, we feel it is useful to retain the classification based on geography for the time being.

III. ALLOPATRIC AND NON-ALLOPATRIC SPECIATION; A CONTRAST IN ASSUMPTIONS

Rejection of non-allopatric speciation is based on assumptions concerning the nature and importance of several biological factors such as gene pool coadaptation, mutations, gene flow, reproductive isolation, reinforcement, and speciation rates. We will explore these assumptions and the evidence invoked to support them, then contrast each with the views advanced by advocates of non-allopatric speciation.

A. Coadaptation of the Gene Pools

The idea of a highly coadapted and cohesive gene pool carefully sculpted by natural selection has been used for many years by advocates of the universality of allopatric speciation as a major factor mitigating against the creation of discontinuities within a gene pool, rapid change in a gene pool, and the evolutionary significance of mutations with major effects. This concept of a coadapted gene pool harkens from a time when organismal development was perceived by evolutionary biologists as proceeding by an intricate interplay among all the genes in a genome. In this view every gene is somehow selected for its ability to interact well with all the other genes in the gene pool and only genes able to coadapt harmoniously at all levels are incorporated (Mayr, 1954, 1963, 1982; Stebbins, 1971; Futuyma and Mayer, 1980). The consequence of this coadaptation is a tight genetic cohesion that resists genetic change, especially dramatic alterations produced by mutations affecting the timing and pattern of developmental events (Mayr, 1963). However, as noted recently by several authors (e.g., Hedrick *et al.*, 1978; Nei, 1980; Bush, 1982) there is little empirical evidence that supports the concept of comprehensive gene pool coadaptation, and the theoretical objections to the concept seem compelling (Kimura, 1974).

Proponents of non-allopatric speciation as well as developmental and molecular biologists have suggested that neither the genome nor the gene pool of a species is highly coadapted in the global sense of Mayr and others, but organized in a way that offers only limited and quite different constraints on genome evolution and the speciation process (Goldschmidt, 1940; Bush, 1975, 1981; Wilson *et al.*, 1975; Bush *et al.*, 1977; Raff and Kaufman, 1983). This view is supported by recent research on organismal development and the molecular organization of the genome.

It is now clear that the control of development and gene expression has hierarchical and compartmental aspects (Hunkapiller *et al.*,1982; Martinez-Arias and Lawrence, 1985) which results in large sets of genetic information being controlled and coordinated by a single genetic element (Raff and Kaufman, 1983; Sternberg and Horvitz, 1984). Although many genes are required for organismal development, only a small minority make developmental decisions. Moreover, because genes can be expressed in a compartmental fashion, alterations in expression in one compartment can result in a significant phenotypic change without drastically affecting developmental processes in other compartments. If genetic interactions impose constraints on genetic substitutions, adaptive evolution, and speciation they may well be limited to those among developmental controlling elements (Raff and Kaufman, 1983).

Molecular genetic studies have also shown that ancestral phenotypes can be cryptically retained long after they have seemingly disappeared because the genes involved can be used and expressed elsewhere in the genome (Hunkapiller *et al.*, 1982). This phenotypic potential can provide the raw material for large-scale phenotypic experiments if it is suddenly expressed as a result of an alteration in the timing of expression or the nature of association between genes. Furthermore, it is now evident that mobile genetic elements can rapidly alter the genetic information stored in an organism through a physical reorganization of the genome (Syvanen, 1984). The hierarchical and compartmental organization, potential for dramatic change, and dynamic properties of the genome have several important implications for both allopatric and non-allopatric race formation and speciation.

First, selection can operate at many different levels of genome organization not simply at the gene. Second, a single or relatively few mutations can have a major effect on the phenotype and provide conditions advantageous for a shift into a new habitat and the development of reproductive isolation. Third, because gene expression is coordinated in developmental systems and most genes do not interact, many of the negative pleiotropic effects of mutations with major phenotypic effects envisioned by Mayr (1963) and others would be greatly reduced or eliminated. It is of course still true that any mutation which results in reduced fitness of the heterozygote could be fixed in nature only under conditions that promote various combinations of inbreeding, drift and disruptive selection (Bush et al., 1977; Lande, 1979; Bush, 1981; Hedrick and Levin, 1984). These conditions may be met anywhere within a species range (i.e., centrally or peripherally; Wright, 1977, 1978).

B. The Genetics of Reproductive Isolation

A prevalent assumption among evolutionary biologists is that many mutations are required to complete reproductive isolation (Mayr, 1963; Futuyma and Mayer, 1980). The empirical foundation of this assumption rests primarily on hybridization studies of closely related *Drosophila* and have been for the most part concerned with sterility (Dobzhansky, 1936, 1973; Spencer, 1944; Patterson and Dobzhansky, 1945; Spieth, 1949; Weisbrot, 1963). For example, in an early study Dobzhansky (1936) demonstrated that interspecific sterility can have a complex, polygenic basis. However, during the 1960's and early 70's the elegant work of Ehrman and her colleagues (Ehrman, 1960, 1962; Ehrman and Williamson, 1969; Ehrman and Kernaghan, 1971) showed that not all sterility in *Drosophila* is under polygenic or even genetic control. More recent work has added to the diversity of sterility mechanisms in *Drosophila* by implicating transposons as a causative agent (Bingham *et al.,* 1982; Rubin and Spradling, 1982; Rubin *et al.,* 1982).

Most models of non-allopatric speciation are based on the assumption that speciation may be initiated by only a few key genetic changes that provide sufficient isolation between parental and daughter populations to ensure continued genetic divergence in response to selection and the eventual evolution of complete reproductive isolation. Empirical evidence supporting the role of such mutations in the speciation process comes from three different sources. Perhaps the most convincing are plant and animal hybridization studies which have demonstrated that major morphological or behavioral changes, including traits affecting reproductive isolation, can be controlled by only a few genes (Weibe,

1934; Danforth, 1950; Stephens, 1950; Saunders, 1952; Gerstel, 1954; Ae, 1959; Chu and Oka, 1972; Heuttel and Bush, 1972; Oka, 1974; Templeton, 1977; Val, 1977; Sano and Kita, 1978; Oliver, 1979; Grula and Taylor, 1980; MacNair and Christie, 1983). Also, allozyme studies have shown that closely related species can be very similar at the level of structural genes (Nevo *et al.,* 1974; Johnson *et al.,* 1977; Craddock and Johnson, 1979; Harrison, 1979; Guttman *et al.,* 1981; Berlocher and Bush, 1982). Finally, molecular and developmental studies have established that mutations affecting development can, in a single step, result in major alterations of the phenotype (Sternberg and Horvitz, 1984; Sanchez-Herrero *et al.,* 1985). Some heterochronic mutations, such as those studied in the nematodes *Caenorhabditis elegans* and *Panagrellus redivivus,* resemble closely the kinds of phenotypic differences observed between nematode species and genera (Sternberg and Horvitz, 1981; Ambros and Horvitz, 1984).

Theoretical considerations and simulation studies have also provided insight into the conditions favoring and limiting speciation by key genetic changes. Although a prominent early model, which was based on single locus polymorphisms (Maynard Smith, 1966), indicated that reproductive isolation could evolve sympatrically under the right conditions, it was rejected by some as too simplistic on the grounds that speciation is a polygenic process involving many loci of small effect. Not only is the latter position less relevant today; but recent, more sophisticated genetic models and simulation studies have largely confirmed Maynard Smith's results (Endler, 1977; Udovic, 1980; Felsenstein, 1981; Lande, 1982; Kondrashov, 1983b; Rice, 1984). The conclusion from these studies is that non-allopatric modes of speciation are highly likely in certain organisms under suitable conditions. These include the following: (1) One or more new habitats, relatively free of competitors, are available for exploitation and colonization. (2) Genetic variation exists or arises in genes for habitat preference thereby permitting colonization of new habitats. Such colonization can occur across an ecological gradient as in the case of parapatric speciation (Endler, 1977), or sympatrically when a new habitat is distributed mosaicly (Bush, 1969a). (3) Habitat selection is under the control of a few major genes. (4) Selection favors different sets of modifier genes in each habitat. (5) Upon dispersal, individuals settle in preferred habitats to which they are best adapted. (6) Mating is non-random with assortative mating based on habitat preference. (7) Population size in each habitat is independently regulated. Other factors will also enhance the process of non-allopatric divergence such as strong sexual selection associated with habitat preference (Lande, 1982), and population subdivision with inbreeding and drift (Wright, 1940).

A major stumbling block in achieving an understanding of the genetics of reproductive isolation has been the failure to appreciate the fact that species continue to accumulate genetic differences after a speciation event. Thus, studies of closely related species, such as interspecific hybridization analyses, will tend to overestimate the number of genetic changes needed for the evolution of reproductive isolation and can give a false impression of the amount of genetic differentiation that is needed for speciation. Evolutionary biologists must distinguish between the genetics of speciation (i.e., the origin and evolution of reproductive isolation) and the genetics of species differences (Templeton, 1981). If we are to unravel the processes involved in speciation, we must focus on the genetics and evolution of reproductive isolation.

C. Reinforcement

Another argument raised against non-allopatric speciation is the claim that reinforcement, the strengthening of incompletely developed reproductive isolating mechanisms in zones of secondary contact owing to selection against individuals that hybridize (Littlejohn, 1981), is extremely unlikely. This is essentially an assertion that natural selection cannot build isolating mechanisms and is also used to argue against non-allopatric speciation (Futuyma and Mayer, 1980; Paterson, 1982). Although it is clear that selection against hybridization is not a prerequisite for non-allopatric speciation (see earlier section on the genetics of reproductive isolation) we feel it is essential to point out certain weaknesses of the case against reinforcement. First, it is maintained that models of hybrid inferiority generally lead to an unstable equilibrium (Paterson, 1978; Templeton, 1981). Therefore, extinction of one of the populations or elimination of factors responsible for the hybrid inferiority are more likely than the development of premating isolating mechanisms. However, this is only true if there has been no substantial ecological divergence between the two interacting taxa and if allele frequencies at loci responsible for the hybrid inferiority are constant throughout the area of overlap. If ecological divergence has taken place, the population sizes of the two taxa may be regulated independently and a stable equilibrium can result (Maynard Smith, 1966). A stable hybrid zone, maintained by the balance between dispersal and selection against hybrids, can also develop if allele frequencies vary spatially; which seems likely in a zone of secondary contact (Bazykin, 1969, 1972, 1973; Barton and Hewitt, 1981). Once stability has been attained, further divergence and the evolution of reproductive isolation is possible.

A second argument against reinforcement is the allegation that if the initial error in mate choice is large, reinforcement is unlikely even with strong selection against hybridization (Templeton, 1981). A model presented by Wilson (1965) has been cited to support this claim. However, this is not what the model indicated. The outcome, fusion or reproductive isolation, was uncertain over a wide range of hybrid fitnesses and frequencies of mate choice error. Evidently, in most situations there is a "race" between the evolution of reproductive isolation and fusion with no sure winner. Certainly, if individuals tend to mate in their preferred habitats, selection can be much weaker and still be effective.

The most powerful evidence mustered against reinforcement is the paucity of examples of reproductive character displacement (Futuyma and Mayer, 1980; Paterson, 1982). It is true that well documented cases of reproductive character displacement are rare. But it is important to remember that a phenomenon may lack documentation for reasons other than the scarcity of its occurrence. If a phenomenon does not attract the attention of researchers it may appear rare, simply because it is rarely studied. Additionally, if the scientific community imposes very rigid standards for identification and verification of a particular phenomenon it may also appear rare, because of the problems associated with meeting the standards. In fact, it is very difficult to document reproductive character displacement (Grant, 1972; Arthur, 1982). To do so entails demonstrating that the trait under study mediates reproductive isolation, that it varies geographically or ecologically in a manner consistent with displacement rather than clinal or mosaic variation, and that displacement was the direct result of selection against hybridization rather than a response to some other environmental factor. Thus, the dearth of accepted examples may not reflect the actual frequency with which selection against hybridization drives the evolution of reproductive isolation. It should be noted that, although small in number, some good illustrations of reproductive character dispalcement do exist (e.g., Litteljohn, 1965; Fouquette, 1975; Wasserman and Koepfer, 1977; Waage, 1979). Moreover, much laboratory (Koopman, 1950; Knight *et al.*, 1956;

Kessler, 1966; Crossley, 1974) and theoretical work (Maynard Smith, 1966; Crosby, 1970; Caisse and Antonovics, 1978; Kondrashov, 1983a,b; Rice, 1984) supports the idea that selection can build isolating mechanisms in nature.

As a final note, we would urge caution in using broad surveys of potential cases of reproductive character displacement as a means of assessing the frequency with which the process of reinforcement drives the evolution of sexual isolation in nature. The problems with such surveys are manifold, as pointed out by Walker in a 1974 survey of character displacement in male cricket calling songs. Walker concluded that although reproductive character displacement appeared to be rare in this group of organisms, few of the cases studied involved sufficient detail that patterns of geographic variation could be adequately described. Such a survey also presupposes that the traits responsible for reproductive isolation between particular species pairs have been correctly identified. Making such identifications requires detailed analyses on a case by case basis, a requirement scarcely ever fulfilled. The need for a careful analysis is underscored by the work of Waage (1975) and Wasserman and Koepfer (1977), which indicates that selection for improved reproductive isolation may result not in morphological or signalling system divergence but in a change in the species discrimination ability of sympatric individuals. Thus, reinforcement may occur without leading to obvious reproductive character displacement.

If surveys limit their scope to well analyzed areas of allopatry and contact, they can provide meaningful information on the occurrence of reproductive character displacement. But even then, it should be kept in mind that character divergence in sympatry can be obscured by shifting boundaries of the contact zone (Waage, 1979) and by gene flow between populations inside and those outside the zone of sympatry leading to the formation of a smooth cline (Walker, 1974).

D. Gene Flow

The ability to move and hence to disperse is an almost universal property of animals. Mayr (1963) has argued that a consequence of this mobility is the intermingling of populations and the prevention of genetic divergence. This would only be true if animals always moved long distances in a random fashion and if gene flow could always overcome selection pressure for genetic divergence imposed by different physical and biotic environments. Undoubtedly, some animals, such as certain marine invertebrates with a long pelagic phase, disperse randomly over great areas (Srathmann, 1974). However, in many animals intrinsic factors such as the maintenance of territories, homing tendencies, parental care, and very specific mate and habitat preferences lead to directed dispersal which can strongly restrict the level of gene flow between populations. Perhaps the most general and important of these phenomena is habitat selection (Rausher, 1984), a characteristic which has been experimentally verified in many organisms, including rodents (Schroder and Rosenzweig, 1975), crickets (Howard and Harrison, 1984), moths (Kettlewell, 1955; Kettlewell and Conn, 1977), *Drosophila* (Rockwell and Seiger, 1973; Parsons, 1975, 1978; Kawanishi and Watanabe, 1978) and *Rhagoletis* (Prokopy *et al.,* 1971). Furthermore, mating is frequently confined to the preferred habitat either before or after dispersal. This is particularly true of parasites and parasitoids. Therefore, to make a blanket statement that dispersal will result in the intermingling of populations is to hopelessly simplify the complexity of natural systems. The importance of dispersal in preventing the divergence of populations will change from one group to another and will have to be assessed independently for every system under study.

Mayr (1942, 1947, 1963) has forcefully promoted the idea that gene flow is the major factor responsible for the genetic cohesion of the populations of a species. In his earlier papers and books, he alleged that selection would not lead to genetic differentiation in the face of gene flow. As accumulating evidence from theoretical (Wright, 1940; Levene, 1953; Cook, 1962) and laboratory studies (Thoday and Boam, 1959; Thoday and Gibson, 1962) made this position less tenable, he admitted that selection might produce differentiation at a single locus but argued that the "inertia of the total epigenotype" (Mayr, 1963, p. 521) would preclude any major differentiation and the development of reproductive isolation. Futuyma and Mayer (1980, p. 256) employed the same tactic by postulating that reproductive isolation requires changes in many genes and that in the early stages of speciation "divergent selection will be quite weak for most of the genome." As we have already pointed out, unequivocal evidence for global genome coadaptation, epigenetic inertia, as well as polygenic control over reproductive isolation, is lacking. It is also misleading to assume that in the initial stages of divergence the selective regime at most loci will be negligibly different. If a population begins to exploit a new ecological niche (which would be the case for a monophagous insect shifting on to a new host), the ecological requirements of the daughter and parental populations may differ greatly; and the selective regime for many loci may be drastically altered.

E. Speciation Rate

The assertion has been made that speciation is gradual (Mayr, 1947, 1963). This is taken by some biologists to mean that speciation is a slow process (e.g., Futuyma and Mayer, 1980), requiring the passage of many generations for two populations to move from the stage of initial divergence to the acquisition of irreversible reproductive isolation. However, Mayr (1963) characterized gradual speciation as an alternative to "instantaneous speciation," the production of a single individual that was reproductively isolated from the parental stock and capable of establishing a new species population. Even rapid speciation events, as long as they do not occur in a single generation, were considered gradual by Mayr. In fact, the theory of speciation, via the founder principle or peripatric speciation (Mayr, 1954), is a model of rapid speciation. It seems likely that the tempo of speciation can be highly variable and that in some cases new species arise in only a few generations. How fast speciation is completed will depend upon conditions of selection, gene flow, drift, population structure and the nature of the mutations involved (Bush, 1975; Bush *et al.*, 1977).

IV. DISCUSSION

The evidence for non-allopatric speciation continues to be rejected by many evolutionary biologists on the grounds that the models are theoretically implausible and the evidence can be accommodated within the framework of the allopatric model. We have shown in this paper that the assumptions used to question non-allopatric speciation are either outdated, irrelevant, or unsupported by the evidence at hand. As to the accommodation of the non-allopatric predictions within the orthodoxy of the allopatric model, it is well to remember that mere belief in a theoretical model does not give it logical primacy. It is equally true that by the appropriate use of tortuous schemes and post hoc scenarios all the predictions of the allopatric model can be accommodated within the framework of non-allopatric speciation. If present day distribution patterns represent inadequate support for sympatric speciation, then the same evidence must be denied to the theory of allopatric speciation.

One of the reasons that the debate over modes of speciation has persisted unabated for so long is that the field has been dominated by taxonomists and biogeographers who have traditionally concentrated on the description of pattern rather than the elucidation of process. With no clear knowledge of process, each school of thought has been relatively free to construct its own ecology, genetics, and developmental biology of speciation. This allows the identical evidence to be interpreted differently and results in repeated controversy over the same issues.

In the future, our efforts must be concentrated on the genetics, ecology and developmental biology of specific traits responsible for reproductive isolation. This will require in-depth studies of closely related taxa in various stages of divergence in which such traits can be unambiguously identified. Among the questions that will need to be addressed are how many and what kind of mutations control the expression of the traits, and how does genome organization influence the traits' development and expression? We must also establish how such traits are fixed in the process of speciation. What was the relative contribution of selection and drift? Did extrinsic or intrinsic barriers limit gene flow between the diverging populations? Once these parameters have been worked out for a variety of speciation events in a number of different taxa then more realistic theortical mopdels of speciation can be constructed and more meaningful laboratory experiments can be carried out. Altogether, the above studies will provide a process oriented framework for evaluating the relative importance of various models of speciation.

ACKNOWLEDGMENTS

We thank Alan Templeton and Eviatar Nevo for their constructive criticisms of an earlier draft of this chapter.

REFERENCES

Ae, S. (1959). A study of hybrids in *Colias* (Lepidoptera, Pieridae), *Evolution* **13**, 64–88.

Ambros, V., and Horvitz, H. R. (1984). Heterochronic mutants of the nematode *Caenorhabditis elegans*. *Science* **226**, 409–416.

Antonovics, J. (1971). The effects of a heterogeneous environment on the genetics of natural populations. *Amer. Sci.* **59**, 593–599.

Antonovics, J., Bradshaw, A. D., and Turner, R. G. (1971). Heavy metal tolerance in plants. *Advanc. Ecol. Res.* **7**, 2–58.

Arthur, W. (1982). The evolutionary consequences of interspecific competition. *Advanc. Ecol. Res.* **12**, 127–187.

Barendse, W. (1984). Speciation in the genus *Crinia* (Anura: Myobatrachidae) in southern Australia: a phylogenetic analysis of allozyme data supporting endemic speciation in southwestern Australia. *Evolution* **38**, 1238–1250.

Barton, N. H., and Hewitt, G. M. (1981). Hybrid zones and speciation. *In* "Evolution and Speciation" (W. R. Atchley and D. Woodruff, eds.), pp. 109–145. Cambridge University Press, Cambridge.

Bazykin, A. D. (1969). Hypothetical mechanism of speciation. *Evolution* **23**, 685–687.

Bazykin, A. D. (1972). The disadvantage of heterozygotes in a system of two adjacent populations. *Genetika* **8**, 155–161.

Bazykin, A. D. (1973). Population genetic analysis of disruptive and stabilizing selection. Part II. Systems of adjacent populations and populations within a continuous area. *Genetika* **9**, 156–166.

Berlocher, S. H., and Bush, G. L. (1982). An electrophoretic analysis of *Rhagoletis* (Diptera: Tephritidae) phylogeny. *Syst. Zool.* **31**, 136–155.

Bingham, P. M., Kidwell, M. G., and Rubin, G. M. (1982). The molecular basis of P-M hybrid dysgenesis: the role of the P element, a P strain specific transposon family. *Cell* **29**, 995–1004.

Bradshaw, A. D. (1952). Populations of *Agrostis tenuis* resistant to lead and zinc poisoning. *Nature* **169**, 1098.

Bradshaw, A. D., McNeilly, T. S., and Gregory, R. P. G. (1965). Industrialization, evolution, and the development of heavy metal tolerance in plants. *Brit. Ecol. Soc. Symp.* **6**, 327–343.

Brown, W. J. (1959). Taxonomic problems with closely related species. *Ann. Rev. Entomol.* **4**, 77–98.

Bush, G. L. (1969a). Sympatric host race formation and speciation in frugivorous flies of the genus *Rhagoletis. Evolution* **23**, 237–251.

Bush, G. L. (1969b). Mating behavior, host specificity, and the ecological significance of sibling species in frugivorous flies of the genus *Rhagoletis* (Diptera, Tephritidae). *Amer. Nat.* **103**, 669–672.

Bush, G. L. (1974). The mechanism of sympatric host race formation in the true fruit flies (Tephritidae). *In* "Genetic Mechanisms of Speciation in Insects" (M. J. D. White, ed.), pp. 3–23. Australia and New Zealand Book Co., Sydney.

Bush, G. L. (1975). Modes of animal speciation. *Ann. Rev. Ecol. Syst.* **6**, 339–364.

Bush, G. L. (1981). Stasipatric speciation and rapid evolution in animals. *In* "Evolution and Speciation" (W. R. Atchley and D. Woodruff, eds.), pp. 201–218. Cambridge University Press, Cambridge.

Bush, G. L. (1982). What do we really know about speciation? *In* "Perspectives on Evolution" (R. Milkman, ed.), pp. 119–128. Sinauer Associates, Inc., Sunderland.

Bush, G. L., Case, S. M., Wilson, A. C., and Patton, J. L. (1977). Rapid speciation and chromosomal evolution in mammals. *Proc. Natl. Acad. Sci. USA* **74**, 3942–3946.

Caisse, M., and Antonovics, J. (1978). Evolution in closely adjacent plant populations. IX. Evolution of reproductive isolation in clinal populations. *Heredity* **40**, 371–384.

Chu, Y., and Oka, H. (1972). The distribution and effects of genes causing F_1 weakness in *Oryza breviligalata* and *O. glaberina. Genetics* **70**, 163–173.

Claridge, M. F., and Nixon, G. A. (1981). *Oncopsis* leafhoppers on British trees: polymorphism in adult *O. flavicollis* (L.). *Acta. Entomol. Fenn.* **38**, 15–19.

Clarke, C. A., Dickson, C. G. C., and Sheppard, P. M. (1963). Larval color pattern in *Papilio demodocus. Evolution* **17**, 130–137.

Cook, L. M. (1961). The edge effect in population genetics. *Amer. Nat.* **95**, 295–307.

Craddock, E. M., and Johnson, W. E. (1979). Genetic variation in Hawaiian *Drosophila.* V. Chromosomal and allozymic diversity in *Drosophila silvestris* and its homosequential species. *Evolution* **33**, 137–155.

Crosby, J. L. (1970). The evolution of genetic discontinuity: computer models of the selection of barriers to interbreeding between species. *Heredity* **25**, 253-297.

Crossley, S. A. (1974). Changes in mating behavior produced by selection for ethological isolation between ebony and vestigial mutants of *Drosophila melanogaster*. *Evolution* **28**, 631-647.

Danforth, C. H. (1950). Evolution and plumage traits in pheasant hybrids *Phasianus* × *Chrysolophus*. *Evolution* **4**, 301-315.

Diehl, S. R. (1984). The role of host plant shifts in the ecology and speciation of *Rhagoletis* (Diptera: Tephritidae). Ph.D. thesis, University of Texas, Austin.

Dobzhansky, T. (1936). Studies on hybrid sterility. II. Localization of sterility factors in *Drosophila pseudoobscura* hybrids. *Genetics* **21**, 113-135.

Dobzhansky, T. (1973). Genetic analysis of hybrid sterility within the species *Drosophila pseudoobscura*. *Hereditas* **77**, 81-88.

Ehrman, L. (1960). The genetics of hybrid sterility in *Drosophila paulistorum*. *Evolution* **14**, 212-223.

Ehrman, L. (1962). Hybrid sterility as an isolating mechanism in the genus *Drosophila*. *Quart. Rev. Biol.* **37**, 279-302.

Ehrman, L., and Kernaghan, R. P. (1971). Microorganismal basis of infectious hybrid male sterility in *Drosophila paulistorum*. *J. Hered.* **62**, 66-71.

Ehrman, L., and Williamson, D. L. (1969). On the etiology of the sterility of hybrids between certain strains of *Drosophila paulistorum*. *Genetics* **62**, 193-199.

Endler, J. A. (1977). "Geographic Variation, Speciation, and Clines," Princeton University Press, Princeton.

Felsenstein, J. (1981). Skepticism towards Santa Rosalia, or why are there so few kinds of animals? *Evolution* **35**, 124-138.

Fouquette, M. J., Jr. (1975). Speciation in chorus frogs. I. Reproductive character displacement in the *Pseudacris nigrita* complex. *Syst. Zool.* **24**, 16-22.

Futuyma, D. J., and Mayer, G. C. (1980). Non-allopatric speciation in animals. *Syst. Zool.* **29**, 254-271.

Gerstel, D. V. (1954). A new lethal combination in interspecific cotton hybrids. *Genetics* **39**, 628-639.

Goldschmidt, R. (1940). "The Material Basis of Evolution," Yale University Press, New Haven.

Grant, D. R. (1972). Convergent and divergent character displacement. *Biol. J. Linn. Soc.* **4**, 39-68.

Grula, J. W., and Taylor, O. R., Jr. (1980). The effect of X-chromosome inheritance on mate-selection behavior in the sulfur butterflies, *Colias eurytheme* and *C. philodice*. *Evolution* **34**, 688-695.

Guttman, S. I., Wood, T. K., and Karlin, A. A.(1981). Genetic differentiation along host plant line in the sympatric *Enchenopa binotata* Say complex (Homoptera: Membracidae). *Evolution* **35**, 205-217.

Harrison, R. G. (1979). Speciation in North American field crickets: evidence from electrophoretic comparisons. *Evolution* **33**, 1009-1023.

Hedrick, P., Jain, S., and Holden, L. (1978). Multilocus systems in evolution. *In* "Evolutionary Biology," Vol. II(M. K. Hecht, W. C. Steere, and B. Wallace eds.),pp.101-184. Plenum Publishing Corp., New York.

Hedrick, P. W., and Levin, D. A. (1984). Kin-founding and the fixation of chromosomal variants. *Amer. Nat.* **124**, 789-797.

Howard, D. J., and Harrison, R. G. (1984). Habitat segregation in ground crickets: the role of interspecific competition and habitat selection. *Ecology* **65**, 69-76.

Huettel, M. D., and Bush, G. L. (1972). The genetics of host selection and its bearing on sympatric speciation in *Procecidochares* (Diptera: Tephritidae). *Entomol. Exp. Appl.* **15**, 465-480.

Hunkapiller, T., Huang, H., Hood, L., and Campbell, J. H. (1982). The impact of modern genetics on evolutionary theory. *In* "Perspectives on Evolution" (R. Milkman, ed.), pp. 164-189. Sinauer Associates, Inc., Sunderland.

Johnson, M. S., Clarke, B., and Murray, J. (1977). Genetic variation and reproductive isolation in *Partula*. *Evolution* **31**, 116-126.

Jordan, D. S. (1905). The origin of species through isolation. *Science* **22**, 545-562.

Katakura, H., Hinomizu, H., and Yamamoto, M. (1981). Crossing experiments among three ladybird "species" of *Henosepilachna vigintiocto-maculata* complex (Coleoptera, Coccinellidae) feeding on thistles and/or blue cohosh. *Kontyu* **49**, 482-490.

Kawanishi, M., and Watanabe, T. K. (1978). Difference in photo-preferences as a cause of coexistence of *Drosophila simulans* and *Drosophila melanogaster* in nature. *Jap. J. Genet.* **53**, 209-214.

Kessler, S. (1966). Selection for and against ethological isolation between *Drosophila pseudoobscura* and *Drosophila persimilis*. *Evolution* **20**, 634-645.

Kettlewell, H. B. D. (1955). Recognition of the appropriate backgrounds by the pale and black phases of Lepidoptera. *Nature* **175**, 943-944.

Kettlewell, H. B. D., and Conn, D. L. T. (1977). Further background-choice experiments on cryptic Lepidoptera. *Journal of Zoology* **181**, 371-376.

Khasimuddin, S., and DeBach, P. (1976). Biosystematic and evolutionary statuses of two sympatric populations of *Aphytis mytilaspidis* *(Hym: Aphelinidae)*. *Entomophaga* **21**, 113-122.

Kimura, M. (1974). Gene pool of higher organisms as a product of evolution. *Cold Spring Harbor Symp. Quant. Biol.* **38**, 515-524.

Knight, G. R., Robertson, A., and Waddington, C. H. (1956). Selection for sexual isolation within a species. *Evolution* **10**, 14-22.

Kondrashov, A. S. (1983a). Multilocus model of sympatric speciation I. One character. *Theor. Pop. Biol.* **24**, 121-135.

Kondrashov, A. S. (1983b). Multilocus model of sympatric speciation II. Two characters. *Theor. Pop. Biol.* **24**, 136-144.

Koopman, K. F. (1950). Natural selection for reproductive isolation between *Drosophila pseudoobscura* and *Drosophila persimilis*. *Evolution* **4**, 135-148.

Kreslavsky, A. G. Mikheev, A. V., Solomatin, V. M., and Grizenko, V. V. (1981). Genetic exchange and isolating mechanisms in sympatric races of *Lochmaea capreae* (Coleoptera, Chrysomelidae). *Zool. Zh.* **60**, 62-68.

Lande, R. (1979). Effective deme sizes during long-term evolution estimated from rates of chromosomal rearrangement. *Evolution* **33**, 234-251.

Lande, R. (1982). Rapid origin of sexual isolation and character divergence in a cline. *Evolution* **36**, 213-223.

Levene, H. (1953). Genetic equilibrium when more than one ecological niche is available. *Amer. Nat.* **87**, 331-333.

Littlejohn, M. J. (1965). Premating isolation in the *Hyla ewingi* complex. *Evolution* **19**, 234-243.

Littlejohn, M. J. (1981). Reproductive isolation: a critical review. *In* "Evolution and Speciation" (W. R. Atchley and D. S. Woodruff, eds.), pp. 298-334. Cambridge University Press, Cambridge.

MacNair, M. R., and Christie, P. (1983). Reproductive isolation as a pleiotropic effect of copper tolerance in *Mimulus guttatus* ? *Heredity* **50**, 295-302.

Main, A. R., Lee, A. K., and Littlejohn, M. J. (1958). Evolution in three genera of Australian frogs. *Evolution* **12**, 224-233.

Martinez-Arias, A., and Lawrence, P. A. (1985). Parasegments and compartments in the *Drosophila* embryo. *Nature* **313**, 639-642.

Maxson, L. R., and Roberts, J. D. (1984). Albumin and Australian frogs: molecular data a challenge to speciation model. *Science* **225**, 957-958.

Maynard Smith, J. (1966). Sympatric speciation. *Am. Nat.* **100**, 637-650.

Mayr, E. (1942). "Systematics and the Origin of Species," Columbia University Press, New York.

Mayr, E. (1947). Ecological factors in speciation. *Evolution* **1**, 163-288.

Mayr, E.(1954). Change of genetic environment and evolution. *In* "Evolution as a Process" (J. Huxley, A. C. Hardy, and E. B. Ford, eds.), pp. 157-180. Allen and Unwin, London.

Mayr, E. (1963). "Animal Species and Evolution," The Belknap Press of Harvard University Press, Cambridge.

Mayr, E. (1982). Processes of speciation in animals. *In* "Mechanisms of Speciation" (C. Barigozzi, ed.), pp.1-19. Alan R. Liss, Inc., New York.

Menken, S. B. J. (1981). Host races and sympatric speciation in small ermine moths, *Yponomeutidae*. *Entomol. Exp. Appl.* **30**, 280-291.

Miller, D. R., and Kosztarab, M. (1979). Recent advances in the study of scale insects. *Ann. Rev. Entomol.* **24**, 1-27.

Muller, F. P. (1976). Hosts and non-hosts in subspecies of *Aulacorthum solani* (Kaltenbach) and intraspecific hybridizations (Homoptera: Aphididae). *In* "The Host Plant in Relation to Insect Behavior and Reproduction" (T. Jermy, eds.), pp.187-190. Plenum Press, New York.

Nei, M. (1980). Stochastic theory of population genetics and evolution. *In* "Symposium on Mathematical Models in Biology" (C. Barigozzi, ed.), pp. 17-47. Springer-Verlag, Berlin.

Nevo, E., Kim, Y. J., Shaw, C. R., and Thaeler, C. S., Jr. (1974). Genetic variation, selection and speciation in *Thomomys talpoides* pocket gophers. *Evolution* **28**, 1-23.

Oka, H. I. (1974). Analysis of genes controlling F_1 sterility in rice by the use of isogenic lines. *Genetics* **77**, 521-534.

Oliver, C. G. (1979). Genetic differentiation and hybrid viability within and between some Lepidoptera species. *Amer. Nat.* **114**, 681-694.

Parsons, P. A. (1975). Phototactic responses along a gradient of light intensities for the sibling species *Drosophila melanogaster* and *Drosophila simulans*. *Behav. Genet.* **5**, 17-25.

Parsons, P. A. (1978). Habitat selection and evolutionary strategies in *Drosophila*: an invited address. *Behav. Genet.* **8**, 511-526.

Paterson, H. E. H. (1978). More evidence against speciation by reinforcement. *S. African J. Sci.* **74**, 369-371.

Paterson, H. E. H. (1981). The continuing search for the unknown and unknowable: a critique of contemporary ideas on speciation. *S. African J. Sci.* **77**, 113-118.

Paterson, H. E. H. (1982). Perspective on speciation by reinforcement. *S. African J. Sci.* **78**, 53-57.

Patterson, J. T., and Dobzhansky, T.(1945). Incipient reproductive isolation between two subspecies of *Drosophila pallidipennis*. *Genetics* **30**, 429-438.

Price, P. W., and Willson, M. F. (1976). Some consequences for a parasitic herbivore, the milkwood longhorn beetle, *Tetraopes tetrophthalmus*, of a host plant shift from *Asclepias syriaca* to *A. verticilaata*. *Oecologia* **25**, 331-340.

Prokopy, R. T., Bennett, E. W., and Bush, G. L. (1971). Mating behavior in *Rhagoletic pomonella* (Diptera: Tephritidae) I. Site of assembly. *Can. Ent.* **103**, 1405-1409.

Raff, R. A., and Kaufman, T. C. (1983). "Embryos, Genes, and Evolution," Macmillan Publishing Co., Inc., New York.

Rausher, M. D. (1984). The evolution of habitat preference in subdivided populations. *Evolution* **38**, 596-608.

Rice, W. R. (1984). Disruptive selection on habitat preference and the evolution of reproductive isolation: a simulation study. *Evolution* **38**, 1251-1260.

Rockwell, R. F., and Seiger, M. B. (1973). A comparative study of photoresponse in *Drosophila pseudoobscura* and *Drosophila persimilis*. *Behav. Genet.* **3**, 163-174.

Rubin, G. M., Kidwell, M. G., and Bingham, P. M. (1982). The molecular basis of P-M hybrid dysgenesis: the nature of induced mutations. *Cell* **29**, 987-994.

Rubin, G. M., and Spradling, A. C. (1982). Genetic transformation of *Drosophila* with transposable element vectors. *Science* **218** 348-353.

Sanchez-Herrero, E., Vernos, I., Marco, R., and Morata, G. (1985). Genetic organization of *Drosophila* bithorax complex. *Nature* **313**, 108-113.

Sano, Y., and Kita, F. (1978). Reproductive barriers distributed in *Melilotus* species and their genetic bases. *Can. J. Genet. Cytol.* **20**, 275-289.

Saunders, A. P. (1952). Complementary lethal genes in the cowpea. *S. African J. Sci.* **48**, 195-197.

Schroder, G. D., and Rosenzweig, M. L. (1975). Perturbation analysis of competition and overlap in habitat utilization between *Dipodomys ordii* and *Diposomys merriami*. *Oecologia* **19**, 9-28.

Simpson, G. G. (1961). "Principles of Animal Taxonomy," Columbia University Press, New York.

Spencer, W. P. (1944). The genetic basis of differences between two species of *Drosophila*. *Amer. Nat.* **78**, 183-188.

Spieth, H. T. (1949). Sexual behavior and isolation in Drosophila. I. The interspecific mating behavior of species of the *Willistoni* group. *Evolution* **3**, 67-81.

Stebbins, G. L. (1971). "Processes of Organic Evolution," Prentice-Hall, Inc., Englewood Clifs.

Stephens, S. G. (1950). The genetics of "corky". II. Further studies on its genetic basis in relation to the general problem of interspecific isolating mechanisms. *J. Genet.* **50**, 9-20.

Sternberg, P. W., and Horvitz, H. R. (1981). Gonadal cell lineages of the nematode *Panagrellus redivivus* and implications for evolution by the modification of cell linaege. *Dev. Biol.* **88**, 147-166.

Sternberg, P. W., and Horvitz, H. R. (1984). The genetic control of cell lineage during nematode development. *Ann. Rev. Genet.* **18**, 489-524.

Strathmann, R. (1974). The spread of sibling larvae of sedentary marine invertebrates. *Amer. Nat.* **108**, 29-44.

Syvanen, M. (1984). The evolutionary implications of mobile genetic elements. *Ann. Rev. Genet.* **18**, 271-293.

Tauber, C. A., and Tauber, M. J. (1977a). A genetic model for sympatric speciation through habitat diversification and seasonal isolation. *Nature* **268**, 702-705.

Tauber, C. A., and Tauber, M. J. (1977b). Sympatric speciation based on allelic changes at three loci: evidence from natural populations in two habitats. *Science* **197**, 1298-1299.

Templeton, A. R. (1977). Analysis of head shape differences between two interfertile species of Hawaiian *Drosophila*. *Evolution* **31**, 630-642.

Templeton, A. R. (1981). Mechanisms of speciation – a population genetic approach. *Ann. Rev. Ecol. Syst.* **12**, 23-48.

Thoday, J. M., and Boam, T. B. (1959). Effects of disruptive selection. II. Polymorphism and divergence without isolation. *Heredity* **13**, 205-218.

Thoday, J. M., and Gibson, J. B. (1962). Isolation by disruptive selection. *Nature* **193**, 1164-1166.

Udovic, D. (1980). Frequency-dependent selection, disruptive selection, and the evolution of reproductive isolation. *Amer. Nat.* **116**, 621-641.

Val, R. C. (1977). Genetic analysis of the morphological differences between two interfertile species of Hawaiian *Drosophila*. *Evolution* **31**, 611-629.

Waage, J. (1975). Reproductive isolation and the potential for character displacement in the damselflies, *Calopteryx maculata* and *C. aequabilis* (Odonata: Calopterygidae). *Syst. Zool.* **24**, 24-36.

Waage, J. (1979). Reproductive character displacement in *Calopteryx* (Odonata: Calopterygidae). *Evolution* **33**, 104-116.

Wagner, M. (1868). "Die Darwin'she Theorie und das Migrationsgesetz der Organismen," Duncker und Humbolt, Leipzig.

Walker, T. J. (1974). Character displacement and acoustic insects. *Amer. Zool.* **14**, 1137-1150.

Wasserman, M., and Koepfer, H. R. (1977). Character displacement for sexual isolation between *Drosophila mojavensis* and *Drosophila arizonensis*. *Evolution* **31**, 812-823.

Weisbrot, D. R. (1963). Studies on differences in the genetic architecture of related species of *Drosophila*. *Genetics* **48**, 1121-1139.

White, M. J. D. (1968). Models of speciation. *Science* **159**, 1065-1070.

White, M. J. D. (1978). "Modes of Speciation," W. H. Freeman and Company, San Francisco.

Wiebe, G. A. (1934). Complementary factors in barley giving a lethal progeny. *J. Hered.* **25**, 273-274.

Wiley, E. O. (1978). The evolutionary species concept reconsidered. *Syst. Zool.* **27**, 17-26.

Wiley, E. O. (1981). "Phylogenetics." John Wiley and Sons, New York.

Wilson, A. C., Bush, G. L., Case, S. M., and King, M.-C. (1975). Social structuring of mammalian populations and rate of chromosomal evolution. *Proc. Natl. Acad. Sci. USA* **72**, 5061-5065.

Wilson, E. O. (1965). The challenge from related species. *In* "The Genetics of colonizing Species" (H. G. Baker and G. L. Stebbins, eds.), pp. 7-24. Academic Press, New York.

Wood, T. K., and Guttman, S. I. (1982). The ecological and behavioral basis for reproductive isolation in the sympatric, *Enchenopa binotata* complex (Homoptera: Membracidae). *Evolution* **36**, 223-242.

Wright, S. (1940). Breeding structure of populations in relation to speciation. *Amer. Nat.* **74**, 478-494.

Wright, S. (1977). "Evolution and the Genetics of Populations," Vol. 3, The University of Chicago Press, Chicago.

Wright, S. (1978). "Evolution and the Genetics of Populations," Vol. 4, The University of Chicago Press, Chicago.

MECHANISMS OF ADAPTIVE SPECIATION AT THE MOLECULAR AND ORGANISMAL LEVELS[1]

Eviatar Nevo

Institute of Evolution
University of Haifa
Mount Carmel, Haifa, Israel

ABSTRACT

Subterranean mole rats of the *Spalax ehrenbergi* superspecies in Israel represent an active case of ecological speciation. The complex comprises four chromosomal species (2n = 52, 54, 58 and 60) displaying progressive stages of late chromosomal speciation. Their adaptive radiation in Israel during the last 250,000 years from the Middle Pleistocene to recent times is closely associated with fossoriality and increasing aridity, progressive deforestation and savannization, hence with distinct climatic diversity in both the mediterranean and steppic climate regimes: 2n = 52 radiated in the cool-humid Upper Galilee Mountains; 2n = 54 in the cool-dry Golan Heights; 2n = 58 in the warm-humid Lower Galilee Mountains, Central Yizreel and Coastal Plains; finally, 2n = 60 in the mountains of Samaria, Judea, northern Negev and the southern part of the Jordan Valley and Coastal Plain.

The ecological speciation of *S. ehrenbergi* into increasingly arid environments was initiated by chromosomal evolution followed by genic accumulation of positive assortative mating and the concomitant evolution of a syndrome of climatically coadapted genomic adaptations at all

---

[1] I am indebted to the Israel Discount Bank Chair of Evolutionary Biology and to the "Ansell-Teicher Research Foundation for Genetics and Molecular Evolution" established by Florence and Theodore Baumritter of New York for financial support.

439

organizational levels. These integrate the genetic molecular (DNA and proteins) and the phenotypic organismal (morphological, physiological and behavioral) levels, as multiple adaptive strategies to the following major challenges: (i) subterranean life, (ii) mediterranean climate, (iii) aridity index, and its effect on the energy budget, and (iv) extreme burrow hypoxic-hypercapnic atmosphere.

S. ehrenbergi represent a pluralistic example where chromosomal and genic mutations, genetic drift, migration, isolation, and natural selection, all interact in producing adaptive new species associated with environmental diverse challenges, thereby linking the mechanisms of micro- to macroevolution, corroborating the synthetic theory of Darwinian evolution.

I. OPEN PROBLEMS OF SPECIATION

The origin of species, the process leading to the establishment of reproductive isolation between populations, is a key problem of evolution. Despite its fundamental evolutionary importance, speciation theory is still highly descriptive and speculative, not withstanding the recent dramatic advances in new techniques and theoretical models. These include molecular variation, cytogenetical, ethological, and theoretical innovations (reviewed in White, 1978; Barigozzi, 1982 and references therein). The following are some examples of the still unresolved, yet central, issues in speciation today:

1) What are the dynamics, rates, and time spans of the speciation process? Is speciation punctuational, gradual, or varying in rates in space, time and different taxa? 2) Are there universal laws or principles of biological speciation, or alternatively, are there diverse conceivable modes and scenarios? 3) How much of speciation is deterministic, or stochastic, but ultimately displaying adaptive patterns? 4) How does reproductive isolation arise? Does it develop incidentally in allopatry or rather by reinforcement in sympatry? Does reproductive isolation precede

or follow adaptive divergence of gene pools? What are the genetic mechanisms causing the evolution of reproductive isolation? 5) Does speciation necessarily entail major genomic changes? 6) Does successful speciation always involve an ecological component of newly opened niches? 7) Is phenotypic variation between species at multiple organizational levels, including morphology, physiology and behavior, correlated with, and predictable by, molecular variation? Are both phenotypic and genotypic variation between species predictable by taxonomic, ecological, demographic and life history characteristics? Is it possible to disentangle the causative from the associative and consequential genetic variation responsible for speciation?

The present paper overviews speciation patterns and theory in the actively speciating subterranean mole rats belonging to the superspecies *Spalax ehrenbergi* in Israel. In the present review, I will focus on the relation between molecular and organismal differentiation at the morphological, physiological and behavioral levels. For recent reviews in other perspective see Nevo (1979, 1982, 1985a,b). In particular, I will emphasize the deductions obtainable from this case of active mole rat speciation to major recent criticisms of evolutionary theory involving the adaptationist program (Gould and Lewontin, 1979), evolutionary rate of speciation (Eldredge and Gould, 1972; Gould and Eldredge, 1977), micro- and macroevolution (Gould, 1980, 1982) and the origin and nature of speciation.

II. EVOLUTIONARY HISTORY AND ORIGINS OF SUBTERRANEAN MAMMALS AND MOLE RATS, *SPALACIDAE*

Aridization trends during the Cenozoic led to the evolution of open country biota. Savannization led to rapid evolution of an open country flora and fauna including adaptive radiation of cursorial and fossorial mammals. The latter exploited two nonoverlapping niches, the herbivorous, colonized by rodents and the insectivorous, colonized by the marsupials and insectivores. The evolution of these fossorial mammals comprised burrowing adaptations at all organizational levels (Nevo, 1979). Mammalian speciation underground was prolific on all continents and gave rise to an unusually large number of species. Conservative estimates indicate that 144 of 4,060 mammalian species (about 3.5%) speciated underground. The new discoveries of widespread chromosomal sibling species in these mammals may increase manyfold the number of subterranean species (Nevo, 1979).

The evolution of mole rats (family *Spalacidae*, see Savic´, 1982, and Savic´ and Soldatovic´, 1984) into the subterranean ecological zone occurred during Miocene-Pliocene times from a muroid or cricetid ancestor (Petter, 1961). Originating possibly in Asia Minor and/or southeastern Europe, they radiated into the Russian steppes on the one hand, and into the Near East and North African steppes, on the other. Their evolution involved dynamic chromosomal changes, comprising Robertsonian whole-arm rearrangements and pericentric inversions. To date, over 30 karyotypes have been described for mole rats, *Spalacidae* ($2n=38$-62; $NF=74$-124) from Russia (Lyapunova *et al.*, 1974), the Balkans and Asia Minor (Savic´ and Soldatovic´, 1984; Peshev and Vorontsov, 1982), Israel (Wahrman *et al.*, 1969a,b 1985), and Northern Africa (Lay and Nadler, 1972).

III. THE ACTIVE SPECIATION AND ADAPTATION OF *SPALAX EHRENBERGI* COMPLEX: AN OVERVIEW

Subterranean mole rats of the *S. ehrenbergi* superspecies in Israel involve four, morphologically indistinguishable, chromosomally homozygous species each associated with and adapted to a specific climatic regime. The four chromosomal species, 2n = 52, 54, 58 and 60 (Wahrman *et al.*, 1969a,b, 1985) represent progressive late stages of incipient speciation via chromosomal evolution. Their radiation in Israel occurred during Middle and Upper Pleistocene times as documented by multifaceted records including: fossils (Tchernov, 1968), immunological (Nevo and Sarich, 1974) and allozymic distances (Nevo and Shaw, 1972; Nevo and Cleve, 1978), and major histocompatibility complex differentiation (Nizetic *et al.*, 1984, 1985). Evolutionary divergence time for the *S. ehrenbergi* complex, deduced from electrophoretic data, is within the last 250,000 years (Nevo and Cleve, 1978). These dates are in accordance with the hybrid zones (Nevo and Bar-El, 1976; Nevo, 1985a) and fossil (Tchernov, 1968) evidences.

The evolution of the *S. ehrenbergi* complex represents a remarkable case of ecological speciation in action. The climatic evolution of Israel displays, since Miocene times, aridization proceeding northward (Tchernov, 1968). The open country biota that followed the aridity trend provided the open habitats needed for the radiation and speciation of *Spalax* in Israel. Each chromosomal species is associated with a specific climatic regime characterized by a combination of humidity and temperature variables (Nevo, 1985a): cool-humid (2n = 52), cool-dry (2n = 54), warm-humid (2n = 58), and warm-dry (2n = 60), (Fig. 1).

0 10 20 30 km
LEBANON
3
4
SYRIA
52
2
1
6
8
5
54
6
7
Haifa
58
9
JORDAN
10
Tel Aviv
60
Jerusalem
11
Dead
12
Sea
Beer Sheva

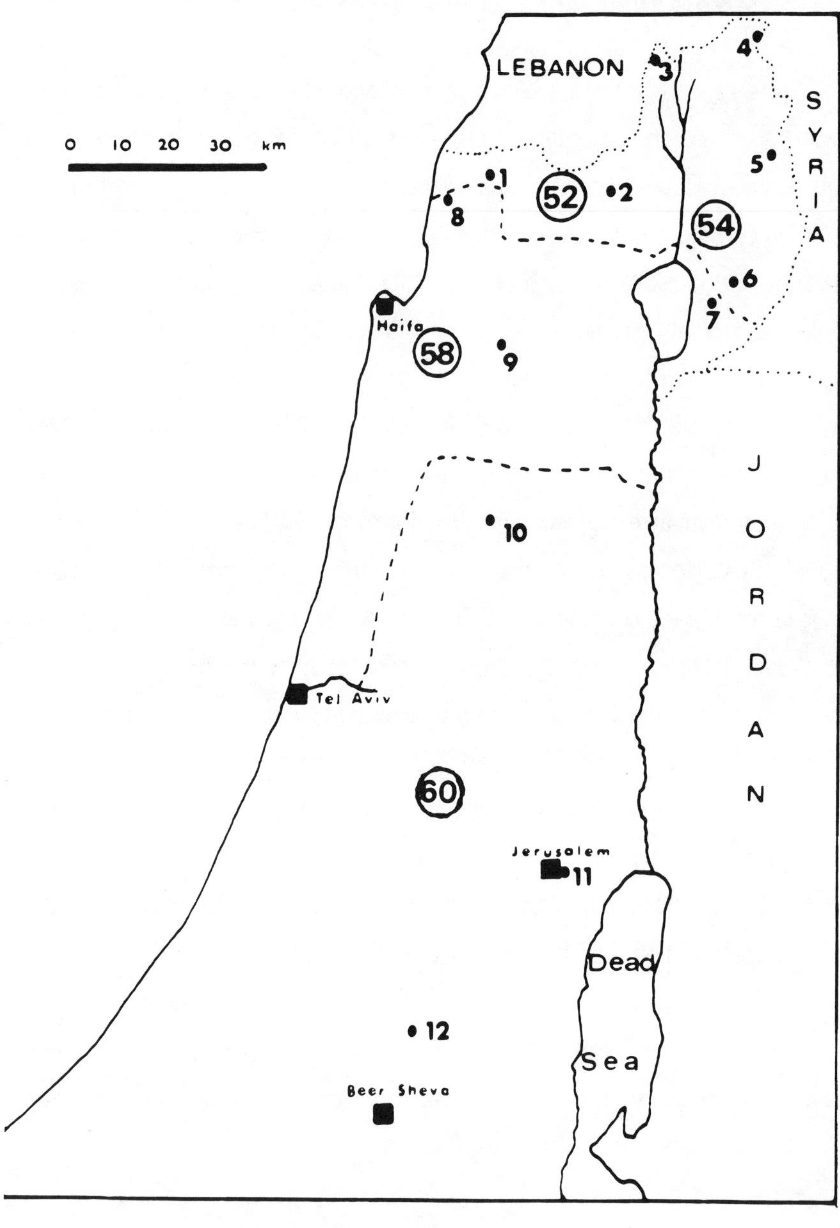

Figure 1

IV. EVOLUTIONARY PATTERNS AND MIDDLE PLEISTOCENE ORIGINS OF THE *SPALAX EHRENBERGI* COMPLEX

Several lines of evidence indicate that the *S. ehrenbergi* complex is geologically recent and is currently under final stages of speciation, as indicated by the following lines of evidence.

A. Immunological Distance

(i) *Microcomplement fixation.* Immunologically, the four chromosomal species display a very close genetic relationship. Microcomplement fixation, using antisera against purified albumin and transferrin of the $2n = 60$ chromosome species, indicates very small (0-5) immunological distances among the four chromosomal species of the complex (Nevo and Sarich, 1974), suggesting Middle to Upper Pleistocene differentiation. By contrast, the immunological distances between the superspecies complexes *S. ehrenbergi* in the Near East and *S. Leucodon* from Yugoslavia is about 20 immunological units, suggesting that they separated in late Miocene or early Pliocene times (i.e., 10 million years ago). Further, the immunological distances between the *S. ehrenbergi* complex and several myomorph rodents outside the genus *Spalax* (e.g., *Mus, Cricetulus, Peromyscus* and *Microtus*) are about 90 units for albumin and 140 units for transferrin. Thus, the results of microcomplement fixation indicate early, possible Oligocene, evolutionary divergence for the family *Spalacidae,* but a very recent one, Middle to Upper Pleistocene, for the differentiation of the *S. ehrenbergi* complex in Israel.

(ii) *Major histocompatibility complex.* The serological and biochemical evidence of the major histocompatibility (Mhc) complex of the mole rats in Israel (Nizetic *et al.,* 1984) and the evidence of the restriction fragment polymorphism (Nizetic *et al.,* 1985) support the notion of relatively recent evolutionary divergence. We have typed over 50 mole rats with mouse monoclonal and polyclonal antibodies specific for class I and class II Mhc molecules. Some of these antibodies were produced against mouse Mhc molecules, others against Mhc molecules of other species. About 25% of the antibodies reacted with mole rat lymphocytes in the cytotoxic test. Some of the serologically positive antibodies precipitated from a glycoprotein pool of mole rat spleen cell molecules that corresponded in size with class I and class II molecules of other species. Thus, mole rats, like other mammals, possess the Mhc which consists of class I and class II loci. We call this Mhc *Spalax* major histocompatibility (*Smh*) complex. The occurrence of a large number of different serotypes among the tested animals suggest that Smh loci are polymorphic. So far no qualitative differences were found between the Smh polymorphism of the four chromosomal species of *S. ehrenbergi*, reinforcing the evidence of their relative evolutionary recency.

The restriction fragment polymorphism of *Smh* also supports the notion of evolutionary recency of the *S. ehrenbergi* complex. The analysis has revealed that the *Smh* complex contains as many class I genes as the mouse does. In the mouse the number of class I Mhc genes, is known to be between 30-40 (Hood *et al.*, 1983). Thus the *Smh* complex contains many class I genes, as is known for other mammals. Although we have no final estimate of class II genes in *Spalax*, the number is less than one-half of the number of class I genes, again close to the rough estimate of eight loci in the mouse (Steinmetz *et al.*, 1982). Neither the cluster of class I or class II genes apparently underwent drastic genome changes since the divergence of mouse and mole rats, presumably in Miocene times. Remarkably, some of the class I genes of *Smh* are highly polymorphic, as are probably the linked loci coding for the complement 4

(C4), agreeing with the results of the serological analysis described previously. The *Smh* polymorphism of the *S. ehrenbergi* complex is primarily quantitative, though few qualitative interspecific variations occur, primarily between the northern (2n = 52, 54) as compared to the central (2n = 58) and southern (2n = 60) species.

B. Genetic Distance

Genic diversity, based on 25 gene loci (Nevo and Shaw, 1972; Nevo and Cleve, 1978) also substantiates the evolutionary recency of the *S. ehrenbergi* complex. The genetic identity (Nei, 1972) between the four chromosomal species is surprisingly high, mean I = 0.966, range 0.931-0.988. These high values indicate that evolutionary divergence times for the *S. ehrenbergi* complex deduced from electrophoretic data (Nei, 1971), are within the last 250,000 years, whereas the 2n = 60, presumably the last derivative of speciation, originated about 75,000 years ago (Nevo and Cleve, 1978).

C. Morphological Distance and the Fossil Record

Morphologically, too, the species are extremely close and indistinguishable. Based on about 40 skull and body variables, we estimated the Mahalanobis distances between the four species. The mean Mahalanobis distance was 2.86, range 2.20-3.29 (Nevo and Tchernov, unpublished). The fossil record of *S. ehrenbergi* in Israel substantiates its Middle to Upper Pleistocene evolutionary history (Tchernov, 1968). The fossil record indicates a dramatic increase in the numbers of *Spalax* during the last 120,000 years of the the Upper Pleistocene. This, presumably, corresponds to the active speciation of the *S. ehrenbergi* complex in Israel.

D. Natural Hybridization

Hybrid zones separate the distribution of the chromosomal species of *Spalax,* decreasing progressively in width northward: (i) 2n=58-60 (2.800 km in width), (ii) 2n = 54-58 (0.750 km), (iii) 2n= 52-58 (0.320 km), and (iv) 2n = 52-54 (width very narrow, but not quantified). These hybrid zones provide telling evidence on the evolutionary dynamics, and stages of speciation in the superspecies (Nevo and Bar-El, 1976; Nevo, 1985a). The first three hybrid zones differ substantially in width, and in the proportion of hybrids and parental types. Field evidence indicates that at least some hybrids are partly fertile. However, the restricted range of the hybrid zones and the nonrandom distribution of hybrids within these zones suggest that hybrids are inferior in fitness to their parental types. We estimated the intensity of selection against hybrids, following Barton's (1979) model. Selection forces, presumably at many loci and/or at whole chromosomes ranging between 0.2% to 19% would be enough to maintain the broader (2n = 58-60) and narrower (2n = 52-58) hybrid zones, respectively, assuming an average dispersal rate of 100 meters per generation. Noteworthy, the position of the hybrid zones appears to be determined mainly by climatic or ecological determinants rather than by geographical barriers, and some zones may be dynamically moving (Nevo, 1985a). Likewise, the rate of disappearance of the hybrid zones, is much longer than originally thought for hybrid zones in general. The progressive narrowness northward of the hybrid zones, is positively correlated with reproductive isolation. Therefore it may monitor the terminalization of the active speciation over time, until effective ethological isolation is established. Regression of the width of the hybrid zones on divergence times (deduced in Nevo and Cleve, 1978), gives r = -0.96, p < 0.001. Accordingly, the process of elimination of the hybrid zones may take 300,000 years, or generations, when the chromosomal species will become closed genetic systems (Heth and Nevo, 1981). This can be hardly defined as a punctuated process, at least not in population genetic terms.

V. THE ORIGIN AND EVOLUTION OF REPRODUCTIVE ISOLATION MECHANISMS

A. Postmating Reproductive Isolation

Cytogenetic barriers to reproduction may exist in the hybrids of the *Spalax ehrenbergi* complex (Wahrman *et al.*, 1985). Noteworthy, hybrid analysis through segregation in litters in the wild (Nevo and Bar-El, 1976) and through cytogenetic analysis (Wahrman *et al.*, 1985) suggests that hybrids are partly fertile. This is true for backcrosses of F1 hybrids and for inter-hybrid crosses. Nevertheless, our recent limited analysis of meiosis, in hybrids from the 2n = 52–58 zone, suggests (i) the presence of unpaired autosomal segments at, presumably, early pachytene, (ii) the formation of nonhomologous associations, and the disappearance of homologue inequalities through synaptic adjustment, and (iii) the proximity between unpaired autosomal segments and the sex chromosome bivalent. If these preliminary findings will be substantiated, they might suggest a reduced fertility for males and partly account for the selection against hybrids. Other chromosomal micro-changes have also been described in *S. ehrenbergi* (Wahrman *et al.*, 1985) which may play a role in the development and maintenance of reproductive isolation (Rosenmann *et al.*, 1985).

The evolution of postmating reproductive isolation in *S. ehrenbergi* seems to have preceded the development of premating isolation. This is evident by the lack of an efficient premating isolation mechanism between the younger divergence of 2n = 58–60, while postmating barriers seem to operate there. By contrast, both post- and premating mechanisms have been substantiated in the older species (2n = 52-54-58, Heth and Nevo, 1981). The chromosomal differences lead to consistent lower hybrid fitness including hybrid breakdown. Hence, they appear to function as primary postmating reproductive isolation that initiated speciation. In other words, in *S. ehrenbergi* chromosome differentiation appears to have been causal not incidental, to the speciation process.

B. Premating Reproductive Isolation

Positive assortative mate selection was revealed in one-choice (Nevo, 1969), and two-choice (Nevo and Heth, 1976) laboratory experiments when estrous females significantly preferred homo-chromosomal mates. Later, an extensive program of two-choice mate selection tests, involving all four karyotypes (Heth and Nevo, 1981), suggested that (i) mate selection was generally similar among females from central and near hybrid zone populations. This evidence suggests, that ethological isolation originated primarily as an incidental by-product of adaptive differentiation rather than by reinforcement in the hybrid zones, and (ii) mate selection efficiency varied among karyotypes in accord with their recency of origin, i.e., 2n = 52, 54 and 58 (when tested against 2n = 52 and 54) select much better than 2n = 60, and 2n = 58 against 2n = 60, which is the last derivative of speciation. This evidence suggests that in mole rats, ethological isolation accumulates slowly and gradually.

C. The Mechanisms of Ethological Isolation

The ethological mechanism of mate selection in *S. ehrenbergi* involves at least three communication systems (olfaction, vocalization and aggression), that contribute to maximize ethological reproductive isolation. Estrous females of the two chromosomal species 2n = 52 and 58, significantly preferred homospecific odors (Nevo *et al.,* 1976). The "mating dialect" of *S. ehrenbergi* has been described (Heth *et al.*, 1985b), as an adaptive feature for communication underground (Heth *et al.*, 1985c). The four species vary in their "mating dialects." Consequently, they provide an effective acoustic isolating mechanisms as revealed in laboratory female discrimination tests (Nevo *et al.*, 1985b). Finally, high interspecific aggression between contiguous species appears also to play a role in premating isolating mechanisms (Nevo *et al.*, 1975, 1985a; Nevo, 1985b).

D. Origin and Evolution of Assortative Mating

The origin and evolution of positive assortative mating in the *S. ehrenbergi* superspecies was deciphered by comparing female mate preference in the laboratory between ancestral and derivative species. Estrous females of the most recent derivative of speciation (2n = 60) showed trimodal mate preference distribution, significantly differing from a normal curve. Females consisted of three phenotypes, comprising negative, low positive, and high positive preference for homospecific mates. By contrast, mate preference in encounters of ancestral species (2n = 52, 54 and 58) was positively homospecific. It is suggested that the three modal distribution is explicable even on the basis of one major gene with three genotypes. The evolution of ethological reproductive isolation proceeded presumably from a high polymorphism in the most recent derivative of speciation, towards increasing monomorphism of positive assortative mating among ancestral species. If an assortative mating locus is combined with sexual selection of the frequent male adapted optimally to the local environment, then speciation and adaptation will be tightly linked in the evolution of mole rats (Beiles *et al.*, 1984).

VI. MULTIPLE ORGANISMAL AND MOLECULAR ADAPTIVE STRATEGIES

The successful speciation of *S. ehrenbergi* in Israel depends on multiple complementary adaptive strategies at all organizational levels to the subterranean ecotope in general (Nevo, 1979), and to the climatic stresses of increasing aridity and temperature, in particular (Nevo, 1982, 1985a,b). Here, I will briefly summarize mainly *new* evidence pertaining to adaptive syndromes to four major ecological stresses, emphasizing primarily organismal and molecular adaptations. These involve adaptations to the: (i) subterranean ecotope, (ii) mediterranean and steppic climate, (iii) aridity index, and (iv) burrow unique atmosphere. The adaptive "stories" detailed below have been documented and explained in detail in the reference attached to each.

A. Adaptations to the Subterranean Ecotope

I will describe here a new morphological finding in *S. ehrenbergi*, related to adaptive speciation.

Evolution of incisor enamel microstructure and speciation. Incisor enamel development is directly related to rodent adaptation and speciation (see several papers in Luckett and Hartenberg, 1985). *Spalax* is a teeth-digger, hence incisor evolution may be related to its speciation patterns (Flynn *et al.*, 1985). The relatively thick outer enamel of *Spalax* appears to be related to its teeth-digging mechanisms. In general, enamel variability does not differ between closely related species, yet it does in *S. ehrenbergi*. The enamel thickness varies as follows: 112, 105, 105, 89.5 microns for 2n = 54, 52, 58 and 60, respectively. Other incisor enamel parameters also show geographical trends. Notably, soil types in the *S. ehrenbergi* range in Israel become lighter southwards. They vary geographically from the largely heavy basalts and terra rossas in northern regions, where 2n = 54 and 52 range, to the largely light rendzinas, marls, sandy-loams, and sands where 2n = 58 ranges, to the primarily rendzinas, sands and loess substrates of 2n = 60. Thus incisor enamel evolution appears to relate to soil texture and structure. Natural selection presumably leads to thicker enamel in heavy soils and thinner enamel in lighter ones.

B. Adaptations to the Mediterranean and Steppic Climates

Interspecific variation in swimming capacity. Adaptation to the mediterranean climate which is characterized by winter floodings and standing water in high precipitation regions is also reflected by the swimming capacity of mole rats. The swimming ability of 251 mole rats involving the four species, was tested in the laboratory (Hickman *et al.*, 1983). The grand mean swimming time for *Spalax* was 28 sec. ± 18.2. Sex, weight, time in captivity and intraspecific populational differences were not significant to swimming performance. Significant differences were found, however, between the species. Individuals from populations of the drier southern regions (2n = 60; mean 14 sec) did not perform as well as those from northern populations (2n = 52 and 2n = 54; mean 35 sec each), with an intermediate performance by individuals of the geographically intermediate species (2n = 58; mean 25 sec). The results seem to support the hypothesis that differential swimming ability of individuals among the species may be associated with the extent and level of flooding and free-standing water, which is highest in the northern regions due to the higher precipitation and heavier soils in these regions. The swimming performance was displayed in a standardized laboratory environment by animals, some of which lived several years in the laboratory. The swimming capacity seems therefore to have a genetic component. It displays an ethological correlate of ecological speciation in accordance with the climatic origins of the species.

C. Adaptation to the Aridity Index and the Energetics Syndrome

Based on the moisture index there are four climatic regions in Israel from north to south: humid, subhumid, semiarid and arid (Atlas of Israel, 1970, maps IV/3). If the thermal patterns are combined, we obtain the four climatic regimes to which the four chromosomal species of mole rats are adapted: cool-humid $(2n = 52)$, cool-dry $(2n = 54)$, warm-humid $(2n = 58)$, and warm-dry $(2n = 60)$. Thus, the aridity index increases southwards (across the ranges of $2n = 52 \rightarrow 58 \rightarrow \rightarrow 60$) and eastwards $(2n = 52 \rightarrow 54)$ (see details in Nevo, 1985a). The major physiological problem across increasingly hotter and drier microhabitats relates to the balance of energetics. Natural selection must optimize at all organizational levels the net energy gain, as it effectively did. Energetic optimization involves all organizational levels: morphology, genetics, ecology, physiology and behavior.

Morphological adaptations. Body size of the four chromosomal species decreases clinally southwards in response to geographic variation in heat load, displaying a positive correlation of weight with latitude in accordance with Bergmann's rule (Nevo and Tchernov, unpublished). Likewise, pelage color displays both local and regional differentiation. The latter involves geographic variation from darker pelage colors in the northern heavier black and red soils to the lighter soils in the south (unpublished results). This accords with incisor enamel differentiation described earlier as an adaptation to lighter soils southwards (Flynn *et al.*, 1985). The smaller body size and paler pelage color, characterizing primarily 2n = 60, contribute to alleviate the heavy heat loads in the hot steppic regions approaching the Negev desert.

Genetical adaptations. Polymorphism P increases progressively towards the southern Negev deserts (P= 0.12 → 0.16 → 0.28 for 2n = 52, 58, 60, respectively), and towards the eastern Syrian desert (P= 0.12 → 0.24 for 2n = 52, 54, Nevo and Cleve, 1978). These trends of increasing genetic diversity towards the climatically unpredictable steppes and deserts are shared by many unrelated species of plants, invertebrates, and vertebrates in Israel. They underline a generalized common adaptive trend in unrelated taxa in accordance with increasing uncertainty in space, and time, i.e., with the niche-width variation hypothesis (Nevo, 1983).

Physiological adaptation. Several adaptations contribute synergistically to energetics optimization. Geographic variation has been found across the stressful southward trend in aridity in the following variables: basic metabolic rates, BMR's (Nevo and Shkolnik, 1974), nonshivering thermogenesis, NST (Haim *et al.,* 1984), and thermoregulation (Haim *et al.,* 1985a). The decrease in BMR towards the desert appears to be adapted to the aridity index by minimizing water expenditure and overheating. Thermoregulation, as well as NST in mole rats, are positively correlated with temperature variation in nature thereby saving energy. Recently we also found adaptive geographic variations in heart and respiratory rates. Heart rate was 152, 137, 112 and 108 for 2n = 52, 58, 54 and 60, respectively, and respiratory rates are higher in the "Humid" (2n = 52, 58) as compared with the "xeric" (2n = 54, 60) species (Arieli *et al.,* 1985a). The combined physiological variation in BMR, NST, thermoregulation, heart and respiratory rates, appears to be adaptive at both the macro- and microclimatic levels, both *between* and *within* species, all contributing to energetic optimization.

Ecological and demographic adaptations. Several adaptive structures, including territory size and population density are evident across the *S. ehrenbergi* range (Nevo *et al.,* 1982b, 1985a). Territory size increases (52 < 54 < 58 < 60) and population density decreases (54 > 52 > 58 > 60) southwards. These trends appear to correlate adaptively with productivity and resource availability which are higher in the northern mediterranean, and lower in the southern steppic and desert regimes.

Biochemical adaptations. Optimal foraging theory assumes that natural selection maximizes fitness by optimizing net energy gained per unit feeding time (Pyke *et al.,* 1977). We have initiated the analysis of the primary structure of ribonucleases of the chromosomal species of *S. ehrenbergi* in an attempt to relate it to the feeding strategies of mole rats across the ecological gradient (Beintema and Nevo, in preparation). The molecular evolution of mammalian ribonucleases and their involvement in digestion processes have been recently reviewed by Beintema and Lenstra (1982). Variation in pancreatic ribonuclease quantity may be a response to the necessity of digesting large amounts of ribonucleic acid. The high rates of ribonuclease evolution in myomorph rodents, to which mole rats belong, may be correlated with the variation in diets and digestive systems of this taxon. Our preliminary results in *S. ehrenbergi* indicate: (i) variation in the amount of ribonuclease between the chromosomal species, and (ii) geographic variation in the proportion of some amino acids (e.g., glutamic acid, lycine, methionine, tyrosine, arginine), some increasing, other decreasing southwards, with increasing aridity. This cline may be correlated with decreasing habitat productivity and food diversity from the richer northern mediterranean to the poorer southern steppic habitats.

Behavioral adaptations including activity, exploratory, habitat selection and aggression patterns also contribute to energy optimization. Activity patterns in both intensity and temporal variation (Nevo *et al.*, 1982a) as well as exploratory patterns (Heth *et al.*, 1985a) vary geographically in accordance with climatic determinants. The species ranging in humid climates (2n = 52, 58) are more active and more exploratory than those living in drier ones (2n = 54, 60). Moreover, species living in a cooler climate (2n = 52, 54) display less rest periods than those living in warm climates (2n = 58, 60). The differential habitat selection of the species found in laboratory testing (Nevo *et al.*, 1979) indicates that individuals from different populations and species select their climatic niche, simulated in the laboratory by a combination of temperature and humidity, in accordance with their climatic origins. Climatic habitat selection was conducted in a standardized environment and showed no correlation with time in captivity, hence, it is presumably largely intrinsic. Aggression intensity also varies geographically, at least partly, in accordance with climatic stresses of aridity and temperature. Behavioral-genetic "militant" phenotypes decrease (2n = 52 > 54 > 58 > 60)), and "pacifists" increase (2n = 52 < 54 < 58 < 60) southwards , in both sexes (Nevo *et al.*, 1985a).

The multifaceted patterns detailed above support the energy budget optimization theory by optimizing net energy gain per unit activity (McNab, 1979; Nevo *et al.*, 1982a), thereby maximizing fitness.

D. Adaptations to the Unique Burrow Atmosphere

Respiratory adaptations. Subterranean mammals face, beside the major problem of energetics, the critical extremes of hypoxia and hypercapnia in their burrow atmospheres. Consequently, they developed multiple respiratory adaptations (Nevo, 1979 and references therein).

Hypoxic-hypercapnic stresses increase in the subterranean burrow atmospheres of mole rats northward in Israel due to the higher precipitation and heavy soil textures in the northern regions of 2n = 52 and 54. We have measured oxygen and carbon dioxide pressures in subcutaneous gas pockets of the four chromosomal species of the *S. ehrenbergi* complex. Oxygen pressures of 11.8, 13.6, 16.9 and 17.2 torr and carbon dioxide pressures of 84.2, 82.9, 80.1 and 64.1 torr were measured for the chromosomal species 2n = 52, 54, 58 and 60, respectively. The differences between the four species in their subcutaneous gas tension appear to reflect adaptive respiratory variation associated with geographic variation in climate (Arieli *et al.*, 1984). It underlines an important respiratory physiological correlate of ecological speciation in the extremely hypoxic and hypercapnic subterranean environment, where, in Israel, these stresses increase northwards.

Urinary adaptations. Another adaptive physiological mechanism to eliminate excess carbon dioxide from the body of a subterranean mammal, caused by the extreme hypercapnic stress in the burrow atmosphere, may be via urine excretion. Supportive evidence to the operation of this mechanism is our recent finding that urine concentration of calcium and magnesium bicarbonates of fossorial mammals, such as the mole rats and the hamster, is higher than that of the white rat (Haim *et al.*, 1985b). Furthermore, a preliminary test of carbon dioxide concentration per liter urine in samples of 10 animals from each of the chromosomal species 2n = 52, 54, 58 and 60 gives the following values: 3.53, 3.06, 2.39 and 2.34, respectively (Nevo *et al.*, unpublished). This result parallels the analysis of carbon dioxide in subcutaneous pockets (Arieli *et al.*, 1984). In other words, the species living in higher hypercapnic burrow atmospheres (2n = 52, 54), due to higher precipitation, flooding, and heavier soil textures, are adapted to higher carbon dioxide pressures in their bodies and apparently developed more efficient mechanisms to eliminate it in the urine.

Hematocrit and hemoglobin concentrations. Hematocrit (HCT) and hemoglobin (Hb) concentrations were measured in the four species (Arieli *et al.,*1985b). HCT was 52.0, 51.4, 50.9 and 47.8%; and Hb was 16.0, 16.6, 16.3 and 14.7g/100ml for 2n = 52, 58, 54 and 60, respectively. The species 2n = 60, which lives in arid habitats, had lower HCT and Hb than the other three species. HCT decreased as aridity increased between the species and within the species 2n = 60. Changes in HCT probably reflect clinal changes in both soil permeability to gases and ambient temperature.

E. DNA Variation and Differentiation

Ribosomal RNA polymorphisms. We have analyzed the chromosomal location of the ribosomal genes in *S. ehrenbergi* (Wahrman *et al.,* 1985) as well as the polymorphisms of the 3'-end of 28S rRNA genes by restriction endonucleases (Suzuki *et al.,* 1984, in preparation). Nucleolus organizing regions (NOR's) are carried by four chromosomes at various positions and are polymorphic within and between populations. Southern blot analysis with a rRNA probe prepared from mouse and two restriction enzymes, Eco-RI and Bam-HI revealed that each of 50 individuals tested, comprising 12 populations and the four species, had various sizes of major restriction fragments, at the 3'-end of 28S rRNA gene, as evidenced by the probe specificity and restriction enzymes. Generally, the NTS polymorphisms appear to be species specific, suggesting their *post* speciational evolution. However, we also detected a common DNA fragment in all species. This common fragment predated speciation. The occurrence of gene flow across hybrid zones between two contiguous species was demonstrated by the restriction patterns.

Mitochondrial DNA polymorphisms. We have analyzed mtDNA variation both between and within the chromosome species of *S. ehrenbergi* by several restriction enzymes (Bam-HI, Hind-II, Eco-RI, Hind-III, and Pst-I, Xba-I and Mbo-I (Yonekawa and Nevo, in preparation; Honeycutt and Nevo, in preparation). Preliminary results showed various levels of polymorphisms within and between species. The mtDNA phenotypes, as the rRNA discussed previously, exhibited both environmental and organismal correlations.

F. Correlations Between Molecular, Climatic and Organismal Diversities

Preliminary analysis indicates that allozymic diversity and DNA polymorphism diversities are intercorrelated, and each is correlated first with climatic variables, and second with organismal diversities at all levels (morphological, physiological and behavioral). Detailed accounts will appear elsewhere. Here, I present the first evidence suggesting that the molecular and organismal diversities are interrelated. Since in the analysis of chromosomal species we had only four entries, only very high correlations (above $r = 0.93$) are significant, and apparently suggest *much lower, but real* correlations. The further analysis of 12 populations (3 for each species) with only some of the molecular variables seem to substantiate the above conclusions. The amount of significant correlations was above the expected 5% by chance.

Correlations of protein and DNA polymorphisms with climate. We found ecological, primarily climatic, correlates with protein (allozymic) and DNA (Smh, rRNA, mtDNA) polymorphisms. For examples, *Polymorphism* (P-1%) based on 25 gene loci is significantly correlated with daily temperature difference; allelic diversity in isocitrate dehydrogenase, *Idh* with rainfall, and 6 phosphogluconate dehydrogenase,

6-Pgd with number of rainy days. Likewise, we found several significant correlations between some DNA polymorphisms and climate. For example, restriction site diversity (Nei and Tajima, 1981) of rRNA correlated with mid-day humidity. Several *Smh* antigens also showed correlations with climatic variables, some with temperature (K548, H-2.160, with mean August temperature), others with rainfall (H-42A, Lam III), r = 0.96*, 0.99** and/or mid-day humidity (H2m8, H-2.159). some of the amino acid frequencies comprising ribonuclease are also correlated with climate (e.g., alanine with evaporation).

Correlations between protein and DNA polymorphisms. Significant correlations have also been found between allozyme and DNA polymorphisms. For example, average number of the combined Eco-RI and Bam-HI rRNA fragments were correlated with *Idh* and *6Pgd* diversities. Likewise, the frequency of some of the six phenotypes of mtDNA showed correlations with average heterozygosity and allele frequency of *Idh*. *Smh* polymorphisms also showed correlations with indices of genic diversity (e.g., P-1% with Y-8KP, H2m8, and *6Pgd* with H-2.110, H-2.115 and H-2.131). Finally, some amino acids of ribonuclease were also correlated with allozyme diversities (e.g., valine with heterozygosity (H), phenylalanine with polymorphisms (*P*). Likewise, allele frequencies of the two *Ldh* loci were correlated with glycine.

Correlations between allozyme and organismal diversities. I will provide here only very few examples to illustrate the general pattern. Allozymic frequencies of *Ldh*-1,5 and *alpha-Gpd, G-6pd* were correlated with morphological, cranial and limb variables, while the frequency of *6Pgd* and *Mdh-1* was correlated with physiological variables such as: BMR with *P*, and *6Pgd;* HCT, with *P;* Hb with *Mdh-1;* thermoregulation with *H, 6Pgd* with *Idh*. Finally, allozymic variation was also correlated with behavioral variables such as swimming capacity.

Correlations between DNA and organismal diversities. DNA polymorphism showed also significant correlations with organismal diversities. For example, rRNA and mtDNA with body size and skull parameters, carbon dioxide pressure, thermoregulation, NST, and swimming capacity; Smh serological variation with body characters, oxygen and carbon dioxide pressures, BMR, carbon dioxide concentration in urine, Hb content, some NST variables, swimming capacity, exploration, activity, aggression and vocalization.

Clearly, some of the above correlations may be spurious, and only our further detailed analysis may eliminate them by testing repetitive new data sets. However, doubtlessly, some of these correlations appear to be genuine and important for highlighting the complex interrelationships between the molecular, organismal and environmental diversities, even if their biological significance needs elucidation.

VII. THEORETICAL OVERVIEW

I will now consider some of the current challenges to evolutionary theory (e.g., Gould, 1980, 1982) in view of the evidence presented above. I argue that nothing learned from *S. ehrenbergi* should change the basic postulates of neo-Darwinism. On the contrary, the evidence presented at all levels, from the molecular to the organismal ones, strongly supports the basic neo-Darwinian concepts. Since the evidence presented above speaks for itself, I will be brief and schematic in discussing three challenges to current evolutionary theory: (i) The adaptationist program, (ii) rate of speciation, and terminating by comments on (iii) genetic differentiation during speciation.

A. The Adaptationist Program

Adaptation is a central concept of evolutionary biology. It was recently reviewed critically by Krimbas (1984), who suggested that it should be excluded from scientific texts in order to avoid confusion in language and thinking. Earlier, Gould and Lewontin (1979) faulted the adaptationist program "for its failure to distinguish current utility from reasons for origins...; for its unwillingness to consider alternatives to adaptive stories; for its reliance upon plausibility alone as a criterion for accepting speculative tales; and for its failure to consider adequately such competing themes as random fixation of alleles, production of non-adaptive structures by developmental correlation with selected features (allometry, pleiotropy, material compensation, mechanically forced correlation), the separability of adaptation and selection, multiple adaptive peaks, and current utility as an epiphenomenon of nonadaptive structure." Of course, these and other criticisms may be true and must be attended to. Does it mean, however, that adaptation becomes obsolete? The evidence presented for the *S. ehrenbergi* complex argues that the adaptationist program is as vital as ever. Our research for evolutionary meaning, i.e., the search for the origin and maintenance of adaptations in nature should not make us uncritical, naive, or pan-selectionists. But as the *S. ehrenbergi* evidence shows, nothing in its multiple structures and functions makes sense unless viewed in terms of the organism-environment relationships. Surely organisms must be analyzed as integrated wholes, with phylogenetic constraints at all levels. It is the combination of past origins and challenges as well as present challenges that make organisms so unique and complex.

The challenge the biologist faces is not to abolish adaptive stories but rather to *critically* substantiate them. This should be done by searching for efficient causes, patterns, and processes which best explain the structures and functions under research out of multiple potential alternative stories. These explanatory models must be testable, falsifiable, and contain at least some level of prediction. Natural selection certainly operates under a variety of phylogenetic and developmental constraints at both the molecular and phenotypic levels, but constraints must not necessarily imply dead ends. Evolutionary tinkering (Jacob, 1977) is operating at the molecular and phenotypic levels. While evolution is indeed constrained by the past, it is nevertheless wildly innovative as molecular evolution reveals (see examples in Dover and Flavell, 1982, and Shapiro, 1983), and as the astounding diversity of organic nature amply demonstrates.

Nonadaptive factors may certainly be important in evolutionary change, due to many obvious reasons. The latter include, inter alia, adaptive compromises and imperfections, historical and epistatic constraints, pleiotropy, genetic hitchhicking, genetic drift, stochasticity, or lack of appropriate genetic variation. Nevertheless, despite the obvious obstacles faced by natural selection, these factors, singly or in combination, do not seem to undermine the major role of natural selection in orienting evolutionary change of whole organisms from the molecular to the phenotypic levels. Otherwise, how can all the multiple structures and functions described in *S. ehrenbergi* be so intimately related to the environment? The massive evidence of *S. ehrenbergi* displays convergent adaptations to the subterranean ecotope in general (Nevo, 1979) and divergent adaptations associated with the speciation and adaptive radiation into four climatic regimes (Nevo, 1982, 1985a,b). While certainly not perfect, the pluralistic adaptive strategies described here are effective in increasing fitness. The significant differential mortality of the four chromosomal species found under a standardized laboratory environment reveals their differential fitness due to their varied climatic origins and adaptations (Nevo *et al.*, 1982c). Thus, the diversity in genetics, physiology, biochemistry, morphology, ecology and behavior of mole rats becomes meaningful *only under the adaptationist program.* Stripped from it, the mole rat story remains unexplained. Why should one avoid that explanatory model which highlights organic nature by critically relating *whole* organisms to their environments?

B. Rates of Speciation

Despite their relatively rapid chromosomal evolution which certainly provided momentum to their speciation, mole rats do not reflect the seemingly non-Darwinian phenomenon of punctuated equilibrium (Gould and Eldredge, 1977). The evidence reviewed here from natural hybridization and the rate of accumulation of ethological assortative mating is discordant with punctuationism. Or, should we regard 300,000 generations as a burst? Similar conclusions for the gradual build up of reproductive isolation is also argued by Barton and Charlesworth (1984). Furthermore, while mole rat morphology appears to be conservative, genetics, physiology, and behavior reflect distinct diversity, far from reflecting any stasis. Mole rat speciation may have been rapid, but not punctuational. It certainly fits into the framework of neo-Darwinian evolution. The latter conceives of a widespread spectrum of speciation and evolutionary rates from the gradual to the punctuational extremes (see Simpson, 1944; and a recent critical review in Bengtsson, 1980, and in Charlesworth *et al.*, 1982).

C. Genetic Differentiation during Speciation

The idea that speciation depends on genetic revolution (Mayr, 1954) has been recently reviewed critically by Barton and Charlesworth (1984) and by Carson and Templeton (1984). Both reviews conclude that genetic revolutions are unlikely to occur during speciation. Earlier, I argued on allozymic evidence that genetic revolution does not accompany the active speciation of mole rats (Nevo and Shaw, 1972; Nevo and Cleve, 1978) pocket gophers (Nevo *et al.*, 1974), or other rodents (Nevo, 1985a). Similar conclusions have been drawn for other organisms by Ayala (1975) and Avise (1976).

Linkage-group conservation appears to be the prevailing rule in vertebrate evolution (Ohno, 1984). Conservation of mammalian x-linkage groups *in toto* origianally proposed by Ohno *et al.* (1964) has now been extended to large blocks of autosomal linkage groups as well (Ohno, 1973, 1984). Chromosomal homologies comprising large blocks of autosomal linkage groups have been found between very remote species such as man and domestic cat (Nash and O'Brien, 1982), and some linkage groups extend from fish to man (Ohno, 1973, 1984). If substantiated in the future, such conservation of large blocks of autosomal linkage groups from fish to mammals may provide indisputable genetic evidence against the idea of genetic revolution. Speciation may indeed be accompanied by relatively few genomic changes. Even future evolutionary divergence, after the establishment of reproductive isolation which accumulates genetic changes does not apparently need dramatic changes at the DNA level. The molecular evidence of *Spalax* to date, involving both protein and DNA, supports this conclusion. A recent discussion on molecular biological mechanisms of speciation appears in Rose and Doolittle (1983).

VIII. CONCLUSIONS AND PROSPECTS

Is a new evolutionary synthesis necessary to explain adaptation and speciation in evolving mole rats? The evidence presented at the molecular and organismal levels, and their interaction in the *S. ehrenbergi* complex, suggests that their evolution is indeed a pluralistic process. The emergence of a new species of mole rats was not concentrated in geologically instantaneous events of branching speciation. The process was certainly gradual in population genetic terms and even geologically was not a burst. Genetic polymorphisms *within* species at all levels of organization extend into differences *between* species suggesting a continuum from micro- to macroevolution. *Spalax* clearly demonstrates the Darwinian concept that evolution implies the conversion of the variation among individuals and populations *within* an interbreeding group into variation *between* groups in space and time. Individual variation is indeed the cornerstone of evolution. Natural selection, either individually or interdemically, and despite its constant interaction with stochastic process, appears to predominate and orient evolutionary change both intra- and interspecifically.

Speciation and adaptation in *S. ehrenbergi* are deeply intermixed. Initial reproductive isolation by Robertsonian chromosomal mutations may have arisen peripatrically by chance in small peripherally isolated populations (Nevo *et al.*, 1982b) followed by adaptation and the gradual accumulation of ethological reproductive isolation. The simplistic sequence of first speciation then adaptation is unrealized. Both processes interact and complement each other along the evolutionary history of the emerging species. Ecological diversity appears to be a major architect of the adaptive speciation of mole rats, efficiently masking nonadaptive traits. The superiority of a descendant over its ancestor depends on whether it succeeded ecophysiologically in the vacant ecological niche it colonized where it met no competition.

The four chromosomal species are morphologically undistinguishable and converge adaptively to their common unique subterranean ecotope. By contrast, they diverge in a variety of integrated adaptations, biochemical, genetical, physiological, ecological and behavioral to their separated microclimatic niches. The adaptive variation syndromes in biochemistry, genetics, morphology, physiology, ecology and behavior differentiating the four chromosomal species are precisely those differentiating populations within a species. Microevolution is perfectly intergrading into macroevolution as suggested by Darwin. Genetic variation between chromosomal species of mole rats in the above mentioned biological determinants arose *directly* from genetic variation within populations of these characteristics. The active ecological speciation of *S. ehrenbergi* into increasingly arid environments is relatively young, and the hybrid zones separating them reflect the progressive terminalization of speciation. Speciation in the *S. ehrenbergi* complex was initiated by chromosomal evolution followed by genic accumulation of positive assortative mating and the concomitant evolution of a syndrome of climatically genomic coadaptations at all organizational levels. *S. ehrenbergi* represents a pluralistic example where chromosomal and genic mutations, genetic drift, migration, isolation and selection all interact in producing adaptive new species associated with environmental diversity and challenges, thereby linking the mechanisms of micro- and macroevolution and corroborating the synthetic theory of Darwinian evolution.

Numerous problems remain open for future research, particularly in relating the molecular and organismal diversities. These involve, among others, the following problems: What are the molecular structures and mechanisms underlying the adaptive phenotypic patterns and environmental correlates of the chromosomal species? What are the changes in DNA structure and differentiation accompanying speciation? Can convergent-divergent patterns be deciphered at the molecular level? Are there any additional evolutionary forces to those outlined above, such as molecular drive, in the speciation of mole rats? The active speciation and adaptive radiation of the *S. ehrenbergi* superspecies provide exciting and stimulating future research at the interface between the molecular and organismal levels. This could highlight major unresolved problems of evolutionary biology.

ACKNOWLEDGMENTS

I wish to thank Giora Heth, Avigdor Beiles, Shimeon Simson, and my many collaborating colleagues for their continuous assistance in the mole rat project.

REFERENCES

Arieli, R., Arieli, M., Heth, G., and Nevo, E. (1984). Adaptive respiratory variation in four chromosomal species of mole rats. *Experientia* **40**, 512-514.

Arieli, R., Heth, G., Nevo, E., Zamir, Y., and Neutra, O. (1985a). Adaptive heart and breathing frequencies in 4 ecologically differentiating chromosomal species of mole rats in Israel. *Experientia* (in press).

Arieli, R., Heth, G., Nevo, E., and Hoch, D. (1985b). Hematocrit and hemoglobin concentration in 4 chromosomal species and some isolated populations of actively speciating subterranean mole rats in Israel. *Experientia* (in press).

Atlas of Israel (1970). Ministry of Labour, Jerusalem and Elsevier Publ. Co., Amsterdam.

Avise, J. C. (1976). Genetic differentiation during speciation. *In* "Molecular Evolution" (F. J. Ayala, ed.). Sinauer Ass. Inc., Sunderland, Mass.

Ayala, F. J. (1975). Genetic differentiation during the speciation process. *Evol. Biol.* **8**, 1-78.

Barigozzi, C. (ed.) (1982). "Mechanisms of Speciation." Alan R. Liss, Inc., New York.

Barton, N. H. (1979). The dynamics of hybrid zones. *Heredity* **43**, 341-359.

Barton, N. H., and Charlesworth, B. (1984). Genetic revolutions, founder effects, and speciation. *Ann. Rev. Ecol. Syst.* **15**, 133-164.

Beiles, A., Heth, G., and Nevo, E. (1984). Origin and evolution of assortative mating in actively speciating mole rats. *Theor. Pop. Biol.* **26**, 265-270.

Beintema, J. J., and Lenstra, J. A. (1982). Evolution of mammalian pancreatic ribonucleases. *In* "Macromolecular Sequences in Systematic and Evolutionary Biology" (M. Goodman, ed.). Plenum Press, New York, London.

Bengtsson, B. O. (1980). Role of karyotype evolution in placental mammals. *Heredity* **92**, 37-47.

Carson, H. L., and Templeton, A. R. (1984). Genetic revolutions in relation to speciation: the founding of new populations. *Ann. Rev. Ecol. Syst.* **15**, 97-131.

Charlesworth, B., Lande, R., and Slatkin, M. (1982). A neo-Darwinian commentary on macroevolution. *Evolution* **36**, 474-498.

Dover, G. A., and Flavell, R. B. (1982). "Genome Evolution." Academic Press, London.

Eldredge, N., and Gould, S. J. (1972). Punctuated equilibria: an alternative to phyletic gradualism. *In* "Models in Paleobiology" (T. J. M. Schopf, ed.), pp. 82-115. Freeman, Cooper and Co., San Francisco.

Flynn, L. J., Carter, M., Nevo, E., and Heth, G. (1985). Incisor enamel microstructure in the *Spalax ehrenbergi* superspecies (Rodentia): phylogenetic implications. (submitted)

Gould, S. J. (1980). Is a new and general theory of evolution emerging? *Paleobiology* **6**, 119-130.

Gould, S. J. (1982). Darwinism and expansion of evolutionary theory. *Science* **216**, 380-387.

Gould, S. J., and Eldredge, N. (1977). Punctuated equilibria: the tempo and mode of evolution reconsidered. *Paleobiology* **3**, 115-151.

Gould, S. J., and Lewontin, R. C. (1979). The spandrels of San Marco and the panglossian paradigm: a critique of the adaptationist programme. *Proc. Roy. Soc. London B* **205**, 581-598.

Haim, A., Heth, G., Avnon, Z., and Nevo, E. (1984). Adaptive physiological variation in nonshivering thermogenesis and its significance in speciation. *Comp. Biochem. Physiol.* **154**, 145-147.

Haim, A., Heth, G., and Nevo, E. (1985a). Adaptive thermoregulatory patterns in speciating mole rats. *Acta Zool. Fenn.* **170**, 137-140.

Haim, A., Heth, G., and Nevo, E. (1985b). Urine analysis of three rodent species with emphasis on calcium and magnesium bicarbonate. *Comp. Biochem. Physiol.* **80A**, 503-506.

Heth, G., and Nevo, E. (1981). Origin and evolution of ethological isolation in subterranean mole rats. *Evolution* **35**, 254-274.

Heth, G., Beiles, A., and Nevo, E. (1985a). Exploratory behaviour patterns of sexes and species of subterranean mammals. (submitted)

Heth, G., Frankenberg, E., and Nevo, E. (1985b). "Courtship" call of subterranean mole rats (*S. ehrenbergi,* Nehring): Physical analysis. (submitted)

Heth, G., Frankenberg, E., and Nevo, E. (1985c). Optimal sound for vocal communication in tunnels of a subterranean mammal (*S. ehrenbergi* Nehring). (submitted)

Hickman, G. C., Nevo, E., and Heth, G. (1983). Geographic variation in the swimming ability of *Spalax ehrenbergi* (Rodentia, *Spalacidae*) in Israel. *J. Biogeography* **10**, 29-36.

Hood, L., Steinmetz, M., and Malissen, B. (1983). Genes of the major histocompatibility complex of the mouse. *Ann. Rev. Immunol.* **1**, 529-568.

Jacob, F. (1977). Evolution and tinkering. *Science* **196**, 1161-1166.

Krimbas, C. B. (1984). On adaptation, neo-Darwinian tautology and population fitness. *In* "Evolutionary Biology," Vol. 17 (M. K. Hecht, B. Wallace and G. T. Prance, eds.), pp. 1-57. Plenum Publ. Corp.

Lay, D. M., and Nadler, C. F. (1972). Cytogenetics and origin of North African *Spalax* (Rodentia: *Spalacidae*). *Cytogenetics* **11**, 279-285.

Luckett, W. P., and Hartenberg, J. L. (1985). Evolutionary analysis of rodents. *In* "Proc. NATO-CNRS Symp." Plenum Press (in press).

Lyapunova, E. A., Vorontsov, N. N., and Martynova, L. Y. (1974). Cytogenetical differentiation of burrowing mammals in the Palaearctic. *In* "Proc. Inst. Symp. Species and Zoogeography of Eur. Mammals," pp. 203-215, Brno, 1971, .

Mayr, E. (1954). Change of genetic environment and evolution. *In* "Evolution as a Process" (J. Huxley, A. C. Hardy and E. B. Ford, eds.), pp. 157-180. Allen and Unwin, London.

McNab, B. K. (1979). The influence of body size on the energetics distribution of fossorial and burrowing mammals. *Ecology* **60**, 1010-1021.

Nash, W. G., and O'Brien, S. J. (1982). Conserved regions of homologous G-banded chromosomes between orders in mammalian evolution: Carnivores and primates. *Proc. Natl. Acad. Sci. USA* **79**, 6631-6635.

Nei, M. (1971). Interspecific gene differences and evolutionary time estimated from electrophoresis data on protein identity. *Amer. Natur.* **105**, 385-398.

Nei, M. (1972). Genetic distance between populations. *Amer. Natur.* **106**, 283-292.

Nei, M., and Tajima, F. (1981). DNA polymorphism detectable by restriction endonuclease. *Genetics* **97**, 145-163.

Nevo, E. (1969). Mole rat *Spalax ehrenbergi*: Mating behaviour and its evolutionary significance. *Science* **163**, 484-486.

Nevo, E. (1979). Adaptive convergence and divergence of subterranean mammals. *Ann. Rev. Ecol. Syst.* **10**, 269-308.

Nevo. E. (1982). Speciation in subterranean mammals. *In* "Mechanisms of Speciation" (C. Barigozzi, ed.), pp. 191-218. Alan R. Liss, Inc., New York.

Nevo, E. (1983). Population genetics and ecology: the interface. *In* "Evolution from Molecules to Men" (D. Bendall, ed.), pp. 287-321. Cambridge Univ. Press, Cambridge.

Nevo, E. (1985a). Speciation and adaptation in subterranean mole rats. *In* "Proc. Symp. Pop. Genetics and Ecology," 49th Italian Congress of Zoology at Bari, October, 1982. *Bool. Zool.* **52**, 69-95.

Nevo, E. (1985b). Evolutionary behavior genetics in active speciation and adaptation of fossorial mole rats. *In* "Proc. of International Mtg. on "Variability and Behavioural Evolution," Rome, November, 1983. Accad. Naz. Lincei (in press).

Nevo, E., and Shaw, C. (1972). Genetic variation in a subterranean mammal. *Biochem. Genet.* **7**, 235-241.

Nevo, E., Kim, Y. J., Shaw, C., and Thaeler, S. C. Jr. (1974). Genetic variation, selection and speciation in *Thomomys talpoides* pocket gophers. *Evolution* **28**, 1-23.

Nevo, E., and Sarich, V. (1974). Immunology and evolution in the mole rat, *Spalax. Israel J. Zool.* **23**, 210-211.

Nevo, E., and Shkolnik, A. (1974). Adaptive metabolic variation of chromosome forms in mole rats, *Spalax. Experientia* **30**, 724-726.

Nevo, E., Naftali, G., and Guttman, R. (1975). Aggression patterns and speciation. *Proc. Natl. Acad. Sci. USA* **72**, 3250-3254.

Nevo, E., and Bar-El, H. (1976). Hybridization and speciation in fossorial mole rats. *Evolution* **30**, 831-840.

Nevo, E., and Heth, G. (1976). Assortative mating between chromosome forms of the mole rat, *Spalax ehrenbergi. Experientia* **32**, 1509-1510.

Nevo, E., Heth, G., and Bodmer, M. (1976). Olfactory discrimination as an isolating mechanism in speciating mole rats. *Experientia* **32**, 1511-1512.

Nevo, E., and Cleve, H. (1978). Genetic differentiation during speciation. *Nature* **274**, 125-126.

Nevo, E., Guttman, R., Haber, M., and Erez, E. (1979). Habitat selection in evolving mole rats. *Oecologia* **43**, 125-138.

Nevo, E., Guttman, R., Haber, M., and Erez, E. (1982a). Activity patterns in evolving mole rats. *J. Mammal.* **63**, 453-463.

Nevo, E., Heth, G., and Beiles, A. (1982b). Population structure and speciation in mole rats. *Evolution* **36**, 1283-1289.

Nevo, E., Heth, G., and Beiles, A. (1982c). Differential mortality of chromosomal species of mole rats in a standardized environment: An unplanned laboratory experiment. *Evolution* **36**, 1315-1317.

Nevo, E., Heth, G., and Beiles, A. (1985a). Aggression patterns in adaptation and speciation of subterranean mole rats. (submitted)

Nevo, E., Heth, G., and Beiles, A.(1985b). Geographic dialects in blind subterranean mammals: The role of vocal communication in active speciation. (submitted)

Nizetic, B., Figueroa, F., Miller, H. J., Arden, B., Nevo, E., and Klein, J. (1984). Major histocompatibility complex of the mole rat I. Serological and a biochemical analysis. *Immunogenetics* **20**, 443-451.

Nizetic, D., Figueroa, F., Nevo, E., and Klein, J. (1985). Major histocompatibility complex of the mole rat II. Restriction fragment polymorphism. *Immunogenetics* **22**, 55-67.

Ohno, S. (1973). Ancient linkage groups conserved in human chromosomes and the concept of frozen accidents. *Nature* **244**, 259-262.

Ohno, S. (1984). Linkage group conservation and the notation of 24 primordial vertebrate groups. *In* "Chromosomes Today," Vol. 8, pp. 268-278. Allen and Unwin, London.

Ohno, S., Becak, W., and Becak, M. L. (1964). X-autosome ratio and the behaviour pattern of individual X-chromosomes in placental mammals. *Chromosoma* **15**, 14-30.

Petter, F. (1961). Affinites des genres *Spalax* et *Brachyuromys* (Rongeurs, Cricetidae). *Mammalia* **25**, 485-498.

Peshev, D. T., and Vorontsov, N. N. (1982). Chromosomal variability in the mole rat *Nannospalax leucodon*. *In* "Abstr. Papers, Third Intern. Theriolog. Congr" (Myllymaki and Pulliainen, eds.). Helsinki.

Pyke, G. H., Pulliam, H. R., and Charnov, E. L. (1977). Optimal foraging: A selective review of theory and tests. *Quart. Rev. Biol.* **52**, 137-154.

Rose, M. R., and Doolittle, W. F. (1983). Molecular biological mechanisms of speciation. *Science* **220**, 157-167.

Rosenmann, A., Wahrman, J., Richler, C., Voss, R., Persitz, A., and Goldman, B. (1985). Meiotic association between the xy chromosomes and unpaired autosomal elements as a cause of human male sterility. *Cytogenetics and Cell Genetics* **39**, 19-29.

Savic', I. R. (1982). Familie Spalacidae Gray, 1821 - Blindmause. *In* "Hanbuch der Saugetiere Europas," Vol. B21 (J. Niethammer and F. Krapp, eds.), pp. 539-584. Akademische Verlagsgesellschaft, Wiesbaden.

Savic′, I., and Soldatovic′, B. (1984). "Karyotype Evolution and Taxonomy of the Genus *Nannospalax* Palmer, 1903, Mammalia, in Europe." Serbian Academy of Sciences and Arts, Beograd.

Shapiro, J. A. (ed.) (1983). "Mobile Genetic Elements." Academic Press, Inc., New York.

Simpson, G. G. (1944). "Tempo and Mode in Evolution." Columbia University Press, New York.

Steinmetz, M., Minard, K., Horvath, S., McNicholas, J., Frelinger, J., Wake, C., Long, E., Mach, B., and Hood, L. (1982). A molecular map of the immune response region from the major histocompatibility complex of the mouse. *Nature* **300**, 35-41.

Suzuki, H., Moriwaki, K., Kominami, R., Murmatsu, M., and Nevo, E. (1984). Restriction fragment length polymorphism of NTS rDNA in mole rats. *Ann. Rep. Nat. Inst. Genetics* **34**. Ann. Mtg., Japan Genetics Congress, 1984.

Takano, T. (1977). Structure of myoglobin resined at 2.0A resolution. *J. Mol. Biol.* **110**, 537-568.

Tchernov, E. (1968). Succession of rodent faunas during the Upper Pleistocene in Israel. *In* "Mammalia depicta" (P. Parey, ed.). Hamburg and Berlin.

Wahrman, J., Goitein, R., and Nevo, E. (1969a). Mole rate *Spalax:* Evolutionary significance of chromosome variation. *Science* **164**, 82-84.

Wahrman, J., Goitein, R., and Nevo, E. (1969b). Geographic variation of chromosome forms in *Spalax,* a subterranean rodent of restricted mobility. *In* "Comparative mammalian cytogenetics" (K. Benirschke, ed.), pp. 30-48. Springer Verlag, New York.

Wahrman, J., Richler, C., Gamperl, R., and Nevo, E. (1985). Revisiting *Spalax:* Mitotic and meiotic chromosome variability. *Isr. J. Zool.* **33**, 15-38.

White, M. J. D. (1978). "Modes of Speciation." W. H. Freeman and Co., San Francisco.

POPULATION STRUCTURE AND SEXUAL SELECTION FOR HOST FIDELITY IN THE SPECIATION OF HUMMINGBIRD FLOWER MITES

R. K. Colwell

Department of Zoology
University of California, Berkeley
Berkeley, California 94720

ABSTRACT

The rigid patterns of affiliation of hummingbird flower mites (Ascidae: *Proctolaelaps* and *Rhinoseius*) with their host plant species may be more a result of selection for mate-finding than for food-finding. A simple model shows that, at low densities, a significant fitness advantage due to differential mating success is enjoyed by individuals that preferentially seek out the host species favored by prospective mates. Selection for host fidelity by this frequency-dependent mechanism can be viewed as a form of sexual selection. The effects of predispersal mating, relative density, and sex ratio alter the quantitative outcome, but not the qualitative effect. Selection for host fidelity (obligate or learned) very likely plays an important role in the extinction and speciation of hummingbird flower mites. Taking into account population structure, conditions appear favorable for local speciation.

I. INTRODUCTION

When resources do not limit population density, mate-finding may be

a more important focus of selection than food-finding, particularly among

mobile species with highly subdivided and ephemeral populations. In such species, differential mating success--the basis of sexual selection--can sharply focus host or habitat affiliation.

The effect of this phenomenon on community structure is the principal subject of a previous paper (Colwell, 1985b), where I discussed the history of ideas about selection for mate-finding in relation to habitat and host fidelity. Here, I present a model that attempts to capture the essence and explore a few of the ramifications of the "host fidelity hypothesis," especially its implications for processes of extinction and speciation.

The inspiration for this work came from my efforts to understand the distribution and patterns of host affiliation among hummingbird flower mites (Colwell, 1973, 1979, 1985a, 1985b). The flowers of many species of hummingbird-pollinated plants are inhabited by mites of the genera *Rhinoseius* and *Proctolaelaps* (Ascidae). On their host plants, these mites feed on nectar and on pollen substances, mate, and oviposit. From all indications, females must mate more than once to realize their maximum fecundity. Males clearly mate numerous times--especially in species with highly female-biased sex ratios (Wilson and Colwell, 1981; Colwell, 1981, 1982).

The mites live in small breeding groups, each group virtually confined to a single inflorescence, which may produce flowers over a period of days, weeks, or months. However, when the inflorescence finishes flowering, all inhabitants must disperse or die, and the group is no more. Meanwhile, other inflorescences, coming newly into flower, represent large potential fitness gains to successful colonists. The mites move between inflorescences in the nostrils of hummingbirds. It appears that considerable emigration occurs during the "life" of the inflorescence, followed by a mass local exodus as flowering nears an end on each inflorescence.

In tropical lowland forest, typically 10-20 species of hummingbird flower mites coexist, each affiliated with different species of host plants. Most of the mite species are monophagous--affiliated with only a single host species. Moreover, as a rule, each host plant species supports only one mite species ("host monopoly"), but in several cases two mite species are regularly affiliated with the same host, in sympatry; in these cases of "host-sharing" the two mite species are invariably of different genera (one *Rhinoseius* and one *Proctolaelaps* [Colwell, 1985b]).

The high incidence of monophagy and host monopoly in these mites cannot be fully accounted for by differential adaptation of morphological, behavioral, or physiological features, nor by interspecific competition (Colwell, 1985b). Densities often remain quite low (Colwell, 1973; Wilson and Colwell, 1981), rarely approaching carrying capacity. The restriction of "host-sharing" exclusively to non-congeners might seem to suggest within-host resource partitioning in these pairs of species, yet we can find no evidence for partitioning, and considerable evidence against it (Colwell, 1985b).

I suggest that sexual selection (differential mating success) may be largely responsible for the evolution of host plant fidelity and monophagy among hummingbird flower mites: individual mites that disembark from birds at the "correct" host plant find more mates. Thus affiliation with a particular host plant species becomes focused by frequency dependent selection--a special case of Fisher's "runaway process" of sexual selection (Fisher, 1958; Lande, 1980; Kirkpatrick, 1982; Arnold, 1983). The affiliation of non-congeners with the same host may be permitted by differences in courtship or morphology sufficient to avoid mistaken courtship or mating, while the two species of the pair exploit the same host plant as a mate-finding cue (Colwell, 1985b).

Host fidelity, however it originates, perpetuates genetic isolation of populations associated with different host species. If host fidelity is indeed an adaptation that promotes mating success *per se*, then selection for host-fidelity becomes self-reinforcing, and may be expected to lead to rapid genetic isolation of errant lineages, under appropriate circumstances. What those circumstance might be, for hummingbird flower mites, and the role that sexual selection for host fidelity may play in the speciation of these mites will be discussed in the final section. First, I will show by means of a simple model that the key principle is valid.

II. A MODEL FOR THE EVOLUTION OF HOST FIDELITY THROUGH DIFFERENTIAL MATING SUCCESS

Envision a habitat with two or more potential host plant species, all equally suitable for habitation by hummingbird flower mites, and all equally visited by hummingbirds. Into this habitat the model sends a population of dispersing adults of a single species of hummingbird flower mite. Let the density of mites in the ith host plant species after dispersal be n_i mites per inflorescence. If there is no host preference (no host fidelity), then n_i will be the same for all host plant species.

If one of the host species is preferred, however, mite density will be greater in inflorescences of the preferred species, and less in other species. The goal of the model is to estimate the component of mite fitness due to mating success among mites, as a function of mite density.

Now suppose that, among dispersers, proportion m are males, and proportion $(1-m)$ are females. Then the mean density of males per inflorescence, for host species i is

$$\overline{y}_i = mn_i \tag{1}$$

and the mean density of females on host species i is

$$\overline{x}_i = (1-m)n_i \, . \tag{2}$$

Given these means, the model assumes that the number of male or female mites per inflorescence in each of the host species is a Poisson variate, with the sexes independently distributed as a function of their mean densities $\overline{y}_i$ and $\overline{x}_i$. The proportion of inflorescences that have y male mites or x female mites in host species i is therefore

$$P_z(\overline{z}) = \frac{e^{-\overline{z}}\,\overline{z}^z}{z!} \, , \tag{3}$$

where $\overline{z} = \overline{x}_i$ or $\overline{y}_i$, and $z = x$ or y.

To keep things simple, an index of fitness is computed separatedly for each sex--V for females and W for males. The present model makes no attempt to combine fitness between sexes or to incorporate genetics.

In the female submodel, each female disperser is assumed to have a maximum potential fitness of V^*, which she will realize if there is at least one male in the inflorescence where she disembarks, regardless of how many other females are present. Of course, there is actually some maximum number of females that a single male mite can fully inseminate (perhaps 10 to 15, allowing for several repeat matings per female). The error introduced by this assumption is negligible, however, because very few inflorescences will have such an extreme sex ratio.

If there are no males in an inflorescence, the realized fitness of a female who disembarks there may nonetheless take on some "baseline" value V_0, resulting from predispersal mating events. If females mate before dispersal, but require additional matings to realize their maximum potential fecundity (as we believe to be true for hummingbird flower mites), then V_0 will be greater than zero, but less than V^*. Thus the mean fitness of females in plant species i is a weighted average of the

fitness of females that land in zero-male inflorescences and the fitness
of females that land in inflorescences with at least one male:

$$\overline{V}_i = P_0(\overline{y}_i)V_0 + [1 - P_0(\overline{y}_i)]V^* . \tag{4a}$$

$$\overline{V}_i = e^{-\overline{y}_i} V_0 + (1 - e^{-\overline{y}_i})V^* . \tag{4b}$$

The male submodel is more complex. The fitness gained by a male
mite after dispersal is a function not just of the local density of females,
but of the local density of competing males as well. This local fitness
function, which I will call $W(x,y)$, could take many forms. The simplest
is just the local number of females per male

$$W(x,y) = x/y, \quad x \geq 0, \quad y \geq 1 . \tag{5}$$

To calculate male fitness in progeny units, this ratio would have to
be multiplied by the mean number of progeny produced per mated female
after dispersal. Here, however, the simple sex ratio is sufficient because
the concern is with relative fitness of males at different mite densities.

Like females, males may arrive at an inflorescence with a baseline
fitness--W_0-- achieved through predispersal matings. To this baseline,
matings in the new inflorescence contribute additional fitness--in
principle, up to some lifetime maximum W^*. In fact, as already
mentioned for the female submodel, the number of females per male in an
inflorescence will so rarely permit W^* to be reached that W^* can safely
be ignored in the male submodel.

Thus, mean male fitness on plant species i is a weighted average of
local male fitness in inflorescences with each combination of x females
($x \geq 0$) and y males ($y \geq 1$):

$$\overline{W}_i = W_0 + \sum_{x=0}^{\infty} \sum_{y=1}^{\infty} [P_x(\overline{x}_i)P_y(\overline{y}_i)yW(x,y)] / \overline{y}_i . \tag{6}$$

After substitution for the Poisson terms and for $W(x,y)$, male fitness can be written as

$$\overline{W}_i = W_0 + \frac{1}{\overline{y}_i} \left[\sum_{x=0}^{\infty} \frac{e^{-\overline{x}_i} \overline{x}_i^{\,x} x}{x!} \sum_{y=1}^{\infty} \frac{e^{-\overline{y}_i} \overline{y}_i^{\,y}}{y!} \right] \qquad (7)$$

The first summation is just $\overline{x}_i$, by definition; the second summation equals $[1 - P_0(\overline{y}_i)]$, so that $\overline{W}_i$ reduces to

$$\overline{W}_i = W_0 + \frac{\overline{x}_i}{\overline{y}_i} (1 - e^{-\overline{y}_i}) . \qquad (8)$$

(Alternatively, the sex ratio $\overline{x}_i / \overline{y}_i$ can be expressed as $[1-m]/m$.)

III. RESULTS OF THE MODEL

Consider, first, the female submodel (Eq. 4), and recall that predispersal female fitness V_0 is assumed to be less than the maximum potential fitness V^*--in other words, there is some potential benefit from dispersal. Clearly, female fitness increases monotonically with increasing male density $\overline{y}_i$, as the proportion of inflorescences with zero males $[P_0(\overline{y}_i)]$ declines. Females do well to disembark at the host species that males prefer.

Likewise, in the male submodel (Eq. 8), for a given sex ratio (which sets $\overline{x}_i / \overline{y}_i$ as a constant), male fitness also increases monotonically with mite density, most simply parameterized by male density $\overline{y}_i$.

Figure 1 shows female fitness $\overline{V}_i$ and male fitness $\overline{W}_i$ as functions of total local density n_i and proportion males m. First of all, notice that the fitness of both males and females is clearly quite

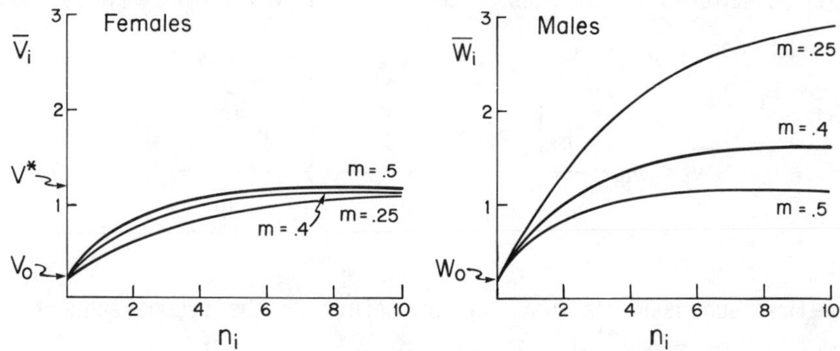

Figure 1. Mean female fitness $(\overline{V})$ and mean male fitness $(\overline{W})$ of mites disembarking from hummingbirds, as a function of mite density on host plant species i (n_i) and as a function of mite sex ratio (proportion males m). For each sex, fitness increases with density as a consequence of greater mating success. Fitness differences between mites that disembark on host species with low mite density and those that disembark on host species with high mite density promote self-reinforcing selection for fidelity to the most-favored host. The density n_i is the average number of mites per inflorescence of host species i, disregarding mite gender. V_0 and W_0 are pre-dispersal fitnesses for females and males, respectively, and V^* is maximum female fitness. See the text for discussion of the effects of sex ratio, scaling of the ordinate, and maximum male fitness.

sensitive to the number of mites per inflorescence (n_i) when the density of mites on at least one host species is fairly low.

Population sex ratio also affects the fitness of both sexes, for a given total mite density. Female fitness declines only slightly with decreasing proportion males (m), for a given value of n_i, because zero-male inflorescences become slightly more common. The effect of sex ratio on female fitness diminishes as overall density increases, and all female fitness curves converge on V^*.

The net effect of increasingly female-biased sex ratios (declining m) on male fitness is much more dramatic, and opposite in direction to the effect on female fitness. For females, only the declining relative density of males affects fitness. For males, on the other hand, both the declining density of competing males and the increasing relative density of females boosts male fitness when the sex ratio is more female-biased, for a given overall density.

In the figure, predispersal fitnesses V_0 and W_0 have both been set arbitrarily to .2, and the maximum fitness for females (V^*) has been set equal to the asmptotic maximum fitness for males ($W_0 + [1-m]/m$) for the case of an unbiased sex ratio ($m = .5$). (Eqs. (4) and (8) are identical for these conditions, and, therefore, so are the curves.)

For both sexes, the figure shows that the curves approach their asymptotes more rapidly for unbiased sex ratios ($m = .5$) than for female-biased sex ratios ($m < .5$)--in other words, density n_i has a significant effect on fitness across a broader range of densities for species with female-biased sex ratios, such as hummingbird flower mites (Wilson and Colwell, 1981), than for species with unbiased sex ratios. Thus female-biased sex ratios amplify selection for host fidelity.

Finally, note that an increased level of predispersal mating would simply raise the intercepts on the fitness axis in Figure 1, without changing the asymptote for either sex, and thereby decrease the effect of postdispersal density on fitness.

IV. DISCUSSION OF THE MODEL

The use of the Poisson distribution in the model carries the assumption that density is low, and that each individual mite disembarks at an inflorescence of its chosen host species completely at random-- without regard to the actions of other individuals.

The assumption of randomness is in fact violated in a variety of ways in real populations of hummingbird flower mites. First, hummingbirds do not visit all inflorescences of a given host species with equal frequency (summed over the "flowering life" of the inflorescence), even on the same individual plant (Colwell *et al.,* 1974). Some inflorescences are less accessible than others, some less defensible (for territorial hummingbirds), some too isolated to be energetically worthwhile. Second, female hummingbird flower mites tend to aggregate--at least in laboratory settings, whereas males tend to avoid one another (S. Naeem and R. Colwell, unpublished data). Third, reproduction within inflorescences is rapid, amplifying initial discrepancies among inflorescences set up at the time of colonization (Colwell, 1973).

With the possible exception of mutual avoidance among males, each of these violations of randomness has the effect of producing an aggregated distribution of individual mites among inflorescences. However, distributions more aggregated than random simply magnify the effects demonstrated by the model--for a given overall density of mites, inflorescences with few or no mites, and thus few or no mates, become more frequent.

In fact, statistical analysis of distributions from field data for several species of hummingbird flower mites from Costa Rica (Fig. 4 in Colwell, 1973) and from Trinidad, W. I. (Colwell, unpublished data), consistently show aggregation for both sexes, with zero-male, zero-female, and completely unoccupied inflorescences forming a substantial porportion of the total (10 to 40%). In principle, these aggregated distributions could be wholly the result of differential visitation by hummingbirds, with mites disembarking at random hummingbird foraging *visits*, but I am quite sure that behavioral aggregation among mites, and certainly, local population growth, are involved as well (Colwell, 1973).

We have so far been unable to demonstrate a role for pheromones that promote aggregation, but they cannot yet be ruled out. Field experiments are underway to yield quantitative estimates of the rates of immigration, emigration, and local population growth for the hummingbird flower mite *Proctolaelaps kirmsei* in its host plant *Hamelia patens* in Costa Rica.

Another assumption of the model is that mites remain where they first disembark. In the real world, fitness losses befalling individuals that land in mateless inflorescences (for both sexes) or mate-poor inflorescences (for males, especially in female-biased mite species), are undoubtedly mitigated to some extent by further episodes of dispersal. On the other hand, mounting a hummingbird, putting one's itinerary completely in the bird's control, and disembarking elsewhere surely is not risk-free, and in any case takes up valuable time. One of the major risks of disembarking in an alien host species, in the real world, is that the inflorescence is very likely to be already occupied by mites of another species of hummingbird flower mite, who may (Colwell, 1985b) or may not (Colwell, 1973) be altogether hospitable, and may further waste one's time with mistaken mating attempts.

The positive fitness effect of host fidelity, through increased mating success, seems certain to be stronger on males than on females in quantitative terms, even though the qualitative effect is quite similar for the two sexes (Fig. 1). In common with classical sexually-selected traits (Fisher, 1958), within-sex variance for mating success due to host fidelity is doutless greater among males simply as a result of their higher reproductive potential. There are as yet no quantitative data for hummingbird flower mites, but female mites of the allied family Phytoseiidae have a lifetime potential fecundity of 50 to 100 eggs, whereas males may produce ten times as many progeny (Amano and Chant, 1978).

Moreover, the expected difference between the sexes in the variance of mating success is magnified by female-biased sex ratios. The matter is complicated by the fact that primary sex ratio in hummingbird flower mites is not only female-biased, but the degree of bias varies systematically among species, as a function of breeding group size (Wilson and Colwell, 1981). The rate of male dispersal, relative to female dispersal, also varies among species of hummingbird flower mites (Colwell, 1985b and unpublished data). Yet a further complication is the likelihood that hummingbird flower mites are parahaploid (a form of functional haplodiploidy [see Hoy, 1977]). My next modelling objective is to work these complexities into the present model, and to combine male and female submodels by incorporating simple population genetics.

In summary, the host fidelity principle illustrated by the model seems qualitatively valid, in spite of diverse simplifications and violation of the assumptions of randomness. The key condition for its applicability, of course, is that populations be at sufficiently low density that host fidelity significantly increases mating success.

V. IMPLICATIONS FOR EXTINCTION AND SPECIATION

A. Extinction

Extensive study of hummingbird flower mites in lowland wet forest areas of Costa Rica and Trinidad (Colwell, 1985b) strongly suggests that each mite species occupies no more host species than necessary to maintain a year-round floral resource base. Thus, in the relatively aseasonal tropical lowland forest, each host plant species typically has its own species of monophagous mite. This one-on-one matchup has two kinds of exceptions. The first, mentioned in the introduction, is the phenomenon of "host-sharing": in six known cases, one species of *Proctolaelaps* and one species of *Rhinoseius* occupy the same host species. The second class of exception concerns host species with seasonal flowering patterns. These are inhabitated by polyphagous hummingbird flower mites that occupy successively flowering alternate hosts seasonally (sequential specialists) or have a year-round "home base" from which seasonal hosts are colonized annually. A detailed account appears (and names are named) elsewhere (Colwell, 1985b).

Extinction threatens monophagous mite species as well as most sequential specialists; all it takes is a flowering failure of a few weeks' duration over the geographical range of a host plant. (The mites have no resting or "resistant" stages.) Thus monophagous mites affiliated with host plants of limited geographical range are the most vulnerable to extinction, and very likely do go extinct frequently, on a geological time scale. We have actually documented a local "temporary extinction" in Trinidad (Colwell, 1985b); the mite *Rhinoseius bisacculatus* disappeared for a period of four years after a ten day flowering failure in its obligate host plant, *Costus scaber*.

Short-term natural selection is of course blind to the long-term risk of extinction from rare extrinsic events. Meanwhile, sexual selection for host fidelity (or any other trait) potentially acts in every generation, at least in a stabilizing mode. In effect, mites are constantly pushed by the selective force of differential mating success towards monophagy, if the yearly resource base permits it, whether it is "good" for them or not in the long run. Extinction--local or complete--is the check on such extreme phenotypes, just as predation often defines the limits of classical sexually-selected characters, such as the colors of male guppies (Endler, 1983), which are otherwise constantly impelled by the agency of female choice toward ever brighter and more contrasting patterns.

B. Host Shifts

The existence of sequential specialist mites presents a challenge to the theory of host fidelity, as well as an opportunity to test it. At our Volcan Colima study site in Mexico, for example, the mite *Rhinoseius epoecus* shifts from its fall-winter host *Castilleja integrifolia* to its local spring-summer host *Lobelia laxiflora* during a one-month period of flowering overlap, and back to *Castilleja* again each fall. In Trinidad, *Proctolaelaps kirmsei* occupies the host plant *Hamelia patens* in the wet season and shifts to *Palicourea crocea* in the dry season; the two plants overlap in flowering period for a few weeks during the two annual transition periods. How do these shifts in host affiliation occur? The host fidelity model predicts that the mites would remain loyal to the current host, whichever it is, even when flowers of the current host are waning in abundance while the alternate host is coming into full flower.

We have carried out many "T-chamber" experiments to investigate preference for various host and non-host nectars and control substances for several species of hummingbird flower mites, among them these two sequential specialists (Colwell 1985b; Heyneman, 1985). In general, mites prefer the nectar of their current host plant to matched sugar solutions, to nectar of the host plants of other sympatric species of mites, and to the nectar of non-host plants pollinated by hummingbirds--presumably, these preferences permit the mites to identify, by olfactory means, flowers appropriate for occupation and mate-finding.

The Mexican sequential specialist mite *Rhinoseius epoecus* was tested during December, when both of its host plant species are in flower, and are occupied by mites of this species. Mites from both host species significantly preferred nectar from the host species from which they were taken for the test (Heyneman, 1985). This result begs the question of how the shift is initiated, but fulfills the expectation that host fidelity may shift rapidly among sequential specialists.

In Trinidad, *P. kirmsei* was tested for its host preference during the dry-season-to-wet-season transition period; although larger sample sizes would be desirable, it appears that *P. kirmsei* prefers its *wet*-season host *Hamelia,* just coming into flower during this transition; indeed, mites are sparse in the dry season host *Palicourea* during transition, and relatively more common in the first flowers of *Hamelia* (Heyneman, 1985). The shift from *Hamelia* back to *Palicourea* has not yet been explored, but it is possible that mites on *Hamelia* when it nears the end of its flowering season may become dense enough to starve, which may relax host fidelity (Hoffman and Turelli, in press), and facilitate the shift to *Palicourea.* It may well be significant that *Hamelia* is the sole host plant of *P. kirmsei* elsewhere in the range of this widespread plant, in areas where it flowers all year (e.g., the Atlantic lowlands of Costa Rica).

Although there may be seasonal genetic shifts in sequential specialists associated with these cyclic host shifts (generation time is about a week), associative learning very likely plays a role--and could in principle be sufficient. We have carried out several very-short-term "transplant" experiments in the field, forcing mites to live in (unoccupied) inflorescences of sympatric host plants normally occupied by other mite species (Colwell, 1985b, and in preparation). We carried out identical preference tests before and after the experiments. *P. kirmsei* was among the species capable of significantly shifting preference to a novel host plant (*Centropogon cornutus* [= *C. surinamensis*]).

Mites at higher latitudes and elevations (such as *Rhinoseius epoecus*) are virtually all sequential specialists--by necessity. A single mite species may occupy four or five seasonal hosts, but have no overlap with other mites in the same habitat, each of which has its own set of hosts spanning several plant families and orders (Colwell, 1985b).   It seems more likely to me that host fidelity in such species is based on learning than on genetic "hard-wiring," whereas monophagous species whose host has a wide geographic range (making complete extinction of the mite highly unlikely) are good candidates for "hard-wired" host fidelity.

The literature on adult conditioning among plant-associated arthropods is growing (Jaenike, 1982, 1983; Prokopy *et al.* 1982; Prokopy and Roitberg, 1984; Rauscher, 1983; but see Hoffman and Turelli, in press). Where densities are low, this phenomenon (where genuine) may be primarily an adaptation for mate-finding, rather than food-finding.

C. Speciation

The annual host shifts in *P. kirmsei* and other sequential specialists may well respresent, in many respects, models for incipient speciation in hummingbird flower mites. If host fidelity does indeed shift seasonally through associative learning, gene frequency changes, or a combination of the two, the process of speciation might very well begin in the same way. How much isolation is required, and how does it occur?

At this time in evolutionary history, the most likely stimulus for a speciation event is probably the extinction of an existing species of hummingbird flower mite, although it is possible that "doubling up" of non-congeners on the same host species may be an ongoing process. Only about one in every two hundred mites is normally found in the "wrong" host species in the field, based on some 12,000 specimens of two dozen species (Colwell, 1985b). Although these errant individuals, had they not been collected, might well have eventually found their way back to their usual host species, these mites probably would not have been among the most successful individuals of their species. On the other hand, the ancestors of any living species that originated by colonization of an "empty" host species were surely "mistake-makers." Heyneman (1985) has shown that hummingbird flower mites tend to accept unfamiliar novel hosts (those not in the foraging repertoire of their usual hummingbird carriers) more readily than "familiar" novel hosts.

When *Rhinoseius bisacculatus* disappeared from our field site in Trinidad after a brief flowering failure in its host plant *Costus scaber* (see above), within weeks the sympatric mite *Proctolaelaps certator*, a "home base" polyphagous species, appeared in numbers in the newly opened flowers of *C. scaber*. Unfortunately the interlopers were not common enough to risk jeopardizing this unusual natural experiment by taking individuals for preference tests. However, the return of the "extinct" mite *R. bisacculatus* four years later did not signal a return to the previous pattern: at last check, in 1981, both species (non-congeners) maintained a strong presence in the host.

I have treated the issue of genetic isolation of colonists on novel host plants in detail elsewhere (Colwell, 1985b). Factors that promote initial isolation from the ancestral population, and that in combination may initiate local speciation include: (1) the frequency of hummingbird "traffic" between the old and the new host plant species, relative to traffic between individuals of the new host plant (Colwell, 1973, Fig. 5); (2) the effectiveness of rapid associative learning of the nectar chemistry of the novel host; (3) founder effects on morphological and behavioral isolating mechanisms (Kaneshiro, 1980; Templeton, 1980, this volume; Lande 1981; Thornhill and Alcock, 1983; Carson, 1978, this volume); (4) founder effects on genetically-specified host preference (if such a preference exists) (Parsons and Hoffman, this volume); and (5) the rate of local population growth relative to immigration.

If these factors together produce a level of genetic isolation sufficient to permit genetic divergence of mites on the novel host plant, then in the longer term, speciation may be consolidated by selection for host fidelity through differential mating success, as well as by classical sexual selection for male morphology or courtship behavior and the correlated preference among females (Templeton, 1980; Lande, 1981; Colwell, 1985; Carson, 1978, this volume). Adaptation to special local conditions (Futuyma and Petersen, 1985; Colwell, 1985b; Parsons and Hoffman, this volume) may then proceed, unhindered by gene flow from the ancestral population. The existence of local sets of extremely similar sibling species, each affiliated with its own host plant (Colwell, 1985b), testifies to the possibility of local speciation, although more biogeographic data are needed for any firm inference.

The effect of host fidelity on fitness in low-density populations plays a two-fold role in this speciation scenario. In ancestral populations, especially of polyphagous hummingbird flower mites, selection for associative learning of host cues may predispose errant individuals and their offspring on novel hosts to disembark from hummingbirds at inflorescences of the new host species, rather than the ancestral one. Meanwhile, rigid host fidelity in the ancestral population prevents swamping of the incipient species by further immigration, which is surely very much smaller than local population growth on the new host.

I have supported these ideas about speciation with a very general model for a mechanism promoting genetic isolation, which in principle could apply to many different kinds of habitat fidelity or host fidelity. My own familiarity with the natural history of hummingbird flower mites has prompted me to use them as a case history in support of these ideas, but of course I hope others will examine carefully the possibility that sexual selection for host or habitat fidelity may play a role in community organization and speciation in other groups of organisms.

As I have discussed elsewhere in some detail (Colwell, 1985), certain groups of flies appear to present many parallels with hummingbird flower mites, and seem good candidates for the study of the evolution of host fidelity through selection for mate-finding, and its possible role in speciation. These groups include certain lineages of Tephritidae (the true fruit flies) and of Drosophilidae. Certain monophagous *Drosophila* in the Neotropics, in fact, very likely share some host plant species with hummingbird flower mites (Pipkin *et al.*, 1966). A vast radiation of flower-breeding *Drosophila* awaits investigation in New Guinea (see Okada and Carson, 1982, and references therein).

ACKNOWLEDGMENTS

I am grateful to Carlo Matessi for his critical review of the manuscript and for his useful suggestions about the model, many of which found their way into the final version. I thank Ron Prokopy, two anonymous reviewers, and the participants in the Conference for their comments and encouragement, especially Guy Bush, Hampton Carson, Peter Parsons, Alan Templeton, and Michael Turelli. Students and faculty at Kellogg Biological Station provided spirited criticism at a critical stage. I thank Robin Chazdon for helping when it counted most, and Sam Karlin for his infinite patience.

REFERENCES

Amano, H., and Chant, D. A. (1978). Some factors affecting reproduction and sex ratios in two species of predaceous mites, *Phytoseiulus persimilis* Athias-Henriot and *Amblyseius andersoni* (Chant) (Acarina: Phytoseiidae). *Can. J. Zool.* **56**, 1593-1607.

Arnold, S. J. (1983). Sexual selection: the interface of theory and empiricism. *In* "Mate Choice" (P. Bateson, ed.), pp. 67-107. Cambridge University Press, Cambridge.

Carson, H. L. (1978). Speciation and sexual selection in Hawaiian *Drosophila*. *In* "Ecological Genetics: The Interface" (P. F. Brussard, ed.), pp. 93-107. Springer-Verlag, New York.

Carson, H. L. (1986). Sexual selection and speciation. *In* "Evolutionary Processes and Theory" (S. Karlin and E. Nevo, eds.). Academic Press, San Diego.

Colwell, R. K. (1973). Competition and coexistence in a simple tropical community. *Amer. Natur.* **107**, 737-760.

Colwell, R. K. (1979). The geographical ecology of hummingbird flower mites in relation to their host plants and carriers. *Recent. Adv. in Acarology* **2**, 461-468.

Colwell, R. K. (1981). Group selection is implicated in the evolution of female-biased sex ratios. *Nature* **290**, 401-404.

Colwell, R. K. (1982). Female-biased sex ratios. (Reply to Charlesworth and Toro, Wildish, and Borgia). *Nature* **298**, 494-496.

Colwell, R. K. (1985a). Stowaways on the Hummingbird Express. *Natural History* **94 (7)**, 56-63.

Colwell, R. K. (1985b). Community biology and sexual selection: Lessons from hummingbird flower mites. *In* "Ecological Communities" (T. Case and J. M. Diamond, eds.), pp. 406-424. Harper and Row.

Colwell, R. K., Betts, B. J., Bunnell, P., Carpenter, F. L., and Feinsinger, P. (1974). Competition for the nectar of *Centropogon valerii* by the hummingbird *Colibri thalassinus* and the flower-piercer *Diglossa plumbea*, and its evolutionary implications. *Condor* **76**, 447-452.

Endler, J. A. (1983). Natural and sexual selection on color pattern in poeciliids. *Environmental Biology of Fishes* **9**, 173-190.

Fisher, R. A. (1958). "The Genetical Theory of Natural Selection," 2nd Ed., Dover, New York.

Futuyma, D. J., and Petersen, S. C. (1985). Genetic variation in the use of resources by insects. *Ann. Rev. Entomol.* (in press).

Heyneman, A. J. (1985). Selective use of floral nectars by hummingbirds and hummingbird flower mites. Ph.D. Dissertation, University of California, Berkeley.

Hoffman, A. A., and Turelli, M. (1985). Distribution of *Drosophila melanogaster* on alternative resources: Effects of experience and starvation. *Amer. Natur.*, in press.

Hoy, M. A. (1977). Inbreeding in the arrhenotokous predator *Metaseiulis occidentalis* (Nesbitt) (Acari: Phytoseiidae). *Intl. J. Acarol.* **3**, 117-121.

Jaenike, J. (1982). Environmental modification of oviposition behavior in *Drosophila*. *Amer. Natur.* **119**, 784-802.

Jaenike, J. (1983). Induction of host preference in *Drosophila melanogaster*. *Oecologia* **58**, 320-325.

Kaneshiro, K. Y. (1980). Sexual isolation, speciation, and the direction of evolution. *Evolution* **34**, 437-444.

Kirkpatrick, M. (1982). Sexual selection and the evolution of female choice. *Evolution* **36**, 1-12.

Lande, R. (1980). Sexual dimorphism, sexual selection, and adaptation in polygenic characters. *Evolution* **34**, 292-305.

Lande, R. (1981). Models of speciation by sexual selection on polygenic traits. *Proc. Natl. Acad. Sci. USA* **78**, 3721-3725.

Okada, T., and Carson, H. L. (1982). Drosophilidae associated with flowers in Papua New Guinea. *Kontû, Tokyo* **50**, 511-526.

Parsons, P. A., and Hoffman, A. A. (1986). Ecobehavioral genetics: Habitat preference in *Drosophila*. *In* "Evolutionary Processes and Theory" (S. Karlin and E. Nevo, eds.). Academic Press, San Diego.

Pipkin, S. B., Rodriguez, R. L., and Leon, J. (1966). Plant host specificity among flower--feeding neotropical *Drosophila* (Diptera: Drosophilidae). *Amer. Natur.* **100**, 135-156.

Prokopy, R. J., Averill, A. L., Cooley, S. S., and Roitberg, C. A. (1982). Associative learning in egglaying site selection by apple maggot flies. *Science* **218**, 76-77.

Prokopy, R. J., and Roitberg, C. A. (1984). Foraging behavior of true fruit flies. *Amer. Sci.* **72**, 41-49.

Rauscher, M. D. (1983). Conditioning and genetic variation as causes of individual variation in the oviposition behavior of the tortoise beetle, *Deloya guttata*. *Anim. Behav.* **31**, 743-747.

Templeton, A. (1980). Theory of speciation via the founder principle. *Genetics* **94**, 1011-1038.

Templeton, A. (1986). The relation between speciation mechanisms and macroevolutionary patterns. *In* "Evolutionary Processes and Theory" (S. Karlin and E. Nevo, eds.). Academic Press, San Diego.

Thornhill, R., and Alcock, J. (1983). "The Evolution of Insect Mating Systems." Harvard University Press, Cambridge, Mass.

Wilson, D. S., and Colwell, R. K. (1981). Evolution of sex ratio in structured demes. *Evolution* **35**, 882-897.

THE RELATION BETWEEN SPECIATION MECHANISMS
AND MACROEVOLUTIONARY PATTERNS[1]

Alan Templeton

Department of Biology
Washington University
St. Louis, MO 63130

ABSTRACT

Evolution can be viewed as a hierarchical process in which the properties or attributes of one evolutionary level emerge from the properties of lower evolutionary levels. Many aspects of macroevolution can be regarded as emergent properties of the process of speciation, which in turn is an emergent property of underlying microevolutionary events such as adaptive divergence, founder events, molecular turnover in the genome, etc. The principle of emergent properties states that the whole is much more than the sum of the parts, so that although many macroevolutionary patterns can be derived from underlying speciation mechanisms, the derivation is not simply an extrapolation. This is illustrated by considering the role of founder-induced speciation in the genus *Drosophila*. It is argued that this speciation mechanism is common in the genus only in the Hawaiian Islands, whereas the geographical and biological situation on the continents makes this an extremely unlikely speciation mechanism over the vast majority of the range of this genus. Thus, from a geographical standpoint, founder-induced speciation is extremely rare and limited within the genus *Drosophila*. Nevertheless, over a quarter of the species in the genus are endemic to the Hawaiian Islands, the genus *Scaptomyza* probably evolved

---

[1] Supported by NIH Grant RO1 GM31571.

497

from *Drosophila* on the Hawaiian Islands and then spread over the globe, and the most derived and extreme morphologies are found in the Hawaiian *Drosophila*. Consequently, this "rare" form of speciation is a major contributor to the genus' species diversity and is the primary determinant of the extremes of morphological diversity within the genus. These extreme consequences of founder-induced speciation are related to the fact that such speciation events often involve genetic architectures consisting of major genes with many pleiotropic effects. Although this genetic architecture may be rare in many microevolutionary processes because of the constraints induced by pleiotropy, this same attribute allows this genetic architecture to be a major contributor to macroevolutionary pattern. Other examples illustrating these principles are given.

I. INTRODUCTION

Much of biology is organized around the principles of hierarchy and emergent properties. The principle of hierarchy recognizes various levels of biological organization (molecular, cellular, tissue, organism, population, community, etc.). The principle of emergent properties further states that certain biological phenomena are only manifest at some specific level or levels of organization. For example, populations evolve, not individuals. However, the principle of emergent properties does *not* imply that lower level processes are irrelevant to the emergent property; rather the phrase emergent property emphasizes the fact that the higher level property emerges from the operation of lower level processes, although the whole is often greater than the sum of the parts. For example, the emergent, population-level property of adaptive evolution arises from the individual-level phenotype being a gene-by-environment interaction and from the mechanisms of inheritance and reproduction. The mechanism by which adaptive evolution emerges from these lower level processes is known as natural selection. Indeed, the mechanisms of microevolution--natural selection, genetic drift, molecular turnover, etc.--

can all be viewed as level crossing mechanisms that explain how lower level processes create evolutionary processes at the population level.

Evolution itself can be regarded as consisting of hierarchies (Templeton, 1982a). Quite commonly, a distinction is made between microevolution, which refers to the evolutionary changes that occur within a species, versus macroevolution, which refers to the evolutionary fates of entire species or assemblages of species through geological time. Although the hierarchical nature of evolution is commonly acknowledged, far less attention has been focused upon applying the principle of emergent properties to evolutionary hierarchies. Indeed, many authors have tried to do just the opposite by claiming that the hierarchies are not true hierarchies at all, but rather completely separate levels that are largely irrelevant to one another. For example, Goldschmidt (1982, reprint of 1940) clearly distinguished between micro- and macroevolution, but regarded the processes and even the material basis (i.e., the types of mutations) of microevolution as being totally separate and irrelevant to macroevolution. Unfortunately, this denial of the importance of emergent properties continues in much of the current writings on evolution. For example, Stanley (1979), like Goldschmidt, regards microevolutionary processes as largely irrelevant to macroevolution (Templeton, 1980a). Likewise, Wilson *et al.* (1984) have recently argued that population-level processes are largely irrelevant to molecular evolution, despite the fact that all theories of molecular evolution are clearly rooted in population-level models (e.g., see Kimura, 1983) and that empirical evidence documents that population-level processes influence the rate and pattern of molecular evolution (Templeton, 1985). Claims about the irrelevancy of one branch of evolutionary biology to another do not strengthen evolutionary theory. Rather, more effort should be devoted to seeking out the interrelations between the various subdisciplines of evolutionary biology with the goal of understanding emergent properties.

With this goal in mind, I wish to focus upon the relationship between microevolution and macroevolution. The trends observed at the macroevolutionary level can be regarded as stemming from two major processes; speciation and extinction. I will limit my discussion to the process of speciation, and only to a limited subset of speciational processes. Given that speciation is one of the determinants of macroevolutionary pattern, it is particularly important to realize that the mechanisms of speciation can be described in terms of such microevolutionary processes as natural selection, genetic drift, molecular turnover, etc. (Templeton, 1981). Hence, speciation is a level-crossing mechanism between micro- and macroevolution, just as natural selection is a level-crossing mechanism from the organismal to population levels. However, because of the principle of emergent properties, crossing from micro- to macroevolution through the process of speciation is not in general a simple extrapolation of microevolutionary trends, as will now be discussed in more detail for a specific speciation mechanism: founder-induced speciation or genetic transilience (Templeton, 1980b).

A. Founder-Induced Speciation and the Genus Drosophila

The idea that founder effects can initiate the process of speciation is an old one (Mayr, 1954), but it still remains controversial (e.g., see Barton and Charlesworth, 1984, versus Carson and Templeton, 1984). Much of the controversy stems over how likely or widespread this speciation mechanism is rather than whether or not is is absolutely impossible. In this regard, I have emphasized in my writings on this subject (Templeton, 1980b, 1980c, 1981; Carson and Templeton, 1984) that genetic transilience is an unlikely mode of speciation and that the vast majority of founder events have no significance whatsoever with regard to the process of speciation.

There are several reasons why genetic transilience is an uncommon mode. First, genetic transilience is likely only when the founder event is extreme and is followed almost immediately by a large increase in population size (Templeton, 1980b; Carson and Templeton, 1984). The biogeographical conditions that will produce small founder populations coupled with the ecological opportunity of immediate and rapid population expansion do not seem to be particularly common. Moreover, given the appropriate biogeographical and ecological conditions, the chances of a genetic transilience still depend upon several innate traits of the ancestral species. These are discussed in more detail in Templeton (1980b), but a couple of factors will be mentioned here. First, the founders must be drawn from a large, panmictic ancestral population. If the ancestral population is inbred or highly subdivided, a founder event has little genetic impact and speciation is unlikely. Second, given that the founder event does have a large genetic impact, the founder population must have an open genetic system recombinationally in order to respond to that impact; otherwise speciation is once again unlikely (Templeton, 1980b). Consequently, genetic transilience requires the correct conjunction of biogeographical, ecological and genetic preconditions to be at all likely, and much more than just a founder event is required for founder-induced speciation. As argued by Templeton (1980b; 1980c; 1981), the simultaneous satisfaction of all these conditions is extremely unlikely, and genetic transilience must be regarded as a rare mode of speciation on a global basis. However, because the chances for genetic transilience depend upon many innate characteristics of the genetic system and population structure of an organism, when the appropriate biogeographical and ecological conditions do exist, genetic transilience can be a very common form of speciation for the groups of organisms (often closely related) that share the appropriate innate traits.

Most of the genus *Drosophila* is found on continental land masses, and the appropriate biogeographic and ecological conditions for genetic transilience are probably rare on most continents. Moreover, many *Drosophila*, such as *D. melanogaster*, do not have the appropriate innate traits (Templeton; 1980b). Consequently, genetic transilience is an unlikely speciation mechanism over the vast majority of the range inhabitated by the genus *Drosophila*. However, one place where the appropriate biogeographic and ecological conditions exist is Hawaii. Due to the presence of a hot spot in the Earth's mantle, new volcanoes and islands are constantly being formed that then drift to the northwest due to plate tectonic movement. As a consequence, new habitats are constantly being produced that are initially uninhabitated, and thus would allow the rapid expansion of any colonizers that make it to a new island or volcano.

These hot spot islands produce founder events within a very different ecological context from the *Drosophila* founder events induced on continents by geological and climatic shifts. Such shifts can certainly create founder events, but only rarely would these continental processes put the founder population into a totally open ecological situation. Indeed, most terrestrial continental founder events occur in the context of peripheral isolates or relictual populations, neither of which will commonly allow rapid population growth immediately after the founder event. Without the founder event being followed by a population flush, there is virtually no chance of founder-induced speciation (Carson and Templeton, 1984). (This is not to say that peripheral and relictual populations never form the basis for a speciation event, merely that the founder event does not *induce* speciation in these contexts. See

Templeton (1980c) for a more detailed discussion of the distinction between speciation that is merely association with a founder event versus founder-induced speciation.) Hence, genetic transilience is rare on the continents *not* because founder events in general are rare on the continents, but because the right type of founder event is rare for continental *Drosophila*.

The opportunity for the appropriate type of founder event is further enhanced in Hawaii due to the fact that the Hawaiian *Drosophila* for the most part live in the higher altitudes, and hence even the volcanoes on the same island are often separated by large distances of inhospitable habitat. Consequently, interisland and, to a lesser extent, intervolcanic, dispersal should be an extremely rare event, a conclusion that has much supporting evidence (Carson and Templeton, 1984). Therefore, the Hawaiian Islands have the biogeographic conditions that foster extreme founder events, probably most often consisting of a single gravid female, and, *most importantly*, have the ecological conditions that allow rapid population expansion into new and relatively unexploited habitats.

In addition, as argued in more detail by Templeton (1980b), the Hawaiian *Drosophila* have many of the innate characteristics that allow genetic transilience to be probable. In particular, the Hawaiian *Drosophila* have one of the most recombinationally open genomes within the genus, and their population sizes are sufficiently large and panmictic to confer upon them levels of isozyme polymorphism and heterozygosity that are comparable to mainland *Drosophila*. Consequently, the Hawaiian Islands represent an area having the appropriate set of biogeographical, ecological and biological conditions that make genetic transilience probable for *Drosophila*. It is difficult to identify other areas of the world where a comparable situation exists, particularly on the continental mainlands where the bulk of the genus lives.

The above arguments imply that genetic transilience is common as a speciation mechanism within the genus *Drosophila* only in the Hawaiian Islands and is rare in mainland situations. Since the Hawaiian Islands collectively have an area only about half that of the Netherlands, and since the genus *Drosophila* is found on all the continents except for Antarctica, one can make the argument that genetic transilience is not a very important speciation mechanism within the genus and represents a very specialized mode of speciation that is severely limited in applicability.

However, the macroevolutionary importance of any speciation mechanism must be judged by what proportion of the total genus is affected by this type of speciation, by what types of morphological diversity are generated by it, and by what potential this speciation mode has for making fundamental breakthroughs that could serve as the basis of adaptive radiation. By these criteria, genetic transilience is of major importance. First, although Hawaii is an extremely small fraction of the Earth's surface, it has produced about a quarter of the species found in the genus *Drosophila*. Consequently, in any quantitative tally of the species composition of the genus *Drosophila*, genetic transilience would have to be regarded as one of the primary modes of speciation.

Genetic transilience would also have to be regarded as a major mode of speciation in any qualitative tally of the genus with respect to morphological diversity and extremes. It has long been recognized (Carson *et al.*, 1970; Hardy and Kaneshiro, 1981) that the Hawaiian *Drosophila* have produced some of the most unusual morphological features found anywhere in the genus. For example, certain Hawaiian *Drosophila* have extremely modified tarsi known as "split-tarsi," "spoon-tarsi" and "bristle-tarsi." One group of Hawaiian *Drosophila* is well known for the extreme elaboration it displays in wing pigmentation patterns, and is therefore known as "picture-winged." Likewise, the mouth-parts have been extremely modified in some groups. The size ranges from flies of comparable size to mainland *Drosophila* to giants that have wing spans of 18-20mm. Moreover, it is known that these extreme morphological changes can arise extremely rapidly; for example, the only hammer-head *Drosophila, D. heteroneura,* is endemic to the Island of Hawaii and is only about 400,000 years old at the maximum. Consequently, in any quantitative or qualitative assessment of morphological evolution in the genus *Drosophila,* genetic transilience would have to be regarded as the major mode that has produced the most extreme morphological deviants found within the genus for a large number of diverse traits.

Finally, genetic transilience has apparently served as the basis for fundamentally new adaptations that allowed adaptive radiation to occur, sometimes on a global scale. For example, some ancestral Hawaiian *Drosophila* evolved a radically new feeding mode in which the larvae are predators on the eggs of spiders. This evolutionary event was the basis of a radiation resulting in the genus *Titanochaeta* (Hardy and Kaneshiro, 1981). Another example is provided by the world-wide genus *Scaptomyza.* More than two-thirds of the species in this genus are found in Hawaii, and the genera *Drosophila* and *Scaptomyza* intergrade in Hawaii although they are very distinct elsewhere. For these and other reasons (Hardy and Kaneshiro, 1981), it is believed that the genus *Scaptomyza* evolved from Hawaiian *Drosophila* and subsequently spread out from the islands to its present-day global distribution. Consequently, the speciation events that occurred in Hawaii laid the basis for the evolution of new genera and adaptive radiations on a global scale.

This example clearly shows that a mode of speciation that is admittedly quite improbable and can only occur under very limited and specialized conditions can nevertheless play an extremely important macroevolutionary role in both a quantitative and qualitative sense. The next section will present another example that illustrates this same point, but over a much longer period of geological time.

B. The Macroevolution of Shallow Sea Communities

Jablonski and Bottjer (1983) and Valentine and Jablonski (1983) have examined the macroevolutionary trends of shallow sea communities from the fossil record. The great advantages with working with these fossil communities are, as with the Hawaiian *Drosophila*, that a great deal is known about the biogeographical constraints influencing their evolution and, most remarkably, information is available about some aspects of their population structure. A major determinant of population structure in these species is whether or not the larvae are planktotrophic. Species with planktotrophic larvae are capable of extensive gene flow over large geographical distances and are prone to panmixia, whereas nonplanktotrophic species have more limited dispersal abilities and are prone to population subdivision. Fortunately, the larval habits of many of these species leave morphological and fossilizable traces in the adult (Jablonski and Bottjer, 1983). Consequently, it is possible to make inferences about the probable population structure of many fossil species from shallow sea communities.

As Jablonski and Bottjer (1983) and Valentine and Jablonski (1983) point out, species with nonplanktotrophic larvae would be very unlikely to undergo genetic transilience under the theory of Templeton (1980b), whereas genetic transilience is at least a possibility for the more panmictic, planktotrophic species. However, as emphasized with the Hawaiian *Drosophila* example, genetic transilience also requires some special biogeographic conditions, and these were considered as well by Jablonski and Bottjer (1983) and Valentine and Jablonski (1983). They point out that the near-shore environment in broad, two-dimensional shelfs, such as the Indo-Pacific shelf, and during marine transgressions over broad expanses of continental platforms provides the necessary pattern of habitat patchiness in which islands, basins, straits and embayments would be produced with high probability and could change rapidly with even minor fluctuations in sea level. They therefore concluded that the near-shore environments provided repeated opportunities for founder-events, and often into new, unexploited shallow marine habitats, whereas off-shore environments did not.

They next showed that most species in near-shore environments are planktotrophic, whereas most species in off-shore environments are nonplanktotrophic. Moreover, planktotrophic species are much more likely to be involved in colonizing and founder events than nonplanktotrophic species for the simple reason that their larvae are more likely to settle in distant and/or newly created habitats. Hence, as with the Hawaiian *Drosophila*, there is a conjunction of the biological and biogeographical characteristics that make genetic transilience probable in the near-shore shallow sea communities.

As mentioned earlier, genetic transilience is a rare mode of speciation on a global basis, and moreover it is one of the few modes of speciation that is fostered by panmixia (Templeton, 1980c). Many more modes of speciation are facilitated by population subdivision and poor dispersal abilities (Templeton, 1980c, 1981). Hence, although genetic transilience is likely only in near-shore communities, it still remains a rare form of speciation, and most speciation should take place in the off-shore, nonplanktotrophic communities where population subdivision is more extreme (Valentine and Jablonski, 1983).

An examination of the shallow sea fossil record confirms this latter prediction; nonplanktotrophic communities display very high rates of speciation and species turnover, with individual species having relatively short durations. In contrast, the near-shore species show much longer species durations and much less speciation (Valentine and Jablonski, 1983). Note that in making this conclusion, Jablonski and Bottjer (1983) and Valentine and Jablonski (1983) are comparing fossil "species" (i.e., morpho-species) from one community to fossil "species" of a second community from the same geological age. Hence, their observed difference in speciation rates is based upon comparable criteria in both communities and is not confounded by comparing paleontological species with neontological species.

One could argue from this observed difference in speciation rates that genetic transilience is therefore an unimportant mode of speciation in shallow-sea communities, and in a quantitative sense this would be true. There is no doubt that the vast majority of speciation events in these communities occur under biogeographic and genetic conditions that virtually eliminate genetic transilience as a speciation mode.

Yet, the studies of Jablonski and Bottjer (1983) revealed what at first seems to be a paradox. Despite the higher speciation and turnover rates in the off-shore environments, the near-shore communities at any given time tend to have the more modern lineages while the off-shore communities have the more archaic ones. The reason for this seeming paradox is, that although speciation is a rare event in the near-shore communities, when speciation does occur in the near-shore environment, it often produces new types that form the basis for a radiation that eventually extends out to the off-shore environment. Thus, the macroevolution of the entire shallow-sea community is driven by the speciation events occurring in the near-shore environments even though speciation is rarer in this environmental type than in offshore environments. Jablonski and Bottjer (1983) explained this pattern in terms of the genetic transilience theory, arguing that although genetic transilience is a rare mode of speciation, that when it does occur, the evolutionary changes are more radical and substantive, and thereby more likely to serve as the basis for a new phase of adaptive radiation. Consequently, although genetic transilience at any given time is a rare mode of speciation in shallow-sea communities and occurs only in the limited context of certain near-shore environments, it could nevertheless be the mode of speciation that dominates the macroevolution of both near-shore and off-shore shallow sea communities under the Jablonski and Bottjer (1983) interpretation.

C. Genetic Transilience and Genetic Architecture

In both the Hawaiian *Drosophila* and the shallow-sea community example, genetic transilience seems to cause more radical evolutionary change than other modes of speciation. This is probably related to the types of genetic architecture that are associated with this mode of speciation. As emphasized by Templeton (1981, 1982b), there is a strong interaction between the mechanisms of speciation and the genetic architectures underlying the traits that evolve during speciation. In particular, genetic transilience preferentially involves traits depending upon only a few major genes with many pleiotropic effects that can be modified by minor polygenes (Templeton, 1980b). Fixation at a major locus not only induces direct evolutionary change, but induces secondary evolutionary processes involving the modifer loci that alter the pleiotropic effects.

Interestingly, Lande (1983) argues that this type of genetic architecture is not very important in natural populations because many of the pleiotropic effects associated with a major gene will have a negative fitness impact. Hence, very strong selection on the advantageous pleiotropic effects is needed, and Lande regards such strong selective forces as rare under natural conditions. Lande views the many natural examples of this architecture -- e.g., sickle-cell anemia in humans, industrial melanism, and insecticide resistance -- as artificial situations induced by human activities. However, as pointed out by Carson and Templeton (1984), there are many other examples of this architecture that cannot be dismissed as artifacts of human activity, including the dramatic change in head shape in *Drosophila heteroneura* mentioned earlier.

However, for the sake of argument, let us assume that Lande is correct and that this type of genetic architecture is rarely involved with microevolutionary adaptation because of its extensive pleiotropic effects. This still does not imply that this type of genetic architecture is unimportant in speciation or macroevolution. First, speciation is a rare and improbable evolutionary event in most situations. Hence, we want something that rarely occurs in microevolution. Second, it is commonly accepted that speciation is often not driven by directly adaptive processes, but rather is the pleiotropic or indirect consequence of other evolutionary processes. Consequently, an evolutionary change that involves a major gene with many pleiotropic effects is far more likely to serve as the basis of speciation than a very precise adaptive process that alters only selected traits under weak selection. Hence, Lande argues against major genes being important in microevolution because they are more likely to produce unforeseen evolutionary consequences through pleiotropy than polygenic systems of minor genes, and I argue that major genes are more important in speciation and macroevolution for exactly the same reason. These two arguments are not mutually contradictory since they depend upon exactly the same attributes of this genetic architecture; rather, they simply refer to different levels in the hierarchy of evolutionary processes.

CONCLUSION

Evolution is a hierarchy in which processes occurring at one level can be amplified or reduced in importance as contributors to emergent properties at a higher level. As the examples given in this paper illustrate, a genetic architecture that is rarely involved in microevolution or a mode of speciation that is improbable except under extremely limited circumstances can still plan an important or even dominant role at the macroevolutionary level.

ACKNOWLEDGMENT

I wish to thank Dr. Eviatar Nevo for his useful comments on an earlier draft of this manuscript.

REFERENCES

Barton, N. H., and Charlesworth, E. (1984). Genetic revolutions, founder effects, and speciation. *Ann. Rev. Ecol. Syst.* **15**, 133-164.

Carson, H. L., Hardy, D. E., Spieth, H. T., and Stone, W. S. (1970). The evolutionary biology of the Hawaiian *Drosophilidae*. *In* "Essays in Evolution and Genetics in Honor of Theodosius Dobzhansky (M. K. Hecht and W. C. Steere, eds.), pp. 437-543. Appleton-Century-Crofts, New York, .

Carson, H. L., and Templeton, A. R. (1984). Genetic revolutions in relation to speciation phenomena: the founding of new populations. *Ann. Rev. Ecol. Syst.* **15**, 97-131.

Goldschmidt, R. (1982). "The Material Basis of Evolution." Yale University Press, New Haven and London.

Hardy, D. E., and Kaneshiro, K. Y. (1981). *Drosophilidae* of Pacific Oceania. *In* "The Genetics and Biology of *Drosophila* ," Vol. 3a (M. Ashburner, H. L. Carson, and J. N. Thompson, Jr., eds.) pp. 307-347. Academic Press, London,.

Jablonski, D., and Bottjer, D. J. (1983). Soft-bottom epifaunal suspension-feeding assemblages in the late *Cretaceous:* Implications for the evolution of benthic paleocommunities. *In* "Biotic Interactions in Recent Fossil Benthic Communities" (M.J.S.Tevesz and P. L. McCall, eds.), pp. 747-812. Plenum Press, New York.

Kimura, M. (1983). "The Neutral Theory of Molecular Evolution." Cambridge University Press, New York.

Lande, R. (1983). The response to selection on major and minor mutations affecting a metrical trait. *Heredity* **50** , 47-65.

Mayr, E. (1954). Change of genetic environment and evolution. *In* "Evolution as a Process" (J. Huxley, A. C. Hardy, and E. B. Ford, eds.), pp. 157-180. Allen & Unwin, London.

Stanley, S. M. (1979). "Macroevolution: Pattern and Process." Freeman Press, San Francisco.

Templeton, A. R. (1980a). Review of "Macroevolution: Pattern and Process," by S. M. Stanley. *Evol.* **34**, 1224-1227.

Templeton, A. R. (1980b). The theory of speciation via the founder principle. *Genetics* **94**, 1011-1038.

Templeton, A. R. (1980c). Modes of speciation and inferences based on genetic distances. *Evol.* **34**, 719-729.

Templeton, A. R. (1981). Mechanisms of speciation -- a population genetic approach. *Ann. Rev. Ecol. Syst.* **12**, 32-48.

Templeton, A. R. (1982a). Review of "The Material Basis of Evolution" by R. Goldschmidt. *Paleobiology* **8**, 474-481.

Templeton, A. R. (1982b). Genetic architectures of speciation. *In* "Mechanisms of Speciation" (C. Barigozzi, ed.), pp. 105-121. Alan R. Liss, New York.

Templeton, A. R. (1985). Genetic systems and evolutionary rates. *In* "Rates of Evolution" (K. S. W. Campbell, ed.), in press.

Valentine, J. W., and Jablonski, D. (1983). Speciation in the shallow sea: general patterns and biogeographic controls. *In* "Evolution, Time and Space: The Emergence of the Biosphere (R. W. Sims, J. H. Price, and P. E. S. Whalley, eds.), pp. 201-226. Academic Press, New York.

Wilson, A. C., Kunkel, J. G., and Wyles, J. S. (1984). Morphological distance: an encounter between two perspectives in evolutionary biology. *Evol.* **38**, 1156-1159.

PART V.   POPULATION GENETICS: OBSERVATION, EXPERIMENT AND THEORY

PHYLOGENETIC RELATIONSHIPS OF MITOCHONDRIAL DNA UNDER VARIOUS DEMOGRAPHIC MODELS OF SPECIATION[1]

Joseph E. Neigel[2]
John C. Avise

Department of Genetics
University of Georgia
Athens, Georgia 30602

ABSTRACT

Recent empirical studies have revealed instances in which the maternally-transmitted mitochondrial DNA (mtDNA) genotypes of some populations are more closely related to those of a distinct biological species than to those of other conspecifics. One plausible explanation involves secondary hybridization and introgression. Here we examine whether phylogenetic sorting of matriarchal lineages, independent of any hybridization, can in principle also give rise to discordancies between biological species boundaries and mtDNA genotype.

Computer models are developed to simulate the evolutionary dynamics of matriarchal lineages during formation of daughter species from an ancestral parent population. The relative probabilities of monophyly, paraphyly, and polyphyly of the daughter species (their *phylogenetic status* with respect to mtDNA) are monitored as a function of time since speciation under a variety of demographic scenarios. Major results are as follows: (1) phylogenetic distributions of mtDNA can lack concordance with species boundaries when species are recently separated; (2) the

[1] Work was supported by an NIH predoctoral fellowship to JEN, and by NSF Grant BSR-8217291.

[2] Present address: Department of Microbiology and Immunology, School of Medicine, University of California, Los Angeles, CA 90024

phylogenetic status of a given pair of species is itself a dynamic evolutionary characteristic, with a common time-course of changes subsequent to speciation being polyphyly —> paraphyly —> monophyly; (3) the demographic mode of speciation will have a major influence on the developing phylogenetic status of related species.

I. INTRODUCTION

Because mitochondrial DNA (mtDNA) is maternally inherited in higher animals (Avise and Lansman, 1983), its evolutionary dynamics can differ from that of nuclear DNA. In contrast to the nuclear genome, which is transmitted biparentally and is subject to recombination at each generation, a given mitochondrial linaege is apparently isolated from genetic exchange with other such lineages. In effect, recombining nuclear genotypes within a population or species have a reticulate evolutionary history, while mitochondrial genotypes have a linear history.

Recent empirical studies have revealed situations in which the phylogenetic distributions of mtDNA lack concordance with biological species boundaries. Thus the mtDNA genotypes of some populations have appeared to be more closely related to those of a distinct biological species than to those of other conspecifics. For example, Powell (1983) found that *Drosophila pseudoobscura* exhibits a mtDNA often indistinguishable from that of *D. persimilis* where these sibling species are sympatric, although the mtDNA in allopatric populations of *pseudoobscura* differed from that of *persimilis*. Similarly, Ferris *et al.* (1983) found that some European mouse populations, classified by morphology and allozymes as *Mus musculus*, possess a mtDNA genotype normally characteristic of a sibling species, *Mus domesticus*. In both of these instances, results were attributed to past hybridization and introgression of mtDNA between species, with the ultimate fixation of an "alien" mtDNA genotype. Takahata and Slatkin (1984) have mathematically modeled the effects of low level mitochondrial gene flow between species, and conclude that these scenarios are plausible. Sterility of male hybrids or a selective advantage for the introduced mtDNA genotype can theoretically contribute to the establishment of a foreign mtDNA without appreciable contamination of the nuclear gene pool.

Without necessarily calling into question the interpretations of results of these previous studies, it may nonetheless be worthwhile to consider whether and under what conditions other evolutionary processes, apart from secondary hybridization and introgression, might account for some cases in which the distributions of mtDNA genotypes do not coincide with species boundaries. There is an additional empirical rationale for this concern. Using restriction-site maps of mtDNA, Avise *et al.* (1983) suggested that some populations of the mouse *Peromyscus maniculatus* are genetically closer to a sibling species *P. polionotus* than they are to each other. In this case, the geographic ranges of *maniculatus* and *polionotus* do not overlap, and there is no compelling reason to suppose they have previously hybridized or experienced introgression. In phylogenetic terminology, *maniculatus* appears to be paraphyletic (Wiley, 1981) with respect to *polionotus* in matriarchal ancestry; *polionotus* is monophyletic with respect to *maniculatus*, but constitutes a subclade within the larger *maniculatus-polionotus* assemblage.

The purpose of this study is to stimulate thinking about possible mtDNA relationships among closely related species by focusing attention on demographically-influenced patterns of phylogenetic sorting of matriarchal lineages during speciation. Other workers have studied the theory of phylogenetic relationships of DNA sequences among individuals within and between populations (e.g., Hudson, 1983; Tajima, 1983). Here we employ computer simulations to examine the relative probabilities of monophyly, polyphyly, and paraphyly of matriarchal lineages belonging to species-pairs generated under various speciation scenarios. Speciations can be associated with an immense array of different demographies; we will present outcomes for a few selected examples.

II. MATERIALS AND METHODS

A. General Outline for the Models

Since only female pedigrees are relevant to transmission of mitochondria, mtDNA lineages can be represented as a non-anastomosing tree (Fig. 1). Each node in the tree marks a female individual, and each branch arising from a node traces that female's lineage to her female progeny. At any given time, a species or population can be thought of as a horizontal cross section through the tree. The last common female ancestor shared by two individuals is the point where their respective branches split from a common node. If genetic differentiation is time-dependent, the genetic distance (D) between two individuals can be defined as the time, in generations, since they last shared a common female ancestor.

As shown schematically in Figure 2, a given pair of biological species (A and B) can in principle exhibit one of four patterns of relationship that determines their *phylogenetic status* with respect to mtDNA lineages: I) A and B are both monophyletic in matriarchal ancestry; II) A and B are both polyphyletic; IIIa) A is paraphyletic with respect to B; or IIIb) B is paraphyletic with respect to A. The phylogenetic status of any pair of species can be defined in terms of maximum genetic distances among conspecifics (max D_{AA}, max D_{BB}) and minimum distances between individuals of the two species (min D_{AB}) by the inequalities presented in Table I. For example, when max D_{AA} < min D_{AB} and max D_{BB}

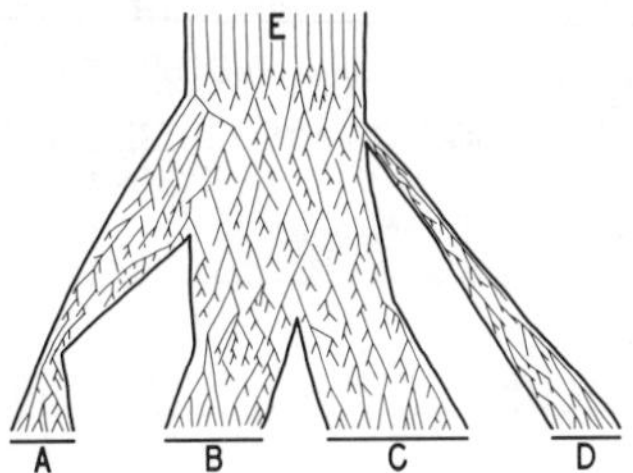

Figure 1. Schematic representation of non-anastomosing mitochondrial lineages, illustrating some possible evolutionary relationships among populations or species A-D. Living members of A trace oldest shared female ancestries to a population bottleneck which postdates the cladogenetic event (separation of A from B). Living members of D trace oldest female ancestries exactly to the separation of D from C. Some living representatives of B and C trace oldest female ancestries to times greatly predating their cladogenetic separation.

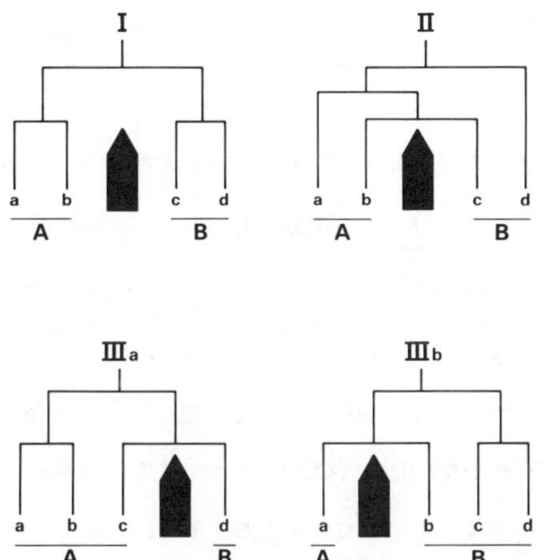

Figure 2. Schematic representations of the *phylogenetic status* of species *A* and *B*. Lower case letters (a-d) are extant mtDNA lineages, each of which can be interpreted to represent an individual animal or an assemblage of related animals sharing a common female ancestor at any time *after* separation from the nearest pictured node. Solid dark arrows indicate onset of reproductive isolation (speciation). See text and Table I for additional explanation.

Table I. Definitions of phylogenetic status in terms of maximum genetic distance within (max D_{AA} or max D_{BB}) versus minimum genetic distance between (min D_{AB}) species A and B. See Figure 2 and text for additional explanation.

Category	Phylogenetic Status	Genetic Distance Relationship
I	A and B monophyletic	max D_{AA} < min D_{AB} and max D_{BB} < min D_{AB}
II	A and B polyphyletic	max D_{AA} > min D_{AB} and max D_{BB} > min D_{AB}
IIIa	A paraphyletic with respect to B	max D_{AA} > min D_{AB} and max D_{BB} < min D_{AB}
IIIb	B paraphyletic with respect to A	max D_{AA} < min D_{AB} and max D_{BB} > min D_{AB}

< min D_{AB}, A and B by definition are monophyletic with respect to each other (phylogenetic status I); conversely, when max D_{AA} > min D_{AB} and max D_{BB} > min D_{AB}, A and B by definition are polyphyletic (status II).

When speciation occurs, individuals from a parental stock become the founders of two or more daughter species. We can envision a continuum of possibilities about how this may occur, but for purposes of tractability in the simulations we will consider two general *modes*: (1) the founders of each daughter species are a random sample of individuals in the parent species; or (2) the parent species may be structured in some way (e.g., geographically) so that individuals that become founders of new species are more likely to share a recent common ancestor than individuals drawn at random. The position of an individual along the parental lineage tree would then correspond to its geographic location. This latter possibility will in turn be considered in two submodes: 2a) individuals founding each daughter species are drawn at random from the tree but are grouped so that they come from opposite sides of the tree; or 2b) individuals founding each daughter species are drawn from end-points of the parental lineage tree, corresponding to extreme geographic regions.

B. Computer Simulations

Our simulations involve the tracing of mtDNA lineages through the process of speciation. To determine the phylogenetic status of a pair of species (with respect to mtDNA) under various speciation modes, it is necessary and sufficient to know: (1) the distribution of genetic distances among the lineages in the parental generation which supplied founders for the daughter species; and (2) the founder lineages which have survived in the daughter species to the time of interest. Space does not permit a complete listing of the computer programs written to provide this information, but the entire programs and all supporting documentation are available from JEN upon request. Programs were written in FORTRAN, and were run on the Digital PDP-11/34 computer. The following is a brief outline of the three major routines.

1. PARDIS

This program (PARent DIStances) generates a vector of mtDNA genetic distances (D) for the individuals in the parental population. A tree represents the development of the population which is assumed to exhibit density regulated growth and a Poisson distribution of female offspring per mother. The mean of the Poisson distribution (AVPROG) is determined by the population size (NPOP) and the carrying capacity (K) with the formula AVPROG = EXP((K-NPOP)/K). The program's inputs are the number of founders for the tree (NFND), the length (in G generations) of the tree, the carrying capacity of the population (K), and two random number seeds. In practice, we used K = NFND = 2500 and let 4K generations pass to generate a vector of distances. In every case the maximum D in the vector was less than $4K$. The program's output (stored in a file named "DIST.DAT") consists of the number of values in the vector of genetic distances and the vector itself.

The most significant feature of this program is its efficient (time-saving) mode of generating and retaining distance information. It is not necessary to generate distances between all possible pairs of extant individuals for a given generation; all relevant information can be summarized in a set of values that specifies the distances (from common female parent) between pairs of adjacent individuals in the tree as it is being developed. The required number of values which need be retained is thus only one less than NPOP (see Fig. 3 for a pictorial representation of the procedure).

2. EXTANT

EXTANT simulates the random extinction of founder lineages through time in the daughter species. As before, each population is assumed to exhibit density regulated growth, with a Poisson distribution of offspring and mean number of female progeny determined by AVPROG = EXP((K-SUM)/K), where in this case SUM and K refer to the population size and carrying capacity of a particular daughter species. The program's inputs are the initial number of lineages founding the daughter species (NLIN1, NLIN2), the carrying capacities (K1,K2) of the respective species, the total

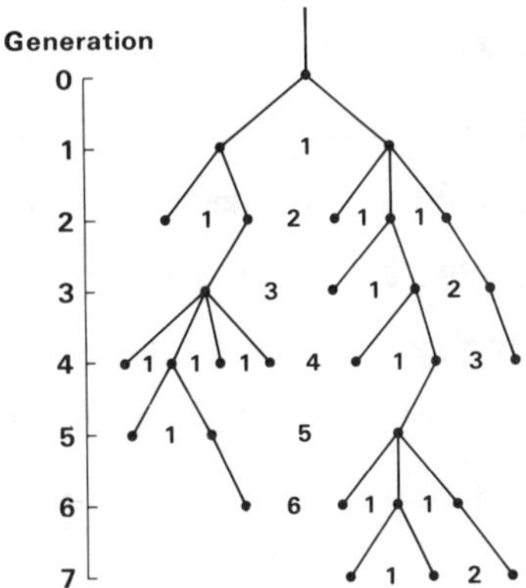

Figure 3. Pictorial representation of the "bookkeeping" involved in the computer program PARDIS. In each generation of the developing computer tree, a vector of distance values is updated as follows: multiple progeny from a given female receive a new distance value of 1; progeny from different but adjacent mothers in the tree receive a distance value one unit greater than the *largest* distance in the line of values between their mothers. The number of distance values which must be retained is thus one less than NPOP. The distance between any non-adjacent individuals in a given generation of the tree is simply the largest distance value between them.

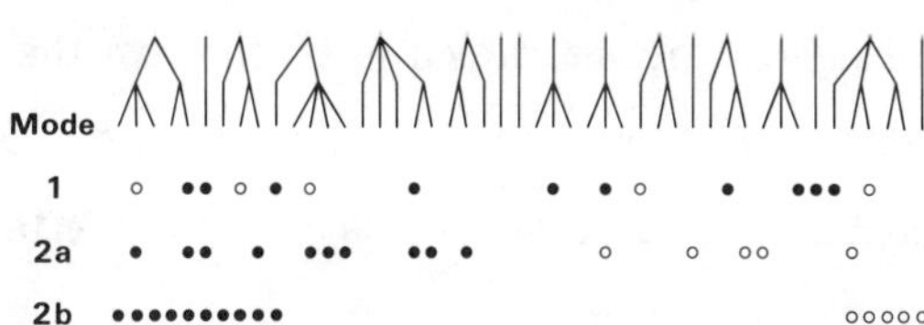

Figure 4. Pictorial representation of alternative speciation modes utilized by the program STATUS for choosing founders for two daughter species. The tree represents the last two generations of a parent population before speciation. In mode 1, founders for daughter species A and B (shown by closed and open circles, respectively) are chosen at random from the parental species. In mode 2a, individuals are chosen at random from separate parts of the parent species tree; and in mode 2b, founders are taken from endpoints of the parent tree. In the simulations, numbers of founders for the daughter species can be varied.

duration (LRUN) of the run in generations, the interval (IVAL) in generations at which the program will output data, the number of desired replicates, and two random number seeds. The program's output (stored in a file named "NLIN.DAT") contains the numbers of original founder lineages still surviving (represented by one or more extant individuals) at each output interval. This program is only a slight modification of one which we used earlier (Avise *et al.*, 1984) to study mtDNA lineage survivorship patterns within a population.

3. STATUS

STATUS selects two sets of lineages from a parent population represented by a vector of genetic distances, and determines the "phylogenetic status" of the two daughter species (Table 1; Fig. 2). The manner in which these lineages are chosen from the parent population can be altered in the program to reflect the *mode* of speciation, as discussed above and pictured in Figure 4. The program's inputs are the vectors of distances for the parent population, which are taken from the file DIST.DAT previously generated by PARDIS, and the number of surviving lineages in each daughter species, which is taken from the file NLIN.DAT previously generated by EXTANT. The output consists of the phylogenetic status of the daughter species at various time intervals after speciation.

III. RESULTS

A. Parent Population

An example of distances in a parent population (consisting in this case of 2458 lineages) is presented in Figure 5. In this figure, each vertical line is the distance (D) to the most recent common ancestor for pairs of *adjacent* individuals along the vector of the final tree. Most such distances (> 96 percent) are less than 50; that is, adjacent individuals have usually shared a common female ancestor within the past

50 generations. Nonetheless, some distances are much larger, and reflect

the chance evolutionary retention of separate lineages for long periods of

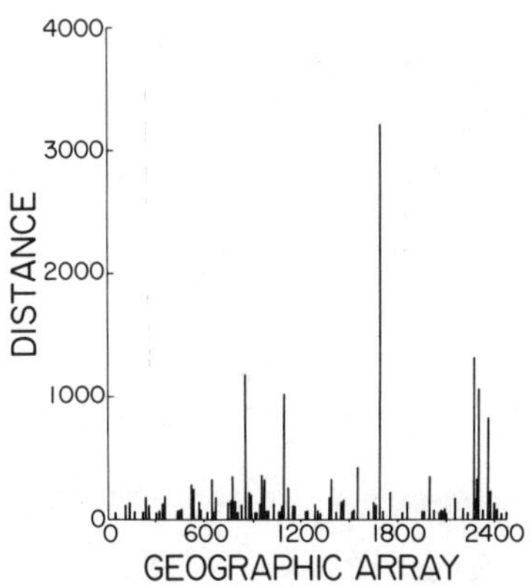

Figure 5. Example of a plotted "DIST.DAT" output, consisting of distances (measured in generations since last shared node) between geographically *adjacent* lineages in a parental population prior to speciation. This population has 2458 extant elements. Distances less than 50 do not stand out above the abscissa.

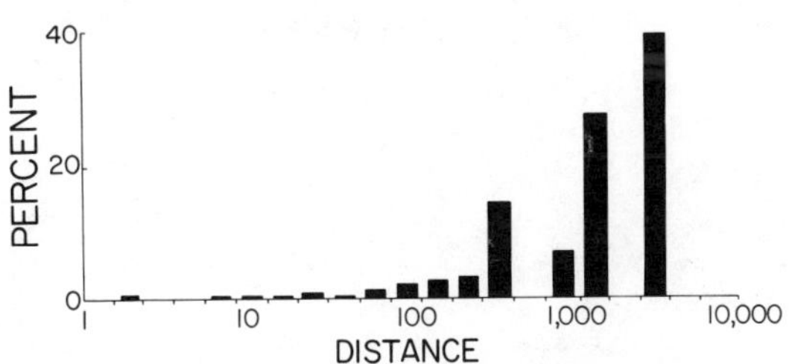

Figure 6. Example of frequency distribution of distances (measured in generations since last shared node) between 1000 *randomly selected* pairs of lineages in a parental population prior to speciation.

Table II. Demographic conditions utilized in the computer simulations. For each of the seven conditions, speciations were allowed to occur by any of the modes pictured in Figure 4.

Simulation file name	# of founders		Carrying capacity		Replicate runs
	Species A	Species B	Species A	Species B	
DAT1	3	2	100	100	400
DAT2	30	20	100	100	400
DAT3	300	200	1000	1000	100
DAT4	300	200	300	200	400
DAT5	300	20	300	200	100
DAT6	300	20	300	20	200
DAT7	300	2	300	200	400

time. For example, two lineages near position 1800 in the array last shared a common female ancestor 3294 generations ago.

The structure of a parent population can also be expressed by distances between *randomly selected* pairs of lineages from the tree. An example is given in Figure 6. The distance categories in the figure were chosen to provide a log scale that covers the observed range with 20 intervals. Note the multi-modal character of this distance distribution, reflecting the major divisions in the parent tree. The multiple-modes also exemplify how random extinction processes can generate discrete clusters of lineages.

B. Daughter Species

The demographic conditions accompanying the simulated speciations are listed in Table II. For each of the seven conditions, speciations were allowed to occur by any of the three modes pictured in Figure 4. Thus a total of 21 simulations (each replicated 100-400 times) was conducted.

In our simulations, all speciations via mode 2b in Figure 4 (in which the founders of the daughter species are selected from the extreme ends of the parent species tree) resulted in monophyletic daughter species (case I, Table I and Fig. 2). However, speciations via modes 1 and 2a (Fig. 4) yielded varied outcomes of monophyly, polyphyly, and paraphyly. The general pattern of the phylogenetic status curves proved to be similar in all cases; computer outputs for eight of the twenty-one sets of simulations are shown in Figures 7-9.

To exemplify the interpretation of these curves, we will describe Figure 7 in some detail. This set of simulations was run with the demographic conditions (DAT4) specified in Table II -- daughter species A and B were initiated with 300 and 200 founders, respectively, and were subsequently maintained at these same carrying capacities. Speciation

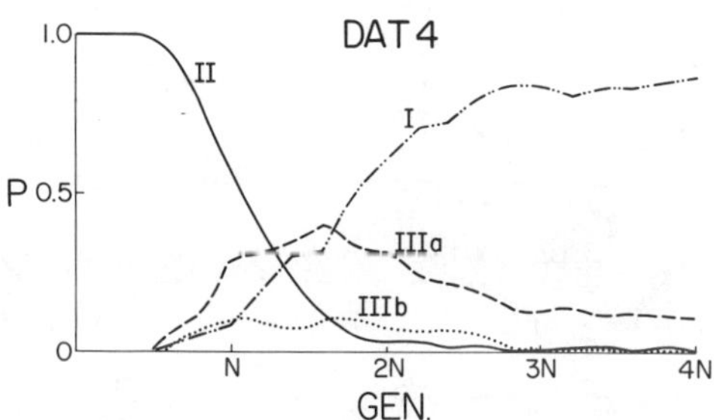

Figure 7. Probability (P) of a given phylogenetic status for two species GEN. generations following simulated speciation. Status I, both species monophyletic; status II, both species polyphyletic; status IIIa, species A paraphyletic with respect to species B; status IIIb, species B paraphyletic with respect to A. Speciation was via mode 1 (see Fig. 4) with demographic parameters DAT4 (see Table II). N is the carrying capacity of the larger daughter species.

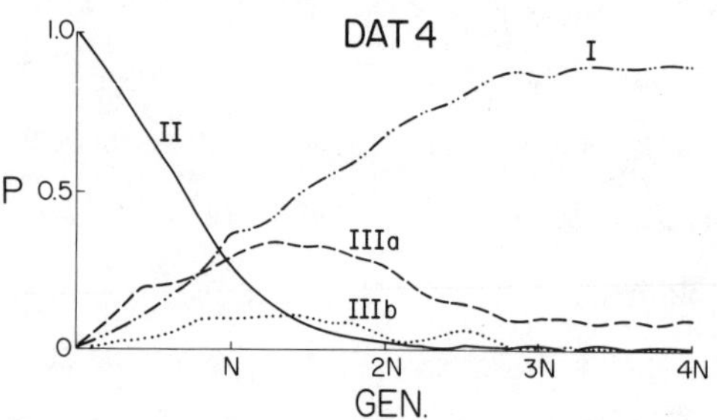

Figure 8. Probability (P) of a given phylogenetic status for two species GEN. generations following simulated speciation. Symbols as in legend to Figure 7. Speciation was via mode 2a (see Fig. 4) with demographic parameters DAT4 (see Table II).

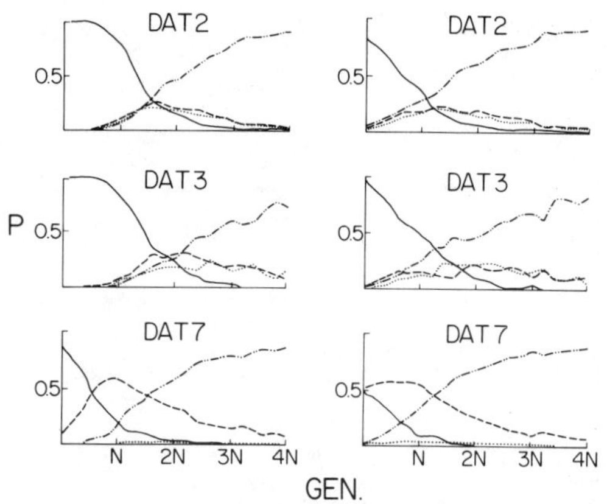

Figure 9. Probability (P) of a given phylogenetic status for two species GEN. generations following simulated speciation. Symbols and line designations as in Figure 7 and its legend. The indicated demographic parameters (e.g., DAT2) are given in Table II. The three graphs on the left are for speciation via mode 1; the graphs on the right are for speciation via mode 2a (Fig. 4). In all cases, N is the carrying capacity of the larger daughter species.

was via mode 1 (foundresses chosen at random from the parental population). The solid line in Figure 7 plots the probability of polyphyly (case II, Table I and Fig. 2) for the daughter species at various times subsequent to speciation. These times are measured in N generations, where N is the carrying capacity of the larger daughter species. The other lines in Figure 7 plot the probabilities of monophyly and of paraphyly (Table I, Fig. 2) of these daughter species. The simulations shown in Figure 8 were conducted under conditions identical to those in Figure 7, but in this case speciation was via mode 2a (Figure 4).

Throughout our simulations, daughter species generated via speciation modes 1 and 2a were polyphyletic, with high probability, for at least N generations following speciation. At intermediate $(N - 4N)$ times after speciation, probabilities of paraphyly typically rise and then fall. In most cases, only after about $4N$ generations is it highly probable that both daughter species will appear monophyletic. The evolutionary change to monophyly is of course attributable to the random extinction of lineages through time in the simulations. In general, probabilities of polyphyly remain high for longer times under speciation mode 1, where foundresses are chosen at random, than under speciation mode 2a, where foundresses are chosen in part by location in the tree (compare Figs. 7 and 8, and the left- versus right-hand columns of Figure 9).

IV. DISCUSSION

Our simulations trace the evolutionary histories of non-recombining, asexually-transmitted genomes (such as mtDNA) across speciation events in sexually reproducing organisms. We assume that mtDNA genome differentiation is strictly time-dependent, so that knowledge of the times of separation of female lineages in a non-anastomosing evolutionary tree is sufficient to specify the phylogenetic status of species-pairs with respect to mtDNA. The particular simulations shown in Figures 7-9 are, of course, far from exhaustive of the possible demographies of speciation, but they do allow several qualitative conclusions about general patterns of distribution of mtDNA sequences among closely related species.

1) *Phylogenetic distributions of mtDNA can lack concordance with species boundaries.* When the time since separation of two biological species is relative short (e.g., less than about $N - 4N$ generations), the probability is high that some mtDNA lineages from one species are more closely related to those from another species than they are to other mtDNA lineages from the same species. Thus in an empirical study, an observed discordance between species affiliation and mtDNA genotype does not necessarily signal effects of secondary introgression, particularly if there is evidence that the species involved could have shared a common ancestor less than about $4N$ generations earlier. This counter-intuitive result is not an artifact of sampling error in choice of individuals or in number of mtDNA genetic characters assayed. It is a biologically meaningful phenomenon attributable to potentially realistic patterns of lineage survivorship across speciation events. However, the probability of such phylogenetically-generated discordancies between species boundaries and mtDNA genotype decreases greatly for species-pairs that have been separated for longer periods of time. Using a somewhat different, mathematical approach, Tajima (1983) obtained these same results and concluded that for recently separated species relationships of particular DNA sequences can differ from the species phylogeny.

Nonetheless, the expected times to monophyly in our simulations are probably very conservative. Particularly under speciation modes 1 and 2a, any factor inhibiting lineage extinction should further extend the times for which daughter species appear poly- or paraphyletic. For example, population growth following speciation extends the expected absolute time to monophyly by temporarily dampening extinction of lineages (compare DAT3 and DAT4 in Figures 7-9). Another very important factor (not modeled here) likely to inhibit lineage extinction in nature is density regulation in subdivided populations (Avise *et al.*,1984). Thus the results presented in this report may in reality represent *minimal* expected times to monophyly.

In an empirical study of mtDNA within and between populations or species, there will of course be sampling errors associated with estimating sequence divergence from a finite number of bases or restriction sites, and from limitations in the number of sampled extant lineages. The former source of sampling error might also account for some situations in which species truly monophyletic in matriarchal genealogy appear at face value polyphyletic, or vice versa. The latter source of sampling error would be more likely to leave unrecognized some true cases of poly- or paraphyly.

2) *The phylogenetic status of species is itself dynamic.* For a given pair of species, phylogenetic relationship is not necessarily fixed at time of speciation, but rather will itself change through time. Thus the time after speciation at which a pair of species is observed will usually be an important factor influencing phylogenetic status. The usual order of status following speciation will be polyphyly ——> paraphyly ——> monophyly, although the polyphyly or polyphyly + paraphyly stages may be bypassed. Once lineage sorting reaches a stage where species are monophyletic, reestablishment of a poly- or paraphyletic appearance can only be achieved through secondary gene exchange, most likely mediated by hybridization and introgression.

3. *Mode of speciation influences the phylogenetic status of related species*. Time (in generations) to expected monophyly for daughter species is clearly strongly influenced by population sizes at and subsequent to the speciation event. It is also strongly influenced by the geographic distributions of lineages among the founders. When founders of daughter species arise from geographically distinct portions of the parent species range (modes 2a and 2b), initial probabilities of monophyly are enhanced relative to the situation in which founders are chosen at random (mode 1). In effect, in geographic speciations much of the lineage sorting eventually leading to monophyly may already have been achieved prior to the actual speciation.

Speciations can certainly be associated with a wide variety of demographic conditions (Powell, 1982). For example, Dobzhansky (1970) summarized a model of speciation involving large, allopatric populations; Mayr (1954) emphasized speciation by founder effect in small peripheral isolates; and Carson (1968) emphasized founder events following by rapid population flushes (expansions) and contractions. Templeton's (1980) model of speciation via founder event from a large panmictic ancestral population would appear to have genetic consequences similar to our "mode 1" speciation (Fig. 4). Results of the models presented in this paper demonstrate that demographies of speciation can have considerable influence on expected relationships of mtDNA genotypes within versus between species.

We have explicitly focused on the evolutionary dynamics of mtDNA across speciations because of empirically motivated concerns about the pattern of mtDNA genetic variation among related species (see Introduction), and because the linear mtDNA transmission simplifies the modeling process. Nonetheless, the results should apply equally well to the distribution of mtDNA within and among conspecific populations. For example, Cann *et al.* (1982) and Nei (1982) showed that the distribution of assayed mtDNA genotypes in humans does not always accord with racial (e.g., white, Orientals, American blacks) designation. The exceptions might be attributable to interracial gene exchange, but the possibility should also be considered that some or all exceptions might be due to phylogenetic sorting independent of introgression. Some portions of the nuclear genome may also exhibit for varying times a linear evolutionary history, unaffected by processes such as interallelic recombination or gene conversion (examples may include the DNA within chromosomal inversions in *Drosophila*--Anderson and Aquadro, personal communication). The principles elucidated by our models may also apply to expected patterns of distribution of such nuclear DNA sequences within and among population or species.

REFERENCES

Avise, J. C., and Lansman, R. A. (1983). Polymorphism of mitochondrial DNA in populations of higher animals. *In* "Evolution of Genes and Proteins (M. Nei and R. K. Koehn, eds.), pp. 147-164. Sinauer, Sunderland, Mass.

Avise, J. C., Shapira, J. F., Daniel, S. W., Aquadro, C. F., and Lansman, R. A. (1983). Mitochondrial DNA differentiation during the speciation process in *Peromyscus. Mol. Biol. and Evol.* 1, 38-56.

Avise, J. C., Neigel, J. E., and Arnold, J. (1984). Demographic influences on mitochondrial DNA linaege survivorship in animal populations. *J. Mol. Evol.* 20, 99-105.

Cann, R. L., Brown, W. M., and Wilson, A. C. (1982). Evolution of human mitochondrial DNA: molecular, genetic, and anthropological implications. *In* "Human Genetics, Part A: The Unfolding Genome" (B. Bonné-Tamir, ed.), pp. 157-165. Alan Liss, New York.

Carson, H. L. (1968). The population flush and its genetic consequences. *In* "Population Biology and Evolution (R. C. Lewontin, ed.), pp. 123-138. Syracuse University Press, New York.

Dobzhansky, T. (1940). "Genetics and the Origin of Species," 2nd ed. Columbia University Press, New York.

Ferris, S. D., Sage, R. D., Huang, C.-M., Nielsen, J. T., Ritte, U., and Wilson, A. C. (1983). Flow of mitochondrial DNA across a species boundary. *Proc. Natl. Acad. Sci. USA* **80**, 2290-2294.

Hudson, R. R. (1983). Testing the constant-rate neutral allele model with protein sequence data. *Evolution* **37**, 203-217.

Mayr, E. (1954). Change of genetic environment and evolution. *In* "Evolution as a Process" (J. Huxley, eds.), pp. 157-180. Macmillan Publishing, New York.

Nei, M. (1982). Evolution of human races at the gene level. *In* "Human Genetics, Part A: The Unfolding Genome" (B. Bonné- Tamir, eds.), pp. 167-181. Alan Liss, New York.

Powell, J. R. (1982). Genetic and nongenetic mechanisms of speciation. *In* "Mechanisms of Speciation" (C. Barigozzi, ed.), pp. 67-74. Alan Liss, New York.

Powell, J. R. (1983). Interspecific cytoplasmic gene flow in the absence of nuclear gene flow: evidence from *Drosophila*. *Proc. Natl. Acad. Sci. USA* **80**, 492-495.

Tajima, F. (1983). Evolutionary relationship of DNA sequences in finite populations. *Genetics* **105**, 437-460.

Takahata, N., and Slatkin, M. (1984). Mitochondrial gene flow. *Proc. Natl. Acad. Sci. USA* **81**, 1764-1767.

Templeton, A. R. (1980). The theory of speciation via the founder principle. *Genetics* **94**, 1011-1038.

Wiley, E. O. (1981). "Phylogenetics." John Wiley and Sons, New York.

ECOBEHAVIORAL GENETICS: HABITAT PREFERENCE IN *DROSOPHILA*

P. A. Parsons

Department of Genetics and Human Variation
La Trobe University
Bundoora, Victoria, Australia

A. A. Hoffmann

Department of Genetics
University of California
Davis, California 95615 USA

ABSTRACT

Ecobehavioral genetics is concerned with the variation of the ecological and behavioral traits important in determining distribution and abundance in the broadest sense. We illustrate this approach in *Drosophila* by habitat preference studies in olfactometers in which responses to odors are tested. Cues from fruit resources from an orchard and responses to other individuals at the adult and larval stages vary among *Drosophila* species and within *D. melanogaster*. Our observations have implications on the interface of ecology, behavioral studies and quantitative genetics concerning the nature of residual odor attraction, sexual selection, sexual isolation, aggregation and interaction at the breeding site level, and on the association between olfactory variation and resource utilization.

I. INTRODUCTION

Genetic variation for habitat preference within natural populations is being increasingly considered. Many studies are basically genotypic in approach, considering gene (in particular electrophoretic) and chromosome polymorphisms as the primary units for study (Parsons, 1983a; Hoffmann, Parsons and Nielsen, 1984). This approach is useful for questions about the association between such polymorphisms and environmental heterogeneity. Studies where variants have been shown to differ behaviorally raise the possibility that genotypes prefer habitats where they are relatively fit. However, a genotypic approach is limited to genetic polymorphisms such as structural enzyme loci which can be easily screened and cannot fully explain the nature of the genetic variation.

An alternative and phenotypic approach (Table I) involves the *a priori* and direct study of quantitative ecobehavioral traits important in the location and utilization of habitats (Parsons, 1983a). Studies on geographic variation in host preference and utilization in insects (reviews, Diehl and Bush, 1984; Futuyma and Peterson, 1985) come under this category. These studies are generally limited to demonstrating geographic variation in host utilization and some preference behaviors, without undertaking detailed quantitative genetic analyses. Host associated variation within populations has only been considered in some cases (e.g., Via, 1984; Rausher, 1983).

In Drosophila, there are numerous laboratory studies on genetic variation in behavioral traits including phototaxis (Rockwell *et al.* 1975; McDonald and Parsons, 1973), larval locomotion and feeding (Sewall *et al.* 1975; Sokolowski, 1980), pupation site location (Ringo and Wood, 1983) and olfaction (Fuyama, 1978; Hoffmann, 1983). Some of these studies, particularly those in *D. melanogaster,* indicate that behavioral traits are

Table I. Approaches to the study of variation at the ecobehavioral level.

1. Ecobehavioral studies + quantitative genetics ⟶	Ecobehavioral phenotypes ⟶	The phenotypic approach
2. Ecobehavioral studies + population genetics ⟶	Gene (electrophoretic) + chromosomal polymorphisms ⟶	The genotypic approach

amenable to fairly sophisticated genetic analysis. However, while researchers often comment on the potential adaptive significance of these traits, their relevance to variation in habitat preference in the wild remains equivocal.

We have therefore initiated an ecobehavioral study of habitat preference in *Drosophila*, examining variation in attraction at the inter- and intraspecific levels and attempting to relate this variation to field behaviors. We consider the following questions:

1. Which behaviors contribute most to variation in habitat preference?

The term "habitat preference" encompasses a number of behaviors involved in locating, examining and utilizing habitats. These behaviors involve responses to environmental cues and also depend on the internal state of the organism (Dethier, 1982) which is affected by factors such as desiccation and starvation. Habitat preference has often been applied to individual behaviors in the evolutionary literature without rigorous definition and any attempt to consider the relative importance of different behaviors. The term "preference" may represent a misconception about the behavioral mechanism for locating and accepting habitats by many organisms. It implies weighing alternatives, a behavior which may not be important in insects (Miller and Strickler, 1984).

2. What component of the variation is genetic?

There are three factors contributing to phenotypic variation in habitat preference: genetic and environmental variation, and experience. Experience effects have been demonstrated in *Drosophila* with laboratory paradigms testing for olfactory aversion to repellants (Thorpe, 1939; Manning, 1967) and oviposition on chemically supplemented media and fruit juices (Jaenike, 1983, 1985). Environmental variation includes those factors which may affect the internal state of individuals. These factors could interact in a complex manner. For example, the expression of variation due to genetic composition or conditioning may depend on internal states. Hungry flies may not discriminate between resources regardless of their prior experience or genotype. Such considerations highlight the necesssity of studying behavioral variation under a range of conditions likely to be encountered in the field.

3. To what extent is there a correlation between habitat location and resource utilization?

This question is central to models of habitat selection (e.g., Taylor, 1976) and sympatric speciation (e.g., Maynard-Smith, 1966). Two mechanisms have been proposed: individuals are positively conditioned by the resources from which they emerge, or there is a genetic correlation between traits affecting resource location and utilization. Data on this question in *Drosophila* are restricted to variation among mutant strains in artificial habitats (e.g., Jones and Probert, 1980; Taylor and Condra, 1983).

This paper addresses these questions by describing some findings on olfactory variation based upon two types of olfactory cues which may affect the distribution of flies in the wild: those emanating from fruit and those related to the presence of other individuals.

II. OLFACTORY VARIATION AMONG *DROSOPHILA* SPECIES

This study was carried out with flies from a mixed orchard at Wandin, 35km. southeast of Melbourne. Four *Drosophila* species are commonly collected from this site: *D. melanogaster*, *D. simulans*, *D. immigrans*, and the endemic *D. lativittata*. These species can be reared from the same fruit resources (peach, apple, plum), although *D. immigrans* tends to be associated with lemons in this area (Prince and Parsons, 1980). Flies of each species were maintained on laboratory medium for several generations before testing. Olfactory response was tested in a vertical wind tunnel modified from a design by Wright (1966) and described in Hoffmann *et al.* (1984). The olfactometer tests mainly for optomotor anemotaxis which is a mechanism whereby *Drosophila* respond to distant plant odors (Kellogg *et al.* 1962; Shorey, 1976). An exhaust fan draws odors (prepared by filtering the liquid fraction of pulp from mature fruits of each fruit type) from flasks via trap cylinders into the middle chamber of the wind tunnel. Flies from the four species were released in the middle chamber. The apple, peach and plum odorants were tested in pairwise combinations, while lemon was tested only in combination with apple and distilled water.

Flies accumulated for 5 hours in the trap cylinders and were then collected. Data are summarized in Table II by the probability estimate P_1 representing the number of flies of a species captured with odorant 1 divided by the total number of flies of that species captured. Statistical analysis was by log-linear models incorporating the sex, odor and *Drosophila* species terms (for details, see Hoffmann, 1985). The term of

Table II. Response of species to lemon, peach, plum and apple odorants (adapted from Hoffmann, 1985).

Odor combination			P_1 estimate[a]				OD interaction[b]
1	2	Trial	*D.mel.*	*D.sim.*	*D.lat.*	*D.imm.*	
lemon	water	1	0.92	0.68	0.82	0.93	+
		2	0.92	0.67	0.85	0.98	+
		3	0.86	0.58	0.92	0.98	+
lemon	apple	1	0.61	0.30	0.52	0.81	+
		2	0.56	0.25	0.52	0.75	+
		3	0.55	0.24	0.39	0.73	+
peach	plum	1	0.82	0.82	0.69	0.68	+
		2	0.56	0.68	0.82	0.73	−
		3	0.68	0.60	0.63	0.64	−
peach	apple	1	0.42	0.45	0.27	0.55	+
		2	0.60	0.62	0.62	0.85	+
		3	0.49	0.40	0.60	0.80	+
plum	apple	1	0.59	0.61	0.62	0.37	+
		2	0.65	0.72	0.67	0.47	+
		3	0.35	0.40	0.34	0.21	−

[a] P_1 = probability of capture with odor 1. [b] Indicates whether odor by *Drosophila* species (OD) interaction included in log linear model.

interest is the odor by species interaction denoted OD in Table II since it tests for interspecific variation for attraction to the odor. Significance ($P < 0.05$) of this term is tested by its inclusion in the log-linear model of "best" fit.

The OD interaction is included in all models for experiments with the lemon odorant. *D. immigrans* was relatively more attracted to lemon than the other species, with *D. simulans* the least attracted, irrespective of the other odorants. The olfactory data are consistent with the field distribution of these species, and moreover, the attractiveness of lemons for *D. melanogaster* relative to *D. simulans* is consistent with the ability of these species to utilize this fruit as a resource (Prince and Parsons, 1980; Atkinson, 1981).

Results for the other fruit combinations indicate consistent OD interactions for the peach-apple combination, but not the peach-plum combination. This interaction was significant for two of the three trials with the plum-apple combination, although trends were consistent for the three trials. For the peach-apple combination, the probability estimates indicate that *D. immigrans* was more frequently caught in the peach trap cylinders, while rankings for the other species are not consistent. For the apple-plum combination, *D. immigrans* was less attracted to plum than the other three species. It is surprising that consistent species differences were not observed for the peach-plum combination, as a preference of the peach odor by *D. immigrans* is expected on the basis of results for the other two combinations. The behavior of *D. immigrans* therefore seems to depend on the combination in which fruit odors are tested. This observation may complicate studies on habitat preference because it suggests that the relative attractiveness of a resource cannot necessarily be determined from tests with individual resouces in all cases.

III. OLFACTORY VARIATION FOR RESOURCES IN *DROSOPHILA MELANOGASTER*

In nature, cosmopolitan colonizing species such as *D. melanogaster* exploit a wide array of resources associated with fermentation. Therefore, an orchard with a variety of fruits is an appropriate habitat for investigating the possibility of genetic differences in *D. melanogaster* associated with fruit types. Adults were collected as they emerged from apples, peaches and plums from adjacent groves of the above orchard, and then maintained in the laboratory for two generations. The three "fruit type" populations were then tested in pairs in the middle chamber of the vertical wind tunnel olfactometer. Flies marked with fluorescent dusts from two "fruit type" populations were exposed to two trap cylinders containing odorants derived from the same two fruits.

The results of these experiments are summarized in Table III in the form of indices of choice difference:

$$\hat{\Delta} = \hat{P}_{1|1} - \hat{P}_{1|2} \, ,$$

where the probability estimate $\hat{P}_{1|1}$ is the number of flies derived from the fruit 1 population captured in the cylinder containing the odors of fruit 1, divided by the total number of flies from the fruit 1 population that were captured, and similarly $\hat{P}_{1|2}$ is the number of flies from the fruit 2 population captured in the cylinder containing the odors of fruit 1, divided by the total number of flies derived from this population that were captured. (See Turelli *et al.* 1984 for derivation of Δ and standard errors.) This index measures the association among flies from given fruit types and attractants, and has a range from -1 to $+1$, with values >0 indicating a positive association.

The indices show that there is a consistent tendency for flies to be attracted to the odors of fruit types from which they originated; indices for the three fruit type combinations taken in pairs are all >0, and for three peach-plum and two apple-plum experiments they are significantly >0. Two points emerge from this study. Firstly, the variation is related to the fruit resources from which the flies originated. This suggests that the behaviors measured in this olfactometer are important in determining the microdistribution of flies in the field. Secondly, the findings suggest

Table III. Indices of choice difference, Δ, in individual experiments for response of *D. melanogaster* collected from different fruits to fruit odors in the wind tunnel olfactometer (calculated from Hoffmann, Parsons and Nielsen, 1984).

Combination	Index of choice difference
(a) Peach-plum	0.16***, 0.07, 0.26***, 0.08*
(b) Plum-apple	0.20***, 0.02, 0.09*
(c) Apple-peach	0.07, 0.05, 0.02

* $P < 0.05$, ** $P < 0.01$, *** $P < 0.001$ for deviation of indices of choice difference from 0.

that this variation is genetic, since all flies were cultured under a uniform laboratory environment.

There are a number of theoretical models relevant to this variation. On the one hand, olfactory variation may be unrelated to resource utilization and maintained by a mutation-selection balance (Lande, 1975) or selection for foraging efficiency (Rausher, 1984). Alternatively, the variation may be associated with resource utilization, and maintained via differential fitness on the resources (e.g., Taylor, 1976). The fruit resources tested are available simultaneously in Melbourne orchards for two to three months annually, with one of the fruits (apple) persisting longer than the other. Hence selection could be quite strong if variation were maintained by resource utilization or foraging efficiency. Further studies could help to decide between these alternatives. Genetic correlations between olfaction and resource utilization could be examined. Additionally, it would be interesting to test whether the variation is detected when only one fruit type is presented as the odor. If individuals from the populations are attracted equally to apple when presented alone, then the variation may be effectively neutral when only this more persistent resource is available.

IV. OLFACTORY VARIATION FOR RESPONSES TO OTHER INDIVIDUALS

Organisms may respond to cues from other individuals of the same or different species. Such behaviors have not been considered much in *Drosophila,* except for those associated with courtship (Ehrman and Parsons, 1981). We investigated the possibility of cues from other individuals modifying olfactory attraction unrelated to courtship.

The wind tunnel provides a paradigm for testing this possibility in the laboratory. A second wind tunnel design was used for testing residual odors from the sibling species *D. melanogaster* and *D. simulans* (see Spence, Hoffmann and Parsons, 1984 for experimental details). In this design, flies of the two species were placed into the middle chamber as before. The attractants consisted of two trap cylinders in which three categories of flies of each species were separately used as marker populations; 600 virgin females, 600 non-virgin females and 700 males. After 8 hours, the flies were discarded and the "marked" trap cylinders were placed in the wind tunnel. Flies were permitted to accumulate overnight from the middle chamber into the "marked" cylinders. We refer to this phenomenon as "marking," although we are not proposing that individuals mark habitats specifically for the purpose of eliciting behaviors in other individuals.

The residual odors from all three categories of "marker" populations attracted more males than females (Table IV). For females as markers in particular, there was a significant tendency of males to locate traps marked with the residual odor of their own species. Indices of choice difference (Δ) suggest that this tendency is somewhat higher for non-virgin females perhaps because of the deposition of eggs. Considering the attraction of females, fewer flies were attracted compared with males, and are inadequate for the detection of trends at the statistical level

Table IV. Attraction of *Drosophila melanogaster* and *D. simulans* to residual odors. (Summarized from Spence, Hoffmann and Parsons, 1984).

Population odor (attractant)	flies captured					
	Males			Females		
	melanogaster	*simulans*	Δ	*melanogaster*	*simulans*	Δ
Non–virgin females						
melanogaster	56	25	0.49***	17	10	0.31
simulans	16	63		10	21	
Virgin females						
melanogaster	43	18	0.38***	17	11	-0.06
simulans	21	44		10	5	
Males						
melanogaster	48	31	0.28***	23	12	0.20
simulans	18	38		15	18	

Δ = index of choice difference, *** P < 0.001

although the data are consistent with the male data. Thus the major phenomenon is the attraction of males by the residual odor of females.

In nature, this implies that flies at a feeding and oviposition site may attract males to that site. Interestingly, unpublished field collection data in Melbourne, Victoria, reveal a frequent excess of males when flies are collected by sweeping, especially when physical conditions are optimal. Given the extremely high fecundity of *Drosophila* females in natural populations, a corresponding attraction of females could be disadvantageous since it would lead to excessive competition for larval resources.

Attraction to "marked" microhabitats is a dimension to habitat preference additional to that based upon the abiotic and biotic factors of the environment as considered by Andrewartha and Birch (1954). A prerequisite is obviously permissive physical conditions and available resources. While the chemicals in resources may act as attractants for both sexes of several species attracted to fermenting fruit and vegetable matter, the marking of habitats may result in the attraction of males of a given species to highly favorable microhabitats previously occupied by that species. In addition, a degree of isolation may occur at the microhabitat level unforeseen at the present time, and calls for studies of species distributions at this level. Extrapolations to the field are of course difficult with *Drosophila* although it is hoped that flies can be assumed to approximate their normal behavior in responding to odors in the olfactometer.

Assuming attraction to favorable microhabitats based upon habitat marking by residual odors, it is a reasonable hypothesis that the same phenomenon should occur at the larval stage. In particular, the more efficient use of underexploited resources could follow from the attraction of adults to resources containing larvae.

We investigated responses to larval cues in *D. melanogaster* and *D. simulans* with the wind tunnel set up as for the fruit attraction experiments. A sucrose-agar-dead yeast medium was placed in the flasks, and ovipositing flies of each species were left on this medium for 24 hours. The attraction of adults to odors in the trap cylinders which came from the flasks was monitored over successive days.

Data from four experiments are summarized in Table V. Numbers of flies attracted fluctuated from day to day, although lower numbers of both species were attracted on day 1 than on later days. Therefore the medium became more attractive following larval activity. We consider first the data from experiments 1, 3 and 4. The indices of choice difference, Δ, are positive for male data from all trials, and significantly >0 in many cases. Hence, males are preferentially attracted to media containing larvae of their own species. The female results are rather less clear cut. Large numbers of females were attracted, unlike the experiments with

Table V. Indices of choice difference, Δ, for attraction of *D. melanogaster* and *D. simulans* to media containing larvae at different stages of development.

	Day 1	Day 2	Day 3	Day 4	Day 5	Day 6	Day 7	Day 8	Day 9
Experiment									
1 ♂♂	a	a	a	a	0.02	0.32**	0.42**	0.26	0.63**
♀♀	a	a	a	a	0.03	0.16**	0.23	0.37*	a
2 ♂♂	0.32*	0.45**	-0.24**	-0.37**	0.29**	0.22**	0.40**	0.28**	0.26**
♀♀	-0.23	0.25	-0.33**	-0.09	-0.06	0.12	0.09	0.27	-0.33
3 ♂♂	0.20	0.11	-0.17	0.12	0.13*	0.38**	0.23**	0.11	0.11
♀♀	a	0.15	0.43	0.23**	0.08	0.41**	0.23*	0.08	0.22
4 ♂♂	0.25*	0.15	0.48**	0.10	0.38**	0.50**	0.21	0.45**	0.21
♀♀	-0.30	0.07	0.17	0.05	0.10	a	-0.36	0.04	0.13

* P < 0.05, ** P < 0.01, a Less than 5 *D. melanogaster* or *D. simulans* attracted.

residual odors. Indices were mostly positive, including all the cases where Δ differed significantly from 0 so that attraction tended towards conspecific larval generated odors. However, the male data showed significant associations in several trials when the females showed no differential response, suggesting that the sexes may respond differently to the larval cues. Data from experiment 2 suggest that the pattern of differential attraction may change in some cases with larval development. The Δ value for male data is significantly positive on days 1 and 2, significantly negative on days 3 and 4, and positive again on the remaining days. Preliminary experiments suggest that larval density needs to be controlled to reduce this variability among experiments.

V. HABITAT MARKING WITHIN SPECIES

Does habitat marking occur within species? A reasonable approach is to commence with populations from habitats that are widely divergent ecologically. Furthermore, this ecological divergence should be reflected in ecologically important characteristics (or ecological phenotypes) from these populations. Melbourne in southern Australia has a temperate climate, while Townsville is in the humid tropics. *D. melanogaster* from these two populations is distinguishable for cold and desiccation resistance, utilization of ethanol and its metabolite acetic acid, and the number of larvae choosing ethanol (Parsons, 1983b). In particular, the ethanol and acetic acid differences may correspond to the very different resources in the temperate zone versus the humid tropics (Stanley and Parsons, 1981). Parallel differences between Melbourne and Townsville populations also apply to *D. simulans,* although less marked for ethanol utilization, which is expected since *D. simulans* is basically an ethanol-sensitive species (McKenzie and Parsons, 1972).

Data were obtained using non-virgin females as marker populations, since this was the experimental situation providing maximum habitat marking differences between *D. melanogaster* and *D. simulans* (Table IV). The summarized results are presented in Table VI, based upon five trials in each species. In both species there is a significant tendency of males to select cylinders marked with the residual odor of females of their own populations. As expected from earlier data, fewer females are attracted, and there is no significant effect at the population level. Hence, the results at the intraspecific level parallel those at the interspecific level, although as would be expected the intraspecific Δ's are < interspecific Δ's (Table IV and VI).

Table VI. Attraction of *Drosophila melanogaster* and *D. simulans* to residual odors from Melbourne and Townsville non-virgin females (summarized from Parsons and Hoffmann, 1985).

Population odor (attractant)	Males			Females		
	Melbourne	Townsville	Δ	Melbourne	Townsville	Δ
D. melanogaster						
Melbourne	132	96		33	54	
Townsville	73	155	0.26***	28	50	0.02
D. simulans						
Melbourne	154	46		87	42	
Townsville	107	77	0.22***	49	31	0.06

Δ = index of choice difference, *** $P < 0.001$

Table VII. Indices of choice difference, Δ, for attraction of Townsville and Melbourne females to larval cues.

		Δ			
Species	Day	Expt. 1	Expt. 2	Expt. 3	Expt. 4
D. melanogaster	5	0.27*	0.08	0.22**	-0.01
	6	-0.14	0.30**	0.43**	0.08
	7	0.11**	0.19**	0.35**	0.03
	8	0.34**	-0.06	0.07	0.15
D. simulans	5	0.11	0.11*	0.03	
	6	0.12	0.20**	0.20**	
	7	0.26*	0.06	0.13*	
	8	0.26**	-0.09	0.08	

* $P < 0.05$, ** $P < 0.01$

We examined the response of Townsville and Melbourne females to larval cues, with 50 larvae from a population placed into a flask. Table VII indicates many significant positive Δ values. Hence females are preferentially attracted to larval odors from their own population at this larval density.

The interpopulation differences are heritable because they were repeatable and detected after culturing flies for several generations under the same conditions. This suggests genetic variation in odor production and recognition. Another possible factor is the differential attraction of flies to yeast or other microorganisms from the collection sites. These microorganisms could persist across generations, being passed on in the eggs or carried by the adults and larvae. An explanation based solely on microorganisms seems unlikely, given the following requirements:

1. *D. melanogaster* and *D. simulans* should harbor different microorganisms, even though they are bred on the same medium and were collected from the same site.

2. Populations of the same species should harbor different microorganisms.

3. Results presented elsewhere (Parsons and Hoffmann, 1985) indicate that isofemale lines from the same site show differential responses to residual female odors. Hence, individuals from the same species and collected at the same site should be associated with different microorganisms.

4. Differential attraction involves variation in the recognition of cues as well as their production. Hence microorganisms associated with flies should induce the flies to differentially respond to cues.

5. Microorganism induction should be sex specific.

Further experiments could help to assess the relevance of this factor if any; these include testing for maternal effects in progeny from crosses between the populations, or using flies cultured under axenic conditions.

VI. DISCUSSION

The results have diverse implications at the interface of ecological and behavioral genetics, and some of these are briefly discussed below:

A. The Nature of Residual Odor Attraction

The interspecific variation in residual odor attraction by males is sex specific, since males showed little attraction to the residual odors of other males. This specificity suggests variation in both odor recognition by males and odor production by females. One alternative explanation is that males may be attracted to substances transmitted during copulation. However, this possibility can be discounted by the fact that the phenomenon was observed with both virgin and non-virgin females, and from the experiments with larvae as attractants.

Chemical stimuli from mature *Drosophila* females are important in initiating courtship of males (Tompkins *et al.,* 1980; Venard and Jallon, 1980; Antony and Jallon, 1982). These stimuli consist of apparently species-specific long chain hydrocarbons in the cuticle of females, and there is the possibility of some genetic variation within species (Tompkins and Hall, 1984). These chemicals may be important in residual odor marking, although many have low volatility and may only affect male behaviors over a few millimeters, probably being detected by contact rather than olfaction (Antony and Jallon, 1982).

On the other hand, following suggestions concerning preferences by rare *Drosophila* males (Ehrman and Probber, 1978), habitat marking could alternatively or additionally be the consequence of a recognition system based on compounds that are normal metabolic products (see Parsons, 1983b). In this way, females or larvae from different populations would leave a characteristic bouquet preferentially attracting flies to the bouquet of a given population. These recognition compounds may be normal fermentation products, attractive over relatively long distances, and in this light the parallel results for the sibling species are expected. It follows that isofemale strains within a population that are extreme for metabolic phenotypes may also show this type of marking, and preliminary data suggest this possibility (Parsons and Hoffmann, 1985).

B. Sexual Selection

Mating has been the subject of many laboratory studies in *Drosophila,* but not usually in an ecological context. Generally, there are greater differences in laboratory mating success among males of *D. melanogaster* than among females, indicating the predominant importance of sexual selection among males (Bateman, 1948; Parsons, 1983a). In natural populations, Anderson *et al.* (1979) found evidence for virility selection in *D. pseudoobscura* and Brittnacher (1981) deduced that virility may dominate the dynamics of *D. melanogaster* and *D. pseudoobscura* polymorphisms. Cues from residual marking and larvae will attract males to particular microhabitats, leading to increased competition among males for mates. This would enhance sexual selection at the microhabitat level, and may contribute to the high levels of multiple insemination usual in populations of *D. melanogaster* and other widespread *Drosophila* species (Anderson, 1974; Milkman and Zietler, 1974).

C. Sexual Isolation

The preferential attraction of flies from Melbourne and Townsville to habitats marked by females and larvae of their own population is suggestive of sexual isolation between these populations, especially as the courtship behaviors of cosmopolitan species tend to be associated with their resources (Spieth, 1952). Habitat marking combined with ecological divergence should therefore be associated with some sexual isolation. Indeed, although many authors regard animal speciation as the development of reproductive isolating mechanisms, an ability to utilize the resources of the environment in such a way as to be protected against niche competitors (Mayr, 1977; Parsons, 1983b) may also be important and needs further emphasis.

Tests of sexual isolation among *Drosophila* populations are usually carried out in mating chambers, and these often indicate that isolation is weak or absent especially in *D. melangoaster* (Spieth and Ringo, 1983). However, the behaviors exhibited in wind tunnels are unlikely to be important in mating chamber tests, and it seems worthwhile to further investigate isolation with paradigms incorporating resources and long distance attraction.

D. Aggregation and Interaction at the Breeding Site Level

Both olfactory responses to resources and habitat marking suggest that individuals with similar olfactory phenotypes tend to aggregate at particular fruit types. Since resources are mainly ephemeral, the genetic effect of attraction to fruit odors would be rather transitory in cosmopolitan species such as *D. melanogaster* that use a wide array of resources, while habitat marking and responses to larval cues would presumably be a less transitory phenomenon. These effects would lead to a tendency for similar phenotypes to coexist together, further developed by aggregation from a limited number of females ovipositing on each site (Hoffmann *et al.*, 1984; Hoffmann and Nielsen, 1985). Therefore flies from different resource items would tend to have genotypic distributions with greater variability than under random mating, which leads to decreased interactions among widely divergent genotypes.

Among *Drosophila* species, aggregation by responses to adult and larval cues limits contact among species in microhabitats, so that there would be fewer interactions among species compared with a model where resource exploitation occurs indiscriminantly. Additional results suggestive of reduced interspecific-level interactions include the seasonal changes in the *melanogaster/simulans* numbers, whereby *melanogaster* is dominant at a different time of the year to *simulans* (McKenzie and Parsons, 1974).

In summary, interactions among *Drosophila* species utilizing similar resources may be relatively unimportant. This is consistent with Atkinson's (1984) observations on field-collected rainforest flies where he found no evidence for interspecific competition even though there is demonstrable larval competition within species. The coexistence of species in this rainforest community could best be explained by aggregated distributions over discrete breeding sites. In a consideration of decomposer insects (which include *Drosophila*), Strong (1983) argues that the theoretical model that most accurately describes these organisms is not one of homogeneous niches at some equilibrium but one with highly dissected, rapidly changing, resource patches. Under these circumstances interactions among species must, at best, be transitory.

E. Implications For Habitat Preference Studies

Considering the questions raised in the "Introduction," we have shown that olfactory attraction to natural resources contributes to variation in habitat preference and that this behavior may affect the field microdistribution of *Drosophila*. Additionally, we have demonstrated a genetic component to this variation since it was detected after culturing flies in a uniform environment.

The approach of sampling variable habitats in the field could be applied to other behaviors considered mainly in the laboratory. One example is from Kekic *et al.* (1980) who reared *D. subobscura* collected from dark and light habitats and provided suggestive evidence that offspring differ in their phototactic response. An advantage of this approach is that laboratory traits can in principle be related to field behaviors. It should also be possible to separate genetic and other components by comparing phenotypic variation among flies obtained from fruit in the field with variation after laboratory culturing. This may be difficult for fruit attraction in the wind tunnel because the Δ values varied across trials (Table III). The work of Fuyama (1978) indicates that olfactory response can be analyzed genetically, although these experiments with individual chemicals in a Y-tube olfactometer are far removed from field behaviors.

We have shown that responses to other individuals at the adult or larval stages may be important in habitat preference, and demonstrated heritable variation in this behavior. The relative importance of these cues compared to those from resources will need to be considered in future studies. Our findings suggest that habitat discrimination in *Drosophila* may be complex, depending on chemical cues from resources and individuals. The presence of different odors may also affect olfactory response, as suggested by the effect of odor combination in *D. immigrans* behaviors.

We have little data on the association between olfactory variation and resource utilization, with the exception of the interspecific data for lemons. The population and species specific attraction to larval cues may be related to resource utilization, especially if there is a facilitation effect. It would be interesting to compare the quality of resources with larvae from different populations as breeding sites for those populations.

In conclusion, a prerequisite for understanding variation in habitat preference in *Drosophila* will be an increased knowledge of the genetics of ecobehavioral traits under an array of environments (e.g., varying temperatures) especially those simulating the situation in the wild. These need to be combined with the increasing number of studies on microhabitats occupied in the wild. Progress in this latter aim is being achieved, although alowly, by a combination of laboratory experiments on ecobehavioral traits together with related field studies.

REFERENCES

Anderson, W. W. (1974). Frequent multiple insemination in a natural population of *Drosophila pseudoobscura. Am. Nat.* **108**, 709-711.

Anderson, W. W., Levine, L., Olvera, O., Powell, J. R., de la Rosa, M. E., Salceda, V. M., Gaso, M. I., and Guzman, J. (1979). Evidence for selection by male mating success in natural populations of *Drosophila pseudoobscura. Proc. Natl. Acad. Sci. USA* **76**, 1519-1523.

Andrewartha, H. G., and Birch, L. C. (1954). "The Distribution and Abundance of Animals." University of Chicago Press.

Antony, C., and Jallon, J. M. (1982). The chemical basis for sex recognition in *Drosophila melanogaster. J. Insect. Physiol.* **28**, 873-880.

Atkinson, W. D. (1981). An ecological interaction between citrus fruit, *Penicillium* moulds and *Drosophila immigrans* Sturtevant (Diptera: Drosophilidae). *Ecol. Entomol.* **6**, 339-344.

Atkinson, W. D. (1984). Coexistence of Australian rainforest Diptera breeding in fallen fruit. *J. Anim. Ecol.* (in press).

Bateman, A. J. (1948). Intra-sexual selection in *Drosophila. Heredity* **2**, 349-368.

Brittnacher, J. G. (1981). Genetic variation and genetic load due to the male reproductive component of fitness in *Drosophila. Genetics* **97**, 719-730.

Dethier, V. G. (1982). Mechanism of host-plant recognition. *Ent. Exp. Appl.* **31**, 49-56.

Diehl, S. R., and Bush, G. L. (1984). An evolutionary and applied perspective of insect biotypes. *Ann. Rev. Entomol.* **29**, 471-504.

Ehrman, L., and Parsons, P. A. (1981). "Behavior Genetics and Evolution." McGraw-Hill, New York.

Ehrman, L., and Probber, J. (1978). Rare *Drosophila* males: the mysterious matter of choice. *Am. Scientist* **66**, 216-222.

Futuyma, D. J., and Petersen, S. C. (1985). Genetic variation in the use of resources by insects. *Ann. Rev. Entomol.* (in press).

Fuyama, Y. (1978). Behavior genetics of olfactory response in *Drosophila*. II. An odorant-specific variant in a natural population of *Drosophila melanogaster. Behav. Genet.* **8**, 399-414.

Hoffmann, A. A. (1983). Bidirectional selection for olfactory response to acetaldehyde and ethanol in *Drosophila melanogaster. Genet. Sel. Evol.* **15**, 501-518.

Hoffmann, A. A. (1985). Interspecific variation in the response of *Drosophila* to chemicals and natural fruit odors. *Aust. J. Zool.* (in press).

Hoffmann, A. A., and Nielsen, K. M. (1985). The effect of resource subdivision on genetic variation in *Drosophila. Am. Nat.* **125**, 421-430.

Hoffmann, A. A., Nielsen, K. M., and Parsons, P. A. (1984). Spatial variation of biochemical and ecological phenotypes in *Drosophila;* electrophoretic and quantitative variation. *Devel. Genetics* **4**, 439-450.

Hoffmann, A. A., Parsons, P. A., and Nielsen, K. M. (1984). Habitat selection: olfactory response of *Drosophila melanogaster* depends on resources. *Heredity* **53**, 139-145.

Jaenike, J. (1983). Induction of host preference in *Drosophila melanogaster. Oecologia* **58**, 320-325.

Jaenike, J. (1985). Genetic and environmental determinants of food preference in *Drosophila tripunctata. Evolution* **39**, 362-369.

Jones, J. S., and Probert, R. F. (1980). Habitat selection maintains a deleterious allele in a heterogeneous environment. *Nature* **287**, 632-633..

Kekic, V., Taylor, C. E., and Andjelkovic, M. (1980). Habitat choice and resource specialization by *Drosophila subobscura. Genetika* **12**, 219-225.

Kellogg, F. E., Frizel, D. E., and Wright, R. H. (1962). The olfactory guidance of insects IV. *Drosophila. Canad. Ent.* **94**, 844-888.

Lande, R. (1975). The maintenance of genetic variability by mutation in a polygenic character with linked loci. *Genet. Res. Camb.* **26**, 221-235.

Manning, A. (1967). Pre-imaginal conditioning in *Drosophila*. *Nature* **216**, 338-340.

Maynard Smith, J. (1966). Sympatric speciation. *Am. Nat.* **100**, 637-650.

Mayr, E. (1977). The study of evolution, historically viewed. *In* "The Changing Scenes in Natural Sciences 1776-1976," Special Publ. 12, pp. 39-58, Academy of Natural Sciences, Philadelphia.

McDonald, J., and Parsons, P. A. (1973). Dispersal activities of the sibling species *Drosophila melanogaster* and *Drosophila simulans*. *Behav. Genet.* **3**, 293-301.

McKenzie, J. A., and Parsons, P. A. (1972). Alcohol tolerance: an ecological parameter in the relative success of *Drosophila melanogaster* and *Drosophila simulans*. *Oecologia* **10**, 373-388.

McKenzie, J. A., and Parsons, P. A. (1974). Numerical changes and environmental utilization in natural populations of *Drosophila*. *Aust. J. Zool.* **22**, 175-187.

Milkman, R. D., and Zeitler, R. R. (1974). Concurrent multiple paternity in natural and laboratory populations of *Drosophila melanogaser*. *Genetics* **78**, 1191-1193.

Miller, J. R., and Strickler, K. L. (1984). Insect/plant interactions – role of chemicals in finding and accepting hosts. *In* "Chemical Ecology of Insects" (W. J. Bell and R. T. Carde, eds.), Chapman and Hall.

Parsons, P. A. (1983a). Ecobehavioral genetics: habitats and colonists. *Ann. Rev. Ecol. Syst.* **14**, 35-55.

Parsons, P. A. (1983b). "The Evolutionary Biology of Colonizing Species." Cambridge University Press, New York.

Parsons, P. A., and Hoffmann, A. A. (1985). Habitat marking: parallel genetic divergence in two *Drosophila* species. *Heredity* **54**, 203-207.

Prince, G. J., and Parsons, P. A. (1980). Resource utilization specificity in three cosmopolitan *Drosophila* species. *J. Nat. Hist.* **14**, 559-565.

Rausher, M. D. (1983). Conditioning and genetic variation as causes of individual variation in the oviposition behaviour of the tortoise beetle, *Deloyala guttata*. *Anim. Behav.* **31**, 743-747.

Rausher, M. D. (1984). The evolution of habitat preference in subdivided populations. *Evolution* **38**, 596-608.

Ringo, J., and Wood, D. (1983). Pupation site selection in *Drosophila simulans*. *Behav. Genet.* **13**, 17-27.

Rockwell, R. F., Cooke, F., and Harmsen, R. (1975). Photobehavioral differentiation in natural populations of *Drosophila pseudo-obscura* and *D. persimilis*. *Behav. Genet.* **5**, 189-202.

Sewall, D., Burnet, B., and Connolly, K. (1975). Genetic analysis of larval feeding behaviour in *Drosophila melanogaster. Genet. Res.* **24**, 163-173.

Shorey, H. H. (1976). "Animal Communication by Pheromones." Academic Press, New York.

Sokolowski, M. B. (1980). Foraging strategies of *Drosophila melanogaster*: A chromosomal analysis. *Behav. Genet.* **10**, 291- 302

Spence, G. E., Hoffmann, A. A., and Parsons, P. A. (1984). Habitat marking: males attracted to residual odors of two *Drosophila* species. *Experientia* **40**, 763-765.

Spieth, H. T. (1952). Mating behavior within the genus *Drosophila* (Diptera). *Bull. Am. Mus. Nat. Hist.* **99**, 401-474.

Spieth, H. T., and Ringo, J. M. (1983). Mating behavior and sexual isolation in *Drosophila*. *In* "The Genetics and Biology of *Drosophila* " (M. Ashburner, H. L. Carson, and J. N. Thompson Jr., eds)., Vol. 3c, pp. 223-284, Academic Press, London.

Stanley, S. M., and Parsons, P. A. (1981). The response of the cosmopolitan species *Drosophila melanogaster* to ecological gradients. *Proc. Ecol. Soc. Aust.* **11**, 121-130.

Strong, D. R. (1983). Natural variability and the manifold mechanisms of ecological communities. *Am. Nat.* **122**, 636-660.

Taylor, C. E. (1976). Genetic variation in heterogeneous environments. *Genetics* **83**, 887-894.

Taylor, C. E., and Condra, C. (1983). Resource partitioning among genotypes of *Drosophila pseudoobscura*. *Evolution* **37**, 135-149.

Thorpe, W. H. (1939). Further studies on pre-imaginal olfactory conditioning in insects. *Proc. Roy. Soc. Lond.* **77B**, 424-433.

Tompkins, L., and Hall, J. C. (1984). Sex pheromones enable *Drosophila* males to discriminate between conspecific females from different laboratory stocks. *Anim. Behav.* **32**, 349-352.

Tompkins, L., Hall, J. C., and Hall, L. M. (1980). Courtship stimulating volatile compounds from normal and mutant *Drosophila. J. Insect. Physiol.* **26**, 689-697.

Turelli, M., Coyne, J. A., and Prout, T. (1984). Resource choice in orchard populations of *Drosophila*. *Biol. J. Linn. Soc.* **22**, 95-106.

Venard, R., and Jallon, J. M. (1980). Evidence for an aphrodisiac pheromone of female *Drosophila. Experientia* **36**, 211-212.

Via, S. (1984). The quantitative genetics of polyphagy in an insect herbivore. *Evolution* **38**, 881-905.

Wright, R. H. (1966). An insect olfactometer. *Canad. Entomol.* **98**, 282-285.

POPULATION BIOLOGY OF SUEZ CANAL MIGRATION –
WHICH WAY, WHAT KIND OF SPECIES AND WHY[1]

Uriel N. Safriel

Department of Zoology
The Hebrew University of Jerusalem
Jerusalem 91904 Israel

Uzi Ritte

Department of Genetics
The Hebrew University of Jerusalem
Jerusalem 91904 Israel

ABSTRACT

Since the opening of the Suez Canal, about 130 Red Sea species have become successful colonizers of the Mediterranean Sea, while only 10-15 Mediterranean immigrants colonized the Red Sea. We found that the currents in the Suez Canal were instrumental in this directionally biased migration. The question remains, though, what distinguishes the 130 immigrants from the other hundreds of Red Sea species that have not colonized the Mediterranean? To answer it, we studied demographic parameters and genetic variability in two groups of intertidal molluscs. Within each group all species were taxonomically related and ecologically similar, and each consisted of two Red Sea species, a colonizer and a non-

---

[1] This paper is No. 16 in the series "Colonization of the Eastern Mediterranean by Red Sea species immigrating through the Suez Canal," which was supported in part by a grant from the United States-Israel Binational Science Foundation (BSF), Jerusalem, Israel.

colonizer. Electrophoretic enzyme variability of colonizers was not significantly different from that of their non-colonizing relatives, and for the colonizers the Canal did not serve as a genetic bottleneck. On the other hand, the innate capacity for increase (r) of the colonizer was consistently higher than that of the non-colonizer. Each studied group included also a third, closely related, indigenous Mediterranean species. Relative to the colonizer, the indigenous mussel shows a stronger r-strategy, whereas the indigenous cerithiid is a definite **K**-strategist. Thus, the Mediterranean Sea does not constitute an environment necessarily favoring r-strategists; rather, an r-strategy constitutes a colonization pre-adaptation. The successful colonizer is the one that arrives at the colonized area first, increases fast and then competitively excludes late-arriving species.

I. INTRODUCTION

The Suez Canal, which is 115 years old, is about 150km long. It connects the Gulf of Suez with the eastern basin of the Mediterranean Sea (Fig. 1). Although these two seas have been completely isolated for about five million years, regional geological events through the last 50 million years (Neev *et al.* 1976) were responsible for the marked differences between their biotas; that of the Gulf of Suez (and the Red Sea) is tropical, Indo-Pacific, whereas that of the eastern Mediterranean is temperate-Atlantic, with certain subtropical affinities. Unlike the Panama Canal, the Suez Canal is at sea-level, and it allows the exchange of species between the two seas.

Indeed, soon after the opening of the Canal both Mediterranean and Red Sea species have been found in it (for references see Por, 1977), and new records for the Canal have been added ever since (Safriel *et al.*, 1980). Furthermore, many species have been found in both seas, which means that they had migrated via the Suez Canal (Por, 1977), although, due to lack of systematic research, only for some of them individuals have been found in the Canal itself. The objective of this paper is to evaluate in what way the Suez Canal, rather than the two seas that it connects, determines the kinds and numbers of colonizing species.

II. THE UNIDIRECTIONALITY OF MIGRATION

The identification of definite colonizers depends on whether their actual spread in the colonized area has been documented, or on biogeographical considerations (e.g., a Red Sea species is a colonizer of the Mediterranean if its original affinities have been strictly indo-Pacific; Por, 1977). When the appropriate analysis was performed, it was soon realized that the appearance of colonizers in the two seas was not symmetrical. About 130 Red Sea species have by now colonized the Mediterranean, whereas only 10-15 Mediterranean species have colonized the Red Sea; new Red Sea species are being constantly added to the list of the Mediterranean fauna (e.g., Ben-Tuvia, 1983).

Sea surface temperatures of the eastern Mediterranean and the Red Sea cannot be invoked as an explanation for this disparity: due to the low winter temperatures of the eastern Mediterranean (17°C mean monthly temperature for the coldest month; anonymous, 1957), species of the Red Sea (with 21°C average annual minimal temperature at the northern end of the Red Sea; Morcos, 1970) may not be adapted to inhabit the Mediterranean, whereas for Mediterranean species Red Sea winter temperatures may be quite 'comfortable,' let alone summer temperatures which are nearly identical in both seas (27°C and 26°C for the Mediterranean and the Red Sea, respectively; sources as above). Yet, hardly any of the Mediterranean species colonized the Red Sea. Also, the biota of the eastern Mediterranean have been traditionally regarded as 'impoverished' whereas that of the Gulf of Suez are highly diverse, and it has been therefore suggested that Red Sea species are 'diffusing' through the Canal to fill-in 'empty ecological niches' in the Mediterranean (Ben-Tuvia, 1966; Por, 1977). Why had these 'empty niches' not been filled-in earlier by species from the more diverse communities of the western Mediterranean and the Atlantic?

A more plausible interpretation is that Red Sea species have high competitive ability relative to Mediterranean species, on the grounds that species from highly diverse communities are competitively superior to those from less diverse ones (Briggs, 1968). Indeed, it was suggested that an immigrant fish (Ben-Tuvia, 1966) and an immigrant crab (Gilat, 1963) have competitively outnumbered their respective native relatives in the eastern Mediterranean. However, the data on which this statement is based are rather unsatisfactory, and it has not even been quantitatively established whether the immigrant species had indeed originated from highly diverse communities and penetrated less diverse ones.

We have set out to examine the latter interpretation (Safriel and Ritte, 1980), by investigating several groups of intertidal rocky shore animals -- a polychaete family (Ben-Eliahu and Safriel, 1980), and two mollusc families (Safriel and Ritte, 1983). We found (Table I) that in two of these families (Nereidae and Cerithiidae) species richness was indeed 2-6 times greater in the Red Sea than in the Mediterranean, but in one (Mytilidae) it was slightly greater in the Mediterranean than in the Red Sea. When it comes to possible competitive interactions we note (Table II) that numerical relations, between the colonizer in the colonized area and the ecologically most similar and sympatric indigenous species, are in two cases (in which the colonizer is less abundant in the colonized area than in the origin) in favor of the indigenous species, and in one case (in which the colonizer is more abundant in the colonized area than in the origin) in favor of the colonizing species, with no relation to the differences in species richness of the two communities.

Table I. Number of indigenous species and numbers of successful colonizers in the rocky intertidal of the northern gulfs of the Red Sea and in the eastern Mediterranean.[1]

| | Red Sea | | Mediterranean | |
	Indigenous	Colonizers from the Mediterranean	Indigenous	Colonizers from the Mediterranean
Polychaeta				
Nereidae[2]	9	0	4	1
Mollusca				
Mytilidae[3]	3	0	4	1
Cerithiidae[4]	6	0	1	1

[1] Data pertain to the Gulf of Suez, the Gulf of Elat and the Mediterranean coast of Israel. [2] Living in structurally similar intertidal biogenic reef (Ben-Eliahu and Safriel, 1982). [3] Living mainly on intertidal ledges and beachrock slabs (Safriel *et al.*, 1980). [4] Ayal (1978).

Table II. Relative abundance of Red Sea colonizers and their related indigenous species in the Mediterranean habitats.[1]

Species Colonizer (C) / Indigenous (I)	Proportions[2] (C)/(I)
Pseudonereis anomala / Perinereis cultrifera	1:50
Brachidontes variabilis / Mytilaster minimus	1:200
Cerithium scabridum / Cerithium rupestre	10:1

[1] For reference see comments for Table I. [2] For simplification the actual counts are combined, and the final results are rounded.

What does matter is the type of life-history strategy of the colonizer relative to that of its potential competitor, when both are involved in a 'migration-and-extinction competition' (Levins and Culver, 1971; Hanski, 1983) on resources in the patchy and unstable Mediterranean intertidal environment. Thus, the mytilid colonizer *Brachidontes variabilis*, with an (experimentally measured) high competition coefficient (Yiftakh, 1981), would be competitively superior to the indigenous Mediterranean mytilid *Mytilaster minimus,* if their common habitat would have been stable. But the latter species, having an extreme **r**-strategy, is faster than the colonizer in monopolizing newly created patches (Gilboa and Safriel, 1977; Yiftakh *et al.,* 1978). On the other hand, the indigenous cerithiid snail *Cerithium rupestre* is a typical **K**-strategist that has been endowed with a high competitive ability, but is apparently being displaced by the **r**-strategist colonizer *C. scabridum,* from an unstable but significant portion of their common habitat (Ayal, 1978).

III. THE CANAL AS A SALINITY BARRIER

So far we touched on possible ecological reasons for the disparity in colonization success between the two seas.   Alternatively, it is conceivable that there are more colonizers in the Mediterranean than in the Red Sea because Red Sea species can migrate through the Canal better than Mediterranean species.

It has been pointed out, for example, that salinity in the northern end of the Gulf of Suez is 4.4% (Morcos, 1970), and only 3.9% in the eastern Mediterranean (Oren, 1952). Hence, Red Sea species may be more likely to overcome the salinity barrier of the Great Bitter Lake (at the southern part of the Canal, Fig. 1), whereas for eastern Mediterranean species the Bitter Lake constitutes a genuine barrier. However, the salinity of the Great Bitter Lake has been considerably reduced by the flow of the Canal waters (Por, 1977); the Red Sea snail *Cerithium caeruleum,* for example, is more tolerant to high salinities than *C. scabridum,* yet only the latter colonized the Mediterranean (Ayal and Safriel, 1983).

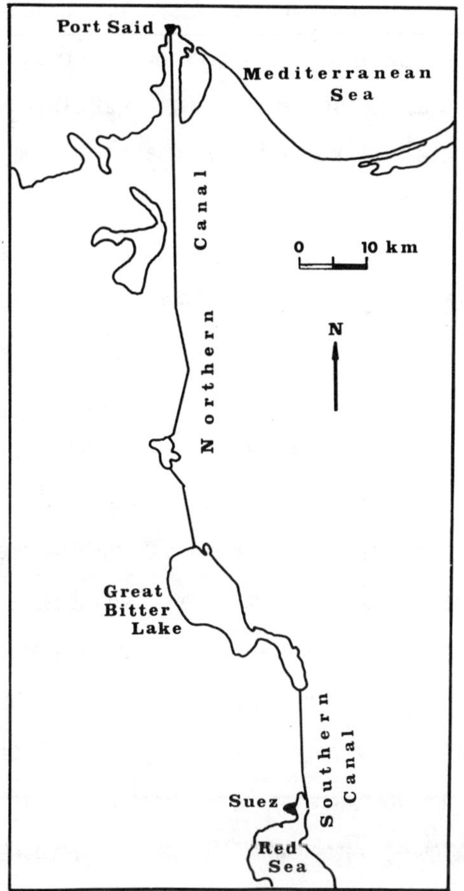

Figure 1. The Suez Canal

IV. THE CURRENTS IN THE CANAL AND THE UNIDIRECTIONALITY OF MIGRATION

Another approach has been to implicate the regime of currents in the canal for migration unidirectionality. This regime is determined primarily by the tidal schedules of the eastern Mediterranean and the Gulf of Suez, that diverge dramatically in their features. For example, based on early measurements (Fox, 1926), the prevalence of a northbound current during most of the year was postulated (Steinitz, 1929; Ben-Tuvia, 1973), but the accuracy of these measurements was challenged, and the existence of such a current was denied (Morcos, 1959).

The regime of the currents in the Suez Canal was resolved elegantly by Stiassnie (1972) who produced a mathematical, hydraulic model which was based on Red Sea and Mediterranean tidal predictions, on seasonal barometric pressure-affected sea-levels, evaporation and salinity, and on run-off and wind data. The model predicts the direction and velocity of the current along the whole length of the Canal, throughout the year. These may vary from one hour to the next, but the predictions of the model have been satisfactorily validated by field measurements.

Agur and Safriel (1981) used this model for simulating the movement of propagules floating in the Canal waters, after being 'introduced' (through a computer program) into either entrance of the Canal, at different times and frequencies. Once introduced, the fate of each propagule was followed, either until it was 'reabsorbed' into the sea of its origin, or until it emerged through the opposite end ('one jump journey'). It was also allowed to perform a 'two-jump' journey, using the Great Bitter Lake as a 'stepping-stone,' due to its favorable temperatures, lower turbidity and less gelatinous mud bottom (Table III).

Table III. Computer simulation of passive movement of floating propagules in the Suez Canal.[1]

| | Northbound migration Red Sea to Mediterranean | | | | Southbound migration Mediterranean to Red Sea | | | |
| | One jump journey (Suez-Port Said) | | Two jump journey (G. Bitter Lake-Port Said) | | One jump journey (Port Said - Suez) | | Two jump journey (Port Said-G. Bitter Lake) | |
	mean passage[2]	percent success	mean passage[2]	percent success	mean passage[2]	percent success	mean passage[2]	percent success
April	321	42	178	63	–	0	–	0
May	456	45	250	90	–	0	–	0
June	1017	7	839	19	–	0	–	0
July	–	0	–	0	–	0	–	0
August	–	0	–	0	–	0	–	0
Sept.	–	0	–	0	692	1.3	606	8.6

[1] Means and percentages for the results of 8784 simulations (a single propagule introduced during the main breeding season, every hour, from 1st April to 30th September, at each entrance of the Canal). After Agur (1982). [2] In hours.

Table IV. The southbound migration of Mediterranean propagules versus the northbound migration of Red Sea propagules, based on the Suez Canal pattern of currents.

	The southbound passage	The northbound passage
Probability for favorable currents[1]	0.402	0.598
Favorable current during peak of reproduction	no	yes
Chances of being 'absorbed' into the canal	1	3
Proportion of successful journeys	0.01	0.15
Duration of successful journeys (days)	30	11
Chances of a successful journey	1	75

[1] The probability of a northbound or a southbound current at any given point in the Canal, calculated as a mean for all model predictions.

These simulations produced an unequivocal result -- Red Sea species should be very successful in producing propagules that complete the northbound passage into the Mediterranean, whereas Mediterranean species are hardly able to complete a southbound passage to the Gulf of Suez (Table IV). Since about one-third of the Red Sea species that have colonized the Mediterranean have not been recorded in the Canal or its lakes, it is plausible that many of them indeed performed the 'one jump' migration. Even a species like *C. scabridum*, whose adults were found living in the Canal five years before the species was reported from the Mediterranean, could have easily performed this migration, since its larval longevity is ca 4-5 longer than the minimal period required for a completion of a successful northbound journey.

Finally, we note that although during the month of September southbound passage is feasible, the recent effect of the Aswan Dam (not incorporated into Stiassnie's model) is to reverse the currents trend during the second half of September so that even then it becomes predominantly northbound (Morcos and Gerges, 1974). Thus, though the ongoing reduction in salinities of the Great Bitter Lake should have improved the likelihood of Mediterranean species crossing to the Red Sea, the recent changes in currents counteract this trend (Agur, 1982). Indeed, the few Mediterranean species that colonized the Red Sea are either fish that swim actively (e.g., *Dicentrarchus punctatus* and *Liza aurata*) or invertebrate species usually transported by boats (e.g., the polychaetes *Scalisetosus fragilis* and *Lumbrineris coccinea*; Por, 1977).

Table V. Relative potential of two closely related Red Sea cerithiid snails to migrate through the Suez Canal.

	Cerithium scabridum (colonized the Mediterranean)	*Cerithium caeruleum* (did not colonize the Mediterranean)
Larval longevity (days)[1]	45	90-120
Range of tolerance to salinity fluctuations[2]	0-3	0-4

[1] Both species have planktonic larvae. Data are from Ayal (1978).

[2] Range is scaled 0-4, in which 0 means no fluctuation and 4 means highest fluctuations (Ayal and Safriel, 1980).

Table VI. Electrophoretic analysis of Red Sea versus Mediterranean populations of colonizers.

	Cerithium scabridum (19 loci)		*Brachidontes variabilis* (17 loci)	
	Red Sea[1]	Mediterranean[2]	Red Sea[3]	Mediterranean[4]
Total number of alleles	75	83	93	92
Loci with exclusive alleles[5]	3 (16%)	8 (42%)	7 (41%)	5 (29%)
Number of exclusive alleles[6]	6 (8%)	14 (17%)	8 (7%)	7(8%)
Mean heterozygosity[7]	0.61	0.63	0.62	0.63
Mean number of alleles per locus	3.95	4.37	5.47	5.41

[1] One site in the Gulf of Suez (Gebel Hammam Sidna Musa). [2] One site in the Mediterranean (Akko). [3] Four sites – Abu Zanima, Gebel Hammam Sidna Musa, Ras Muhammad in the Gulf of Suez and El Kura in the Gulf of Elat. [4] Three sites – Shave Ziyyon, Palmachim and Tlul. [5] Percents are of the total loci examined. [6] Percents are of the total number of alleles found. [7] In *Cerithium* differences between seas are insignificant (one-way ANOVA). In *Brachidontes* there are no significant differences between sites (in each sea, one-way ANOVA), no significant differences in heterozygosities between seas, but there is a significant difference between loci (P<0.001, two-way ANOVA).

V. DOES THE CANAL FILTER POTENTIAL RED SEA COLONIZERS OF THE MEDITERRANEAN?

If the Suez Canal does not constitute a barrier for Red Sea migrants, and its currents allow for a rapid flow of floating propagules from the Red Sea to the Mediterranean, why then out of the hundreds of Red Sea species with pelagic larvae, only 130 did successfully colonize the Mediterranean?

For example, *C. caeruleum,* the Red Sea cerithiid species which exhibits the greatest ecological similarities to the colonizer *C. scabridum* (Ayal and Safriel, 1983), should have migrated through the Suez Canal with much greater ease than the colonizer (Table V). However, though both species have been found in the Canal (Ayal, 1978), only *C. scabridum* has colonized the Mediterranean. Similarly, *Modiolus auriculatus,* the Red Sea mytilid species which exhibits the greatest ecological similarity to the colonizing mytilid *B. variabilis,* has occasionally migrated through the Canal, but has not established successful colonies in the Mediterranean (Safriel *et al.,* 1980).

VI. THE CANAL AS A GENETIC BOTTLENECK

Neither the salinity of the Canal, nor the pattern of its currents, sort out Red Sea species into colonizers versus non-colonizers. Yet, the Canal could constitute a genetic bottleneck that differentially affects the chances of Red Sea species in successfully colonizing the Mediterranean. Namely, it is possible that the Canal samples only a small segment of the genetic variability existing in the Red Sea, hence for some species the propagules that manage to cross the Canal establish founding populations that are genetically inviable.

The extent of the genetic variability in populations of *C. scabridum* (Pashtan, 1978; Pashtan and Ritte, 1977) and of *B. variabilis* (Lavee, 1981) in the Red Sea was compared with that of their colonizing populations in the Mediterranean, through an electrophoretic analysis of enzymes (Table VI). Only 8% and 7% of the Red Sea alleles of *C. scabridum* and of *B. variabilis*, respectively, have not been found in the Mediterranean populations of these species. Thus, we might have inferred that more than 90% of the genetic variability of the source populations have been 'sampled' by the Canal.

However, 17% and 8% of the Mediterranean alleles, of the cerithiids and the mytilids, respectively, are 'unique', i.e., we discovered them only in the Mediterranean but not in the Red Sea samples. Although it is possible that these differences are due to chance, it may be that the unique alleles exist also in the Red Sea, but are very rare there. In spite of their rarity they were sampled by the Canal, and then they were selected by the Mediterranean environment, to become sufficiently common so as to be detected in our Mediterranean samples. Similarly, the 'unique' Red Sea alleles may have been selected against in the Mediterranean, and have become very rare.

The conclusion from this electrophoretic analysis is that the Suez Canal has not acted as a genetic bottleneck in the process of colonizing the Mediterranean. As can be seen in Table VI, there are no significant differences between populations of the source and the colonized area in mean heterozygosity, and in *C. scabridum* the mean number of alleles is even 11% higher in the Mediterranean than in the Red Sea. Thus, at least in the two cases so far studied, the Canal did not distort the genetic structure of the colonizers to a point that will jeopardize their successful colonization.

This should not be necessarily true with regard to the colonization of Mediterranean species in the Red Sea. Since there is a real numeric bottleneck here, caused by the direction of the currents, a genetic bottleneck is not unlikely, and Mediterranean propagules reaching the Red Sea may be unable to establish persistent colonies.

VII. COLONIZATION SUCCESS AND THE EXTENT OF GENETIC VARIABILITY

Although in particular circumstances colonization can be achieved without any genetic variability (Selander and Kaufman, 1973), it is generally postulated that high genetic variability is a prerequisite for successful colonization (Baker and Stebbins, 1965). A comparison of the extent of electrophoretic variability in Red Sea populations of colonizing and non-colonizing species did not reveal significant differences (Table VII), and we thus conclude that at least this variability cannot distinguish between colonizers and non-colonizers.

VIII. DEMOGRAPHIC PARAMETERS AND SUCCESS IN COLONIZATION

Could it be that the two non-colonizing molluscs that we studied had not succeeded in colonizing the Mediterranean simply because they had not found suitable habitats there? At least with regard to *Cerithium* this is not plausible (Ayal and Safriel, 1983): the habitat range of the colonizing cerithiid is a subset of a much wider habitat range of the non-colonizing cerithiid (Fig. 2). Hence *C. caeruleum* could have found a suitable habitat in the Mediterranean, yet it has not colonized. What are then the problems in colonizing the Mediterranean that *C. caeruleum* could not overcome, and are they identical to those encountered by the mytilid *M. auriculatus*, as well as by other Red Sea species that have not colonized the Mediterranean after the opening of the Suez Canal?

Table VII. Electrophoretic analysis of colonizers and non-colonizers in the source area (Red Sea).[1]

	Cerithiids		Mytilids	
	Colonizer	Non-colonizer	Colonizer	Non-colonizer
	(C. scabridum)	*(C. caeruleum)*	*(B. variabilis)*	*(M. auriculatus)*
No. of examined loci	20	21	17	16
Total no. of alleles found	89	81	100	94
Mean no. of alleles per locus[2]	4.45 (1.19)	3.86 (1.06)	5.88 (2.09)	5.87 (1.78)
Mean heterozygosity per locus[2,3]	0.53 (0.12)	0.60 (0.11)		
	0.66 (0.10)	0.68 (0.07)	0.59 (0.23)	0.63 (0.17)

[1] Populations of all species sampled in several sites in the Gulf of Suez and the Gulf of Elat (see footnote 3 in Table VI), including an artificial lagoon at Elat ('Venetzia'). [2] In brackets – standard deviations. No significant differences between colonizers and non-colonizers (one-way ANOVA). When both colonizers are compared with both non-colonizers (mean no. of alleles per locus is 5.11 vs. 4.73), no significant differences between species and loci exist (two-way ANOVA). [3] Heterozygosity as calculated for each population. In cerithiids, the first line gives the mean for the populations exhibiting minimal heterozygosities, and the second line gives the mean for the populations exhibiting maximal heterozygosities. No significant differences were found (one-way ANOVA).

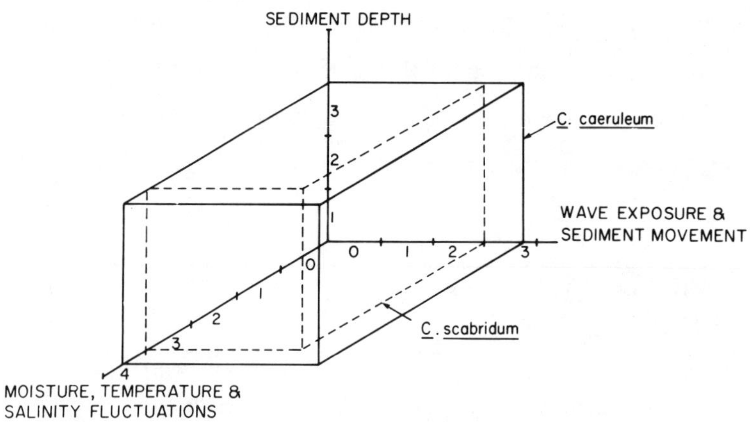

Figure 2. A schematic representation of habitat occupancy of the colonizing and the non-colonizing cerithiid, along ordinally scaled axes of relevant environmental variables (Ayal and Safriel, 1982a). The colonizer's space (delimited by broken lines) is a subset of that of the non-colonizer (delimited by solid lines).

A common feature to all colonization events preceded by migration through a narrow pasageway such as the Suez Canal, is that the size of the founding population is small. It is therefore vulnerable to the danger of random extinction; under all possible environmental regimes, including absolutely uniform and stable ones, 'demographic stochasticity' may critically shorten the persistence time of the founding colony, especially at the initial stage of colonization. Note that environments are often non-uniform and unstable, and that newly arrived colonizers are likely to perceive their new environment as non-uniform and fluctuating more than their source environment, since only in the latter they have become well adapted. Whether the colonized area is perceived as fluctuating regularly in space and time, or randomly ('environmental stochasticity'), a potential for rapid population growth, expressed by high value of the demographic parameter **r** (intrinsic growth rate at low densities), would shorten the time during which a colonizing population, threatened by demographic stochasticity, is exposed to the risks of environmental instability and randomness. Thus, a high **r** (and associated features of the 'r-strategy') is a pre-adaptation for colonizing areas with all types of spatial and temporal heterogeneity, as well as for penetrating communities with potential competitors and predators (for a review see Safriel and Ritte, 1983). Indeed, each of the two colonizing molluscs that we studied is much of an 'r species', relative to its non-colonizing counterpart: judging from their relative size and growth rate (Table VIII), the colonizing mytilid should have a higher **r** than the non-colonizing one, and the same holds true for the cerithiids, for which **r** was actually measured (Table IX).

Both colonizers encounter competition in the communities they penetrated --*B. variabilis* with the mytilid *M. minimus* (Yiftakh *et al.*, 1978), and *C. scabridum* with *C. rupestre* (Ayal, 1978). The presence of these competitors in the Mediterranean enabled us to evaluate whether high **r** is a pre-adaptation for colonization in general, or a pre-adaptation for colonizing just the Mediterranean, or those environments within it which induce **r**-selection. We compared demographic parameters of the indigenous competitors to those of the colonizers, and found that the Mediterranean indigenous mytilid is a typical **r**-strategist (Table VIII), whereas the cerithiid is a typical **K**-strategist (Table IX). Thus, at least in the species studied by us, the Mediterranean selects for both **r**- or **K**-strategy, depending on the specific environment and species, but a high **r** value of the species in the Red Sea communities preadapts them to succeed in colonizing, irrespective of the type of the colonized environment in the Mediterranean.

Table VIII. Some demographic features of mytilids.

| | Red Sea | | Indigenous in the Mediterranean |
	Non-colonizer (*Modiolus auriculatus*)	Colonizer (*Brachidontes variabilis*)	(*Mytilaster minimus*)
Mean size of adults (mm)[1]	35	20	10
Age of attaining half final size (yrs)[2]	?	~ 3.5	~ 1.5
Size of sexual maturation (mm)[3]	18	8	4
Maximal longevity (yrs)[3]	?	~ 13	~ 6

[1] Data from Safriel *et al.* (1980). [2] Data from Safriel and Ritte (1983). [3] Data from Gilboa (1976) and D. Lavee (pers. comm.).

Table IX. Some demogrphic features of cerithiids.

| | Red Sea | | Indigenous in the Mediterranean |
	Non-colonizer (*C. caeruleum*)	Colonizer (*C. scrabridum*)	(*C. rupestre*)
Mean size of adults (mm)[1]	28	18	22
Age of 1st reproduction (yrs)[1]	~ 4	~ 1	~ 3
Maximal longevity (yrs)[1]	~ 10	3	5
Generation time (yrs)[1,2]	7.3	2.1	?
Diameter of egg capsule (μ)[3]	150	130	400
Incubation period (days)[3]	4-5	4-5	14
Maximal no. eggs/ female. days[1]	~ 25000	~ 13000	80
r value [3,4]	1.6	0.9	?

[1] Data from Ayal (1978). [2] of a cohort (T_c). [3] From Ayal and Safriel (1982b). [4] This is r_m (see Ayal and Safriel, 1982b), for larval survivorship of 10^{-3}.

It should be pointed out that the non-colonizing cerithiid has a high fecundity (Table IX). Its life history strategy is that of 'bet hedging' (Stearns, 1976) rather than of a typical **K**-strategy. It is very likely that the great disparity in age of first reproduction between the two species determines their difference in colonization success: a founding colony of *C. scabridum* produces propagules to colonize adjacent habitat patches within a year, whereas that of *C. caeruleum,* if it does not perish, only after three years. By that time most patches in the surroundings are already occupied by *C. scabridum,* which has the upper hand in this type of 'migration and extinction competition' (Levins and Culver, 1971). Thus, it is plausible that *C. caeruleum* could have colonized the Mediterranean were it not living in the source area toegether with a closely related but potentially better colonizing species.

IX. CONCLUSION

The currents of the Suez Canal determine the unidirectionality of migration and the subsequent colonization of the Mediterranean by Red Sea species. The few Mediterranean species that colonized the Red Sea via the Suez Canal had migrated counter-current, either attached to vessels or actively swimming. In the northbound immigration process the Canal does not act as a genetic bottleneck or filter, so that the gene pool of the Red Sea source populations is amply represented in the Mediterranean colonies. However, the Canal does constitute a very narrow passageway that samples only a minute portion of the available propagules and allows the successful passage of only few at a time. Therefore founding colonies are always small, and their persistence time depends on their ability to grow fast. Within each of the two ecologically very different groups of Red Sea molluscs (the vagile, grazing cerithiids, and the sessile filter-feeding mytilids), the successful colonizers indeed have a higher potential for population growth than their non-colonizing related species. High genetic variability may also promote colonization success, but in these two groups genetic variability is exceedingly high in both colonizers and non-colonizers. Yet, at least for the non-colonizing species, it could not substitute for the lack of a high potential for population growth, which is the major predictor for success in colonizing the Mediterranean after migration through the Suez Canal, irrespective of the type of the colonized environment. This should hold true also for the Mediterranean colonizers of the Red Sea, though the issue has not been yet investigated.

ACKNOWLEDGMENT

We would like to thank Z. Agur, Y. Ayal, M. N. Ben-Eliahu, and D. Lavee for unpublished information.

REFERENCES

Agur, Z. (1982). Ph.D. dissertation. The Hebrew University of Jerusalem, Jerusalem, Israel.

Agur, Z., and Safriel, U. N. (1981). Why is the Mediterranean more readily colonized than the Red Sea, by organisms using the Suez Canal as a passageway? *Oecologia* **49**, 359-361.

Anonymous (1957). "Middellandse Zee." Edited by the Koninklijk Neederlands Meteorologisch Institute, s-Gravenhage.

Ayal, Y. (1978). Ph.D. dissertation. The Hebrew University of Jerusalem, Jerusalem, Israel.

Ayal, Y., and Safriel, U. N. (1982a). Role of competition and predation in determining habitat occupancy of Cerithiidae (Gastropoda: Prosobranchia) on the rocky, intertidal Red Sea coasts of Sinai. *Marine Biology* **70**, 305-316.

Ayal, Y., and Safriel, U. N. (1982b). r-curves and the cost of the planktonic stage. *Amer. Natur.* **119**, 391-401.

Ayal, Y., and Safriel, U. N. (1983). Does a suitable habitat guarantee
successful colonization? *Jr. of Biogeography* **10**, 37-46.

Baker, H. G., and Stebbins, G. L. (eds). (1965). "The Genetics of Colonizing
Species." Academic Press, New York.

Ben-Eliahu, M. N., and Safriel, U. N. (1982). A comparison between species
diversities of polychaetes from tropical and temperate structurally
similar rocky intertidal habitats. *Jr. of Biogeography* **9**, 371-390.

Ben-Tuvia, A. (1966). Red Sea fishes recently found in the Mediterranean.
Copeia **2**, 254-275.

Ben-Tuvia, A. (1973). Man made changes in the eastern Mediterranean and
their effect on the fishery resources. *Marine Biology* **19**, 197-203.

Ben-Tuvia, A. (1983). An Indo-Pacific goby *Oxyurichthys papuensis* in
the eastern Mediterranean. *Israel Jr. of Zoology* **32**, 37-44.

Briggs, J. C. (1968). Panama's sea-level canal. *Science* **162**, 511-513.

Fox, M. H. (1926). Cambridge expedition to the Suez Canal. 1924, General
Part. *Trans. of the Zool. Soc. Lond.* **22**, Part 1, No. 1.

Gilat, E. (1963). The macrobenthic animal communities of the Israeli
continental shelf in the Mediterranean. *Ext. Rapp. Proc. Verb.
C.I.E.S.M.M.* **17**, 103-106.

Gilboa, A. (1976). M.Sc. dissertation. Department of Zoology, The Hebrew
University of Jerusalem, Jerusalem, Israel.

Hanski, I. (1983). Coexistence of competitors in patchy environment.
Ecology **64**, 493-500.

Lavee, D. (1981). Ph.D. dissertation. The Hebrew University of Jerusalem,
Jerusalem, Israel.

Levins, R., and Culver, D. (1971). Regional coexistence of species and competition between rare species. *Proc. Natl. Acad. Sci. USA* **68**,
1246-1248.

Morcos, S. A. (1970). Oceanography of the Red Sea. *Oceanogr. Mar. Biol.
Annu. Rev.* **8**, 73-202.

Morcos, S. A. (1974). Circulation and mean sea level in the Suez Canal.
In "L'Oceanography Physique de la Mer Rouge," pp. 267-287. IA
PSC-UNESCO-SCOR Symp. Paris, 1972. CNEXO Publ. Ser. Actes
Colloq. 2.

Neev, D., Almagor, S., Arad, A., Ginsburg, A., and Hall, J. K. (1976). The
geology of the southeastern Mediterranean. *Geol. Surv. Israel
Bull.* **68**, 1-51.

Oren, O. H. (1952). Some hydrographical features observed off the coasts
of Israel. *Bull. Inst. Oceanogr. Monaco* **1017**.

Pashtan, A. (1978). M.Sc. dissertation. Departments of Genetics and
Zoology, The Hebrew University of Jerusalem, Jerusalem, Israel.

Pasthan, A., and Ritte, U. (1977). Genetic changes in *Cerithium scabridum* (Mollusca: Gastropoda) following colonization in the Mediterranean (abstract). *Israel Jr. Zoology* **26**, 258-259.

Por, F. D. (1977). Lessepsian migration - the influx of Red Sea biota into the Mediterranean by way of the Suez Canal. *In* "Ecological Studies," Vol. 223. Springer-Verlag, Berlin, Heidelberg, New York.

Safriel, U. N., and Ritte, U. (1980). Criteria for the identification of potential colonizers. *Biological Bull. of the Linnean Soc. of London* **13**, 287-297.

Safriel, U. N., and Ritte, U. (1983). Universal correlates of colonizing ability. *In* "The Ecology of Animal Movement" (I. R. Swingland and P. J. Greenwood, eds.), pp. 215-239. Clarendon Press, Oxford.

Safriel, U. N., Gilboa, A., and Felsenburg, T. (1980). Distribution of rocky intertidal mussels in the Red Sea coasts of Sinai, the Suez Canal and the Mediterranean coast of Israel, with special reference to recent colonizers. *Jr. of Biogeography* **7**, 39-62.

Sasson, Z. (1981). M.Sc. dissertation. Department of Zoology, The Hebrew University of Jerusalem, Jerusalem, Israel.

Selander, R. K., and Kaufman, D. W. (1973). Self-fertilization and genetic population structure in a colonizing land snail. *Proc. Natl. Acad. Sci. USA* **70**, 1186-1190.

Stearns, S. C. (1976). Life history tactics: a review of the ideas. *Quarterly Review of Biology* **51**, 3-47.

Steinitz, W. (1929). Die wanderung indopazifischer Arten ins Mittelmeer seit Beginn der Quatarperiode. *Int. Rev. Hydrobiol.* **22**, 1-90.

Stiassnie, M. (1972). A mathematical hydraulic model for the current regime of the Suez Canal. Mimeographed handout and a computer printout. Technion Institute of Technology, Haifa, Israel.

Yiftakh, Z., Gilboa, A., and Safriel, U. N. (1978). Can an Indo-Pacific mytilid successfully colonize a Mediterranean mytilid bed? *In* "2nd Int. Cong. Ecol. INTECOL, Jerusalem 1978." Abstracts Vol., p. 419.

EVOLUTIONARY GENETICS: HLA AS AN EXEMPLARY SYSTEM[1]

Philip W. Hedrick[2]
Glenys Thomson
William Klitz

Department of Genetics
University of California
Berkeley, California 94720

ABSTRACT

Comparison to neutrality expectations of the extent and pattern of single-locus and two-locus variation in the HLA region strongly suggests that selection is important at these loci. However, the four complement loci embedded in the HLA region do not show this pattern. The extent of disequilibrium among pairs of loci is map-distance dependent. The pattern of disequilibrium is consistent with selection events in this region.

I. INTRODUCTION

The classic examples of evolutionary genetics include melanism in *Biston betularia*, shell color and pattern in *Cepaea nemoralis*, and sickle-cell anemia in humans. With recent biochemical techniques many more loci have been investigated, but the system on which there is the most data, including the number of loci and number of alleles, is the

[1] Supported by National Institutes of Health grant HD 12731.
[2] Permanent Address: Division of Biological Sciences, University of Kansas, Lawrence, Kansas 66045.

major histocompatibility complex in humans (HLA). Because of this richness of information and the interesting characteristics of the genetic variation at HLA, we feel that HLA has become an exemplary system for understanding evolutionary genetics.

HLA was first investigated in a search for white-blood-cell groups that could form the basis for tissue transplantation matching in humans. It is now known that the HLA system contains a number of closely linked loci on chromosome 6 determining cell surface molecules, immune response, components of the complement system as well as other functions (e.g., Hood *et al.*, 1983; Klein *et al.*, 1983). In addition, a number of antigens of the HLA system show very striking associations with a large array of diseases (e.g., Dausset and Svejjaard, 1977; Ryder *et. al.*, 1979; Ryder *et al.*, 1981).

There exists an extensive amount of data on the genetic variation for HLA genes in different populations (for example, see the Histocompatibility Testing volumes for 1972, 1980 and 1984 (Dausset and Colombani, 1973; Terasaki, 1980; Albert *et al.*, 1984). Two features of these data make the HLA genes of particular importance for examination from an evolutionary perspective. First, many of the loci exhibit extremely high levels of polymorphism. For example, at least 23 and 47 different antigens have been defined at the HLA-*A* and *B* loci, respectively. Because the alleles have rather even distributions, the heterozygosity is around 90% for both these loci. Second, there are non-random associations (gametic disequilibria) between certain pairs of antigens at different loci. For example, the frequencies of antigens *A*1 and *B*8 are 0.172 and 0.127, respectively, in northern Europeans, making the expected frequency of this haplotype (chromosome or gametic type) 0.0218. However, the observed frequency of the *A*1 *B*8 haplotype is 0.0985, giving a normalized disequilibrium value of 0.69.

These observations would tend to imply that selective forces have been operating on the alleles in the HLA region. However, it is important that we objectively examine the variation in this region, and not make *a priori* assumptions about selective factors. The neutrality theory gives a useful starting point for our analysis in that it assumes that different alleles at a locus have equivalent effects on fitness. In this model, the equilibrium heterozygosity and multilocus associations in a population are a function of the combined effects of genetic drift and mutation. The neutral model, therefore, provides theoretical predictions against which the observed genetic variation in a population or a sample may be compared. First, we will compare single locus and then two-locus variation in the HLA region to neutrality expections. Second, we will examine the relationship of two-locus disequilibrium and map distance. Finally, we use an approach called disequilibrium pattern analysis to investigate two-locus associations.

II. HLA HOMOZYGOSITY AND NEUTRALITY EXPECTATIONS

Ewens (1972) developed a sampling theory to predict the distribution of alleles observed in a sample of size n taken from a population at equilibrium under neutrality. Watterson (1978a,b) extended this approach and developed a test that allows the comparison of the observed homozygosity expected in a sample of size n containing k alleles to the homozygosity expected under neutrality. This conditional homozygosity F is defined as

$$F = \sum_{i=1}^{k} p_i^2 \tag{1}$$

where p_i is the frequency of the ith allele in the sample of size n. (Note that this test is not examining deviations from Hardy-Weinberg proportions.)

Using this approach, Hedrick and Thomson (1983) compared the observed conditional homozygosity in 22 samples at both the A and B loci (Terasaki, 1980) to neutrality expectations. In all cases, the observed homozygosity was less than that expected from neutrality and was statistically significantly less at the 0.05 level in 25 of the 44 cases. They then evaluated the evolutionary factors that could be important in influencing the level of homozygosity conditioned on n and k, particularly those factors such as gene flow, bottlenecks, unidentified alleles, and balancing selection that could increase heterozygosity relative to neutrality expectations. After extensive consideration of these factors, Hedrick and Thomson suggested that some form of balancing selection (e.g., frequency-dependent selection) is the explanation most consistent with the level of conditional homozygosity at the A and B loci in the populations studied. The relatively high level of gene conversion in this region should have an effect similar to an elevated mutation rate. As a result, gene conversion should not increase conditional homozygosity which is independent of $4Nu$ (Ewens, 1972; N is the effective population size and u is the mutation rate).

Using the data from the most recent histocompatibility workshop (Albert *et al.*, 1984), Klitz *et al.* (1985b) have applied this approach to samples for nine loci of the HLA region. Figure 1 gives a map of the HLA region indicating the position of these loci: A, B, and C - class I loci; DR - a class II locus (both class I and II loci determine membrane glycoproteins); $C2$, Bf, $C4A$, and $C4B$ - class III loci of the complement system; and Glo-1 that codes for the enzyme Glyoxylase I.

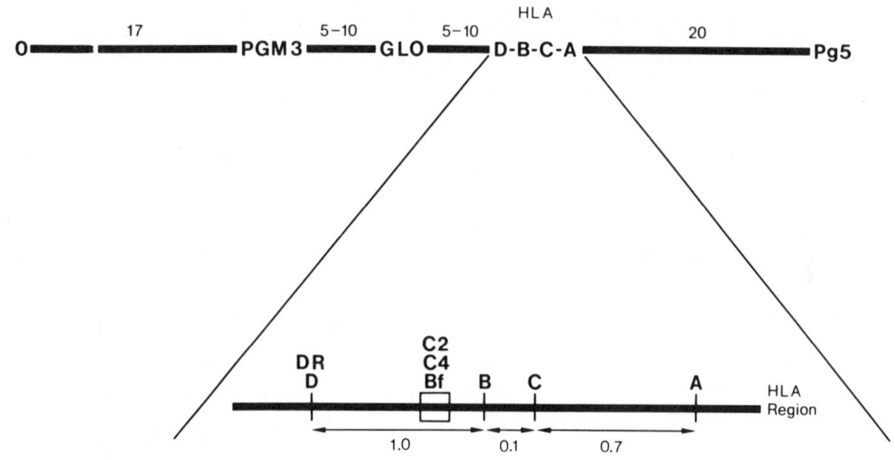

Figure 1. The map of nine loci in the HLA region.

As in the previous samples, loci *A* and *B* had conditional homozygosities statistically significantly less than neutrality expectations (Table I). In addition, the other loci that code for membrane glycoproteins, *C* and *DR*, as well as *Glo*-I had homozygosities significantly less than expected. However, the four complement loci had homozygosities either consistent with (*Bf*, *C4B*, and *C4A*) or exceeding (*C2*) neutrality expectations. These results are particularly interesting because they suggest that the complement loci, which are embedded in the HLA region, and the HLA loci display quite different selection histories despite their close linkage and the background of extensive disequilibrium in the region.

III. DISEQUILIBRIUM AND NEUTRALITY EXPECTATIONS

Under neutrality, a population at equilibrium has an expected association between alleles at different loci. Even though there is no selection among different alleles at a locus, the combined effects of

Table 1. The expected and observed homozygosity for samples from different populations at nine loci in the HLA region (Klitz *et al.*,1985b).

Locus	Sample	k	n	Homozygosity Expected	Observed	P
A						< 0.001
	Chinese	12	197	0.265	0.211	
	English-speaking*	16	392	0.228	0.149	
	French	15	265	0.261	0.133	
	German-speaking*	15	336	0.236	0.147	
	Italian	15	372	0.241	0.143	
	Japanese	9	536	0.395	0.232	
	Scandinavian	14	226	0.233	0.196	
C						< 0.001
	Chinese	9	197	0.340	0.196	
	English-speaking	9	381	0.381	0.168	
	French	9	251	0.357	0.170	
	German-speaking	9	316	0.369	0.166	
	Italian	9	360	0.379	0.181	
	Japanese	8	525	0.429	0.276	
	Scandinavian	9	228	0.352	0.151	
B						< 0.001
	Chinese	24	159	0.112	0.087	
	English-speaking	28	384	0.127	0.090	
	French	27	256	0.118	0.082	
	German-speaking	27	328	0.126	0.075	
	Italian	28	341	0.086	0.057	
	Japanese	22	495	0.174	0.080	
	Scandinavian	23	203	0.133	0.099	
$C4b$						ns
	English-speaking	7	159	0.411	0.514	
	French	8	205	0.379	0.517	
	German-speaking	5	382	0.566	0.584	
	Scandinavian	8	659	0.430	0.346	

(Table I cont'd)

Locus	Sample	k	n	Homozygosity Expected	Homozygosity Observed	P
C4A						ns
	English-speaking	7	159	0.411	0.415	
	French	7	205	0.425	0.556	
	German-speaking	7	310	0.445	0.493	
	Scandinavian	7	659	0.472	0.423	
Bf						ns
	Chinese	2	93	0.809	0.791	
	English-speaking	4	178	0.623	0.596	
	French	4	209	0.608	0.627	
	German-speaking	2	95	0.809	0.700	
	Scandinavian	3	136	0.727	0.696	
C2						< 0.01
	English-speaking	2	92	0.809	0.912	
	German-speaking	2	95	0.809	0.940	
	Japanese	4	521	0.651	0.883	
	Scandinavian	3	500	0.746	0.934	
DR						< 0.001
	Chinese	11	125	0.260	0.136	
	English-speaking	11	389	0.323	0.144	
	French	11	240	0.299	0.120	
	German-speaking	11	336	0.314	0.130	
	Italian	11	356	0.318	0.138	
	Japanese	11	528	0.337	0.167	
	Scandinavian	11	228	0.294	0.129	
Glo-1						< 0.001
	Chinese	2	101	0.809	0.707	
	English-speaking	2	123	0.815	0.500	
	French	2	209	0.831	0.507	
	German	2	98	0.809	0.507	
	Scandinavian	2	180	0.828	0.502	

*English-speaking includes Australian, English, Irish, Scottish, Welsh, and German-speaking includes Austrian, German, Swiss.

genetic drift and mutation result in an interlocus association when there is limited recombination (e.g., Ohta and Kimura, 1969; Hill, 1975). We have extended the approach of Ewens (1972) and Watterson (1978a,b) using the computer simulation approach of Hudson (1983) to determine the extent of disequilibrium expected in a sample of size n with k and l different alleles at two loci (Hedrick and Thomson, 1985). These disequilibrium values also depend upon the amount of recombination as measured by the quantity $4Nc$ where N is the size of the population from which the sample is drawn and c is the rate of recombination between the loci.

The extent of gametic disequilibrium can be measured in several ways for a specific gamete (e.g., Hedrick *et al.*, 1978 for a review). A widely used measure of gametic disequilibrium for a given gamete is

$$D_{ij} = x_{ij} - p_i q_j \qquad (2)$$

where x_{ij} is the observed frequency of gamete $A_i B_j$, p_i and q_j are the frequencies of alleles A_i and B_j at loci A and B, and the expected frequency of gamete $A_i B_j$ is $p_i q_j$, assuming no association between the alleles. The range of this measure of gametic disequilibrium is a function of the allelic frequencies, making a measure that has the same range for all allelic frequencies desirable. For this reason, Lewontin (1964) suggested using the normalized measure

$$D'_{ij} = \frac{D_{ij}}{D_{max}} \qquad (3)$$

where if $D_{ij} < 0$, D_{max} is the lesser of $p_i q_j$ and $(1-p_i)(1-q_j)$ and if $D_{ij} > 0$, D_{max} is the lesser of $p_i(1-q_j)$ and $(1-p_i)q_j$.

An example of the distribution of D'_{ij} values is given in Figure 2 for a sample of size $n = 200$ when there are six alleles at both loci and

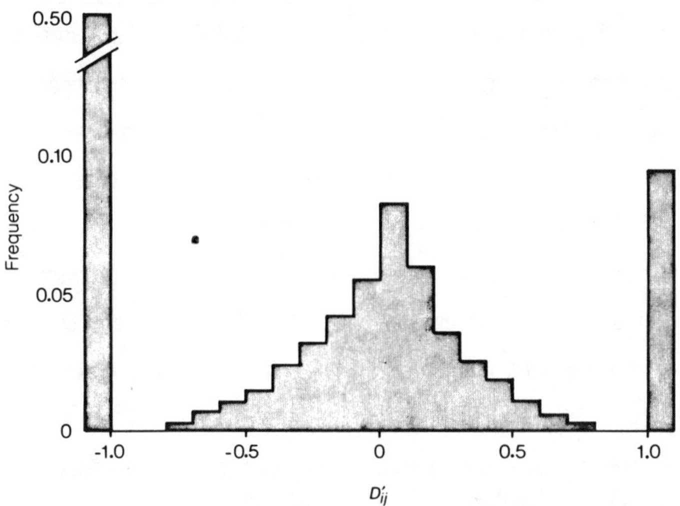

Figure 2. The distribution of D'_{ij} values in a sample of size $n = 200$ and $4Nc = 500$ when there are six alleles at both loci.

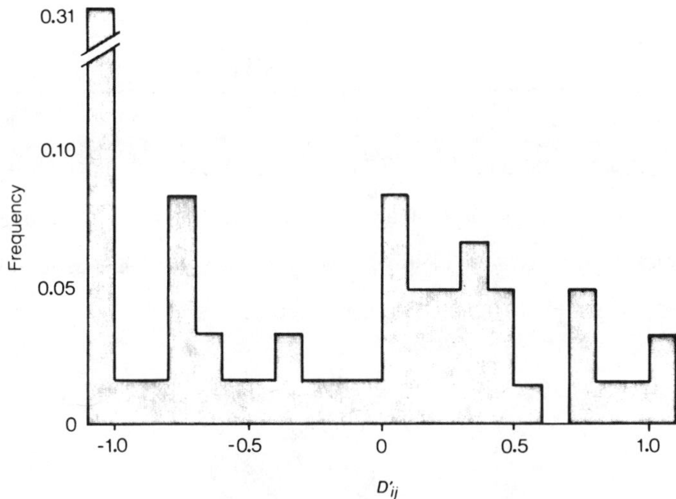

Figure 3. The distribution of D_{ij} values between HLA-A and HLA-B in samples from three South American Indian populations.

$4Nc = 500$, a high level of recombination. The distribution of these conditional D'_{ij} values for those that are not 1.0 or -1.0 has a mean near zero and a nearly symmetrical distribution. There is a larger proportion of -1.0 values, the case where a haplotype is not observed, than of 1.0 values. If $4Nc$ is lower, then the size of the terminal classes is larger.

The composite distribution of D'_{ij} values for the HLA-*A* and *B* loci for three South American Indian samples (Black and Salzano, 1981) is given in Figure 3 (Hedrick and Thomson, 1985). Notice that these observed values are spread out over the whole range, are somewhat asymmetrical and have few $D'_{ij} = 1.0$ values. This pattern appears to be quite different from the neutrality expectations, particularly in the absence of small negative D'_{ij} values.

The total disequilibrium between all the alleles at two loci can be measured in several ways. For example, the total disequilibrium can be expressed as

$$D^2 = \sum_{i=1}^{k} \sum_{j=1}^{\ell} D_{ij}^2 . \tag{4}$$

However, this measure, like D, is highly dependent upon allelic frequencies. As a result, other measures of association such as those given in Hedrick and Thomson (1985), that are not so dependent on allelic frequencies, are preferable. One such measure is

$$Q = n \sum_{i=1}^{k} \sum_{j=1}^{\ell} \frac{D_{ij}^2}{p_i \, q_j} \tag{5}$$

a statistic that gives a normalized measure of total disequilibrium (Pearson, 1900 was the first to describe this type of measure). Q is approximately chi-square distributed under the null hypothesis of $D = 0$ (see Hill, 1975) with $(k-1)(\ell-1)$ degrees of freedom. As a result, if there is no association between alleles at the loci, then Q divided by the degrees of freedom should equal unity within sampling fluctuations. For samples with $D \neq 0$, this measure is sample-size dependent, therefore, to standardize for different sample sizes that influence the chi-square value, the measure

$$Q^* = \frac{Q}{n(k-1)(\ell-1)} \qquad (6)$$

is appropriate.

Table II gives values of Q^* for several combinations of n, k, ℓ, and $4Nc$ (see Hedrick and Thomson, 1985, for a more complete table with 95% intervals). Conditional Q^* is highly dependent upon the level of recombination and decreases as recombination increases, asymptotically approaching 0.0 as $4Nc$ increases. The Q^* values also decrease with increasing sample size, and decrease with increasing numbers of alleles, particularly when $4Nc$ is small.

Let us compare the total disequilibrium between loci A and B from the three groups of South American Indians and a sample of Hutterites from Canada (Morgan *et al.*, 1980) to these theoretical expectations. Table III gives the sample size, number of alleles at both loci, and the observed Q^* values for these samples. If we compare these values to those in Table II, then the observed disequilibrium suggests that $4Nc$ would be quite small. For example, the Q^* value for the Parak is greater than that for $4Nc = 0$ for the same n and k, ℓ values. Because c between A and B is estimated to be 0.008 (Robson and Lamm, 1983), N would have to be extremely small to account for these high values. In other words, there appears to be much greater total conditional disequilibrium among alleles at these loci than predicted for a sample taken from a neutrality population.

Table II. The expected level of disequilibrium between two loci, Q^*, for given values of n, k, ℓ, and $4Nc$ (Hedrick and Thomson, 1985).

	k,ℓ	0	10	500
			$4Nc$	
n = 100	2,2	0.164	0.045	0.011
	4,4	0.093	0.039	0.011
	6,6	0.070	0.035	0.011
	8,8	0.054	0.032	0.011
n = 200	2,2	0.124	0.037	0.006
	4,4	0.075	0.030	0.006
	6,6	0.056	0.027	0.006
	8,8	0.047	0.024	0.006
n = 400	2,2	0.103	0.023	0.003
	4,4	0.066	0.024	0.004
	6,6	0.049	0.022	0.004
	8,8	0.038	0.020	0.004

Table III. The observed level of total disequilibrium between HLA loci A and B for four different samples (Hedrick and Thomson, 1985).

Sample	k,ℓ	n	Q^*
Parak	4,4	196	0.099
Tirizo	4,5	196	0.076
Waiapi	5,5	234	0.065
Hutterites	7,13	406	0.037

IV. HLA DISEQUILIBRIUM AND MAP DISTANCE

The major factors that generate disequilibrium, namely, selection, genetic drift, gene flow, and genetic hitchhiking, are generally more important when there is tight linkage between the loci (see Hedrick *et al.*, 1978 for a review). However, empirical information on the size of the linkage unit in which these factors are influential is limited. There has been some indication that there is more disequilibrium among tightly linked loci than more loosely linked loci for both electrophoretic variants in *Drosophila melanogaster* (Langley, 1977) and for loci in the HLA region (Hiller *et al.*, 1978; Karlin and Piazza, 1981). Here we will examine data from the 1980 and 1984 Histocompatibility Workshops in an effort to more precisely evaluate the size of the evolutionary map unit in the HLA region (see Klitz *et al.*, 1985a, for a more detailed treatment).

To compare the extent of disequilibrium among different populations and map distances, a modification of Q is useful. To reduce the right skew of the chi-square distributed samples, the logarithm of Q^* is taken (after multiplying by 10^4 to eliminate negative logarithms). In other words, the modification of the overall association between two loci is

$$Q' = \log\left(\frac{10^4 Q}{n(k-1)(\ell-1)} \right) . \qquad (7)$$

In Caucasians, there are three independent estimates of disequilibrium available for the genes in the HLA region. European and North American samples are present in the 1980 workshop and a composite sample of Caucasians from the 1984 workshop. The four HLA loci, *A, B, C* and *DR*, can be arranged in six different locus pairs with recombination distances ranging from 0.001 for *B - C* to 0.018 for *A - DR*. All of these pairs have significant levels of disequilibrium. In contrast, locus pairs of the four HLA loci with *Glo*-1 located four map units away, each have non-significant disequilibrium. When the Q' values of the four HLA loci (all having different sample sizes and numbers of alleles) are arranged according to map distance (Table IV), it is clear that disequilibrium decreases with map distance over our range of points from 1% to 1.8% recombination. Between 1.8% and 4% recombination the disequilibrium drops to zero. Approximately two map units, then,

Table IV. The disequilibrium Q' between pairs of loci for three difference Caucasian samples.

Locus pair	Recombination	8th Workshop		9th Workshop
		European	North American	
B-C	0.001	2.02	2.11	2.23
A-C	0.007	1.37	1.38	1.48
A-B	0.008	1.45	1.44	1.49
B-DR	0.010	1.52	1.48	1.62
C-DR	0.011	1.49	1.59	1.73
A-DR	0.018	1.23	1.35	1.30

effectively characterizes the size of the evolutionary map unit in the HLA region. The possible existence of hot spots of recombination in the HLA region should not affect these results as they will be accounted for in the estimation of map distance.

V. DISEQUILIBRIUM PATTERN ANALYSIS

In the last two sections, we have illustrated that there is more disequilibrium between loci in the HLA region than expected from neutrality and that there is greater disequilibrium for pairs of closely linked loci. However, from these data, we cannot directly determine the cause of these associations. In this section, we will discuss an approach designed to examine the distribution of disequilibrium and identify patterns that are consistent with past selective events (Thomson and Klitz, 1985; Klitz and Thomson, 1985). A strength of this approach is that disequilibrium in multilocus gametic frequencies may be retained for a number of generations, and decays in time only as a function of the recombination rate between two loci.

The disequilibrium values for a pair of loci are constrained by the following relationships

$$\sum_{i=1}^{k} D_{ij} = \sum_{j=1}^{\ell} D_{ij} = 0 . \qquad (8)$$

In other words, the disequilibrium values for a given allele at one locus and all other alleles at the other locus sum to zero so that the negative disequilibrium values for a given allele are exactly balanced by the positive disequilibrium values.

Let us assume that selection favors a particular haplotype $A_1 B_1$ and examine the resulting pattern of disequilibrium values. (The results obtained apply whether the haplotype $A_1 B_1$ increases in frequency via a hitchhiking event or via selection for this haplotype.) The theoretical development is given in Thomson and Klitz (1985) and is an extension to multiple alleles of the hitchhiking models of Thomson *et al.* (1976) and Thomson (1977). If initially all disequilibrium values between the alleles at the two loci of interest are zero, or small (as they will be in the case of newly arisen mutants), then the following simple relationships for the disequilibrium values resulting from the selection event can be given.

First, the positive disequilibrium of the haplotype $A_1 B_1$, which increases in frequency as a result of the selection or hitchhiking event, and the $A_i B_j$ haplotypes, $i = 2,...,k$, $j = 2,...,\ell$, will be balanced by the negative disequilibrium values of the $A_i B_1$, $i = 2,...,k$ and $A_1 B_j$, $j = 2,...,\ell$ haplotypes. Further, the negative disequilibria of the $A_i B_1$ haplotypes, $i = 2,...,k$ will be proportional to the A_i allele frequency, while the normalized disequilibrium values D' for all these haplotypes will be equal when $p_{A_i} + p_{B_1} \leq 1$. For the $A_1 B_j$ haplotypes, $j = 2,...,\ell$, the disequilibria will be proportional to the B_j allele frequency, and the normalized disequilibria values will be equal for all these haplotypes, when $p_{A_1} + p_{B_1} \leq 1$, but in most cases these will be different from the constant normalized disequilibrium value for the $A_1 B_1$ haplotypes.

These features are illustrated in Figures 4 and 5, which give plots of the disequilibria and normalized disequilibria values, from the iterations of a selection model. This example considers three alleles at the A locus, and four alleles at the B locus, giving twelve haplotypes in the population. The model is strictly deterministic in that the effects of genetic drift are not considered. Selection favors one haplotype, $A_1 B_1$, with directional selection giving a 50% selection advantage for the favored haplotype. A large selection coefficient was chosen in order to observe rapid movement of haplotypes in the disequilibrium space but the results are not qualitatively different when other selection values are used.

The increase and then decrease in the value of the disequilibrium for the $A_1 B_1$ haplotype with time relates to the fact that the maximum value that D can take is a function of the allele frequencies. Note that the normalized disequilibrium increases over time. The disequilibria of each of the classes of related haplotypes, that is, $A_1 B_2$, $A_1 B_3$ and $A_1 B_4$, and $A_2 B_1$ and $A_3 B_1$, respectively, form a line in the negative space passing through the point $D = 0$, $pq = 0$ in the D space (Fig. 4) and assume a single D' value (Fig. 5). Unrelated haplotypes ($A_2 B_2$, $A_2 B_3$, $A_2 B_4$ and $A_3 B_2$, $A_3 B_3$, $A_3 B_4$) fall on a line in the D space but form complicated alignments in the D' space because the value of D_{max} varies with the relative allelic frequencies.

Let us compare these expectations to that observed for a large (5,202 individuals) and relatively homogeneous sample for the A and B loci from Denmark (Hansen *et al.*, 1979). Figures 6 and 7 give the observed distribution of disequilibrium and normalized disequilibrium values for allele $A1$ and all alleles at the B locus. Haplotype $A1 B8$, which has the highest positive disequilibrium (and normalized disequilibrium) value in

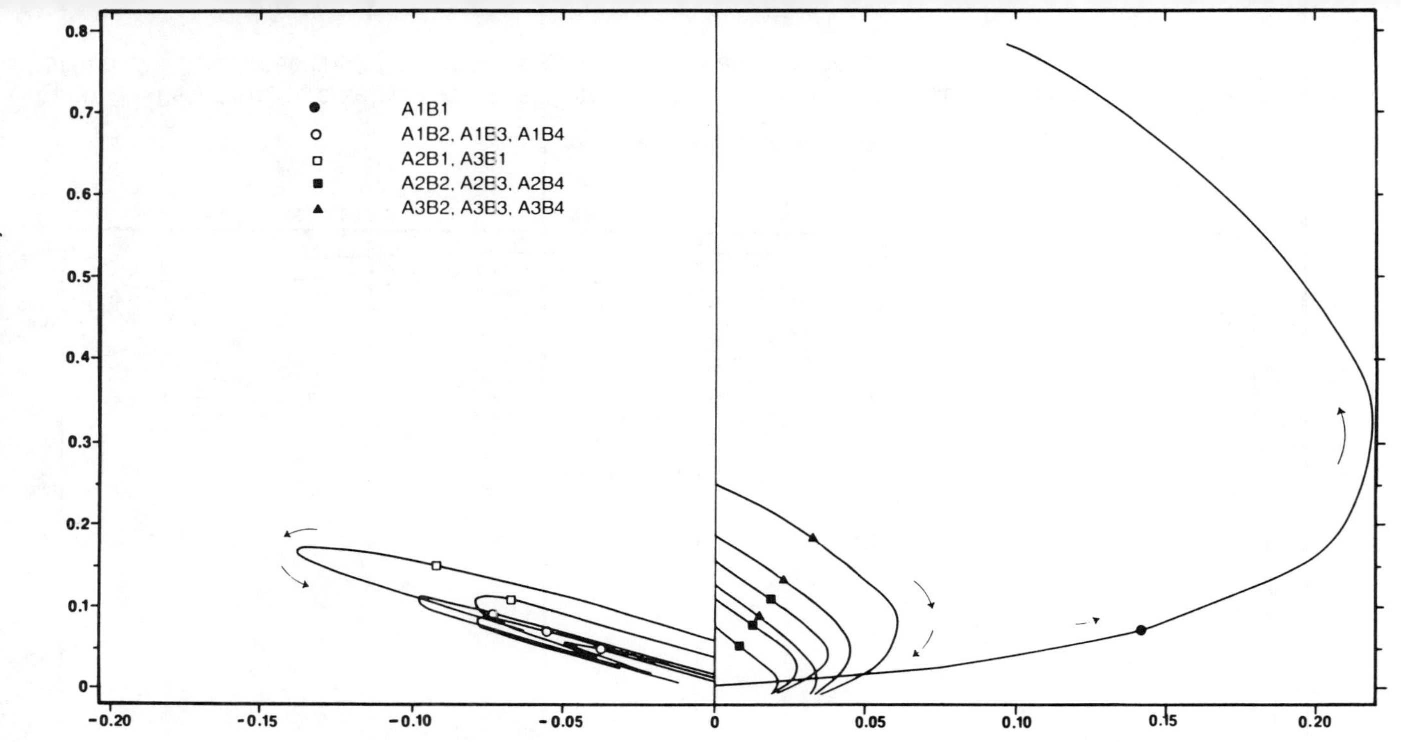

Figure 4. Results of a deterministic two-locus selection model having three alleles at locus A and four alleles at locus B showing the change in D of the twelve haplotypes after 25 generations of selection favoring the A_1B_1 haplotype. The initial generation is in gametic equilibrium and the location of haplotypes at the eighth generation is indicated. The initial frequencies for alleles A_1, A_2 and A_3 are 0.03, 0.37 and 0.60, and for alleles B_1, B_2, B_3 and B_4 are 0.10, 0.20, 0.30 and 0.40, respectively.

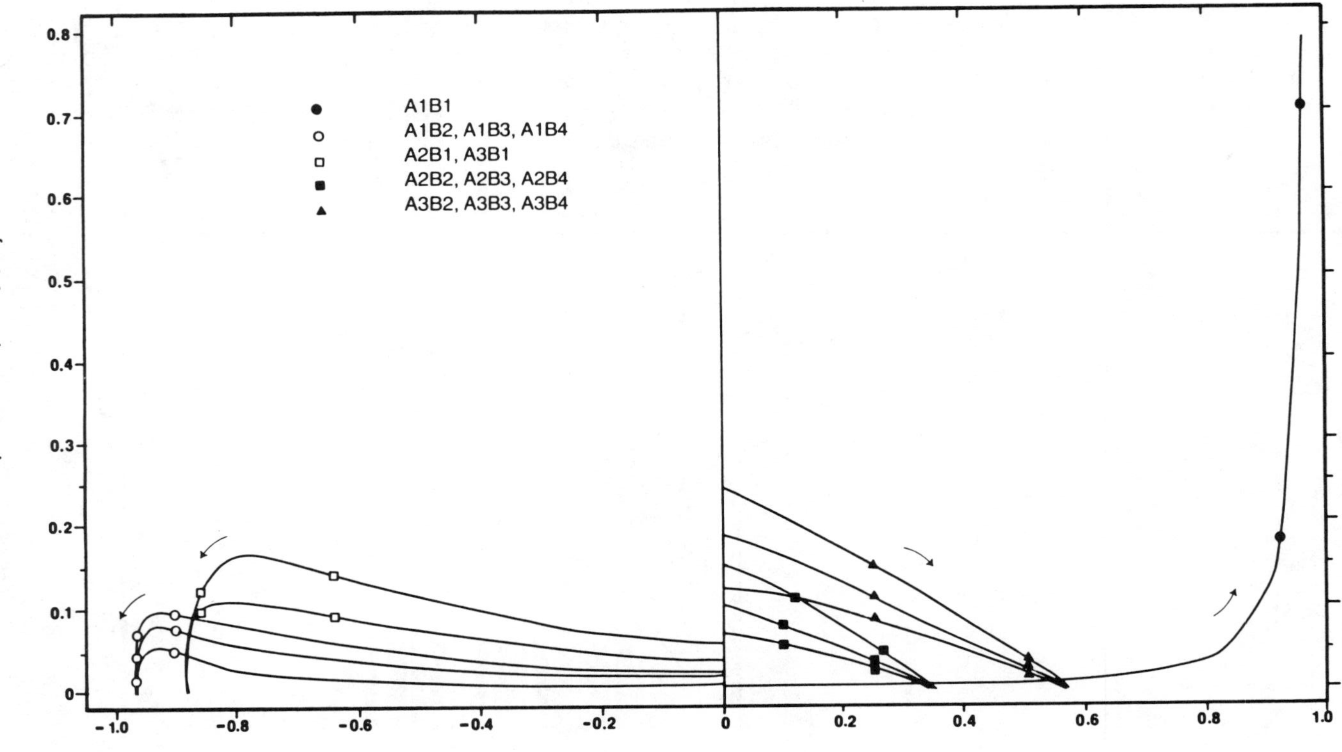

Figure 5. Deterministic selection model identical to that in Fig. 4, plotted against the normalized disequilibrium D'. The locations of haplotypes at the eighth and sixteenth generations are indicated.

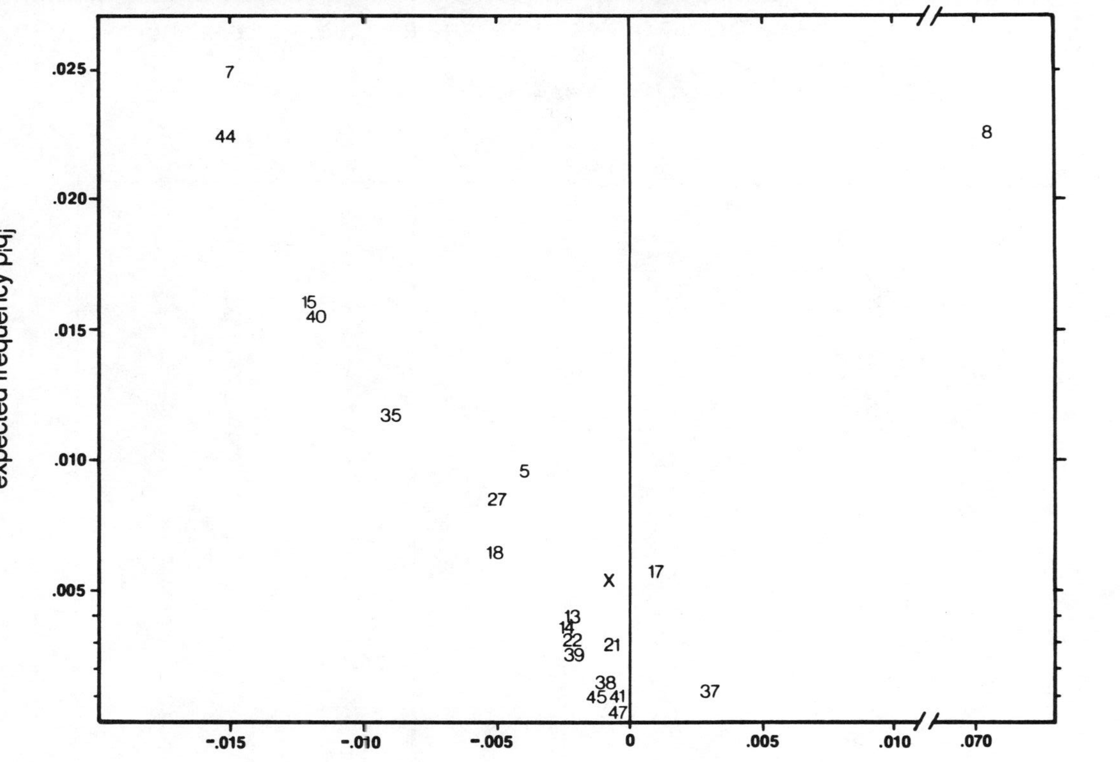

Figure 6. All haplotypes containing the allele *A1* taken from the Danish sample and plotted against *D*. The *B* allele designation is used to indicate the position of each haplotype.

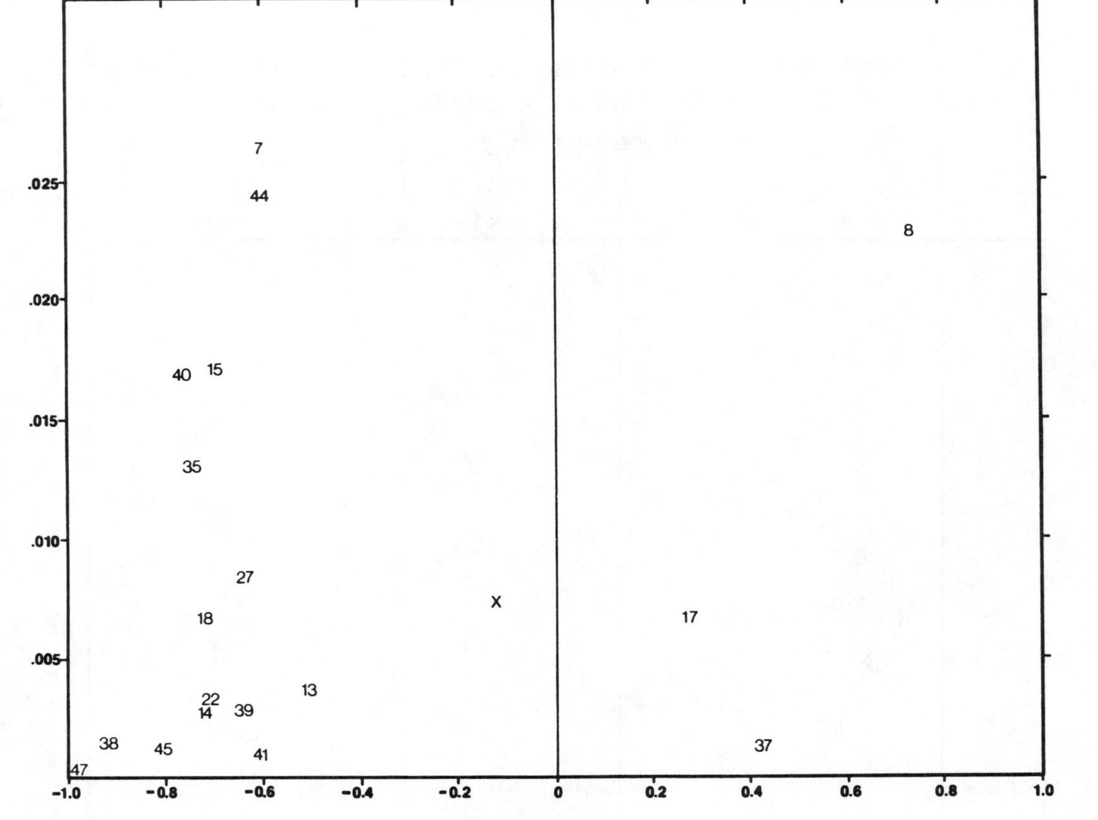

Figure 7. All haplotypes containing the allele A_1 taken from the Danish sample plotted against D' (otherwise like Fig. 6).

the population (D = 0.0766, D' = 0.728) reveals a pattern indicative of selection. In Figure 6, $A1\,B8$ is the main haplotype in the positive space with two low frequency haplotypes having low positive disequilibrium, while the rest of the related haplotypes ($B8$ not $A1$) fall in a linear array in the negative space with disequilibrium values approximately proportional to the frequency of the unshared A allele. The graph of normalized disequilibrium values shown in Figure 7 for $A1$ haplotypes reveals an alignment of the negative values for the commoner haplotypes. The values fall between -0.6 and -0.8. Rarer haplotypes, for example, $A1\,B47$, $A1\,B38$ and $A1\,B13$ depart furthest from this alignment apparently due to sampling effects. The misalignment of $A1\,BX$ probably occurs because the blank allele X is an unidentified mixture of B locus alleles.

There are eight haplotypes that have disequilibrium patterns consistent with a past selective event, the most striking being $A1\,B8$ and $A3\,B7$. Others have disequilibrium distributions completely different from these expectations. The theoretical distribution of disequilibrium from a neutrality population or from admixture is quite different from that expected from a selective event.

VI. CONCLUSIONS

An examination of the pattern and extent of single-locus and two-locus variation in the HLA region strongly suggests that selection has played an important role on HLA variants for the following reasons:

1) The conditional heterozygosity for five loci in the region, HLA-A, B, C, and DR and Glo-1, are significantly greater than neutrality expectations. On the other hand, four complement loci embedded in the region have heterozygosities less than or equal to neutrality expectations.

2) The extent of disequilibrium between loci A and B for samples of South American Indians and Hutterites is larger than neutrality expectations.

3) The disequilibrium between more closely linked loci in the HLA region is higher than for loosely linked loci and suggests that the size of the evolutionary map unit is about two map units.

4) The pattern of disequilibrium for HLA-A and B locus haplotypes is consistent with a selection model. Eight haplotypes have been identified as being favored by selection.

From these observations, it is obvious that evolutionary forces are greatly influencing the amount and pattern of genetic variation in the HLA region. As a result, we feel that HLA should be included along with other classic examples for investigation of the principles of evolutionary genetics.

REFERENCES

Albert, E., Baur, M., and Mayr, W. (eds.) (1984). "Histocompatibility Testing 1984." Springer Verlag, Berlin.

Black, F. L., and Salzano, F. M. (1981). Evidence of heterosis in the HLA system. *Amer. J. Hum. Genetic.* **33**, 894-899.

Dausset, J., and Colombani, P. J. (eds.) (1973). "Histocompatibility Testing 1972." Munksgaard, Copenhage.

Dausset, J., and Svejgaard, A. (eds.) (1977). "HLA and Disease." Munksgaard, Copenhagen.

Ewens, W. J. (1974). The sampling theory of selectively neutral alleles. *Theor. Pop. Biol.* **3**, 87-112.

Hansen, H. E., Larsen, S. E., Ryder, L. P., and Nielsen, L. S. (1979). HLA-A, B haplotype frequencies in 5,202 unrelated Danes by a maximum-likelihood method of gene counting. *Tissue Antigens* **13**, 143-153.

Hedrick, P. W., Jain, S., Holden, L. (1978). Multilocus systems in evolution. *Evol. Biol.* **11**, 101-182.

Hedrick, P. W., and Thomson, G. (1983). A neutrality test for HLA. *Genetics* **104**, 449-456.

Hedrick, P. W., and Thomson, G. (1985). A two locus neutrality test:
Applications to humans, *E. coli,* and lodgepole pine. *Genetics* (in press).

Hill, W. G. (1975). Linkage disequilibrium among multiple neutral alleles produced by mutation in finite population. *Theor. Pop. Biol.* **8,** 117-126.

Hiller, C., Bischoff, M., Schmidt, A., and Bender, K. (1978). Analysis of the HLA-ABC linkage disequilibrium: decreasing strength of gametic association with increasing map distance. *Hum. Genet.* **41,** 301-312.

Hood, L., Steinmetz, M., and Malissen, B. (1983). Genes of the major histocompatibility complex of the mouse. *Ann. Rev. Immunol.* **1,** 529-568.

Hudson, R. R. (1983). Properties of a neutral allele model with intragenic recombination. *Theor. Pop. Biol.* **23,** 183-201.

Karlin, S., and Piazza, A. (1981). Statistical methods for the assessing linkage disequilibrium at the HLA-A, B, C loci. *Ann. Hum. Genet.* **54,** 79-94.

Klein, J., Figueroa, F., and Nagy, Z. A. (1983). Genetics of the major histocompatibility complex: The final act. *Ann. Rev. Immunol.* **1,** 119-142.

Klitz, W., Hedrick, P., and Thomson, G. (1985a). Measuring the unit of selection using the HLA region (manuscript).

Klitz, W., and Thomson, G. (1985). Disequilibrium pattern analysis II. Application to Danish HLA-A and B locus data (manuscript).

Klitz, W., Thomson, G., and Baur, M. P. (1985b). Contrasting selection histories in tightly linked loci (manuscript).

Langley, C. (1977). Nonrandom associations between allozymes in natural populations of *Drosophila melangoaster. In* "Measuring Selection in Natural Populations" (F. B. Christiansen and T. M. Fenchel, eds.), pp. 205-273. Springer-Verlag, Berlin.

Lewontin, R. C. (1964). The interaction of selection and linkage. I. General considerations; heterotic models. *Genetics* **49,** 49-67.

Morgan, K., Holmes, T. M., Schlaut, J., Marchuk, L., Kovithavongs, T., Pazderka, F., and Dossetor, J. B. (1980). Genetic variability of HLA in the Dariusleut Hutterites. A comparative genetic analysis of the Huterrites, the Amish, and other selected Caucasian populations. *Amer. J. Hum. Genet.* **32,** 246-257.

Ohta, T., and Kimura, M. (1969). Linkage disequilibrium due to random genetic drift. *Genet. Res.* **13,** 47-53.

Pearson, K. (1900). On the criterion that a given system of deviations from the probable in the case of a correlated system of variables is such that it can be reasonably supposed to have arisen from random samples. *Philos. Mag.* **50**, 157–175.

Robson, E., B., and Lamm, L. U. (1983). Report of the Committee on chromosome 6. *In* "International Human Gene Mapping Workshop No. 7," Los Angeles.

Ryder, L. P., Svejgaard, A., and Dausset, J. (1981). Genetics of HLA disease association. *Ann. Rev. Genet.* **15**, 169–187.

Ryder, L. P., Anderson, E., and Svejgaard, A. (eds.)(1979). "HLA and Disease Registry, Third Rport." Munksgaard, Copenhagen.

Terasaki, P. I. (ed.)(1980). "Histocompatibility Testing 1980." University of California, Los Angeles.

Thomson, G. (1977). The effect of a selected locus on linked neutral loci. *Genetics* **85**, 753–788.

Thomson, G., Bodmer, W. F., and Bodmer, J. (1976). The HLA system as a model for studying the interaction between selection, migration, and linkage. *In* "Population Genetics and Ecology" (S. Karlin and E. Nevo, eds.), pp. 465–498. Academic Press, New York.

Thomson, G., and Klitz, W. (1985). Disequilibrium pattern analysis. I. Theory (manuscript).

Watterson, G. A. (1978a). An analysis of multia-allelic data. *Genetics* **88**, 171–179.

Watterson, G. A. (1978b). The homozygosity test of neutrality. *Genetics* **88**, 405–417.

GAUSSIAN VERSUS NON-GAUSSIAN GENETIC ANALYSES OF POLYGENIC MUTATION-SELECTION BALANCE[1]

Michael Turelli

Department of Genetics
University of California
Davis, California 95616

ABSTRACT

Like allozyme variation, heritable variation for polygenic traits is maintained by unknown processes. In an influential paper that combined mathematical analysis with a review of relevant data, Lande (1975) proposed that high levels of variation could be maintained by mutation in the face of stabilizing selection. In addition to presenting an appealing hypothesis for the persistence of heritable variation, Lande's paper spawned a new industry of mathematical analyses for the genetic dynamics of polygenic traits. These analyses are based on a model that assumes only additive allelic effects and a continuum of alleles at each locus. In addition, the analyses assume that at each locus the distribution of allelic effects segregating in a population is approximately Gaussian. The mathematical tractability of the resulting *Gaussian genetic model* permits analytical treatment of formidable multilocus problems. Despite its widespread use, relatively little attention has been devoted to the biological and mathematical assumptions underlying this model. Using both mathematical and empirical evidence, I demonstrate that the Gaussian genetic model relies on the unsubstantiated assumption that per locus mutation rates typically exceed 10^{-4}. I present alternative (non-Gaussian) approximations that apply for lower per locus mutation rates

[1] This work was supported by National Institutes of Health Grant GM22221 and National Science Foundation Grant BSR 84-15847.

607

and compare the resulting predictions to those obtained from Gaussian analyses. The primary conclusion is that qualitatively different predictions can arise from conventional analyses that *derive* rather than *assume* the distribution of allele effects. In particular, I argue that it will be very difficult to test the hypothesis that significant levels of heritable variation are maintained by mutation-selection balance.

I. INTRODUCTION

The past ten years have seen a major resurgence of theoretical and experimental work in quantitative genetics. Many experimentalists have turned away from allozymes and towards quantitative characters because their evolutionary importance is more apparent and fitness differences between phenotypes can be readily documented. The papers of Lande beginning in 1975 and 1976 have strongly influenced the modeling of phenotypic evolution and the underlying genetic changes. Both his phenotypic and genotypic models are based on Gaussian distributions, but they describe different phenomena and rely on different biological assumptions.

Lande's analyses of phenotypic evolution (e.g., Lande, 1976, 1979) assume that the distribution of phenotypes in a population is approximately Gaussian. This is a classical, empirically based assumption that traces back through Fisher (1918) to pre-Mendelian biometrical analyses (e.g., Galton, 1889). In Lande's Gaussian phenotypic analyses, the genetic parameters, specifically additive genetic variances and covariances, enter as time-invariant constants. Only population means evolve and the equations for their dynamics follow from standard properties of multivariate Gaussian distributions. Simplifying genetic assumptions (described in Bulmer, 1980) are implicit in the interpretation of the variance parameters as covariances between relatives. (See

Feldman and Cavalli-Sforza, 1979, and Karlin, 1979, for alternative analyses of phenotypic evolution.)

These Gaussian phenotypic analyses are not intrinsically related to the Gaussian analyses for genotypes that Lande (1975, 1980) has also popularized. Although superficially no less plausible than the Gaussian phenotypic analyses, the Gaussian genetic analyses can lead to erroneous conclusions unless restrictive biological assumptions are valid. This paper is intended to: 1) review the assumptions underlying the Gaussian genetic analyses and argue that they are not empirically supported; 2) present alternative simplifying assumptions applicable to polygenic mutation-selection balance; and 3) compare the predictions of Gaussian and non-Gaussian analyses. In addition to reviewing in a relatively nontechnical form the results of Turelli (1984, 1985), I will also present some new data analyses and theoretical results.

II. EXAMINATION OF ASSUMPTIONS UNDERLYING THE GAUSSIAN GENETIC MODEL

What I will call the "Gaussian genetic model" has two components. The first is a set of explicit genetic assumptions originally proposed by Crow and Kimura (1964), then generalized by Lande (1980). The second is a mathematical assumption concerning the statistical distribution of phenotypic effects of alleles segregating in a population. This assumption is based on a mathematical derivation by Kimura (1965) that was generalized by Fleming (1979). Lande (1975), following Latter (1970), spawned a new industry in quantitative genetics theory by converting Kimura's *result* into an *assumption* that greatly simplifies the analysis of multilocus dynamics and equilibria.

In Lande's (1980) generalization that includes pleiotropy, the Gaussian genetic model assumes that a multivariate phenotype $P = (P_1,...,P_k)^T$ is codetermined by genetic and environmental factors according to

$$P = G + E \qquad (2.1)$$

with the distribution of environmental effects, E, independent of G and multivariate Gaussian with mean O and variance-covariance matrix Σ_e. The model assumes diploidy but no dominance or epistasis so that G can be decomposed as

$$G = \sum_{i=1}^{n} (x_{i,\male} + x_{i,\female}), \qquad (2.2)$$

with n the number of loci contributing variance to the characters in P, x_i the effect of an allele at locus i, and the subscripts $\male$ and $\female$ denoting the parent from whom the allele was inherited. A central assumption is that there is a continuum of alleles at each locus, with each allele labeled by its contribution to P, i.e., the x_i in (2.2). A continuum of effects is generated by assuming that each allele at locus i mutates with probability μ_i to a new allele of effect

$$y_i = x_i + M_i \qquad (2.3)$$

with the mutational increment M_i independent of the pre-mutation effect x_i and multivariate Gaussian with mean O and variance-covariance matrix $\Sigma_m^{(i)}$. The variance contributed to character j is denoted $m_{j,i}^2$.

In modeling mutation-selection balance, nonoverlapping generations and Gaussian stabilizing selection are generally assumed. For simplicity, the optimal phenotype may be translated to O so that phe⁻ type P has relative fitness (viability)

$$w(P) = \exp(P^T \Sigma_W^{-1} P) \, . \qquad (2.4)$$

Together with (2.1), this implies that genotypes also experience a Gaussian fitness function, namely

$$w(G) = C \exp(G^T \Sigma_S^{-1} G) \qquad (2.5)$$

with the constant C independent of G and

$$\Sigma_S = \Sigma_W + \Sigma_e \qquad (2.6)$$

(Lande, 1980). As selection on characters j weakens, the jth diagonal element of Σ_S, denoted $V_{s,j}$, increases. These explicit genetic assumptions will be referred to as the "continuum-of-alleles model."

Once the linkage relationships among the loci are assigned, the problem is to describe the genetic dynamics and equilibria produced by recombination, mutation, mating and selection. The continuum of alleles at each locus precludes a standard description of the population in terms of allele frequencies. Instead, the composition of the population at locus i in generation t must be described by a multivariate probability density function $p_{i,t}(x_i)$; similarly gametes are described by a multilocus density $p_t(x_1,...,x_n)$. Even without pleiotropy, it is difficult to describe the dynamics of p_t. No explicit equilibrium is known and only Fleming (1979) has attempted to *derive* an equilirium directly from the multilocus recursions (see Nagylaki (1984) for a simpler exposition). In his original single-character analysis, Kimura (1965) ignored linkage disequilibrium and analyzed each locus separately. He found that under specific assumptions concerning the relative magnitude of the mutation and selection parameters (i.e., μ_i, m_i^2 and V_s), there is an equilibrium for $p_{i,t}(x_i)$ that is approximately Gaussian. This was confirmed by Fleming (1979), who also presented a more accurate non-Gaussian approximation, and the accuracy of both approximations was examined numerically by Turelli (1984).

The central simplification of the Gaussian genetic model is to forego a direct analysis of the dynamics of $p_t(x_1,...,x_n)$. Unlike traditional population genetic analyses that *derive* allele frequencies directly from recursions, the Gaussian genetic model *assumes* that p_t is Gaussian. The dynamics of mutation, meiosis, mating and selection are modeled by deriving recursions for the means, variances and covariances of effects within and between loci. Given the Gaussian assumptions for mutation and selection, plus Kimura's (1965) equilibrium result, this critical assumption may appear reasonable, at least for large t. However, as demonstrated below, it relies on implicit biological assumptions that are difficult to justify.

As noted by Lande (1975) and elaborated by Turelli (1984, 1985), the assumption that p_t is approximately Gaussian is valid only if the variance of mutation effects is small relative to the variance of effects for currently segregating alleles. This must apply to each character considered. The reason is that after mutation, the population is a mixture. A fraction $(1-\mu_i)$ of the alleles at locus i have not mutated and their effects have variance $V_g^{(i)} + m_i^2$, where m_i^2 is the variance of mutation-induced effects in (2.3). Analytical considerations suggest that this mixture will be nearly Gaussian at equilibrium only if

$$m_i^2 \ll V_g^{(i)} \tag{2.7}$$

(Lande, 1975; Fleming, 1979). Numerical analyses in Turelli (1984, 1985) suggest that $3m_i^2 < V_g^{(i)}$ suffices for reasonable agreement; but once $m_i^2 > 3V_g^{(i)}$, the shape of the equilibrium distribution is markedly non-Gaussian, e.g., $\kappa = E(x^4)/3[E(x^2)]^2 > 2$ whereas $\kappa = 1$ for a Gaussian distribution with mean 0.

Data are available to evaluate the plausibility of (2.7). Clayton and Robertson (1955) were the first to quantify mutation rates for polygenic traits by estimating the amount of new heritable variance added per zygote per generation. Using the parameters of the single-character continuum-of-alleles model, this quantity is

$$\sigma_m^2 = 2 \sum_{i=1}^{n} \mu_i m_i^2. \tag{2.8}$$

Data from *Drosophila melanogaster* sternopleural and abdominal bristles, nine characters in maize, and several in mice all yield estimates on the order of

$$\sigma_m^2 / V_e \cong 10^{-3} \tag{2.9}$$

or larger, with V_e denoting environmental variance (Lande, 1975; Hill, 1982). Because each of these characters has a moderate heritability in the environments used (i.e., $.25 < h^2 < .75$), (2.9) can be translated into

$$\sigma_m^2 \cong 10^{-3} V_g , \tag{2.10}$$

with

$$V_g \cong 2 \sum_{i=1}^{n} V_g^{(i)}$$

assuming low levels of linkage disequilibrium.

To convert (2.10) into a statement about m_i^2 and $V_g^{(i)}$, note that both sides involve summation over loci. If we assume that $V_g^{(i)}$ is proportional to $\mu_i m_i^2$, we obtain

$$10^3 \mu_i m_i^2 \cong V_g^{(i)} . \tag{2.11}$$

Thus, if traditional per locus mutation rates apply, i.e., $\mu_i \leq 10^{-5}$, or even if $\mu_i \leq 10^{-4}$, we expect

$$m_i^2 > 10 V_g^{(i)} , \qquad\qquad (2.12)$$

which contradicts assumption (2.7) underlying the Gaussian approximation. *This suggests that the Gaussian genetic model will only be accurate if the per locus mutation rates relevant to quantitative traits are much higher than available data indicate* (see Turelli (1984, Sec. 5) for an elaboration).

Two new estimates of σ_m^2 can be obtained from published data on *D. melanogaster*. In an experiment aimed at describing selection on loci contributing to variation in sternopleural bristle number, Linney *et al.* (1971) mated full-sibs for 42 generations. Their data on "within line heterozygosity" can be analyzed by the techniques of Clayton and Robertson (1955) to yield

$$\sigma_m^2 \cong 4.8 \times 10^{-3} V_g , \qquad\qquad (2.13)$$

where V_g is the amount of genetic variation in their initial outbred population. This is comparable to, but slightly higher than, the estimates summarized by Hill (1982). Mukai *et al.* (1984) reported mutation accumulation experiments for ADH activity. After 300 generations in balancer stocks, replicates of two independent second chromosomes yielded 0.709 and 0.731 as estimates for V_g/V_e. Making the conservative assumptions that only second chromosome loci contribute to ADH activity variation and that V_g/V_e in generation 300 estimates 300 σ_m^2, we obtain

$$\sigma_m^2 \cong 2.5 \times 10^{-3} V_e . \qquad\qquad (2.14)$$

Like previous estimates of σ_m^2, (2.13) and (2.14) support (2.11).

In addition to the implicit assumption of very high per locus mutation rates, the Gaussian genetic model makes the explicit assumption that each locus can produce an unbounded continuum of effects on every character it influences. Unless the variance–covariance matrix for mutation effects in (2.3) is singular, this model assumes that locus i can contribute any effect to each character it influences, irrespective of its effects on other characters. This level of genetic flexibility is a strong assumption. As shown in Turelli (1985), it requires over 100 alleles at a locus to accurately mimic a two-character continuum of alleles at mutation-selection equilibrium. If less genetic flexibility is assumed, dramatically different quantitative predictions can result.

III. ALTERNATIVE SIMPLIFYING ASSUMPTIONS

The attraction of the Gaussian genetic model is that it drastically simplifies analysis of the multilocus genetics underlying quantitative traits. By extending Latter's (1970) Gaussian analysis for individual loci to gametes, Lande (1975) popularized the Gaussian treatment of linkage disequilibrium via covariance of allele effects at different loci. Previous analyses of mutation-selection balance (Latter, 1960; Kimura, 1965; Bulmer, 1975) had ignored linkage disequilibrium. They assumed that each locus could be analyzed in isolation and that linkage equilibrium would be closely approximated so that

$$p(x_1,...,x_n) \cong \prod_{i=1}^{n} p_i(x_i) \qquad (3.1a)$$

and

$$\hat{V}_g \cong 2 \sum_{i=1}^{n} \hat{V}_g^{(i)} . \qquad (3.1b)$$

This simplifying approximation, when supplemented with numerical analysis, remains valuable and is frequently quite accurate as documented by extensive numerical calculations in Turelli (1984). In contrast, the usefulness of the Gaussian treatment of linkage disequilibrium is brought into question by two observations. The first is that for typical parameter values, Lande's (1975) multilocus Gaussian prediction for equilibrium genetic variance differs only trivially from one obtained by analyzing each locus separately then using (3.1) (Turelli, 1984, Sec. 3). The second is that Lande's multilocus prediction is slightly *less* accurate than Kimura's (1965) prediction based on (3.1) (Nagylaki, 1984; Turelli, 1984, Sec. 4).

When analyzing equilibria under mutation-selection balance, the empirically based inequality

$$m_i^2 \gg V_g^{(i)} \tag{3.2}$$

leads to a major simplification. The mutation-selection equilibrium can then be approximated by assuming that the phenotypic effect of a new mutant is essentially independent of its effect before mutation, i.e., (2.3) can be approximated by

$$y_i = \bar{x}_i + M_i \tag{3.3}$$

where $\bar{x}_i$ denotes the average effect of alleles segregating at locus i. This approximation works because (3.2) ensures that the magnitude of $x_i - \bar{x}_i$ is likely to be much smaller than the magnitude of $\bar{x}_i + M_i$. It is analogous to the classical strong selection, low mutation rate approximation for mutation-selection balance in which essentially all deleterious mutants arise from "wild type" alleles (cf. Haldane, 1927). Approximation (3.3) is called the house-of-cards approximation and is analyzed in detail in Turelli (1984, 1985). It is generally quite accurate for low per locus mutation rates, unless selection is extremely weak. Although the approximation is named after a model proposed by Kingman (1978), it is not being proposed as an alternative model for polygenic mutation. Instead it is an *alternative approximation* for the same continuum-of-alleles genetic model to which Lande (1975, 1980) has applied the Gaussian approximation.

IV. COMPARISON OF GAUSSIAN AND NON-GAUSSIAN GENETIC PREDICTIONS

I will compare the predictions from alternative approximations for the continuum-of-alleles model described in Section II and discuss the implications of assuming only a finite number of alleles rather than a continuum. All of the models assume Gaussian stabilizing selection and assign allele effects so that $\hat{V}_g = 0$ in the absence of mutation. Rather than repeat the analyses in Turelli (1984, 1985), I will simply summarize the models, methods of analyses and results. Extensions of earlier results appear in B and C below.

A. Single Character with Random Mating

Latter (1960) presented the first complete analysis of polygenic mutation-selection balance. Instead of a continuum of alleles, he assumed only two alleles per locus, with additive effects $- c/2$ and $c/2$, and equal forward and back mutation rates. His analysis treated each locus in isolation and relied on (3.1) to produce an n-locus prediction which was independently derived by Bulmer (1972). Both concluded that

$$\hat{V}_g \cong 4UV_s \tag{4.1}$$

with $U = \sum_{i=1}^{n} \mu_i$ and V_s a measure of the intensity of stabilizing selection [see (2.5)]. Relatively low per locus mutation rates are implicit in their analyses so that $\hat{V}_g(i) \ll c^2$. This inequality is analogous to (3.2) which motivates the house-of-cards approximation. When it holds, (4.1) is quite accurate, at least for identical loci, as long as recombination rates are not extremely low (Turelli, 1984). Note that according to (4.1), $\hat{V}_g$ is independent of the parameter c, which describes the phenotypic effects of mutation, and it depends on the number of loci and their mutation properties only through U.

Apparently unaware of Latter's (1960) work, Kimura (1965) addressed the same question but assumed a continuum of alleles at each locus. Unlike Latter (1960) and later workers, Kimura used a continuous time selection model and assumed a quadratic optimum rather than Gaussian fitness function. Fleming (1979) showed, however, that the same answer emerges with discrete generations and Gaussian selection. Like Latter, Kimura relied on (3.1) for n-locus predictions. He found that if the product $\mu_i m_i^2$ is small for each locus and comparable in magnitude to V_s^{-1}, the equilibrium distribution of allele effects at each locus is approximately Gaussian and

$$\hat{V}_g \cong \sqrt{2V_s}\ \sqrt{n_E \sigma_m^2} \tag{4.2}$$

with

$$\sigma_m^2 = 2 \sum_{i=1}^{n} \mu_i m_i^2 \quad \text{and} \quad n_E = 2\left(\sum_{i=1}^{n} \sqrt{\mu_i m_i^2} \right)^2 / \sigma_m^2 .$$

The effective number of loci, n_E, was defined by Lande (1975) and satisfies $n_E \leq n$ with equality only if $\mu_i m_i^2$ is independent of i. Kimura's approximation for $\hat{V}_g$ differs from (4.1) in three important respects: 1) It is proportional to $\sqrt{V_s}$ instead of V_s; 2) it is proportional to the m_i, which are analogous to c in the diallelic model, whereas (4.2) is independent of c; and 3) for fixed σ_m^2, (4.2) is proportional to $\sqrt{n_E}$ whereas (4.2) is independent of the number of loci for fixed U. As noted by Crow (pers. comm.), the last property reflects the fact that the genetic load per locus is proportional to μ_i in the diallelic model (as in the classic mutation load theory reviewed by Crow and Simmons, 1983) but proportional to $\sqrt{\mu_i}$ under the Gaussian approximation for a continuum of alleles.

Fleming (1979) obtained (4.2) from the leading term in an asymptotic expansion for the equilibrium of an n-locus recursion assuming discrete generations and Gaussian selection. His second order, non-Gaussian expansion produced a correction term for (4.2) that reduces $\hat{V}_g$ for realistic parameter values. His refined approximation for $\hat{V}_g$ is more accurate than (4.2) for moderate mutation rates but becomes negative as per locus mutation rates decrease (see Turelli (1984, Sec. 4) for numerical comparisons). This degeneration is reasonable because relatively high per locus mutation rates are assumed in Fleming's derivation as they are in Kimura's (1965).

Lande's (1975) analyses also attempts to account for linkage effects but it ignores departures from normality. It produces

$$\hat{V}_g \stackrel{\sim}{=} \sqrt{2V_s + n_E\sigma_m^2} \; \sqrt{n_E\,\sigma_m^2} + n_E\sigma_m^2 \qquad (4.3)$$

with σ_m^2 and n_E as in (4.2). Because $n_E\,\sigma_m^2$ is estimated as 0.1 or less, whereas V_s is estimated as 10 or larger (Lande, 1975), the numerical difference between the Kimura (4.2) and Lande (4.3) approximations is generally negligible.

The discrepancy between the two-allele prediction (4.1) and continuum-of-allele predictions (4.2) and (4.3) was noted by Bulmer (1980). Its "obvious" source is the explicit difference in the number of alleles assumed. However, Turelli (1984) demonstrated that the discrepancy is actually attributable to implicit assumptions concerning the relative magnitude of the mutation and selection parameters.

Specifically, the Kimura (1965), Lande (1975) and Fleming (1979) analyses all assume high per locus mutation rates, e.g., $\mu > 10^{-4}$, because they rely on (2.7). Among them, Fleming's non-Gaussian analysis is the most accurate. If the continuum-of-alleles model is reanalyzed for low per locus mutation rates using the house-of-cards approximation, prediction (4.1) again emerges for $\hat{V}_g$. Thus the critical biological assumption is the relative magnitude of mutation effects and equilibrium genetic variance, i.e., (2.7) versus (3.2), rather than the number of alleles per locus. This conclusion is supported by analytical and numerical analyses of a triallelic model (Turelli, 1984). My low mutation rate analyses rely on the single-locus approach used by Latter (1960), Kimura (1965) and Bulmer (1972). As shown numerically in Turelli (1984), their accuracy is impaired by linkage only if it is extremely tight.

B. Single Character with Inbreeding

A remarkable prediction of the Gaussian genetic model is that $\hat{V}_g$ is independent of the system of mating (Lande, 1977). Lande (1977) treated inbreeding by assuming a fixed inbreeding coefficient f which is interpreted as "a correlation between the effects of alleles at the same locus in uniting gametes." It is easy to show that the Gaussian prediction is violated by a diallelic model with relatively low per locus mutation rates. If inbreeding is approximated by a constant departure from Hardy-Weinberg proportions, a simple extension of the single-locus analysis that produces (4.1) leads to

$$\hat{V}_g(f) = 4UV_s(1+f)/(1+3f) \tag{4.4}$$

with U as in (4.1). Thus as f increases towards one, $\hat{V}_g$ decreases toward half the value maintained in a random mating population. This effect has been verified by six-locus numerical calculations analagous to those presented in Turelli (1984). Thus the invariance of $\hat{V}_g$ predicted by the Gaussian genetic model may well be an artifact of assuming high per locus mutation rates.

C. Multiple Characters and Pleiotropy

Lande (1980) introduced a multicharacter extension of the Crow and Kimura (1964) continuum-of-alleles model to describe the interaction of mutation, selection and pleiotropy. Using the Gaussian approximation for allele effects, he determined conditions under which linkage disequilibrium would be negligible. With this simplification, he presented a single-locus approximation that generalizes Kimura's (1965). If locus i affects only one selected character, Kimura (1965) predicts

$$\hat{V}_g(i) \cong \sqrt{\mu_i m_i^2 V_s} \ . \tag{4.5}$$

If locus i affects several selected characters, Lande (1980) predicts

$$\hat{\Sigma}_g(i) \cong \Sigma_s^{1/2} \left(\mu_i \Sigma_s^{-1/2} \Sigma_m(i) \Sigma_s^{-1/2} \right)^{1/2} \Sigma_s^{1/2} \tag{4.6}$$

with positive semidefinite matrix square roots throughout. In (4.6), $\hat{\Sigma}_g(i)$, Σ_s and $\Sigma_m(i)$ denote the variance-covariance matrices for the equilibrium genetic effects at locus i, the multivariate Gaussian fitness function for genotypes (2.5), and the Gaussian distribution of pleiotropic mutation effects described by (2.3).

Analysis of a special case reveals a general qualitative difference between the Gaussian (high mutation rate) prediction (4.6) and non-Gaussian predictions basd on lower per locus mutation rates. According to (4.6), if Σ_s and $\Sigma_m(i)$ are diagonal matrices, then (4.5) applies for each character with m_i^2 and V_s given by the appropriate diagonal elements from $\Sigma_m(i)$ and Σ_s . In particular, for character one,

$$\hat{V}_{g,1}{}^{(i)} \cong \sqrt{\mu_i m^2_{1,i} V_{s,1}} \ . \tag{4.7}$$

Thus, without correlation in the selection or mutation schemes, the equilibrium variance contributed to character one by a pleiotropic locus is independent of the effects of that locus on other characters, even if they are also under selection.

Turelli (1985) used the house-of-cards approximation to reanalyze this model for two pleiotropically connected characters. No compact form was found for the general solution. However, in the simple case without correlation in selection or mutation, the equilibrium variance contributed to P_1 by locus i is approximately

$$\hat{V}_{g,1}{}^{(i)} \cong 2\mu_i V_{s,1} \ / \ (1+\beta_i) \tag{4.8}$$

with

$$\beta_i{}^2 = s_{2,i} \ / \ s_{1,i} \quad \text{and} \quad s_{j,i} = m^2_{j,i} \ / \ V_{s,j} \ . \tag{4.9}$$

In contrast to the Gaussian prediction (4.7), (4.8) predicts that the presence of selection on the pleiotropically connected character P_2 reduces $\hat{V}_{g,1}{}^{(i)}$ even through the mutation effects are uncorrelated and the selection pressures independent. The parameter β_i that quantifies this reduction has a simple interpretation. When an allele of optimal effect at locus i mutates, fitness is reduced in proportion to $s_{j,i}$ by selection on character j. Thus, under the house-of-cards approximation, in which most mutants arise from nearly optimal alleles, $\beta_i{}^2$ measures the relative intensity of selection on new mutants at locus i which comes from their effect on P_2 versus their effect on P_1. This description also clarifies the cause of the pleiotropy-induced reduction of $\hat{V}_{g,1}{}^{(i)}$. Under the house-of-cards assumptions, new mutants experience stabilizing selection associated with all the characters they affect, whether or not those characters, or the mutation effects, are correlated. Although this extreme situation with no correlation is biologically unlikely, it clearly illustrates a qualitative discrepancy between the Gaussian and non-Gaussian predictions that persists with correlation but becomes less obvious.

When the house-of-cards analysis is extended to three pleiotropically connected characters, then even without correlation in selection and mutation, $\hat{V}_{g,1}^{(i)}$ can in general only be expressed in terms of special functions known as elliptic integrals. However, if we also assume that selection acts equally on P_2 and P_3, i.e.,

$$s_{2,1} = s_{3,1} \qquad (4.10)$$

with $s_{j,1} = m^2_{j,i} / V_{s,j}$ as in (4.9), then the approximations illustrated in Appendix 2 of Turelli (1985) yield

$$V_{g,1}^{(1)} = 2\mu_1 V_{s,1}\{1-\beta_1(1-\beta_1^2)^{-1/2} \tan^{-1}[(\beta_1^{-2}-1)^{1/2}]\}/(1-\beta_1^2)$$
$$\text{if } \beta_1 < 1 \qquad (4.11a)$$

$$V_{g,1}^{(i)} \cong 2\mu_1 V_{s,1} / 3 \qquad \text{if } \beta_1 = 1 \qquad (4.11b)$$

and

$$\hat{V}_{g,1}^{(1)} \cong 2\mu_1 V_{s,1}\{\beta_1(\beta_1^2-1)^{-1/2} \tanh^{-1}[(1-\beta_1^{-2})^{1/2}]-1\}/(\beta_1^2-1)$$
$$\text{if } \beta_i > 1, \quad (4.11c)$$

with $\beta_1^2 = s_{2,1}/s_{1,1}$ as in (4.9). These expressions allow us to determine the effect on $\hat{V}_{g,1}^{(1)}$ of distributing a fixed amount of pleiotropic selection over two characters, P_2 and P_3, instead of just one. If the pleiotropic selection $s \equiv s_{2,1}$ in (4.9) is equally distributed over two characters, i.e., set $s_{2,1} \equiv s_{3,1} = s/2$, then it can be shown from (4.8) and (4.11) that $\hat{V}_{g,1}^{(1)}$ is reduced and the magnitude of the reduction increases with s. This suggests that the pleiotropy-induced reduction of $\hat{V}_{g,1}^{(1)}$

predicted by the two-character analysis may be conservative and further supports the conclusions reached in Section IVD below.

A major result from the low mutation rate analyses for a single character was that the same $\hat{V}_g$ prediction emerged for two-allele, three-allele and continuum-of-alleles models. Unfortunately, this robustness is destroyed by pleiotropy. Using a two-character, five-allele extension of the three-allele model in Turelli (1984), I have shown (Turelli, 1985) that without correlation

$$\hat{V}_{g,1}{}^{(i)} \stackrel{\sim}{=} 2\mu_1 V_{s,1} / (1 + \beta_1{}^2) \tag{4.12}$$

with $\beta_1{}^2$ as in (4.9) but the variance parameter $m_{j,1}$ replaced by a fixed allele effect parameter analogous to the constant c in Latter's (1960) diallelic model. Unlike the two-character continuum-of-alleles model, this five-allele model constrains the relative effects of mutation on the two characters. This deterministic coupling accounts for the difference between (4.8) and (4.12). Because over 100 alleles are needed at each locus to accurately approximate a two-dimensional continuum (Turelli, 1985), deterministic constraints, such as those in the five-allele model, are likely to characterize pleiotropic loci. Rose (1982) has shown that particular patterns of constraints, termed "antagonistic pleiotropy," can produce overdominant selection and thus preserve variation in the absence of mutation.

D. Hidden Pleiotropic Effects

All of the models above make the simplifying assumption that selection acts on the relevant loci only through the character(s) under study. A more realistic analysis must recognize that these loci will generally affect other, unknown characters which are also under selection. Turelli (1985) analyzed the mathematical and biological consequences of these hidden pleiotropic effects. An idealized model was considered in which each locus affects two characters, P_1 and P_2. P_1 is the character under study, and P_2 represents the unknown effects of the loci contributing variation to P_1. This model was analyzed from two perspectives. The first assumes that all mutation and selection parameters relevant to P_1 and P_2 are known so that $\hat{V}_{g,1}$ can be predicted from a complete bivariate analysis. The second mimics empirically obtainable predictions by assuming that only P_1 is observed so that $\hat{V}_{g,1}$ must be predicted by an extrapolation, denoted $\tilde{V}_{g,1}$.

Obviously hidden effects will diminish the accuracy of $\tilde{V}_{g,1}$. What is not obvious is how the relative magnitudes of $\tilde{V}_{g,1}$ and $\hat{V}_{g,1}$ depend on the genetic model and analysis used. Surprisingly, $\hat{V}_{g,1}$ and $\tilde{V}_{g,1}$ are relatively close to one another for a wide range of parameters under the Gaussian genetic model. In contrast, $\tilde{V}_{g,1}$ tends to overestimate $\hat{V}_{g,1}$ when the house-of-cards analysis applies, and this effect is accentuated by assuming only five alleles per locus rather than a continuum. The cause of the overestimation is illustrated by (4.8) and (4.12) which show that selection on a hidden character can decrease $\hat{V}_{g,1}$ without creating phenotypic correlations that would influence $\tilde{V}_{g,1}$. However, $\tilde{V}_{g,1}$ may also underestimate $\hat{V}_{g,1}$, even under the house-of-cards analysis, so that no simple relationship exists between empirical extrapolations and actual equilibria. This uncertainty makes it extremely difficult to test Lande's (1975) hypothesis that much of the additive variance observed for quantitative characters can be explained by mutation-selection balance.

V. CONCLUSIONS

Despite its widespread use, relatively little attention has been paid to the biological and mathematical assumptions underlying the Gaussian genetic model. I expect that most biologists will find those assumptions surprising and implausible. As indicated in Section II, the Gaussian genetic model is likely to be accurate only if per locus mutation rates are much higher than available data suggests. In principle, long-term experiments could distinguish between the alternative predictions for mutation-selection equilibria without identifying parameters for individual loci. As shown in Section IVA, the high-mutation-rate (Gaussian) and low-mutation-rate (house-of-cards) analyses predict qualitatively different dependence of $\hat{V}_g$ on selection intensity and mutation rates. This discrepancy is preserved when pleiotropy is introduced. Thus if empirical equilibria could be obtained under different selection or mutation regimes, the competing hypotheses could be tested. Unfortunately, the time scale for reaching mutation-selection equilibria is generally quite long.

Given the lack of empirical support for the assumptions underlying the Gaussian genetic model, its predictions should be viewed with some skepticism. Some investigators, notably Felsenstein (1979) and Kirkpatrick (1982), have used numerical analyses to show that their Gaussian-based predictions can be recovered from standard genetic recursions. Most have not. It remains a challenge to develop polygenic models that are both analytically tractable and biologically believable.

ACKNOWLEGMENTS

I thank A. A. Hoffmann, S. Karlin and J. B. Walsh for their comments on an earlier draft.

REFERENCES

Bulmer, M. G. (1972). The genetic variability of polygenic characters under optimizing selection, mutation and drift. *Genet. Res.* **19**, 17-25.

Bulmer, M. G. (1980). "The Mathematical Theory of Quantitative Genetics." Clarendon Press, Oxford.

Clayton, G. A., and Robertson, A. (1955). Mutation and quantitative variation. *Am. Natur.* **89**, 151-158.

Crow, J. F., and Kimura, M. (1964). The theory of genetic loads. *Proc. XI Int. Congr. Genet.* **2**, 495-505.

Crow, J. F., and Simmons, M. J. (1983). The mutation load in *Drosophila*. *In* "The Genetics and Biology of Drosophila," Vol. 3C (M. Ashburner, H. L. Carson, and J. N. Thompson, eds.), pp. 1-35. Academic Press, London.

Feldman, M. W., and Cavalli-Sforza, L. L. (1979). Aspects of variance and covariance analysis with cultural inheritance. *Theor. Pop. Biol.* **15**, 276-307.

Felsenstein, J. (1979). Excursions along the interface between disruptive and stabilizing selection. *Genetics* **93**, 773-795.

Fisher, R. A. (1918). The correlation between relatives on the supposition of Mendelian inheritance. *R. Soc. (Edinburgh) Trans.* **52**, 339-433.

Fleming, W. H. (1979). Equilibrium distributions of continuous polygenic traits. *SIAM J. Appl. Math.* **36**, 148-168.

Galton, F. (1889). "Natural Inheritance." Macmillan, London.

Haldane, J. B. S. (1927). A mathematical theory of natural and artificial selection. Part V. Selection and mutation. *Proc. Camb. Phil. Sco.* **23**, 838-844.

Hill, W. G. (1982). Predictions of response to artificial selection from new mutations. *Genet. Res.* **40**, 255-278.

Karlin, S. (1979). Models of multifactorial inheritance: I. Multivariate formulations and basic convergence results. *Theor. Pop. Biol.* **15**, 308-355.

Kimura, M. (1965). A stochastic model concerning the maintenance of genetic variability in quantitative characters. *Proc. Nat. Acad. Sci. USA* **54**, 731-736.

Kingman, J. F. C. (1978). A simple model for the balance between selection and mutation. *J. Appl. Prob.* **15**, 1-12.

Kirkpatrick, M. (1982). Quantum evolution and punctuated equilibria in continuous genetic characters. *Am. Nat.* **119**, 833-848.

Lande, R. (1975). The maintenance of genetic variability by mutation in a polygenic character with linked loci. *Genet. Res.* **26**, 221-235.

Lande, R. (1976). Natural selection and random genetic drift in phenotypic evolution. *Evolution* **30**, 314-334.

Lande, R. (1977). The influence of the mating system on the maintenance of genetic variability in quantitative characters. *Genetics* **86**, 485-498.

Lande, R. (1979). Quantitative genetic analysis of multivariate evolution, applied to brain: body size allometry. *Evolution* **34**, 402-416.

Lande, R. (1980). The genetic covariance between characters maintained by pleiotropic mutations. *Genetics* **94**, 203-215.

Latter, B. D. H. (1960). Natural selection for an intermediate optimum. *Aust. J. Biol. Sci.* **13**, 30-35.

Latter, B. D. H. (1970). Selection in finite populations with multiple alleles. II. Centripetal selection, mutation, and isoallelic variation. *Genetics* **66**, 165-186.

Linney, R., Barnes, B. W., and Kearsey, M. J. (1971). Variation for metrical characters in *Drosophila* populations. III. The nature of selection. *Heredity* **27**, 163-174.

Mukai, T., Harada, K., and Yoshimaru, H. (1984). Spontaneous mutations modifying the activity of alcohol dehydrogenase (ADH) in *Drosophila melanogaster*. *Genetics* **106**, 73-84.

Nagylaki, T. (1974). Selection on a quantitative character. *In* "Human Population Genetics: The Pittsburgh Symposium" (A. Chakravarti, eds.), pp. 275-306. Van Nostrand Reinhold, New York.

Rose, M. R. (1982). Antagonistic pleiotropy, dominance, and genetic variation. *Heredity* **48**, 63-78.

Turelli, M. (1984). Heritable genetic variation via mutation-selection balance: Lerch's zeta meets the abdominal bristle. *Theor. Pop. Biol.* **25**, 138-193.

Turelli, M. (1985). Effects of pleiotropy on predictions concerning mutation-selection balance for polygenic traits. *Genetics* **111**, 165-195.

THE GAUSSIAN APPROXIMATION FOR RANDOM GENETIC DRIFT[1]

Thomas Nagylaki

Department of Molecular Genetics and Cell Biology
The University of Chicago
920 East 58th Street
Chicago, Illinois 60637

ABSTRACT

A Gaussian approximation is established for the evolution of the gene frequencies at a multiallelic locus under selection, mutation, and random genetic drift. Generations are discrete and nonoverlapping; the diploid, monoecious population mates at random. All evolutionary forces are weak, and the deterministic ones are much stronger than random drift. Explicit formulae are presented for the variance of the Gaussian fluctuations about the deterministic trajectory in three diallelic special cases: pure mutation, pure selection without dominance, and mutation and overdominant selection at equilibrium.

I. INTRODUCTION

Feller (1951) described three qualitatively different behaviors of the

Wright-Fisher model in the limit of large population number. For two

alleles without selection, he showed that (i) if mutation is much weaker

than random drift, the limiting diffusion is the one for pure random drift;

---
[1] Supported by National Science Foundation Grant DEB81-03530.

(ii) if mutation and random drift are comparable, the classical diffusion of Fisher and Wright applies; and (iii) if mutation is much stronger than random drift, there are small, Gaussian fluctuations around the deterministic equilibrium. Karlin and McGregor (1964) explored in detail the possible limiting diffusions of the haploid, neutral, diallelic Moran (1958a) model with overlapping generations. Norman (1968, 1972, 1974, 1975a,b) has shown very generally that if deterministic forces dominate random ones, a stochastic process exhibits small, Gaussian fluctuations around a deterministic trajectory, and he has established this Gaussian approximation for the multiallelic Wright-Fisher model and for a diallelic model of a dioecious population with discrete, nonoverlapping generations (Moran, 1958b). Thus, not only has Norman treated more general models more rigorously than has Feller (1951), but he has also extended Feller's approximation from the neighborhood of the equilibrium to the entire trajectory.

No convincing biological rationale appears to have been advanced for the Wright-Fisher model. In his model with discrete, nonoverlapping generations, Moran (1958b) introduced selection by assigning different constant fertilities to the three genotypes. If matings occur, however, even in an infinite population this formulation requires multiplicative fertilities (Penrose, 1949; Bodmer, 1965; Nagylaki, 1977, pp. 51–55); the necessary detailed biological assumptions in a finite population are unknown. In this paper, we shall use Norman's general theorems to deduce the Gaussian approximation for the monoecious multinomial-sampling model of Ethier and Nagylaki (1980). We shall obtain explicit formulae for the variance of the Gaussian fluctuations in three diallelic special cases: pure mutation, pure selection without dominance, and mutation and overdominant selection at equilibrium.

II. THE MODEL

The life cycle starts with N monoecious adults. We focus attention on a single locus with r alleles and denote the frequency of the unordered genotype A_iA_j, $i \leq j$, just before reproduction by P_{ij}. It is essential to use *unordered* frequencies as the basic variables because random variation would destroy the symmetry of the ordered frequencies. The frequency of A_i at this stage reads

$$p_i = P_{ii} + \frac{1}{2} \sum_{j:j>i} P_{ij} + \frac{1}{2} \sum_{j:j<i} P_{ji} . \tag{1}$$

Reproduction is panmictic, including selfing, and without fertility differences. The adults produce an infinite number of gametes, which fuse at random to form zygotes in Hardy-Weinberg proportions with unordered genotypic frequencies $(2-\delta_{ij})p_ip_j$, in which δ_{ij} represents the Kronecker delta and p_i is still given by (1).

Selection acts through viability differences. If w_{ij} represents the viability of A_iA_j individuals, after selection the genotypic frequencies become

$$P^*_{ij} = \frac{(2-\delta_{ij})w_{ij}p_ip_j}{\sum_{k \leq m} (2-\delta_{km})w_{km}p_kp_m} . \tag{2}$$

The population size remains infinite.

Mutation is next. Let u_{ij} designate the probability that A_i mutates to A_j; by convention, $u_{ii} = 0$. Then

$$R_{ij} = \left(1 - \sum_k u_{ik}\right)\delta_{ij} + u_{ij} \tag{3}$$

is the probability that an A_i allele in a zygote appears as A_j in a gamete. Assuming that the two genes carried by an individual mutate independently, we obtain after mutation the germ-line genotypic frequencies

$$P_{ij}^{**} = \frac{1}{2}(2-\delta_{ij}) \sum_{k \leq m} (R_{ki}R_{mj} + R_{kj}R_{mi})P_{km}^{*} . \qquad (4)$$

The population number is, of course, unaltered.

Random drift operates through population regulation, which reduces the population to N adults with unordered genotypic frequencies P_{ij}', thereby completing the life cycle. Therefore, given the genotypic frequencies P, the distribution of the genotypic numbers NP_{ij}', $i \leq j$, is multinomial with index N and parameters P_{ij}^{**}.

We summarize the above information in the following formal scheme.

reproduction   selection   mutation   regulation

adult ————> zygote ————> adult ————> adult ————> adult
N,P_{ij},P_i $\infty,(2-\delta_{ij})P_iP_j,P_i$ $\infty,P_{ij}^{*},P_i^{*}$ $\infty,P_{ij}^{**},P_i^{**}$ N,P_{ij}',P_i'

To some extent, the order of the evolutionary forces in our life cycle is arbitrary. While selection acts on the phenotype, which develops from the zygotic genotype, the germ cells mutate with no phenotypic effect. Consequently, in any formal scheme, selection must always precede mutation. We are left with two possible sequences in addition to ours: reproduction, selection, regulation, mutation; and reproduction, regulation, selection, mutation. Whereas in our model selection and mutation occur in an infinite population and hence can be treated deterministically, the first of the above alternatives would entail a much more complicated probabilistic formulation of mutation, and the second would necessitate this for both mutation and selection. If all the evolutionary forces were weak, it is plausible, but unproved, that the dynamics of the three models would be quite close.

It is convenient to express the transition probabilities of the Markov chain of genotypic frequencies in terms of probability-generating functions:

$$E\left[\prod_{i \le j} \xi_{ij}^{NP'_{ij}} \Bigg| P\right] = \left(\sum_{i \le j} P^{**}_{ij}\xi_{ij}\right)^N . \tag{5}$$

Computations *within generations* are facilitated by employing ordered (or symmetrized) genotypic frequencies

$$\tilde{P}_{ij} = \frac{1}{2}(1 + \delta_{ij})P_{ij} , \qquad i \le j ; \qquad \tilde{P}_{ji} = \tilde{P}_{ij} . \tag{6}$$

Then (1) reduces to

$$P_i = \sum_j \tilde{P}_{ij} ; \tag{7}$$

with the definition $w_{ji} = w_{ij}$ for $i < j$, (2) becomes

$$\tilde{P}^*_{ij} = w_{ij}P_iP_j / \overline{w} , \qquad \overline{w} = \sum_{i,j} w_{ij}P_iP_j ; \tag{8}$$

and (4) reads

$$\tilde{P}^{**}_{ij} = \sum_{k,m} R_{ki}R_{mj}\tilde{P}^*_{km} . \tag{9}$$

Summing (9) over j gives

$$P^{**}_i = \sum_k R_{ki}P^*_k , \tag{10}$$

where

$$P^*_i = P_iw_i / \overline{w} , \qquad w_i = \sum_j w_{ij}P_j . \tag{11}$$

Most of our interest centers on the evolution of the gene frequencies. Equations (1) to (5) reveal that the transition probabilities depend on the initial genotypic frequencies $\mathbf{P}$ only through the allelic frequencies $\mathbf{p}$. Therefore, the vector of gene frequencies $\mathbf{p}(n)$, where $n \ (= 0,1,2,...)$ is time in generations, is Markovian. To derive the generating function of the transition probabilities of this Markov chain, take $\xi_{ij} = \zeta_i \zeta_j$ in (5). We find

$$
\prod_{i \leq j} (\zeta_i \zeta_j)^{NP'_{ij}} = \prod_{i,j} (\zeta_i \zeta_j)^{N\tilde{P}'_{ij}}
$$

$$
= \left(\prod_{i,j} \zeta_i^{N\tilde{P}'_{ij}} \right) \left(\prod_{i,j} \zeta_i^{N\tilde{P}'_{ji}} \right)
$$

$$
= \prod_{i,j} \zeta_i^{2N\tilde{P}'_{ij}}
$$

$$
= \prod_i \zeta_i^{2Np'_i} .
\tag{12}
$$

Substituting (12) into (5) and noting that the right-hand side depends only on $\mathbf{p}$ yields

$$
E\left[\prod_i \zeta_i^{2Np'_i} \,\middle|\, \mathbf{p} \right] = \left(\sum_{i,j} \tilde{P}^{**}_{ij} \zeta_i \zeta_j \right)^N .
\tag{13}
$$

In the Wright-Fisher model, given $\mathbf{p}$, the allelic numbers $2Np'_i$ are multinomially distributed with index $2N$ and parameters p^{**}_i. This is equivalent to (13) if and only if the fitnesses are multiplicative, i.e., $w_{ij} = v_i v_j$ for every i and j for some $\mathbf{v}$ (Ethier and Nagylaki, 1980).

The diffusion limit (Ethier and Nagylaki, 1980) and the Gaussian approximation in the next section, however, are the same for the two models.

III. THE GAUSSIAN APPROXIMATION

To enforce the weakness of selection and mutation, we posit that these evolutionary forces have intensity ϵ_N,

$$w_{ij} = 1 + \epsilon_N \sigma_{ij} , \qquad u_{ij} = \epsilon_N \mu_{ij} ; \tag{14}$$

we assume that $\epsilon_N \to 0$, $N\epsilon_N \to \infty$ (which implies that the deterministic forces dominate random drift), and σ_{ij} and μ_{ij} are fixed as $N \to \infty$. To establish the Gaussian approximation, we verify the three moment conditions of Norman's (1975a) Theorem 2.

Either from the fact the genotypic numbers are multinomially distributed (Ethier and Nagylaki, 1980) or directly from (13), we obtain

$$E(\Delta p_i \mid P) = \epsilon_N M_i^N(\mathbf{p}) , \tag{15a}$$

$$\text{Cov}(p_i', p_j' \mid P) = (2N)^{-1} V_{ij}^N(\mathbf{p}) , \tag{15b}$$

where

$$\epsilon_N M_i^N(\mathbf{p}) = p_i^{**} - p_i , \tag{16a}$$

$$V_{ij}^N(\mathbf{p}) = p_i^{**}(\delta_{ij} - p_j^{**}) + \tilde{P}_{ij}^{**} - p_i^{**} p_j^{**} . \tag{16b}$$

With the exact functions M_i^N and V_{ij}^N , Norman's (1975a) first two moment conditions [his (2.9) and (2.10)] are satisfied identically. Furthermore, M_i^N and V_{ij}^N converge to the usual drift and diffusion coefficients (Ethier and Nagylaki, 1980, and references therein) as $N \to \infty$:

$$M_i^N(\mathbf{p}) \to M_i(\mathbf{p}) = p_i \left(\sum_{j=1}^{r} \sigma_{ij} p_j - \sum_{j,k=1}^{r} \sigma_{jk} p_j p_k \right)$$

$$+ \sum_{j=1}^{r} p_j \mu_{ji} - p_i \sum_{j=1}^{r} \mu_{ij} , \tag{17a}$$

$$V_{ij}^N(\mathbf{p}) \to V_{ij}(\mathbf{p}) = p_i(\delta_{ij} - p_j) . \tag{17b}$$

To verify Norman's (1975a) simplified third-moment condition [his (2.12')], we employ a standard moment inequality (Feller, 1971, pp. 153–155), appeal twice to Jensen's inequality, condition on P, note that NP_{ij}' is binomially distributed with index N and parameter P_{ij}^{**} for each i and j, $i \le j$, and finally utilize the fact that the fourth central moment of the binomial distribution is $O(N^2)$:

$$\left\{ E \left[\left(\sum_{i=1}^{r-1} (p_i' - p_i^{**})^2 \right)^{3/2} \right] \right\}^{1/3} \le \left\{ E \left[\left(\sum_{i=1}^{r-1} (p_i' - p_i^{**})^2 \right)^2 \right] \right\}^{1/4}$$

$$\le (r-1)^{1/4} \left\{ E \left(\sum_{i=1}^{r-1} (p_i' - p_i^{**})^4 \right) \right\}^{1/4}$$

$$= (r-1)^{1/4} \left\{ E \left[\sum_{i=1}^{r-1} \left(\sum_{j=1}^{r} (\tilde{P}_{ij}' - \tilde{P}_{ij}^{**}) \right)^4 \right] \right\}^{1/4}$$

$$\le [r^3(r-1)]^{1/4} \left\{ E \left(\sum_{i=1}^{r-1} \sum_{j=1}^{r} (\tilde{P}_{ij}' - \tilde{P}_{ij}^{**})^4 \right) \right\}^{1/4}$$

$$= O(N^{-1/2}) . \tag{18}$$

The exact and approximate deterministic gene-frequency trajectories $\mathbf{x}^N(n)$ and $\mathbf{x}(t)$, $n = [t/\epsilon_N]$, satisfy

$$\Delta x_i^N = \epsilon_N M_i^N(\mathbf{x}) , \tag{19a}$$

$$\dot{x}_i = M_i(\mathbf{x}) , \tag{19b}$$

where the superior dot designates the time derivative. Fix any $T_0 < \infty$, suppose that $n \le T_0/\epsilon_N$, and define the matrix K by

$$[K(\mathbf{x})]_{ij} = \frac{\partial M_i}{\partial x_j} (\mathbf{x}) , \tag{20}$$

$1 \le i,j \le r-1$. The vector

$$Z^N(n) = (2N\epsilon_N)^{1/2} [\mathbf{p}(n) - \mathbf{x}^N(n)] \tag{21}$$

has $r-1$ components and represents the scaled deviation of the allelic frequencies from the exact deterministic trajectory. As $N \to \infty$, the distribution of $Z^N(n)$ is asymptotic to the multivariate Gaussian with mean vector $\mathbf{0}$ and covariance matrix

$$g(t) = \int_0^t \exp\left\{\int_{t'}^t K^T[\mathbf{x}(t'')]dt''\right\} V[\mathbf{x}(t')] \exp\left\{\int_{t'}^t K^T[\mathbf{x}(t''')]dt'''\right\}dt' , \tag{22}$$

in which the superscript T signifies transposition.

Observe that the integral (22) is along the approximate deterministic trajectory. According to our proof, the exact moments M_i^N and V_{ij}^N should appear in (22); the result (17), however, allows us to replace these by the much simpler functions M_i and V_{ij}. Although a differential equation specifies g in Norman (1975a), the integral (22) appears in Norman (1968; 1972, p. 118). If $N\epsilon_N^3 \to 0$ as $N \to \infty$, i.e., for fairly weak selection and mutation, we may replace $\mathbf{x}^N$ in (21) by the much simpler trajectory $\mathbf{x}$ (Norman, 1975a). See Norman (1974, 1975b) for some equilibrium results.

The important qualitative conclusion from (21) and (22) is that the gene-frequency fluctuations are of order $(N\epsilon_N)^{-1/2}$.

We specialize now to two alleles.

A. Two Alleles

To simplify the notation, put $p = p_1$ and

$$\mu_1 = \mu_{12}, \quad \mu_2 = \mu_{21}, \quad \mu_0 = \mu_1 + \mu_2, \tag{23a}$$

$$\sigma_1 = \sigma_{11} - \sigma_{12}, \quad \sigma_2 = \sigma_{22} - \sigma_{12}, \quad \sigma_0 = \sigma_1 + \sigma_2. \tag{23b}$$

Then the drift and diffusion coefficients become

$$M(p) = p(1-p)(\sigma_0 p - \sigma_2) + \mu_2 - \mu_0 p, \tag{24a}$$

$$V(p) = p(1-p). \tag{24b}$$

The variance of $Z^N(n)$ reduces to (Norman, 1968, 1972, p. 191)

$$g(t) = \{M[x(t)]\}^2 \int_p^{x(t)} \frac{V(x')}{[M(x')]^3} \, dx', \tag{25}$$

where p represents the initial frequency of A_1.

There is a simple equilibrium theorem if the approximate deterministic equation $\dot{x} = M(x)$ has a unique, exponentially stable, polymorphic equilibrium $\hat{x}$, i.e., if $M(x)$ has a unique zero $\hat{x}$ in $(0,1)$ such that

$$\frac{dM}{dx}(\hat{x}) < 0. \tag{26}$$

In this case, as $N \to \infty$ and $t \to \infty$, the distribution of $Z^N(n)$ is asymptotic to the Gaussian with mean 0 and variance

$$\hat{g} = -\frac{V(\hat{x})}{2\,\dfrac{dM}{dx}(\hat{x})} \qquad (27)$$

(Norman, 1974, 1975a,b).

We discuss three examples.

1. *Pure Mutation*

Let

$$y^N(n) = x^N(n) - \hat{x}, \quad y(t) = x(t) - \hat{x} \qquad (28)$$

denote the deviations from the equilibrium $\hat{x} = \mu_2/\mu_0$ and put $y_0 = p - \hat{x}$. Then

$$y^N(n) = y_0(1 - \epsilon_N\mu_0)^n, \quad y(t) = y_0\, e^{-\mu_0 t}, \qquad (29a)$$

$$g(t) = \frac{y^2}{\mu_0}\left[\frac{\mu_1\mu_2}{2\mu_0{}^2}\left(\frac{1}{y^2} - \frac{1}{y_0{}^2}\right) + \frac{\mu_1 - \mu_2}{m_0}\left(\frac{1}{y} - \frac{1}{y_0}\right) + \ell n\,\frac{y}{y_0}\right], \quad (29b)$$

where $y = y(t)$. If $\mu_1, \mu_2 > 0$, we may let $t \to \infty$ to obtain

$$g(t) \to \mu_1\mu_2 / (2\mu_0{}^3). \qquad (30)$$

Although it is easy to derive (29) from the exact expression for the variance of the gene frequency (Nagylaki, 1979), that procedure does not prove the normality of the fluctuations. The analysis of Feller (1951) applies close to the equilibrium $\hat{x}$.

If $\mu_2 = 0$, (29) becomes

$$x^N(n) = p(1 - \epsilon_N\mu_1)^n, \quad x(t) = pe^{-\mu_1 t}, \qquad (31a)$$

$$g(t) = \frac{x^2}{\mu_1}\left(\frac{1}{x} - \frac{1}{p} + \ell n \frac{x}{p}\right) . \tag{31b}$$

which should be accurate if n is much less than the mean extinction time.

2. *Pure Selection without Dominance*

We have $\sigma_1 = -\sigma_2 = \sigma$,

$$x(t) = p/[p + (1-p)e^{-\sigma t}] , \tag{32a}$$

$$g(t) = \frac{x^2(1-x)^2}{\sigma}\left\{2\,\ell n\left[\frac{x(1-p)}{p(1-x)}\right] - \frac{1}{x} + \frac{1}{p} + \frac{1}{1-x} - \frac{1}{1-p}\right\} . \tag{32b}$$

We expect (32b) to be accurate if n is much less than the mean absorption time. No closed-form expression is known for $x^N(n)$.

3. *The Overdominant Equilibrium*

Suppose

$$\sigma_i = -\rho_i , \quad \rho_i > 0 , \quad i = 1,2 ; \quad \rho_0 = \rho_1 + \rho_2 . \tag{33}$$

If $\mu_1, \mu_2 > 0$, the conditions for the validity of (27) are satisfied (Norman, 1974). Therefore,

$$\hat{g} = \frac{\hat{x}(1-\hat{x})}{2[\mu_0 + \rho_0\hat{x}(1-\hat{x}) + (2\hat{x}-1)(\rho_2 - \rho_0\hat{x})]} , \tag{34a}$$

in which $\hat{x}$ designates the unique root in (0,1) of

$$\hat{x}(1-\hat{x})(\rho_2 - \rho_0\hat{x}) + \mu_2 - \mu_0\hat{x} = 0 . \tag{34b}$$

If mutation is much weaker than selection, $\mu_0 \ll \rho_1,\rho_2$, we can obtain an approximation for $\hat{x}$, and hence one for $\hat{g}$:

$$\hat{x} \approx \frac{p_2}{p_0} - \frac{\mu_1}{p_1} + \frac{\mu_2}{p_2} , \tag{35a}$$

$$\hat{g} \approx \frac{1}{2}\left(\frac{1}{p_0} - \frac{\mu_1}{p_1^2} - \frac{\mu_2}{p_2^2} \right) . \tag{35b}$$

The leading (or zeroth-order) terms in (35) correspond to pure selection; mutation serves only to maintain the overdominant equilibrium. Note that the mutation terms in (35) have intuitively sensible signs.

To return to the original units, let r_1 and r_2 represent the selection coefficients against A_1A_1 and A_2A_2, respectively:

$$r_1 = \epsilon_N p_1 , \quad r_2 = \epsilon_N p_2 , \quad r_0 = \epsilon_N p_0 ; \tag{36a}$$

u_1 and u_2 denote the mutation rates from A_1 to A_2 and *vice versa*:

$$u_1 = \epsilon_N \mu_1 , \quad u_2 = \epsilon_N \mu_2 . \tag{36b}$$

From (21), (35), and (36) we deduce the approximate mean and variance of the equilibrium gene frequency $\hat{p}$:

$$E(\hat{p}) \approx \frac{r_2}{r_0} - \frac{u_1}{r_1} + \frac{u_2}{r_2} , \tag{37a}$$

$$\mathrm{Var}(\hat{p}) \approx \frac{1}{4N}\left(\frac{1}{r_0} - \frac{u_1}{r_1^2} - \frac{u_2}{r_2^2} \right) . \tag{37b}$$

The zeroth-order term in (37a) is due to Fisher (1922).

ACKNOWLEDGMENT

I thank Professor M. Frank Norman for helpful discussion.

REFERENCES

Bodmer, W. F. (1965). Differential fertility in population genetics. *Genetics* **51**, 411-424.

Ethier, S. N., and Nagylaki, T. (1980). Diffusion approximations of Markov chains with two time scales and applications to population genetics. *Adv. Appl. Prob.* **12**, 14-49.

Feller, W. (1951). Diffusion processes in genetics. *In* "Proc. 2nd Berkeley Symp. Math. Stat. Prob." (J. Neyman, ed.), pp. 227-246. University of California Press, Berkeley.

Feller, W. (1971). "An Introduction to Probability Theory and Its Applications," Vol. II, 2nd ed. Wiley, New York.

Fisher, R. A. (1922). On the dominance ratio. *Proc. Roy. Soc. Edinb.* **52**, 321-341.

Karlin, S., and McGregor, J. (1964). On some stochastic models in genetics. *In* "Stochastic Models in Medicine and Biology" (J. Gurland, ed.), pp. 245-271. University of Wisconsin Press, Madison.

Moran, P. A. P. (1958a). Random processes in genetics. *Proc. Camb. Phil. Soc.* **54**, 60-71.

Moran, P. A. P. (1958b). A general theory of the distribution of gene frequencies. II. Non-overlapping generations. *Proc. Roy. Soc. Lond.* **B, 149**, 113-116.

Nagylaki, T. (1977). "Selection in One- and Two-Locus Systems." Springer-Verlag, Berlin.

Nagylaki, T. (1979). The island model with stochastic migration. *Genetics* **91**, 163-176.

Norman, M. F. (1968). Slow learning. *Brit. J. Math. Stat. Psych.* **21**, 141-159.

Norman, M. F. (1972). "Markov Processes and Learning Models." Academic Press, New York.

Norman, M. F. (1974). A central limit theorem for Markov processes that move by small steps. *Ann. Prob.* **2**, 1065-1074.

Norman, M. F. (1975a). Approximation of stochastic processes by Gaussian diffusions, and applications to Wright-Fisher genetic models. *SIAM J. Appl. Math.* **29**, 225-242.

Norman, M. F. (1975b). Limit theorems for stationary distributions. *Adv. Appl. Prob.* **7**, 561-575.

Penrose, L. S. (1949). The meaning of "fitness" in human populations. *Ann. Eugen.* **14**, 301-304.

PART VI. POPULATION GENETICS OF ECOLOGICAL AND BEHAVIORAL INTERACTIONS

INSTABILITY AND CYCLING OF TWO
COMPETING HOSTS WITH TWO PARASITES

W. D. Hamilton

Department of Zoology
Oxford University
Oxford, England

ABSTRACT

Competition between hosts complicates but does not remove the tendency to oscillation in host-parasite systems. A model on lines of standard community ecology using host-parasite dynamics from May (1973) plus host competition shows that, as competition increases: (1) The fixed point of the system becomes less likely to be stable; (2) the departure from the fixed point remains as likely to oscillate; and (3) permanent cycling (or irregular quasi-cycling) readily follows.

Presence of parasites complicates but does not eliminate the "competitive exclusion principle" for a pair of hosts. When competition between hosts is strong, the double system easily loses either one parasite or one host-parasite pair through extinction. Low levels of immigration are very effective in steadying cycles and preventing extinction.

These findings are discussed in relation to the theory that sex is an adaptation to combat the rapid coevolution of parasites. Extinctions of genes (or local extinctions of population) due to parasites do not necessarily undermine the quasi-repetitive selection that may support sex.

645

I. COEVOLUTIONARY PURSUIT AND FLEEING BY SEX

Among possible interactions of species in an ecosystem there is a general kind that may be called coevolutionary pursuit (CP). Parasite with host (p-h) and predator with prey (p-p) are examples of it. In a square matrix which shows for a local community the effects that current abundance of each species has on the growth rate of every other, pairs of species involved in coevolutionary pursuit produce pairs of entries of opposite sign: a CP is in fact characterized in a "community matrix" (May, 1973) by this "+\-" interaction, which simply reflects that abundance of the pursued species (host or prey) is positive -- increases growth rate -- for the pursuer, while abundance of the pursuer species (parasite or predator) is negative -- decreases growth rate -- for the pursued.

At least since the work of Lotka and Volterra (Scudo and Ziegler, 1970) it has been realized that the +\- interaction causes oscillatory population phenomena. The oscillations following perturbation may be damped and disappear or may grow. If they grow, the species system may end in some compact finite cycle where both species persist in repetitively changing numbers. Or else the oscillations may grow without limit until one species, most likely the host or prey, hits zero. Even when this happens, however, the system may still allow all species to exist if the local deme where extinction occurs is part of a larger population that from time to time reseeds patches with migrants.

On the basis of its abundance and inherent tendency to cycle, it has been suggested that the CP situation, and especially the p-h variant, can explain the existence of a very widespread and seemingly inefficient system of reproduction - namely, sexuality (Jaenike, 1978; Bremermann, 1980, and *in press*, Hamilton 1980, 1982). In introducing this idea, of course, we at once step outside the confines of the community matrix approach to ecology because the model underlying that approach makes no provision for variation within species. However, it is clear that the dynamical findings must still apply in a broad way even when there is variation. Also, reinterpreting "species" in the community model as "varieties" can make interesting suggestions about the advantages sex might have in a broader model -- a model where sex was being really tested in course of *both* interspecies and varietal changes. Therefore this paper is going to look in the community dynamics model for robust sources of cycling that might support sex over parthenogenesis.

The effects of actually having sex will not be addressed in the models of this paper. The cycles discussed will only involve changes in population numbers. Really, of course, it is clear that sexuality is not primarily concerned with numbers of individuals, it is concerned rather with numbers of genes an gene combinations. Therefore there are going to be crucial questions left unanswered in this paper about what changes of combination need to arise during population cycles for sex to be useful, whether such changes as do arise increase or reduce the mean density of population, whether the changes reduce the cycling tendency or exaggerate it, and so on. None of this will be addressed directly. Following May (1983) it will be assumed that if cycles can come out of population density dependence they can also, in parallel ways, come out of genotype frequency dependence. Following a paper of my own (Hamilton, 1980; but for cautions on the specific model of this see May and Anderson, 1983), it will also be assumed that when a fair amount of dominance and epistatic interaction contributes to the fitness of genotypes, not only does cycling remain possible, but, when it occurs, it gives a sexual population a good chance to outcompete any mixture of asexual clones that might be derived from it, despite the advantage that clones have in sheer reproductive efficiency. The chance is especially good when the selection is strong, as can easily be the case with disease.

Of course, if clones are *not* efficient in reproduction, as might be expected with first mutations to parthenogenesis from long-established sexuality, and as seems often to be the case (e.g., Lamb and Willey, 1979), the chances for sex to resist inroads of parthenogenesis is better at the start. However, there would seem to be many marginal situations where, if low initial fertility were their only handicap, clones could establish themselves (e.g., by colonizing islands where competition is lacking). In such places they could then improve efficiency by further selection (as has sometimes happened with asexual lines under artificial maintenance). Finally, they could reinvade the sexual areas from which their matriarchs came. The rarity of such courses argues, in my opinion, that a fairly short-term strong selection is supporting sexuality in most species.

In my model of 1980 there were no fluctuations of population numbers at all. Cycling of genotype frequencies arose from pure frequency dependent selection of a somewhat artificial kind. In contrast, one could also imagine density dependent dynamics following roughly the lines of the model given below, but with the feedback based on numbers of genotypes taken separately, rather than on numbers of all individuals taken together; further one could imagine that the cycles of p-h-selected genotypes "drag" the frequencies at other polymorphic loci along with them, so producing continual changes in parts of the genome that have nothing to do directly with the CP and the parasites. If epistatic fitness effects are involved, the repetitive changes can give sex an advantage. Unless prevented by the structure of the model, overall population numbers would in such a case cycle in some rather complex fashion but would probably not vary so widely as they did before genotype variation was brought in. This would be due to a tendency to dephase automatically the genotype cycles within each species.   Smoothing of overall changes through such dephasing is revealed in the simulation runs of the model below (Section IV) if we start thinking of the two hosts as genetic varieties of a single species.

In wider perspective, and looking in particular at Jaenike's and my own version of the parasite resistance idea about sex, as long as the cycling tendency is not too strongly damped, or stable cycles that occur are not too closely confined, the *numerous* p-h interactions which must apply to almost all eucaryotic species are to be seen as a recipe for eternal turbulent genotype change.   In effect, the pursued species, if sexual, is using segregation and recombination to "run away from" the coevolutionary advances of the various pursuers.   Such advances come particularly fast in the case of pursuing parasites, it is suggested, because of the usual smallness and their consequent shorter generation time (Hamilton, 1982; Bremermann, 1980). To combat their transmutations sexual species are continually putting back together the yet unextinguished components of defense systems that served them in similar phases in the past.   Doing so, they can "flee" more effectively than can asexual species: hence comes an advantage to sex that may exceed the advantage that parthenogenesis has in efficiency.

"Fleeing" and "pursuit" as described so far are largely a matter of running in circles. However, even if old alleles sometimes go extinct and completely new advantageous ones arise by mutation, it is still probably true that pursuit situations are more apt to make gene transiences *common,* as is needed to protect sex, than are other agencies of natural selection. This may amount to saying that some older ideas about sex, such as "Muller's ratchet," may look stronger if attention is focused particularly on pursuit situations, especially the host-parasite one. Even if not quite running in circles here, the system is still most of the time dodging around in a small evolutionary area. Sex on this view has a basically defensive conservative function and the macroevolutionary developments that arise from it are initially side effects.

The reasons why the parasite-host relation, of the two major classes of CP, is more likely to be important for sex can be summarized as follows:

(1) Parasites tend to be short-lived and consequently can rapidly evolve new modes of attack.

(2) Parasites tend to be specialists and predators to be generalists; it is harder to imagine how a generalist can get into cycles.

(3) For any one host or prey, p-h relations are much more numerous than p-p. (On the severity side, predation events may seem more decimating and parasites often very trivial; but in fact, debility due to parasitism probably precipitates capture by a predator much more often than is recorded (e.g., Jenkins *et al.,* 1964; Murton, 1965; Holmes, 1982).)

(4) The coevolution of host and parasite is more likely to take place in discrete orthogonal steps (e.g., create a new "password" surface molecule), whereas p-p relations are more likely to be a matter of adjustment of position in a continuum (e.g., choose a particular compromsie between fleetness and strength). Cycles are more likely to occur when evolution to an intermediate position between two morphs is not possible.

None of these four factors leans unequivocally in favour of the p-h rather than the p-p variant. For factor (2), for example, it may be objected that the possibility of cycling when there is a generalist predator (or parasite) has hardly been studied theoretically. First glance at what has been done, however, is not at all encouraging. Lotka (1956) already pointed out that a predator with two prey can maintain its numbers on one of them while it drives the other extinct. Extinction is harder to achieve (to the benefit of both pursuer and pursued) when dependence is on a single species. Thus we do not seem on our way towards a recipe for eternal cycles. However, given again metapopulations made up of many demes with reseeding by migrants following extinctions, or given special refuges from parasitism or predation when at low density (perhaps through genotypic or learned switching of the pursuer's attention to the more abundant prey or host), it is possible to imagine systems with generalists that at least preserve all their species. Of course, it is another step to demand that on top of that they also *cycle,* or at least give conditions of selection causing frequent *gene transiences* in the way that would be favourable to sex, but this is a possibility that might repay study.

Such a line towards dynamics of generalist pursuers will not be followed here. The simple connected community that I propose to look at does potentially cover the one-paraiste-two-hosts situation as a special case. However, the primary objective is to understand more about the cycling induced by specialist parasites. It is hoped to show that they still cause cycles when their hosts are not ecologically isolated but interact among themselves--in other words, that Kolmogorov's limit cycle (Scudo and Ziegler, 1970) arising from CP does not disappear when put into a more speciose and interconnected situation. In the course of the same analysis I hope to show reason to believe that a general result on cycling that is implied in Eshel and Akin (1983) is not substantially weakened when *absolute* densities of morphs replace *relative* densities (i.e., replace gene frequencies) as the fitness-determining variables. The particular relation holding between hosts in the model studied will be competition.

II. THE MODEL

Let there be two host species, Ho_1 and Ho_2. Let each have a parasite which has at least a minor degree of specialization. Call these Pa_1 and Pa_2.

If the population numbers of these four species in a particular generation are H_1, H_2, P_1, P_2, let those of the next generation be given by the relations

$$H_1' = H_1F_1, \quad H_2' = H_2F_2, \quad P_1' = P_1G_1, \quad P_2' = P_2G_2 \qquad (1)$$

where all the factors F_1, F_2, G_1, G_2 are, in the most general case, functions of all four population sizes: for example

$$H_1' = H_1F_1(H_1,H_2,G_1,G_2) \qquad (1a)$$

Suppose that the system has an interior equilibrium point $(H_1{}^*,H_2{}^*,P_1{}^*,P_2{}^*)$, that is, a point such that $H_1{}^* = H_1{}^*F_1$ with $H_1{}^* > 0$ and likewise for the three others.

The first step towards describing the stability of this equilibrium is to evaluate the Jacobian matrix of the linear approximation of the system that applies near to the equilibrium point. This matrix, formed from evaluations of all the partial derivatives at the equilibrium point, will be written:

$$J = \begin{bmatrix} \overset{-}{a_1} & \overset{-}{b_1} & \overset{-}{c_1} & \overset{(-)}{d_1} \\[1em] \overset{-}{b_2} & \overset{-}{a_2} & \overset{(-)}{d_2} & \overset{-}{c_2} \\[1em] \overset{+}{e_1} & \overset{(+)}{f_2} & \overset{(-)}{g_1} & \overset{(-)}{h_1} \\[1em] \overset{(+)}{f_2} & \overset{+}{e_2} & \overset{(+)}{h_2} & \overset{(-)}{g_2} \end{bmatrix} \qquad (2)$$

In this, we have by definition, for example, that

$$e_2 = \left.\frac{\partial\{P_2(G_2-1)\}}{\partial H_2}\right|_*$$

where $|_*$ means evaluation using $H_1 = H_1^*$, etc. Note that the particular slope instanced here refers to the effect of the second host on the second parasite and, this being appropriate host for that parasite, the slope will be positive. This is indicated by the $+$ sign just above left of e_2 in the matrix. The expected sign is given likewise for all the other symbols in the matrix, with those which might reasonably be zero bracketed.

The next step is to form another matrix $\mathbf{J} - \lambda\mathbf{I}$ (i.e., to subtract λ from every term in the main diagonal); the next then is to solve the equation formed by setting the determinant of the new matrix equal to zero:

$$|\mathbf{J} - \lambda\mathbf{I}| = 0 . \tag{3}$$

Assume this done and the equation solved numerically or otherwise. It then follows that if one or more of the roots of this equation is such as to imply $\mathrm{mod}\,(\lambda+1) > 1$, then the fixed point $(H_1^*,H_2^*,P_1^*,P_2^*)$ is unstable; if on the contrary the opposite inequality holds for *all* of the roots, then the fixed point is stable (see May, 1973). Any case where the largest modulus is equal to 1 would need further analysis to determine stability; however, such a case is exceptional and even where it holds we can see that stability and instability must be both near at hand. Changes of the parameters of the model in such a state are likely to have large effects for or against stability, and any stochastic event can be expected to have longer aftermath than under other conditions. Even if damping, the aftermath changes will probably be similar to effects of cycling from the present point of view. This case will not be discussed further.

In general, the expansion of (3), the characteristic equation of the system, is a quartic, and this will not factorise or solve neatly. One case where it does factorize into a product of quadratic expressions, however, is general enough to give guidance on some interesting questions. This is the case where the two subsystems, Ho_1 with Pa_1 and Ho_2 with Pa_2, are "identical twins" in all their parameters, including those that imply cross effects onto the other twin of the system. Because the subset having this symmetry is more restricted than one could wish, cases where roots are found numerically from the general Jacobian will be discussed and also some simulations of dynamics of the complete system will be reported. In this way it is hoped to give some idea of how robust the conclusions drawn from the symmetrical case may be. Simulation is especially needed, of course, in those situations--of great interest here-- where instability holds since no useful analytical tools for showing the existence of simple cycles seem to be available once more than two species are involved.

Formally, the intended symmetrical subset of matrices is defined as having $a_1 = a_2 = a$, $b_1 = b_2 = b$, . . . , $h_1 = h_2 = h$; in other words, we drop all subscripts in **J**.

When this is done, the characteristic equation derived from **J**, after factorization, is

$$\{\lambda^2 - (a_+ + g_+)\lambda + a_+ g_+ - c_+ e_+\}\{\lambda^2 - (a_- + g_-)\lambda + a_- g_- - c_- e_-\} = 0 \quad (4)$$

where

$$
\begin{aligned}
a_+ &= a + b, & a_- &= a - b, \\
g_+ &= g + h, & g_- &= g - h, \\
c_+ &= c + d, & c_- &= c - d, \\
e_+ &= e + f, & e_- &= e - f .
\end{aligned}
\quad (4a)
$$

Signs and relative magnitudes in these expressions and their implications for the eigenvalues (the roots of (3)) will be discussed. It is worthwhile to look first, however, at another subset where simple quadratic factors can be found and where symmetry of the subsystems is *not* required. Suppose the subsystems are not linked with one another at all, i.e., all those elements in **J** with letters b, d, f, and h, of whichever subscript, are set at zero. Then obviously a quadratic characteristic equation appears for each of the now separated subsystems:

$$\lambda^2 - (a_1 + g_1)\lambda + a_1 g_1 - c_1 e_1 = 0 , \tag{5a}$$

$$\lambda^2 - (a_2 + g_2)\lambda + a_2 g_2 - c_2 e_2 = 0 . \tag{5b}$$

Comparing these to the factors in the previous equation it can be seen that there is a generic similarity. Instability depends on the relative magnitude of certain sums and differences of effects on growth rates in the linked systems, and on simple within-subsystem effects on growth rates in the individual subsystems when unlinked. In (4) a_+ , for example, can be described as the effect on *joint* growth rate of hosts of perturbing *one* of the host populations alone, while in (5a), a_1 is implying the effect of change of the first host population on its own growth rate--in this case there is no effect from the other. Similar observations hold for all other symbols in the first factor.

Similarity of form allows the conditions for stability to be discussed at least in preliminary stages by selecting any one of the four equations given or implied above and dropping all the subscripts. Subscripts can be reinserted when there is need to go into more detail.

In this light, the roots are all of the form:

$$\lambda = \{(a+g) + \sqrt{(a+g)^2 + 4(ag+ce}\ \}/2 \tag{6a}$$

or equivalently:

$$\lambda = \{(a+g) + \sqrt{(a-g)^2 + 4ce}\;\}/2 \; . \tag{6b}$$

Thus there is a pair of complex roots, with implied potential for oscillatory behavior at least near to the fixed point, if

$$(a-g)^2 < -4ce \; . \tag{7}$$

Clearly ce must be negative if there is to be any chance of oscillatory departure. A glance at the matrix **J** shows that this is just what the CP relationship tends to ensure, except for some situations that might arise from the composite nature of c and e in the linked case (i.e., the case arising from (4); note that, unlike with (5), dropping subscripts in (4) is not equivalent to dropping them in matrix **J**).

We now make some assumptions about magnitudes of the various gradients set out in the Jacobian:

1. With the exception of a and b (where any relation of magnitudes may hold), the within-subsystem effects (measured by c_i, e_i, g_i; i = 1,2) are much larger than the across-subsystem effects (measured by d_i, f_i, h_i; i = 1,2). This means that c_+ and c_- both have the same sign as c_1, and likewise for the other letters with the sign subscripts except for a.

2. Effect on growth from one's own species numbers is much less in parasites than it is in hosts, so that a >> g. This is usually reasonable because for parasites host findability is more crucial than the parasite's own density.

3. Now a repeat of the almost definitional assumption about parasitism: hosts benefit their parasites but parasites harm their hosts. Given 1, this means c < 0 and e > 0, and therefore ce < 0. This is a necessary condition for complex roots as noted above.

Under these assumptions, looking at (6a), (6b), and (7), we see that:

(a) If p-h interactions are generally powerful compared to

competitive effects, then oscillations are likely. This, of course, is the expected generalization of Lotka-Volterra theory.

However, if parasite species interact strongly within and between themselves in much the *same way* that host species do (i.e., if the submatrix of **J** that has g and h is broadly similar to the submatrix that has a and b), contrary to no. 2 above, then p-h interactions have less need to be strong for oscillation to occur. Then with a = g we can see that quite small host parasite effects could cause oscillatory events. This seems unlikely to apply to most parasites. It could more reasonably apply to predators apart from the fact that predators are unlikely to be confined to such a simple two-prey scenario.

(b) In any reasonable model the fitness of hosts on their own will show negative density dependence in the following sense. At low host numbers we may expect that a will be positive (numbers grow faster the more there are to reproduce), while if the equilibrium is at high density, then a will be negative. The latter case, in making a_+ likely to be negative, favors stability. An equilibrium anywhere near to a condition where a is zero, is almost certain, with the expected smallness of g, to imply complex roots. The equilibrium moreover is likely to be unstable if -ce, measuring the power of the p-h interaction, is relatively large. As host density at equilibrium falls below that which gives a = 0, a rises, making strong p-h effects less necessary to attain instability.

Summarizing (a) and (b), if parasites have large effects on mortality of hosts and especially if they depress equilibrium levels far below carrying capacity--or if they interfere competitively among themselves, as discussed in the next section--then the internal equilibria of the system(s) are unlikely to be stable, and any departure will be oscillatory. In such cases, with reasonable assumptions, limit cycles or more complex attracting sets are expected to follow, as discussed by May (1973) for the case of the unlinked subsystems.

III. EFFECTS OF HOST COMPETITION

In this section to help fix ideas we depart from previous generality and make specific assumptions about how each host population affects the other. Later a little thought will show that similar results should hold under other reasonable assumptions.

Assume that in the absence of parasites the recurrence relations for the two hosts are

$$H_1' = H_1[1 + s\{1 - (uH_1 + vH_2)/K\}]$$

$$H_2' = H_2[1 + s\{1 - (vH_1 + uH_2)/K\}] \tag{8}$$

where $0 < v < 1$ and $u+v = 1$. Here v is the parameter linking the subsystems, and it can be called the coefficient of allocompetition. Its value shows the proportion by which each host's growth rate responds to numbers of the other species relative to its responding to numbers of its own. If $v = 0$, the equations are simply those determining discrete logistic growth: this case of course has the numbers of each host growing towards K as asymptotic upper limit. As v is raised, each population becomes more responsive to inhibitory pressure from the other host. At $v = 0.5 = u$ each species is responding to all individuals equally without discrimination of species. If $v > u$, each is more inhibited by numbers of other species than by numbers of its own: this could happen with interference competition.

Increasing allocompetition is intuitively destabilizing. One way of seeing this formally is to note that as v passes upwards through 0.5 the two straight isoclines, $uH_1 + vH_2 = K$ with zero growth for H_1, and

$vH_1 + uH_2 = K$ with zero growth for H_2, switch across each other, changing a stable (improper) node at $H_1 = H_2 = K$ into a saddle. The species with greater numbers then drives the other one out. At $v = 0.5$ the species coexist but do so at best in neutral equilibrium.

Increasing allocompetition is likewise sure to be destabilizing in the model where each host has a parasite. This is easily shown formally using the specific competition model implied in Eq. (8). Parasites being present, these equations each need an additional term which will be a function of parasite numbers and, probably, of host numbers. It will represent the reduction of host numbers or host reproduction that the parasites cause. We need not specify the form of such terms and will simply call them $-Pterm_1$ and $-Pterm_2$.

To complete the system we would also need to describe the functions G_1 and G_2 in (1). The following argument does not require that these be specified. However, one thing will be said about them. It is simplifying and reasonable to assume that v does not appear in either of the Pterm expressions or of the G expression. These are not concerned with interhost competition and therefore presence of v would not be expected. The point of this assumption will appear shortly.

We may abbreviate (8) by replacing the square-bracket expressions by symbols $\tilde{F}_1$ and $\tilde{F}_2$; then after including the Pterms we may form the partial derivatives like (2a) that give a_1, b_1, a_2, b_2 for the Jacobian:

$$a_1 = \left.\frac{\partial\{H_1(\tilde{F}_1 - 1)\}}{\partial H_1}\right|_* - \left.\frac{\partial Pterm_1}{\partial H_1}\right|_*$$

$$= \left.\frac{\partial\{H_1\tilde{F}_1 -\}}{\partial H_1}\right|_* - \left.\frac{\partial Pterm_1}{\partial H_1}\right|_*$$

For compactness the last term of these expressions will be given the symbol $\text{Pslope}_1{}^*$. Then, performing the differentiation in the first term:

$$a_1 = s\{1 - (uH_1{}^* + vH_2{}^*)/K\} - usH_1{}^*/K - \text{Pslope}_1{}^* \ .$$

With the assumption of symmetry we can drop subscripts and obtain:

$$a = s(1 - H^*/K) - usH^*/K - \text{Pslope}^* \ .$$

A like procedure for b_1, assuming that Pterm_1 contains H_1 but not H_2, and after assuming symmetry, gives:

$$b = -vsH^*/K \ .$$

Hence, $a_+ = a + b = s - \text{Pslope}^*$, which is independent of v.

Recalling the assumption made above which implies that v is not involved in any other elements of the Jacobian, we can turn back to (4) and see that the "+" quadratic factor as a whole is also independent of v, so there is one pair of eigenvalues which does not change as the degree of interhost competition is changed. If either of these roots gives $\text{mod}(\lambda + 1) > 1$, the system is unstable irrespective of interhost competition or lack of it. A similar argument can be applied regarding independence of interparasite competition, if any exists.

Note that the constancy of a pair of roots depends on the assumed symmetry and is lost if this symmetry is dropped. The reason for constancy is easy to see if we think of the model being started symmetrically, i.e., with $H_1 = H_2$ and $P_1 = P_2$. Obviously these equalities will continue to hold as the model runs, and so it behaves as if each subsystem was running on its own; and with the assumptions above the way it runs does not depend on v. Thus one of the necessary stability conditions, that given by the "+" factor, is the same as the stability condition for an isolated subsystem.

If both "+" roots are inside a unit circle centered at (-1,0) in the complex plane, then the "-" roots must be examined to determine stability.

As v increases, starting at zero, b decreases also starting from zero, while a increases by an exactly corresponding amount. The trace of the matrix **J** increases, making it generally more likely that there will be an eigenvalue with modulus greater than one. In fact, under reasonable assumptions, it is virtually certain that by the time v = 0.5 (full competition) the system is unstable. Under symmetry this follows because

$$a_- = a-b = s(1 - H^*/K) + (v-u)sH^*/K - Pslope^* .$$

The first term is always positive, and the second is positive as soon as v > 0.5. The last term, concerned with the detrimental effects of parasitism, is also positive under fairly reasonable assumptions--these being to the effect that the ability of parasites to take advantage of increasing host numbers shows diminishing returns as host numbers grow, so that the damage rates from a fixed number of parasites diminishes. If all three are positive, then a_- is positive. Instability then follows from (6), and likelihood of oscillatory departure is implied by (7).

In summary, competition always moves to destabilize the dual host-parasite system when symmetrical and we can reasonably suppose that it does so also in asymmetrical cases. Upon destabiliization, as noted in the last section, there is an oscillatory departure.

IV. FATE OF OSCILLATIONS: PROTECTION FROM EXTINCTION

The main method I have used to investigate the non-linear dynamics has been simulation. The usual device, once away from the linearized region, of plotting the patterns of flow in relation to isoclines is not easy to apply in this case because of the dimensionality. For the simulation I used CP equations that are discrete versions of the differential equations 4.4 given in May (1973). In these equations the host's response to density was generalized to the form of Eq. (8), i.e., the term H/K in May's system became in mine $(uH_1 + vH_2)/K$ for the $H_1{}'$ equation and $(vH_1 + uH_2)/K$ for the $H_2{}'$ equation.

If the subsystems had stable cycles or equilibria when considered alone, then at low levels of the coefficient of allocompetition in the situation was not much changed except that cycles might occur where previously they had not at all, and that a stable point could be pushed into small cycles by the other subsystem if that one cycled intrinsically. Other interesting but also expectable features were that the cycling in the two systems tends to be out of phase, perhaps permanently and at least most of the time, and that when two periods synchronize out of phase they do so at some period intermediate between the periods of the separate parts.

As allocompetition is increased, at least one of the cycles moves to a state where most of the time both host and parasite occur at very low numbers, and where brief upsurges of numbers are followed by crashes. Eventually a crash may, with the formulae of the discrete system, bring a host number below zero -- in effect cause an extinction. Or it may stay positive but go so low that the population would in practice have to die out. Of course extinction of host is followed by extinction of the specific parasite. It also occurs commonly in the simulations that the host numbers do not go to inviable lows but the parasite numbers do so. Then the parasite must be considered extinct. Relieved of it, the host recovers strongly.

Following an extinction various things may happen, but obviously none of them are entirely satisfactory if the hope is that cycling may be common in complex ecosystems as they actually exist. The difficulty of one or another subsystem losing its parasite or going wholly extinct under the added destabilizing influence of competition becomes rapidly worse as v approaches 0.5 At and above that value extinction appears inevitable. All this amounts to saying that parasitism does not enable the model to escape from the competitive exclusion principle even when parasites are most of the time depressing numbers far below the joint carrying capacity. When niches overlap only partially ($v < 0.5$) we know from the previous findings that the model can create cycles where none would exist if the host species did not compete. Simulations show that these cycles can be very complex. Often they are apparently permanently irregular especially when of large amplitude. Complexity in itself is nothing to run away from in a theory that cycling backs sexuality; it is probably good in the sense that it must make it harder for the parasite to adopt any regular system of morph changes to track its host. However, the fact that competition can easily destabilize a situation too much and cause extinctions is prejudicial to the theory.

What might dampen the cycles and, without extinguishing them, hold numbers from going too low? An obvious possibility is any *constant* input of migrants into the cycling population. Simulation shows that this is indeed very effective in damping and simplifying cycles. In the specific model derived from May's 4.4 (May, 1973) a constant small input of the hosts alone, or of hosts and parasites, pushes the cycles away from both zero lines very effectively. Counterintuitively, the hosts-alone input is more effective in raising minimum and mean parasite levels than it is in raising minima and means for the hosts. As the constant input is increased the cycles reduce their amplitude and also their complexity further. Finally, and of course simultaneously, they become a single point--the stable equilibrium.

The moderating effect of immigration is not just a theoretical possibility. A laboratory demonstration that a steady input of "clean" hosts could stabilize a cycling p-h system and prevent extinctions was given by Scott and Anderson (1984), using guppies involved with the monogenean parasite *Gyrodactylus*. Compared to my simulation reality turns out to be complex, as one would expect, and it was found that some levels of immigration in the experiment could increase the amplitude of cycling. However, conditions are far from being those of the simulation and it is not clear that a comparison is valid. For example, generation times of host and parasite are very different, and the period of cycling is actually shorter than the normal generation of hosts.

Either in theory or reality, where could a steady influx of migrants come from? And, in order to stabilize, how steady must it be? On the second question, the only direct answer that will be given here is that making the migrant input a Poisson variate instead of a constant does not much affect the issue. This is seen in the simulations. But this doesn't address the problem of a likely *periodic* inconstancy. An attractive answer to the first question where the system can be thought of as a set of demes is that the influx to individual demes comes from the population as a whole--the immigrants are settling from some sort of migrant cloud. For this to work in the sense of fairly constant input it would have to be that the cycles in the various demes are out of phase. Therefore an important further question emerges as to whether migration in itself will have a synchronizing or desynchronizing effect. Mostly this question will have to left for future work, but some very speculative comments are given below.

1. Physical conditions may set the parameters of population behavior differently in different demes. From this it will result that: (a) Demes have different natural periods of oscillation and will not readily

synchronize; (b) Some demes will be favorable to one species and some to another. High fluctuation correlates with high reduction by parasites and not with high productivity. Thus the main demes supplying the migrant cloud are likely to be relatively constant. (The factor of protection of species of genes from extinction implied here has obvious kinship to the multiple-niche polymorphism of Levene and others; see, for example Hedrick *et al.*, 1976).

2. Migration rates of the various species are all different from one another. This implies that an influx of hosts, say, will not tend to be followed by an influx of parasites at a time appropriate to reinforce the way the deme has reacted to the extra hosts. Hence the recipient deme may go into a minicycle or an extended one and so be brought more out of phase with the deme that is donor of the influx. However, this deme may be doing the same thing back, and it is not quite clear that the effect necessarily dampens cycles: experimentation or, better, analysis is needed.

3. Factors such as time taken to travel, iteroparous reproduction, and deposition of dormant propagules all deserve to be discussed, but generally it is again not clear how their influence will go. They are left as too speculative at present.

Altogether there seems reason to hope that a deme structure spread over a diversity of habitats and with migration between will be able to (a) prevent overall extinctions, (b) prevent even local extinction, and (c) still allow intrinsic cycling in some demes which in turn, through their migrants, will impose cyclical pressures on those demes more intrinsically stable.

Cycling in these statements may refer to numbers or to gene frequencies or to both.

V. "CYCLES" THAT INVOLVE EXTINCTION

If in the course of cycles the numbers of a species go extremely low, then there is much more variability about what will happen next. The situation has become stochastic. One possibility, as already discussed, is that the element goes extinct. If it does, uncertainty as to the future -- including the future of the vanished element -- continues. Two agencies might bring the lost element back: (a) immigration from elsewhere, and (b) mutation. Agency (b) cannot apply in a case of a lost species but it can revive a lost allele or at least produce one that is effectively the same.

Stochastic sequences of extinctions and revivals of lost elements are qualitatively different from deterministic cycling. However, they are repetitive and non-progressive in rather the same way. The period of change would probably be longer and less regular. This would not give such consistent short-term support to sexuality as cycles do (Hamilton, 1982). However, changes of selection, emphasising gene combinations, probably still come very fast under this regime of extinctions and re-establishments concerned with resistance compared to what is expected in the non-CP processes of adaptation. Judging from the literature of parasite and parasitoid ecology, local extinctions and establishments of species occur commonly.

In the case where we are concerned with the extinctions of alleles and their reappearance later by mutation, the suggestion that the CP, and especially the p-h relation, is the main supporter of sex may be considered as a version of the long prevalent Weissmann-Muller-Fisher view in which the principal achievement of sexual reproduction is bringing "good mutations" together into the evolving stock as fast as possible. In the present version the abundance of pursuits coming from parasitism would supply a sufficiently rapid succession of "good mutations" and consequent gene transiences for that theory to work. Likewise, once originally "good mutations" had become outmoded due to spreading innovations in the coevolutionary partner, sex could be considered to be working for efficient elimination of bad elements, both outmoded old and mis-mutated new, much in the way it is supposed to work in the "Muller's Ratchet" idea (for review of this and other sex theories, see Bell, 1982).

In an extension of this view of the role of mutation, it is not even required that the new mutation in a host is like the last one to go extinct. In the current environment it may be that a completely new allele may serve as well as or even better than a revival of the old. All that is needed is that it confers resistance against the upcoming types of parasite. With this suggestion we seem to leave anything that could be described as cycling behind. However, it remains true that there is unlikely to be any other pressure for novelty that is as constant and prolific as the coevolution of parasites. It is arbitrary novelty that arrives for the most part, but that does not matter; in fact, the present theory delights in a general aimlessness of evolution. If it is a case of something like new passwords being needed (Hamilton, 1980), for example, we are still in an evolutionary realm that is essentially unprogressive and not too unlike the case of true cycles. At the same time this is a realm where it is much easier to think of *frequent* suitable variants occurring than it is with most trends of evolution. So here, as in the cases of true cycles, we see possibilities for the kinds of gene combinations that sex could be busy making up and taking apart and at least can imagine it working on the scale of generations with *combinatorial* efficiency that would make it able to compete with the intrinsic, purely *reproductive* efficiency of parthenogenesis.

REFERENCES

Bell, G. (1982). "The Masterpiece of Nature: The Evolution and Genetics of Sexuality," University of California Press, Berkeley, Calif.

Bremermann, H. J. (1980). Sex and polymorphism as strategies in host-pathogen interactions. *J. Theor. Biol.* **87**, 671-702.

Bremermann, H. J. (in press). The adaptive significance of sexuality. *In* "The Evolutionary Significance of Sex" (S. Stearns, ed.). Birkhauser, Basle.

Eshel, I., and Akin, E. (1983). On the evolutionary instability of inner Nash solutions of two coevolving populations. *J. Math. Biol.* **18**,123-133.

Hamilton, W. D. (1982). Pathogens as causes of genetic diversity in their host populations. *In* "Population Biology of Infectious Diseases" (R. M. Anderson and R. M. May, eds.). Dahlem Konferenzen, 1982, Springer, Berlin.

Hedrick, P. W., Ginevan, M. E., and Ewing, E. P. (1976). Genetic polymorphism in heterogeneous environments. *Ann. Rev. Ecol. Syst.* **7**,1-32.

Holmes, J. C. (1983). Impact of infectious disease agents on the population growth and geographical distribution of animals. *In* "Population Biology of Infectious Diseases" (R. M. Anderson and R. M. May, eds.). Dahlem Konferenzen, 1982, Springer, Berlin.

Jaenike, J. (1978). An hypothesis to account for the maintenance of sex within populations. *Evol. Theory* **3**, 191-194.

Jenkins, D., Watson, A., Miller, G. R. (1964). Predation and red grouse populations. *J. Appl. Ecol.* **1**, 183-195.

Lamb, R. V., and Willey, R. B. (1979). Are parthenogenetic and related bisexual insects equal in fertility? *Evolution* **313**, 774-775.

Lotka, A. J. (1956). "Elements of Mathematical Biology." Dover, N.Y.

May, R. M.(1973). "Stability and Complexity in Model Ecosystem." Princeton University Press, Princeton, N.J.

May, R. M. (1983). Non-linear problems in ecology and resource management. *In* "Chaotic Behaviour of Deterministic Systems" (G. Looss, R. H. G. Helleman, and R. Stora, eds.). North Holland, Amsterdam.

May, R. M., and Anderson, R. M. (1983). Epidemiology and genetics in the coevolution of parasites and hosts. *Proc. Roy. Soc. Lond. B* **219**, 283-313.

Murton, R. K. (1965). "The Wood Pigeon." Collins, London.

Scott, M. E., and Anderson, R. M. (1984). The population dynamics of *Gyrodactylus bullatarudis* (Monogenea) within lab populations of the fish host *Poecilia reticulata. Parasitology* **89**, 159-194.

Scudo, F. M., and Ziegler, J. R. (1970). "The Golden Age of Theoretical Ecology: 1923-1940." Springer, Berlin.

RESTRICTION-MODIFICATION IMMUNITY AND THE MAINTENANCE OF GENETIC DIVERSITY IN BACTERIAL POPULATIONS[1]

Bruce R. Levin

Department of Zoology
University of Massachusetts
Amherst, Massachusetts 01003

ABSTRACT

I present a simple model of the population dynamics of phage and bacteria with restriction-modification immunity. The analysis of this model suggests that: i) in phage-limited communities of bacteria, there are broad conditions under which this type of immune system will maintain clonal diversity via frequency-dependent selection that favors rare restriction-modification types; and ii) when phage-resistant bacteria are present and the community is limited by resources, the conditions for this mechanism to maintain clonal diversity are much narrower. Reviewing that which is known about the genetic structure of natural populations of *E. coli* and the population biology of bacteriophage, I evaluate the proposition that this type of phage-mediated, frequency-dependent selection is responsible for maintaining the considerable chromosomal gene and plasmid diversity found in this bacterial species. Finally, I briefly consider the analogy between this type of phage-mediated selection on restriction-modification diversity and that postulated earlier for parasite-mediated, frequency-dependent selection maintaining antigenic diversity in vertebrates.

[1] This research was supported by a grant from the National Institutes of Health, GM 33782.

I. INTRODUCTION

Natural populations of *E. coli* abound with chromosomal and extra-chromosomal genetic diversity. The genic diversity in this bacterial species, H = 0.5 (estimated by enzyme electrophoresis), considerably exceeds that found in virtually all eukaryotic species (Selander and Levin, 1980). More than 95% of freshly isolated *E. coli* carry at least one plasmid and the majority of these carry more than one (Caugant *et al.*, 1982). In this report, I consider the possibility that some of this chromosomal and extrachromosomal genetic diversity is maintained by frequency-dependent selection mediated by immunity to bacteriophage infection.

II. RESTRICTION-MODIFICATION IMMUNITY TO BACTERIOPHAGE INFECTION

Restriction-modification acts as a two-component immune system that protects bacteria from invasive DNAs, such as those injected by bacteriophage. One component, the *restriction endonuclease,* destroys invading foreign DNA molecules by cutting them at specific sequences of bases (restriction sites). The second component, the *modification enzyme,* tags self-DNA at specific sites (most commonly by the methylation of cytosines and adenines). DNA modified in this way is "passed over" by the restriction endonucleases produced by that bacterium. Occasionally, unmodified or 'wrongly' modified phage DNA evades the host restriction endonucleases and the phage completes its lytic cycle. The DNA of the resulting progeny phage is modified (by the host enzymes), and these viruses are able to attack bacteria of that restriction-modification type without losses due this immune system. For a review of the basic biology of restriction-modification, see Arber (1965).

III. RESTRICTION-MODIFICATION AND THE MAINTENANCE OF CLONAL DIVERSITY

A. A Simple Model for the Population Dynamics of Phage and Bacteria with Restriction-Modification

Restriction-modification can promote the maintenance of clonal diversity in bacterial populations by i) assuring that a *necessary* condition for the stable maintenance of multiple populations of bacteria is met; and ii) by imposing a frequency-dependent selection regime that favors bacteria of rare restriction-modification types. This can be seen from the Levin, Stewart and Chao (1977) model of the population dynamics of phage and bacteria that has been extended to incorporate restriction and modification.

Let n_i be the density of phage-sensitive bacteria of the ith restriction-modification state and p_j be the density of phage of the jth modification state; $I = 1,2,...,I,$ $J = 1,2,...,J.$ The ith bacterial population grows at a rate, $\psi(r)(1-\alpha_i)$ which is a monotonic increasing function of the concentration of a limiting resource, r; and α_i, is the *selection coefficient.* Resources are taken up by the bacteria at a rate proportional to the standard growth rate, $\psi(r)$, the total density of bacteria and a *conversion efficiency* parameter, e. Phage adsorb to sensitive bacteria at random at a rate proportional to their density, that of the bacteria and an *adsorption rate constant,* δ. Phage-resistant bacteria, of density, n_x, that are refractory to the virus grow at a rate $\psi(r)(1-\alpha_x)$. The bacteria and phage are maintained in an "equable" (chemostat) habitat in which the limiting resource, from a reservoir where it is held at a concentration C, enters at a constant rate, ρ, the *dilution rate,* which is the same as the rate at which excess resource, bacteria and phage are removed.

To incorporate restriction and modification, I make the following assumptions:

1) When phage are of the same modification state as the restriction state of the bacteria system, $j = i$, all adsorptions are lethal to the bacteria and β, the *burst size,* phage particles of modification state i are produced.

2) When the phage are unmodified or are of the 'wrong' modification state, $j \neq i$, a constant fraction ω_i, $(0 \leq \omega_i \leq 1)$, the *restriction coefficient,* of the $j \neq i$ adsorption are lethal to the bacteria, but may or may not produce progeny phage.

3) Another constant fraction of the adsorptions of phage of the wrong modification state, σ_i $(0 \leq \sigma_i \leq \omega_i)$, the *modification coefficients,* are lethal to the bacteria and produce β phage particles of modification state i.

4) All adsorptions result in the loss of the infecting phage particle.

With these definitions and assumptions and neglecting time delays, the rates of change in the concentration of resource and the component phage and bacterial populations are given by:

$$\dot{r} = \rho(C-r) - \Psi(r)e(N_T + n_x) \tag{1}$$

$$\dot{n}_i = \Psi(1-\alpha_i)n_i - \delta n_i p_i - \omega_i \delta n_i (P_T - p_i) - \rho n_i \tag{2}$$

$$\dot{p}_j = \delta n_j p_j \beta + \sigma_j \delta n_j \beta (P_T - p_j) - \delta N_T p_j - \rho p_j \tag{3}$$

$$\dot{n}_x = \Psi_i(1-\alpha_x)n_x - \rho n_x \tag{4}$$

where a dot ($\cdot$) denotes differentiation with respect to time, $N_T = \Sigma n_i$ and $P_T = \Sigma p_j$.

B. The Maintenance of Multiple Restriction-Modification States

A necessary condition for the maintenance of stable communities in this model is that the number of bacterial populations, I, be less than or equal to the number of resources (one) plus the number of discrete phage populations, J (Levin *et al.*, 1977). As a consequence of modification, the number of phage populations (of different modification states) would be at least equal to the number of restriction-modification types in the bacterial population. Stated another way, restriction-modification diversity in the bacterial population would generate the phage diversity needed for maintaining that variation in the bacteria.

Bacteria of novel restriction states will be able to invade established communities of bacteria and phage, if their growth rate exceeds their rate of loss due to infection by wrongly modified phage and flow:

$$\psi(r)(1-\alpha_i) > \omega_i \delta_i P_T + \rho . \tag{5}$$

As long as bacteria of each restriction-modification state can invade communities of every other restriction-modification state, then it seems reasonable to assume that multiple populations of bacteria of different restriction-modification states could persist as a stable community. Presumably, the number of distinct bacterial clones maintained in these communities (with a single resource) would be limited only by the number of restriction-modification states. Even greater clonal diversity could be maintained by adding additional resources (Levin *et al.*, 1977).

In the absence of a phage resistant population ($n_x = 0$), these invasion conditions are liberal. The resident bacterial population(s) would be limited by phage and the internal resource concentration, r, would be high. With realistic parameters, r could be very close to the resource concentration in the reservoir, C (Levin *et al.*, 1977; Chao *et al.*, 1977;

Lenski and Levin, 1985). The phage in the habitat would be modified for growth on the established bacterial population(s). While these bacteria would be dividing at near their maximal rate (due to the super-abundance of resources), as a consequence of phage predation, their net rate of reproduction would be lower. At equilibrium, the net growth rate of the bacteria (and phage) in the community would be equal to the dilution rate, ρ. Due to restriction immunity, an invading bacterial population of a novel restriction-modification state would be relatively less affected by the resident population(s) of phage than would the established bacterial population(s), for whom there would already be population(s) of modified phage limiting their growth. Thus, initially the invading population of bacteria of the novel restriction state could have a net rate of reproduction that exceeds that of the established bacteria. This can obtain even when the invading bacteria have intrinsic growth rate and restriction efficiency disadvantages, relative to the established bacteria. As can be seen from (5), if the new bacterial population is entering the community at the phage-limited equilibrium, all that is necessary for their successful invasion is that their near maximal growth rate, $\psi(C)(1-\alpha_i)$, exceeds their rate of loss by flow and mortality due to infection by wrongly modified phage.

As long as phage-resistant bacteria are less fit than sensitive ($\alpha_x > \alpha_i$), their presence will not upset the persistence of established phage-bacterial communities (Campbell, 1961; Levin *et al.*, 1977). The phage will be maintained on the minority population of sensitive bacteria and the community will be limited by resources rather than phage. However, the concentration of available resource in these communities with resistant bacteria would be much lower than in phage-limited communities. As a result of the lower r, the exponential growth rates of invading populations of bacteria would be lower than in the phage-limited case and the conditions for establishment by bacteria of novel restriction states are going to be more narrow. Nevertheless, even in these resource-limited communities there are still conditions where this type of phage-mediated, frequency-dependent selection will allow for the stable maintainance communities of bacteria of multiple restriction-modification states.

C. Simulation Results

To illustrate the major points made in the intuitive arguments presented above, I use numerical solutions to the system of equations (1)– (4). For the growth function, $\psi(r)$ in these simulations, I employ the Monod equation $\psi(r) = Vr/(K+r)$ with values of V and K in a range similar to that for *E. coli* B in glucose minimal medium (Monod, 1949; Levin *et al.*, 1977). The numerical solutions were obtained by the Euler method with a finite step size of $dt = 0.02$ (Boyce and DiPrima, 1977). The parameter values used are given in the legends of Figures 1 and 2.

The stable maintenance of multiple restriction-modification states of bacteria and modification states of phage are illustrated in Figures 1 and 2. In these simulation runs, I assume that the restriction and modification coefficients are equal, $\omega_i = \sigma_i$. In frames b and c of these figures, low densities of bacteria of novel restriction-modification states are introduced into the communities of bacteria and phage at the densities present at the 200th hour of the preceding frame. The bacterial populations of novel restriction-modification states invading in frames b and c of these figures have successively lower fitnesses ($\alpha_3 > \alpha_2 > \alpha_1$) and less efficient restriction-modification systems ($\omega_3 > \omega_2 > \omega_1$).

In Figure 1 resistant bacteria are not present and the community is phage-limited. The density of the invading population of bacteria of novel restriction states increase at an initially high rate but levels off when

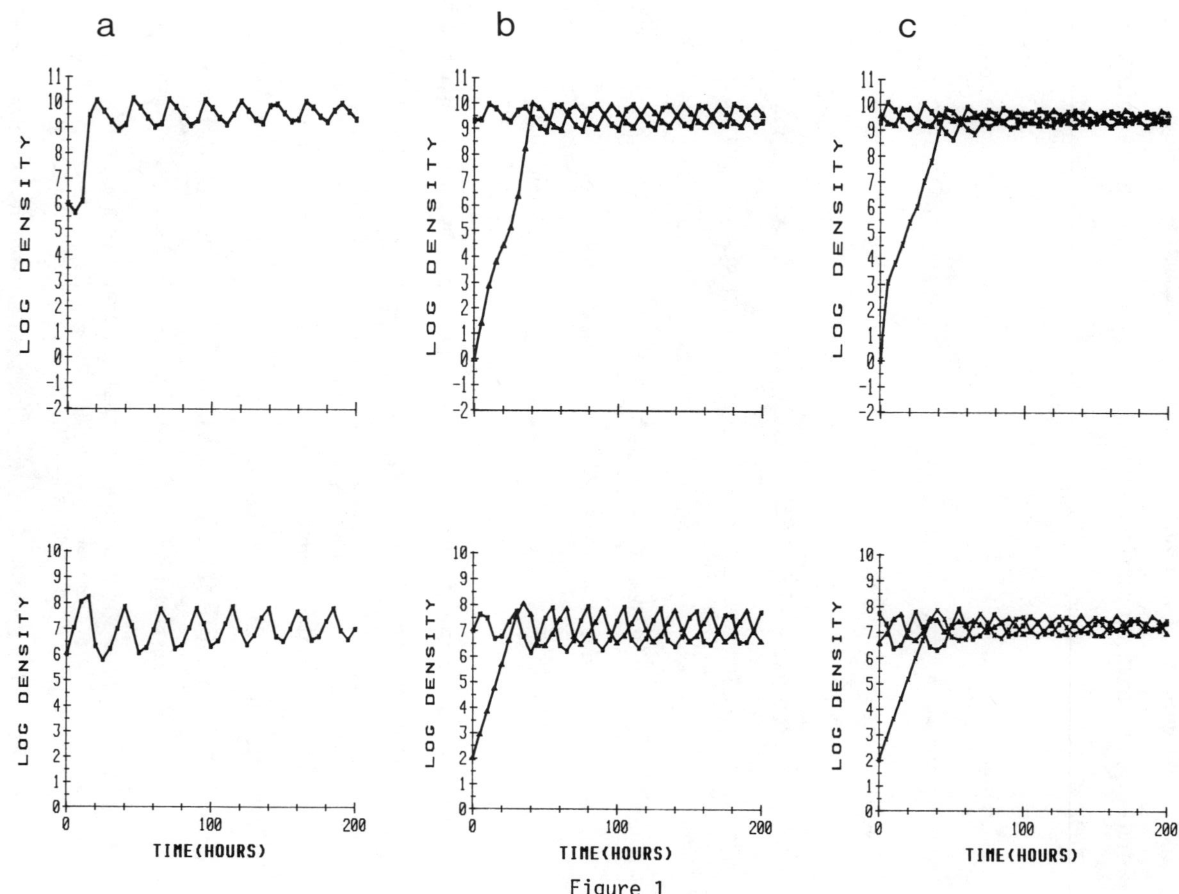

a
b
c
LOG DENSITY
LOG DENSITY
LOG DENSITY
LOG DENSITY
LOG DENSITY
LOG DENSITY
TIME(HOURS)
TIME(HOURS)
TIME(HOURS)

Figure 1

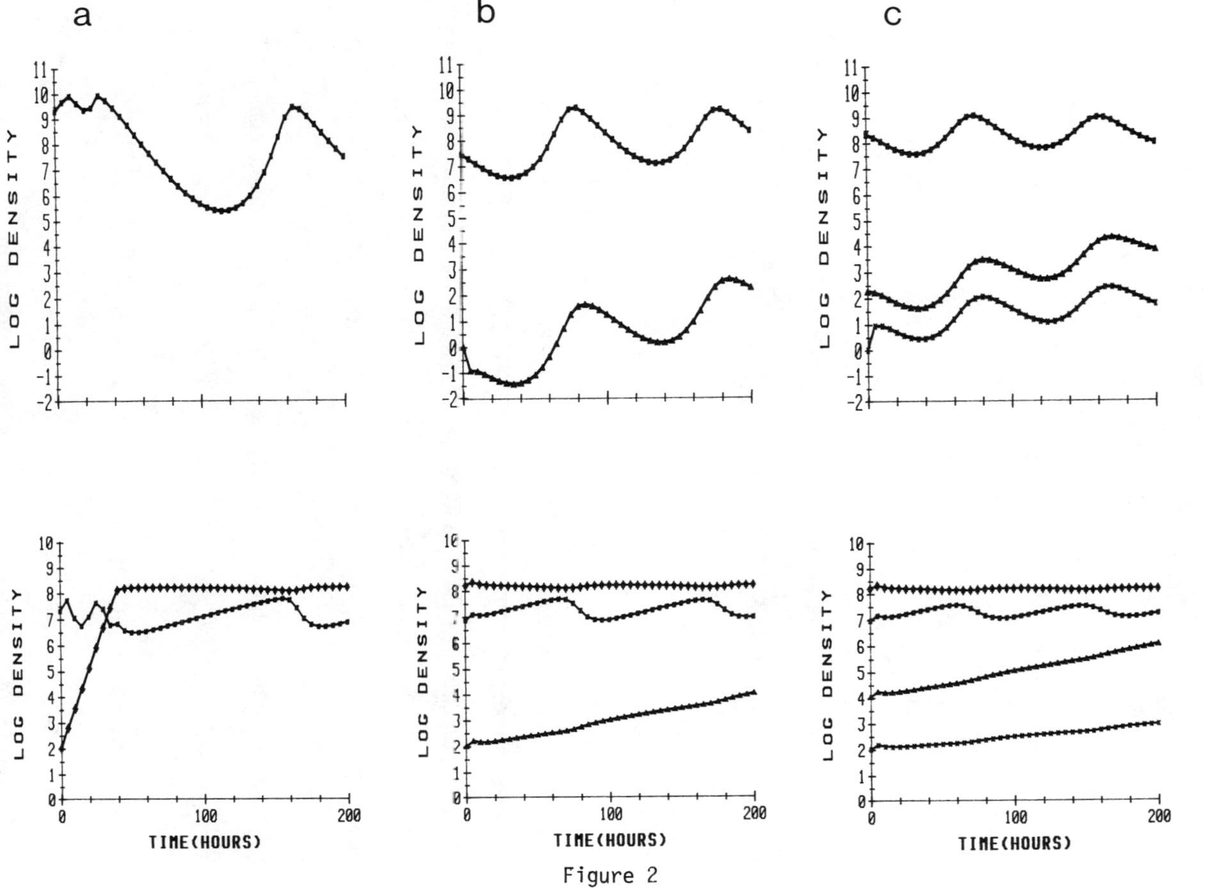

a
b
c
LOG DENSITY
LOG DENSITY
LOG DENSITY
LOG DENSITY
LOG DENSITY
LOG DENSITY
TIME(HOURS)
TIME(HOURS)
TIME(HOURS)

Figure 2

Figure 1 legend. Frequency-dependent selection for restriction-modification in a phage-limited community.

Standard parameter values: $\psi(r) = .70/(4+r)$; $e = 5 \times 10^{-10}$; $p = 0.20$; $C = 100$; $r(0) = 100$; $\delta = 10^{-10}$, $\beta = 100$; $\alpha_1 = 0.00$, $\alpha_2 = 0.05$, $\alpha_3 = 0.10$; $\omega_1 = .0001$, $\omega_2 = .001$, $\omega_3 = .01$ ($\sigma_i = \omega_i$).

a) Establishment of a phage-limited community. b) Invasion of a phage-sensitive population, n_2, into the community depicted at $t = 200$ in Figure 1a. c) Invasion of a phage-sensitive population, n_3, into the community depicted at $t = 200$ in Figure 1b.

Figure 2 legend. Frequency-dependent selection for restriction-modification in a resource-limited community.

The standard parameters are the same as those for Figure 1.

a) Invasion of a phage-resistant population, n_x , into the community depicted at $t = 200$ in Figure 1a; $\alpha_x = 0.15$. b) Invasion of the phage-sensitive population, n_2, into the community depicted at $t = 200$ in Figure 2a. c) Invasion of the phage-sensitive population, n_3, into the community depicted at $t = 200$ in Figure 2b.

phage of the appropriate modification state achieve adequate densities to limit their growth. The consequences of the presence of a phage-resistant populations are illustrated in Figure 2a, where a lower fitness resistant clone is introduced into the community depicted at $t=200$ in Figure 1a. In spite of its growth rate disadvantage, this resistant population invades and is maintained along with established phage and sensitive bacterial populations. However, the rate of ascent of clones of novel restriction-modification states (Figures 2b and 2c) is substantially lower in this resource-limited community than it is in the phage-limited community depicted in the corresponding frames of Figure 1.

D. A Potentially Important Limitation of the Model

In the development of this model, I assumed a constant value for the effectiveness of restriction in protecting cells from mortality due to infection by phage of the 'wrong' modification state, the restriction coefficient, ω_i . I also assumed a constant value for the probability of a productive infection following adsorption by wrongly modified phage, the modification coefficient, σ_i . There is in fact, some evidence that the biology of restriction-modification is somewhat more complex than this. The magnitude of these coefficients are very likely to vary with the physiological state of the cells, the specific phage and restriction-modification systems and the number of phage adsorbing to individual bacteria (multiplicity of infection).

For example, Bertani and Weigle (1953) found that over a multiplicity range of 0.36 to 7.1 there was no significant mortality of *E. coli* B (B-restricting) when infected with the phage P2 that were modified for growth on *Shigella*. The *Shigella*-modified P2 adsorbed equally well on both bacterial species, but had a five order of magnitude lower plating efficiency on *E. coli* B. On the other hand, Paigen and Weinfield (1963) found that the frequency of productive infections (and killed hosts) by unmodified Lambda (grown on *E. coli* C), infecting *E. coli* K-12 (K-modified) varied directly with the multiplicity of infection. With a multiplicity of 1.0, the frequency of productive cells was on the order of 10^{-3}, but with a multiplicity of 8.0 or 9.0, fraction of *E. coli* K-12 yielding phage exceeded 0.10, nearly 0.25 that on the *E. coli* C control.

Although the above observation of multiplicity-dependent kill of restricting bacteria certainly questions the precision of this simple model, it need not alter the qualitative conclusion. As long as restriction-modification augments the survival rate of individual bacteria with this immune systems when phage are present, the potential for this form of frequency dependent selection will exist. However, the conditions for it to operate would be somewhat more limited than the situation considered in the basic model.

IV. DISCUSSION AND CONCLUSIONS

A. The Maintenance of Chromosomal and Extrachromosomal Genetic Diversity by Phage-Mediated Immunity Selection

As presented, this frequency-dependent selection mechanism could account for the maintenance of allelic variation at the specific genes determining restriction-modification specificity. But, without assuming other processes, it is not able to account for the maintenance of genetic diversity at other loci. As a result of recombination, the distribution of alleles at these selected loci would eventually be at random (linkage equilibrium) with those on the rest of the chromosome. However, if the rate of chromosomal gene recombination is extremely low, then as a consequence of local population extinction and recolonization (Maruyama and Kimura, 1980) and periodic selection (Levin, 1981), the genetic structure of asexually reproducing populations would be more that of an array of distinct clones than a randomly distributed gene pool. Permanent linkage disequilibria would exist, and the selection responsible for maintaining the Variation at the restriction-modification loci would, at the same time, result in the stable persistence of clones with different chromosomal genotypes at loci other than those coding for this immunity system.

In cases where restriction-modification is determined by plasmid-borne genes, phage-mediated selection could account for the maintenance of these extrachromosomal genetic elements. This frequency-dependent selection could also maintain a variety of plasmid "species" that differ in regions other than those coding for restriction-modification. The necessary condition for this is analogous to that for phage-mediated selection to maintain chromosomal gene variation, i.e., the rate of inter-plasmid recombination (and transposition) has to be low enough for these genetic molecules to maintain permanent linkage disequilibria between their restriction-modification genes and the rest of their genomes. Since plasmids maintain stable associations with their host bacteria, phage-mediated selection operating on variation in plasmid-determined restriction-modification could also contribute to the maintenance of clonal diversity in bacterial populations.

B. Does Restriction-Modification Account for the Maintenance of Genetic Diversity in *E. coli* ?

It is tempting to suggest that the phage-mediated immunity selection mechanism considered here accounts for much of the enzyme and plasmid variation observed in *E. coli*. However, at this time, there is little justification for a strong advocacy on this issue.

Empirical results certainly support the view that the genetic structure of *E. coli* is that of an array of distinct, and genetically nearly independent clones (Selander and Levin, 1980; Achtman *et al.*, 1982; Ochman and Selander, 1984). Although the genic diversity in this bacterial species is large, genotype variation is relatively modest. Strains of the same enzyme electrophoretic or outer membrane protein type are found in geographically different places and maintain their genetic identity for extended periods. The genetic variation in the *E. coli* flora of individual mammalian hosts can be more readily attributed to clonal migration than in situ recombination (Caugant *et al.*, 1981).

While the absolute extent is not known, there is certainly some variation in chromosomal gene-determined restriction-modification systems in *E. coli*. The laboratory strains *E. coli* B and *E. coli* K12 have different alleles at the loci coding for their Type I restriction-modification systems (Arber,1965), and *E. coli* C is restriction-modification negative. There are also plasmid-borne restriction-modification systems in this bacterial species. The famous type II restriction endonuclease, EcoRI, is coded for by the plasmid RI, which was isolated from *E. coli* (Yoshimori *et al.*,1972). The *E. coli* lactose plasmid pDXX1, described in the appendix to Caugant *et al.* (1981), codes for a restriction-modification system of the Type I class (Goguen, pers. comm.).

Phage have evolved a variety of mechanisms to evade the restriction-modification systems of bacteria. In fact, the majority of laboratory species of coliphage (e.g., T2, T4, T7...) have mechanisms to evade host restriction systems (Kruger and Bickle, 1983). The existence and maintenance of these restriction-evading mechanisms certainly suggests that this type of immune system played an important role in the population dynamics of phage in the past and continues to do so in the present. However, the apparent ubiquity of restriction-evading mechanisms in coliphage also questions whether restriction-sensitive phage are sufficiently frequent in natural communities of *E. coli* to maintain clonal variation by this frequency-dependent selection process.

To obtain an idea of the frequency of naturally occurring bacterial viruses that are sensitive to restriction and modification coded for by *E. coli*, we isolated phage from the influent of a sewage treatment plant in Amherst, Massachusetts (Levin and Laursen, work in progress). Samples were taken on three occasions (February, April and July, 1985) chloroformed and plated on lawns of a restriction-modification negative strain of *E. coli* B (B/6) (Lederberg, 1966). In an effort to maximize the diversity of the sample, phage that produced plaques with a variety of different morphologies were chosen. The 25 phage clones thus isolated were tested for plating efficiency on four lawns: the r^-m^-B/6; an $r^+_B m^+_B$ strain of *E. coli* B (WA251 from B. Bachmann), an $r^+_k m^+_k$ *E. coli* K-12 (CSH 50, Miller, 1972); and an $r^+_x m^+_x$ strain constructed by putting the pDXX1 plasmid (appendix to Caugant *et al.*, 1981) on to a K-12 r⁻m⁻ host (1228, from L. Bullock).

Of the 25 wild phage clones examined, 23 had a 10^{-2} or lower plating efficiency on one or more of the above lawns (relative to that obtained on the r⁻m⁻ B/6 strain it was isolated on). For 12 of these wild phage clones, the reduced plating efficiency can be attributed to restriction rather than resistance. That is, the phage host-range could be modified by growing on r⁺m⁺ hosts and that host-induced modification is lost by growing modified phage on r⁻m⁻ hosts. From the qualitative patterns of restriction sensitivity (some phage did and some did not respond to K restriction), the variation in the extent of the reduction in plating efficiency, and host range on *E. coli* B and K-12 resistant to laboratory phages, it is clear that the 12 restriction sensitive phage represent more than one type. However, at this time we have not yet determined the absolute extent on this diversity. Nevertheless, since 12 of the 25 phage examined were sensitive to host restriction and modification, this preliminary survey certainly supports the proposition that a substantial proportion of naturally occurring coliphage will respond to some restriction-modification systems.

In another way, the results of this and other sewage studies are inconsistent with the hypothesis that phage-mediated selection is important for maintaining variability in bacterial populations. The estimated density of phage in sewage is relatively low, and perhaps too low for these viruses to impose a sufficient selective pressure. In the influent samples considered above, the phage densities estimated on the B/6 lawns were on the order of 10^3 particles per ml. Similarly low concentrations were estimated by other investigators studying the coliphage communities of sewage (Scarpino, 1978; Lerner, 1985). To be sure, laboratory strains of bacteria may not be the appropriate lawn for estimating the densities of wild phage, and sewage may not be the appropriate habitat for this phage-mediated immunity selection. However, free-coliphage are apparently even rarer in fecal samples of humans and domestic animals (Dhillon *et al.*, 1976; Lerner, 1985).

As noted in the preceding theoretical consideration, the breadth of the conditions for restriction-modification immunity to maintain clonal diversity in bacterial populations depend on whether bacterial populations are limited by phage or resources. While in the former case, these conditions appear to be quite broad, in the latter, they are likely to be much narrower. That which is known about the population biology of *E. coli* and virulent phage supports the view that even when virulent viruses are present, they are unlikely to limit the densities of natural populations of this bacterial species. The evidence and arguments for this interpretation are presented in a review by Lenski (1985). The three most compelling lines of evidence are: i) experiments sequentially selecting for resistant *E. coli* and host-range phage tend to terminate with populations of bacteria for which no host-range mutants can be generated (Lenski and Levin, 1985); ii) while phage-limited communities can obtain for extended periods in chemostat culture, in the published sutdies of continuous culture populations of *E. coli* and T phage, these experimental communities eventually become resource-limited due to the presence of phage-resistant bacteria; and iii) as would be anticipated in a resource-limited community, in sewage populations the densities of bacteria exceed those of phage (Scarpino, 1978). Finally, even frequency-dependent selection did favor rare restriction-modification types, the genetic composition of the population may be more determined by the relative fitness of the resistant genotypes of clones than their restriction-modification type.

As a consequence of the caveats implicit in the above consideration, at this time it seems prudent to conclude that, while phage-mediated immunity *could* play a role in the maintenance of the genetic diversity in *E. coli* populations, it is not at all clear how significant this role is. It is also important to emphasize that this phage-mediated immunity selection process is only one of a large number of potential mechanisms for maintaining clonal diversity in bacterial populations.

V. GENERALIZATIONS

The mechanism for phage-mediated frequency-dependent selection considered here is analogous to that postulated for parasite-mediated selection maintaining genetic variation in vertebrates (Bodmer and Bodmer, 1978; Clarke, 1979). Both are consequences of the window of vulnerability against parasites that mimic host cells. For vertebrates this vulnerability is a consequence of the elimination, during the course of development, of clones of antibody-producing cells that generate antibodies that attack self-antigens. In the case of bacterial restriction, this open window is the modification of self-DNA. As long as restriction-sensitive phage represent a significant source mortality, bacteria of rare restriction-modification states will be favored. Similarly, as long as host-mimicking parasites represent a significant source of mortality or morbidity, rare antigenic types will be favored,

> "A parasite in an immunologically competent host could protect itself from attack by reducing its antigenic disparity with the host, and 'mimicking" the hosts antigens. This would produce selection in favour of antigenically variant hosts that could damage the parasite." (Clarke, 1979).

This immunological mechanisms for frequency-dependent selection is particularly appealing because of its parsimony and universality. It is a necessary consequence of the existence of parasites and an intrinsic flaw in generalized immune systems defending against them, i.e., the need to distinguish self from non-self. This type of parasite-mediated frequency-dependent selection can provide a sufficient explanation for the establishment and maintenance of the diversity of restriction-modification systems in bacteria and their plasmids as well as antigenic diversity in vertebrates and their parasites. However, considerably more work will have to be done before we can properly evaluate its importance for maintaining genetic variability in natural populations of bacteria. While this mechanism is appealing as an explanation for the maintenance of the polymorphisms at histocompatibility genes, such as the HLA system of humans (Bodmer and Bodmer, 1978), at this juncture, its generality as a mechanism for maintaining allelic polymorphisms in vertebrates is not at all clear.

ACKNOWLEGMENTS

I am grateful to Allan Campbell and Richard Lenski for critical comments and to Carol Laursen for technical assistance. I also want to acknowledge useful discussions with the EcLF luncheon group.

REFERENCES

Achtman, M., Mercer, A., Busecek, B., Pohl, A., Heuzenroeder, H., Aronson, W., Sutton, A., and Silver, R. P. (1983). Six widespread bacterial clones among *Escherichia coli* K1 isolates. *Infect. and Immun.* **39**, 315-335.

Arber, W. (1965). Host-controlled modification of bacteriophage. *Ann. Rev. Microbiol.* **19**, 365-378.

Bodmer, W. F., and Bodmer, J. G. (1978). Evolution and function of the HLA system. *Br. Med. Bull.* **34**, 309-316.

Boyce, W. E., and DiPrima, R. C. (1977). "Elementary Differential Equations," 3rd Edition. John Wiley, New York.

Campbell, A. M. (1961). Conditions for the existence of bacteriophage. *Evol.* **15**, 153–165.

Caugant, D., Levin, B. R., and Selander, R. K. (1981). Genetic diversity and temporal variation in the *E. coli* populations of a human host. *Genetics* **98**, 467–490.

Chao, L., Levin, B. R., and Stewart, F. M. (1977). A complex community in a simple habitat: an experimental study with bacteria and phage. *Ecology* **58**, 369–378.

Clarke, B. C. (1979). The evolution of genetic diversity. *Proc. Roy. Soc. Lond., Ser. B* **205**, 453–474.

Dhillon, T. S., Dhillon, E. K. S., Chau, H. C., Li, W.-K., and Tsang, H. C. H. (1976). Studies of bacteriophage distribution: virulent and temperate bacteriophage content of mammalian feces. *Appl. Environm. Microbiol.* **32**, 68–74.

Kimura, M. (1983). "The neutral theory of molecular evolution." Cambridge University Press, Cambridge.

Kruger, D. H., and Bickle, T. A. (1983). Bacteriophage survival: multiple mechanisms for avoiding deoxyribonucleic acid restriction systems of their hosts. *Microbiol. Revs.* **47**, 345–360.

Lenski, R. E. (1985). Dynamics of interactions between bacteria and virulent phage. *In* "Microbial Interactions in Communites," Vol. 2 (A. T. Bull and J. H. Slater, eds.). Academic Press, London (in press).

Lenski, R. E., and Levin, B. R. (1985). Constraints on the coevolution of bacteria and virulent phage: a model, some experiments and predictions for natural communities. *Amer. Natur.* (in press).

Lerner, F. L. (1985). The population biology of male-specific bacteriophages: existence conditions. Ph.D. dissertation. University of Massachusetts, Amherst.

Levin, B. R. (1981). Periodic selection, infectious gene exchange and the genetic structure of *Escherichia coli* populations. *Genetics* **99**, 1–23.

Levin, B. R. (1985). The population biology of restriction-modification immunity to bacteriophage infection: A theoretical analysis and evolutionary speculations (submitted manuscript).

Levin, B. R., and Lenski, R. E. (1983). Coevolution in bacteria and their viruses and plasmids. *In* "Coevolution" (D. J. Futuyma and M. Slatkin, eds.), pp. 99–127. Sinauer Assoc., Sunderland.

Levin, B. R., Stewart, F. M., and Chao, L. (1977). Resource limited growth, competition and predation: a model and some experimental studies with bacteria and bacteriophage. *Amer. Natur.* **111**, 3–24.

Ochman, H., and Selander, R. K. (1984). Evidence for clonal population structure in *Escherichia coli*. *Proc. Natl. Acad. Sci. USA* **81**, 198-201.

Maruyama, T., and Kimura, M. (1980). Genetic variability and effective population size when local extinction and recolonization of subpopulations are frequent. *Proc. Natl. Acad. Sci. USA* **77**, 6710-6714

Miller, J. H. (1972). "Experiments in Molecular Genetics." Cold Spring Harbor Laboratory.

Monod, J. (1949). The growth of bacterial cultures. *Ann. Rev. Microbiol.* **3**, 371-394.

Scarpino, P. V. (1978). Bacteriophage indicators. *In* "Indicators of Viruses in Water and Food" (G. Berg, ed.), pp. 201-227. Ann Arbor Science, Michigan.

Selander, R. K., and Levin, B. R. (1980). Genetic diversity and structure in populations of *Escherichia coli*. *Science* **210**, 545-547.

Yoshimori, R., Roulland-Dussiox, D., and Boyer, H. W. (1972). R factor controlled restriction and modification of deoxyribonucleic acid: restriction mutants. *J. Bact.* **112**, 1275-1279.

ALTRUISTIC BEHAVIOR IN SIBLING GROUPS WITH UNRELATED INTRUDERS[1]

Carlo Matessi

Istituto di Genetica Biochimica ed Evoluzionistica
Consiglio Nazionale delle Ricerche
via Abbiategrasso 207, 27100 Pavia, Italy

Samuel Karlin

Department of Mathematics
Stanford University
Stanford, California 94305

ABSTRACT

The conditions for initial increase and fixation of a gene which affects the expression of altruist behavior in social groups comprised of resident siblings and unrelated intruders are studied. In groups of this structure the relatedness of recipients to donors of altruism increases with the extent of relatedness between resident siblings, the proportion of residents in the group and the residents' degree of penetrance of the altruist gene, but decreases with the intruders' level of penetrance. Central to the model is a pair of local fitness functions which prescribe the fitness of the altruist and selfish phenotypes as functions of the composition of local groups into which prereproductives are subdivided. We find that the conditions for the evolution of altruism are sensitive to

[1]Supported in part by NIH Grants GM10452 and HL30856, and NSF Grant MCS 82-15131, and by a grant of the Italian Ministry of Education as part of the national research project on "Island Biogeography".

EVOLUTIONARY PROCESSES
AND THEORY

689

the shape of the fitness functions. Thus, the classical Hamilton Rule (rb > c, where b = benefit to recipient, c = cost to donor and r = recipient to donor relatedness) gives the correct conditions for initial increase and for fixation of altruism only when the fitness functions are linear. The principle which we label as the Hamilton Property, distinguished from the Hamilton Rule, asserts that evolution of altruism is facilitated as relatedness of recipients to donors increases. We find that the Hamilton Property, rather than the Hamilton Rule, is substantially robust over a wide class of fitness functions. However, if the fitness functions are strongly non-linear, the Hamilton Property may fail. In particular, for strongly concave fitness functions, fixation of altruism is easiest when the groups include a moderate fraction of intruders who do not express altruism. On the other hand, if the fitness functions are strongly convex, a moderate fraction of intruders in the group and a high propensity among them to perform altruism provide the most favorable conditions for altruist fixation.

1. INTRODUCTION

The concept of altruism is used in population biology to denote any kind of interaction between a donor and a recipient, where the donor suffers a positive cost in reproductive fitness in exchange for a positive benefit to the recipient's fitness. The behavioral repertoire of social animals in many cases appears to include altruistic components of some form. As a contribution toward understanding the origins of sociality, much theoretical research has been directed to circumscribe the conditions under which altruism can arise in a species by natural selection, and to identify the factors which facilitate its evolution.

The classical theory of altruism, formalized by Hamilton (1964, 1972; cf. Fisher, 1930 and Haldane, 1932; see Michod, 1982, for a recent review and further references), considers genetic relatedness as the decisive factor. Thus, altruism cannot evolve unless recipients are related to donors and its evolution is facilitated by increasing relatedness. These qualitative conclusions were originally formalized by the **Hamilton Rule**, according to which altruism can evolve if and only if

$$cost < benefit \times relatedness.$$

Charlesworth (1978) and Cavalli-Sforza and Feldman (1978) have proposed specific theoretical models where the conditions for evolution of altruism cannot be reduced to the Hamilton Rule. Nevertheless, these different conditions share with the Hamilton Rule the property that if evolution of altruism is feasible at a certain level of relatedness then it will also be feasible at all higher levels. Thus, in order to distinguish this qualitative property from the various quantitative determinations of particular models, we shall say that a given kind of altruism enjoys the **Hamilton Property** if and only if increasing relatedness facilitates its evolution, irrespective of whether it formally satisfies the Hamilton Rule.

In Karlin and Matessi (1983) and Matessi and Karlin (1984), we set forth a general mathematical formulation of the evolution of altruism (which extends a model of Matessi and Jayakar, 1976; also, cf. Levitt, 1975, and Boorman and Levitt, 1980), and have used it to analyze the fate of a single gene which fosters altruism in sibling groups. In this case the average relatedness of a recipient to a donor only depends on the relatedness among siblings and takes the values: 3/4 in the case of haplodiploid sisters, 1/2 for diploid sibs, and 1/4 for haplodiploid half sisters or diploid half sibs. A result of the analysis is that the Hamilton Rule applies only for quite restrictive models, but the Hamilton Property has a considerably wider scope of validity. However, there are cases where even the Hamilton Property can fail, such that altruism can evolve more easily with reduced relatedness.

These results raise new questions. What kinds of altruistic behaviors are endowed with the Hamilton Property? How can they be characterized?

Relatedness, in a social group, may be a function of several factors such as sex ratio (notably in haplodiploid species), forms of polygamy, mating pattern, competitive interactions among potential group members. How do the requisites of the Hamilton Property relative to different components of relatedness compare? And if genetic modification of relatedness is available, in what directions should it evolve?

In this paper we consider the evolution of altruism in local groups which are mixtures of siblings (*residents*) and unrelated individuals (*intruders*). In this context, the average relatedness among group members increases with the degree of relatedness among resident siblings, but also with the proportion of residents in a group, since intruders are unrelated to all group members. The tendency of intruders to perform altruism may differ from that of residents of the same genotype. Thus, the average relatedness of a recipient to a donor is also expected to increase as the tendency of intruders to perform altruism decreases, since in this case the proportion of donors to which residents are related increases.

The classical theory would predict that most social groups include very few intruders which would generally refrain from expressing altruism. Contrary to these expectations we demonstrate the existence of a substantial class of altruistic behaviors that fail to obey the Hamilton Property relative to any of the components of relatedness, so that the predictions of the classical theory are violated in these cases.

In Section II we review our general formulation of the evolution of altruistic traits. Section III details the sibling-intruder model. The conditions for evolution of altruism in this model are presented in Section IV. We identify sufficient conditions for the Hamilton Property to hold, and provide examples which violate the property. Various consequences and biological interpretations are discussed in Section V. Before leaving this introductory section it is useful to define with greater precision some concepts used above in a purely intuitive manner.

A. Criteria for Evolution of Altruism

Evolution of a trait is a dynamic process which cannot be fully characterized by any single event. If however a single event is required as a criterion by which to assess whether evolution of the trait has occurred or not, various natural choices are possible. For example, we could say that evolution of the trait has occurred if the trait successfully invades a population, whatever positive frequency it might ultimately achieve. This is the criterion of **Initial Increase** (I) which is satisfied if each population equilibrium with the trait absent is *locally unstable*. Thus under (I) the genes which are responsible for the trait are protected, i.e., increase in frequency when rare. Alternatively, we may decide that evolution of the trait occurs if the gene or genes for the trait can take over the population. This is the criterion of **Fixation** (F) which obtains whenever an equilibrium at which all individuals are homozygous for the genes associated to the trait is *locally stable*. Other criteria might relate to the existence of stable polymorphisms at one or more loci, where the alleles which are responsible for the given trait segregate with alleles responsible for an alternative trait. We restrict our attention to the criteria of (I) and (F).

B. Factors Which Facilitate Evolution of Altruism

Let π be a parameter, such as relatedness, on which the conditions for Initial Increase (Fixation) of altruism depend. Then the statement

"increasing π facilitates evolution of altruism, with
respect to the criterion of Initial Increase (Fixation)"

signifies that

if I (F) holds for π', then I (F) holds for all $\pi > \pi'$.

In other words, conditions for the fulfillment of I (F) are less stringent for larger values of π.

II. LOCAL GROUPS AND FITNESS FUNCTIONS

We consider an infinite population reproducing in discrete generations. As soon as the population of newborn is produced, it is divided into local groups of finite fixed size N (the results for variable size are qualitatively the same; details will be presented elsewhere). Individuals express one of two possible phenotypes, A-phenotype (altruist) and S-phenotype (selfish), which differ in social behavior. Social interactions take place within the local groups. Prior to reproduction, local groups dissolve and mating occurs in the population at large. The genotypic structure of the adult population and of the local groups of newborn is given in Table I.

Table I. Genotypic structure of the adult population and of a typical local group of newborn.

		Structure	Totals
Genotypes		$G_1 \ldots G_g$	--
Frequencies in adult population		U: $U_1 \ldots U_g$	I
Numbers by phenotype	A	X: $X_1 \ldots X_g$	X
in a local group	S	Y: $Y_1 \ldots Y_g$	Y
Total no. in a local group		Z: $Z_1 \ldots Z_g$	N

Since N is finite, the vector $(\mathbf{X},\mathbf{Y})$ is a random variable. The probability distribution $P(\mathbf{X},\mathbf{Y}|\mathbf{U})$ of $(\mathbf{X},\mathbf{Y})$, given the vector $\mathbf{U}$, is determined by the nature of the process through which local groups are formed. This process, which we designate as **Sampling Scheme,** can be specified in terms of four sets of rules: (1) Mating rules that govern how breeding pairs are formed. (2) Segregation rules that underlie the genotypic composition of the brood of any given pair. These embrace effects of ploidy, multiple loci and alleles, etc. (3) Sampling rules that specify how group members are chosen from the aggregate population of newborn. (4) Expression rules that allocate group members to the A and S phenotypes. Different sampling schemes arise from different determinations for any one of the four sets of rules.

There are various alternative indices that summarize the full properties of $P(\mathbf{X},\mathbf{Y}|\mathbf{U})$ by a single measure of relatedness among group members. A simple, widely used measure (Crozier, 1970; Charlesworth, 1980) is defined in the context of a local group structure as follows.

Definition 1. Let I,J be two random members of a group sampled from distribution P. The relatedness $r(J,I)$ of J to I counts the proportion of alleles at a given locus in I which are identical by descent to alleles in J (note the asymmetry with respect to J and I in this definition). The average relatedness among group members is

$$r_{\text{Mem,Mem}} = E\{r(J,I)\} \, .$$

where $E(\cdot)$ denotes expectation with respect to the sampling of individuals from a group subject to the distribution P. Furthermore, if group members can be assigned to distinct classes B,C,... (e.g., donors and recipients, males and females), then the average relatedness of class C to class B is

$$r_{\mathbf{CB}} = E\{r(J,I)|I \in B, J \in C\} \, .$$

A. Fitness Functions

We assume that A and S phenotypes have different viabilities that in general depend on the number of A and S individuals in a group. Accordingly we represent fitness by a pair of functions (**Local Fitness Functions**):

$$f_A(x), \quad x = 1,...,N \quad \text{and} \quad f_S(x), \quad x = 0,...,N-1 \;,$$

which, for a group with xA and $(N-x)S$, give the probability of surviving to reproduction of A and S individuals, respectively. The specific character of altruistic interactions occurring in different ecological contests (e.g., food provisioning, predator avoidance) may be embodied by different determinations of f_A and f_S .

The concept of local fitness functions leads to a natural definition of cost and benefit of altruism. Thus, we consider the cost to an individual donor A, of behavior A as compared to S, to be the difference between the fitness that the donor would attain, if it were to behave like an S, and the fitness it actually has. The benefit to a recipient (A or S) accruing from a particular donor A, is the difference between the actual fitness of the recipient and the fitness it would attain if the donor were instead to behave like an S. Notice that if a particular A, in a group of xA and $(N-x)S$, should behave like an S, then there would be, effectively, $(x-1)A$ in the group, which would have fitness $f_A(x-1)$, while the others would have fitness $f_S(x-1)$. Accordingly we define, for a group with xA and $(N-x)S$, $x = 1,...,N$, the *individual cost to a donor* and the *cumulative benefit due to an individual donor*, respectively, by

$$c(x) = f_S(x-1) - f_A(x),$$

and

$$b(x) = (x-1)[f_A(x) - f_A(x-1)] + (N-x)[f_S(x) - f_S(x-1)] \;.$$

These we call the **Local Cost** and **Benefit Functions**, respectively. For an altruistic trait we require

$$c(x), b(x) > 0 \quad \text{for} \quad x = 1,...,N .$$

Generally $c(x)$ and $b(x)$ are not constants. In fact, $c(x) \equiv c$ and $b(x) \equiv b$ if and only if $f_S(x) = f_S(0) + (b/(N-1))x$ and $f_A(x) = f_S(0) + (b/(N-1))(x-1) - c$. Accordingly constant cost and benefit correspond to linear fitness functions of the same slope. Notice that the Hamilton Rule implicitly postulates that cost and benefit can be represented by a pair of constants.

B. Recurrence Equations

Given the sampling scheme or equivalently the probability distribution $P(\mathbf{X},\mathbf{Y}|\mathbf{U})$, and the fitness structure $\{f_A,f_S\}$, the equations describing the evolution of the population over successive generations have the form

$$T'U_i' = E\{X_i f_A(X) + Y_i f_S(X)|\mathbf{U}\}, \quad i = 1,....,g ,$$

where T' is a normalization factor such that $U_1' + ... + U_g' = 1$ and $E\{\cdot|\mathbf{U}\}$ denotes the operation of taking expectations under the probability distribution P.

For a wide class of biologically relevant sampling schemes, referred to henceforth as **Conditionally Multinomial**, the distribution P has the form

$$P(\mathbf{X},\mathbf{Y}|\mathbf{U}) = \sum_{\lambda} Q(\lambda|\mathbf{U}) \frac{N!}{X_1!...X_g! \, Y_1!...Y_g!} \lambda_{A,1}^{X_1} \cdots \lambda_{A,g}^{X_g} \lambda_{S,1}^{Y_1} \cdots \lambda_{S,g}^{Y_g} ,$$

where the vector λ has components $\lambda_{A,i}$ and $\lambda_{S,i}$, $i=1,....,g$, which denote the *expected proportions* of AG_i and SG_i individuals respectively (altruist and selfish individuals of genotype G_i) in a given group, and $Q(\lambda|U)$ is a probability distribution for λ which depends on U. Thus, given λ, (X,Y) is multinomially distributed. The total number X of A individuals is binomially distributed with parameter $\theta = \lambda_{A,1} + ... + \lambda_{A,g}$, equal to the *expected proportion* of A in a given group.

With sampling schemes of this kind, the recurrence equations can be expressed in term of the transforms of the local fitness functions f_A and f_S defined by

$$\phi_A(\theta) = \sum_{x=1}^{N} \binom{N}{x} \theta^x (1-\theta)^{N-x} x f_A(x)/N\theta, \qquad 0 \leq \theta \leq 1,$$

$$\phi_S(\theta) = \sum_{x=0}^{N-1} \binom{N}{x} \theta^x (1-\theta)^{N-x}(N-x) f_S(x)/N(1-\theta), \quad 0 \leq \theta \leq 1,$$

which are functions of θ such that the recurrence equations reduce to

$$TU_i' = \sum_{\lambda} Q(\lambda|U)[\lambda_{A,i}\phi_A(\theta) + \lambda_{S,i}\phi_S(\theta)], \quad i = 1,...,g, \qquad (1)$$

where T is a normalization factor. We shall designate ϕ_A and ϕ_S as the **Global Fitness Functions** of A and S individuals, respectively.

The conversion of local into global fitness functions induces an equivalent transformation of the local cost and benefit functions $c(x)$, $b(x)$ into the corresponding global functions $\gamma(\theta)$, $\beta(\theta)$, viz.,

$$\gamma(\theta) = \sum_{x=1}^{N} \binom{N}{x} \theta^x (1-\theta)^{N-x} x c(x)/N\theta,$$

$$\beta(\theta) = \sum_{x=1}^{N} \binom{N}{x} \theta^x (1-\theta)^{N-x} x b(x)/N\theta.$$

The global cost and benefit functions can be expressed in terms of the global fitness function as follows:

$$\gamma(\theta) = \phi_S(\theta) - \phi_A(\theta) \, ,$$

$$\beta(\theta) = \theta\phi_A'(\theta) + (1-\theta)\phi_S'(\theta) \, . \tag{2a}$$

These relations can be inverted yielding

$$\phi_S(\theta) = \phi_S(0) + \theta\gamma(\theta) + \int_0^\theta [\beta(u) - \gamma(u)]du \, ,$$

$$\phi_A(\theta) = \phi_S(\theta) - \gamma(\theta) \, . \tag{2b}$$

Notice that the constant $\phi_S(0)$ must be assigned to determine ϕ_A, ϕ_S uniquely from γ, β. But since the recurrence (1) depends on relative fitness only, without loss of generality we can put $\phi_S(0) = 1$. The one-one correspondence between $\{\phi_A,\phi_S\}$ and $\{\gamma,\beta\}$ allows the evolution of the altruistic trait to be studied in terms of cost and benefit functions.

C. Some Examples

Examination of equations (2a) and (2b) reveals that the global cost and benefit functions are constants, $\gamma(\theta) \equiv \gamma$, $\beta(\theta) \equiv \beta$, if and only if the global fitness functions are linear and parallel: $\phi_S(\theta) = 1 + \beta\theta$, $\phi_A(\theta) = 1 + \beta\theta - \gamma$. This case corresponds to the additive model of Cavalli-Sforza and Feldman (1978). Their additive and multiplicative models can be incorporated into a wider class of models which also includes Charlesworth's models I and II (1978). In this **Generalized CCF-Model** (generalized Charlesworth, Cavalli-Sforza, Feldman model) both ϕ_A and ϕ_S are defined in terms of a single general function as follows:

$$\phi_S(\theta) = 1 + bh(\theta) \, , \tag{3a}$$

$$\phi_A(\theta) = 1 + (1-a)bh(\theta) - c \, , \tag{3b}$$

where $h(\theta)$ is an increasing function with $h(0) = 0$, $h'(0) = 1$, and a, b, c are constants obeying $0 \leq a \leq 1$, $0 < c \leq 1$, $b > 0$.

This model can be interpreted as follows. The presence of altruists in the proportion θ in a local group, increases the fitness of recipients by the amount $bh(\theta)$. Among the altruists, only a fraction $(1-a)$ receive the benefits of altruism while all altruists suffer a constant additive loss of fitness c.

The prescription $a = 1$, $h(\theta) = \theta/(1-\theta)$, produces Charlesworth's model I. The specification $a = c$, $h(\theta) = \theta$, corresponds to Charlesworth's model II, as well as to the Cavalli-Sforza and Feldman multiplicative model. The determination $a = 0$, $h(\theta) = \theta$, yields the Cavalli-Sforza and Feldman additive model.

The cost and benefit functions associated to model (3) are

$$\gamma(\theta) = c + abh(\theta) , \qquad \beta(\theta) = (1-a\theta)bh'(\theta) .$$

Thus, the requirement that h be nonnegative and increasing is a necessary concomitant of the conditions for altruism that $\gamma(\theta)$, $\beta(\theta)$ be positive for all $0 \leq \theta \leq 1$. Obviously c and b indicate the limiting cost and benefit, respectively, as $\theta \to 0$.

III. SIBLING GROUPS WITH UNRELATED INTRUDERS

For the sibling-intruder group structure the sampling scheme is as follows. Matings occur panmictically. A sibship of newborn is selected at random. Then each of the N members of a typical group is chosen at random either from the selected sibship, with probability α, or from the aggregate population of newborn, with probability $(1-\alpha)$. The group members from the sibship are called *residents* and those coming from the population at large are called *intruders*. On average, a group consists of a proportion α residents and a proportion $(1-\alpha)$ intruders.

We focus on three kinds of sibships:

(1) *Haplodiploid sisters*. These consist of the female offspring of a diploid mother and a haploid father. In this case, we assume that males, which are haploid, do not express altruism and are unaffected by social interactions. This model will be designated by [0].

(2) *Diploid sibs*. These consist of the male and female offspring of diploid parents. In this case, social interactions involve all siblings independent of sex. This model will be designated by [1].

(3) *Diploid half sibs*. These consist of the male and female offspring of a common mother but each has a different father. Both sexes are diploid and engage in social interactions. This model will be designated by [2].

Expression of the phenotype, A or S, is governed by the individual's genotype at a single locus admitting two alleles A_1, A_0. The altruist gene A_1 is assumed to be recessive to A_0, in model [R], or dominant, in model [D]. The probabilities of expressing A (penetrance), in the two models [R] and [D] are displayed in Table II. The penetrance of A may depend on the intruder versus resident status of an individual. This could reflect either a nonspecific difference between the two roles such as a general tendency for intruders to be subordinate, which might alter their propensity to express altruism, or the effect of modifier genes which modulate the altruist gene in intruders. The case $p = q$ indicates no differential effects of this kind.

Table II. Probabilities of expressing A, by genotype and by resident-intruder status, when the altruist gene A_1 is recessive [R], and dominant [D] to the selfish gene A_0.

		A_0A_0	A_0A_1	A_1A_1
[R]	Residents	0	0	p
	Intruders	0	0	q
[D]	Residents	0	p	p
	Intruders	0	q	q

A. Recurrence Equations

The sib-intruder sampling scheme is conditionally multinomial. In fact the composition of a local group is clearly multinomial, conditional on the genotypes of the parents of the sibship on which the group is based, in models [0] and [1], or on the genotype of the mother in model [2], provided that intruders and residents are kept distinct.

Table III gives, in the context of a dominant altruist gene with diploid sibship (model [D1]), the expected proportions of the different kinds of individuals (i.e., the different resident and intruder genotypes, by phenotype, A or S) in the local groups for each type of mating. In this listing, the frequencies of genotypes $\{A_0A_0, A_0A_1, A_1A_1\}$ in the parental population are denoted by $\{U_1, U_2, U_3\}$. $u = U_1 + (^1/_2)U_2$ and $v = 1-u = U_3 + (^1/_2)U_2$ indicate the frequencies of the selfish gene A_0 and the altruist gene A_1, respectively.

The entries in the "residents" and "intruders" rows, for each mating type, indicate the components of the vector $\boldsymbol{\lambda}$ in the recurrence equations (1), while the frequencies of the mating types correspond to the probability distribution $Q(\boldsymbol{\lambda}|\mathbf{U})$ for $\boldsymbol{\lambda}$. In agreement with (1), the fitness of group members is given by $\phi_S(\theta)$ or $\phi_A(\theta)$, depending on their

Table III. Expected genotypic-phenotypic composition of local groups in model [D1].

Mating Types		$A_0A_0 \times A_0A_0$	$A_0A_0 \times A_0A_1$	$A_0A_0 \times A_1A_1$	$A_0A_1 \times A_0A_1$	$A_0A_1 \times A_1A_1$	$A_1A_1 \times A_1A_1$
Frequencies		U_1^2	$2U_1U_2$	$2U_1U_2$	U_2	$2U_2U_3$	U_3^2
RESIDENTS A_0A_0	S	α	$\alpha/2$	0	$\alpha/4$	0	0
	A	0	0	0	0	0	0
A_0A_1	S	0	$\alpha(1-p)/2$	$\alpha(1-p)$	$\alpha(1-p)/2$	$\alpha(1-p)/2$	0
	A	0	$\alpha p/2$	αp	$\alpha p/2$	$\alpha p/2$	0
A_1A_1	S	0	0	0	$\alpha(1-p)/4$	$\alpha(1-p)/2$	$\alpha(1-p)$
	A	0	0	0	$\alpha p/4$	$\alpha p/2$	αp
INTRUDERS A_0A_0	S	$(1-\alpha)u^2$	$(1-\alpha)u^2$	$(1-\alpha)u^2$	$(1-\alpha)u^2$	$(1-\alpha)u^2$	$(1-\alpha)u^2$
	A	0	0	0	0	0	0
A_0A_1	S	$2(1-\alpha)(1-q)uv$	$2(1-\alpha)(1-q)uv$	$2(1-\alpha)(1-q)uv$	$2(1-\alpha)(1-q)uv$	$2(1-\alpha)(1-q)uv$	$2(1-\alpha)(1-q)uv$
	A	$2(1-\alpha)quv$	$2(1-\alpha)quv$	$2(1-\alpha)quv$	$2(1-\alpha)quv$	$2(1-\alpha)quv$	$2(1-\alpha)quv$
A_1A_1	S	$(1-\alpha)(1-q)v^2$	$(1-\alpha)(1-q)v^2$	$(1-\alpha)(1-q)v^2$	$(1-\alpha)(1-q)v^2$	$(1-\alpha)(1-q)v^2$	$(1-\alpha)(1-q)v^2$
	A	$(1-\alpha)qv^2$	$(1-\alpha)qv^2$	$(1-\alpha)qv^2$	$(1-\alpha)qv^2$	$(1-\alpha)qv^2$	$(1-\alpha)qv^2$
Total	A	$(1-\alpha)qv(1+u)$	$\alpha p/2 +$ $(1-\alpha)qv(1+u)$	$\alpha p +$ $(1-\alpha)qv(1+u)$	$3\alpha p/4 +$ $(1-\alpha)qv(1+u)$	$\alpha p +$ $(1-\alpha)qv(1+u)$	$\alpha p +$ $(1-\alpha)qv(1+u)$

Legend. U_1, U_2, U_3: population frequencies of genotypes A_0A_0, A_0A_1, A_1A_1, respectively. u: frequency of selfish gene A_0. v: frequency of altruist gene A_1. A,S: altruist and selfish phenotype respectively.

phenotype S or A, respectively, where $\{\phi_S, \phi_A\}$ is a general pair of global fitness functions and θ is the expected proportion of altruists in a group. The values of θ, for each mating type, are indicated in the last row of Table III.

Tables of this kind contain all the information required to derive, in line with the general formula (1), the recurrence system connecting adult genotype frequencies over two successive generations. There are six systems of equations, one for each of the models corresponding to the various combinations $[R,D] \times [0,1,2]$. These recursions cannot be reduced to equations in the gene frequencies only.

B. Relatedness

With the sib-intruder sampling scheme, any two members of a local group are related if and only if both are residents, since intruders are taken at random from the population. The extent of relatedness, s, among residents depends on the kind of sibship. Among the models [0], [1] and [2], s is largest in the case of haplodiploid sisters and smallest in the case of diploid half sibs. The *average* relatedness among group members is expected to increase with s and with the proportion α of residents in a group. Note that when the penetrance of altruism for intruders, q, is increased while the corresponding penetrance for residents, p, is unchanged, the frequency of intruders among donors of altruism certainly increases. On the other hand, if p is increased while q stays constant, the proportion of residents among donors increases. Thus we expect that the average relatedness of a recipient to an altruist (donor) in a local group increases with s, α and p and decreases with q. The Hamilton Property, in the sib-intruder group structure, therefore posits that evolution of altruism is facilitated with increasing s, α and p and with decreasing q.

We determine next a measure of the relatedness $r_{Mem,Alt}$ of general group members to altruist members based on Definition 1. Let π denote the probability that a random member of a local group is a resident, conditional on it being an altruist. The probability that an arbitrary group member is a resident is obviously α. It follows that the probability that, in a random donor-recipient pairing, both individuals are residents is $\alpha\pi$. Therefore $r_{Mem,Alt} = \alpha\pi s$, since the relatedness between any two residents is s. Accounting for a general genotypic composition of the parental population, we have

$$\pi = \frac{\alpha p}{\alpha p + (1-\alpha)q} \tag{4}$$

and we find that

$$r_{Mem,Alt} = \alpha\pi s = \frac{\alpha^2 s p}{\alpha p + (1-\alpha)q}, \tag{5}$$

where, according to Definition 1,

$$s = \begin{cases} 3/4 & \text{for } [0] \\ 1/2 & \text{for } [1] \\ 1/4 & \text{for } [2] \end{cases} . \tag{6}$$

Note that $r_{Mem,Alt}$ of (5) is increasing with s, α and p and decreasing in q.

IV. THE HAMILTON PROPERTY IN THE SIB-INTRUDER MODEL

Determining the conditions for the evolution of altruism on the criteria of Initial Increase (I) and Fixation (F) is equivalent to establishing from the recurrence equations, respectively, the *instability* of the equilibrium state where the selfish gene A_0 is fixed and the *stability* of the equilibrium state of a fixed altruistic gene A_1. Table IV presents these conditions in terms of the global cost and benefit function $\gamma(\theta)$ and $\beta(\theta)$. The following notation is used

$$B(\theta',\theta'') = \frac{1}{\theta''-\theta'} \int_{\theta'}^{\theta''} \beta(u)du, \qquad\qquad C(\theta',\theta'') = \frac{1}{\theta''-\theta'} \int_{\theta'}^{\theta''} \gamma(u)du$$

and

$$D(\theta',\theta'') = \frac{\gamma(\theta'') - \gamma(\theta')}{\theta''-\theta'} .$$

Thus $B(\theta',\theta'')$, $C(\theta',\theta'')$ and $D(\theta',\theta'')$ are average values of $\beta(\theta)$, $\gamma(\theta)$ and $\gamma'(\theta)$ over the interval $\theta' \leq \theta \leq \theta''$. The parameter π in Table IV is defined by (4).

The conditions for Initial Increase in the $[D] \times [0,1,2]$ models and for Fixation in the $[R] \times [0,1,2]$ models affirm that the leading eigenvalue of the linear approximation of the recurrence equations, near the appropriate equilibrium point, is in magnitude greater and smaller than one, respectively. In the other six cases, $[FD] \times [0,1,2]$ and $[IR] \times [0,1,2]$, the study of the linear approximation is not informative because its leading eigenvalue is equal to one, and an analysis of the quadratic terms is required (see Lessard and Karlin, 1982 and Karlin, 1977). These latter cases involve a rare recessive gene, namely the selfish gene A_0 for $[FD]$, or the altruist gene A_1 for $[IR]$. In discussing the results of Table IV we begin with some special cases.

A. Pure Random Groups

For the sib-intruder sampling scheme with $\alpha = 0$, all members of local groups are chosen at random from the aggregate population of

Table IVa. Conditions for Initial Increase of the altruist gene A , in the models $[R,D] \times [0,1,2]$

$$[I]: \text{Initial Increase}$$

[R0]	$$\dfrac{\dfrac{3\alpha^2 p}{2} C(0, \tfrac{\alpha p}{2}) + \alpha(2 - \tfrac{3\alpha}{2}) p\gamma(\tfrac{\alpha p}{2}) + 2(1-\alpha)q\gamma(0)}{2[\alpha p + (1-\alpha)q]} < \dfrac{3}{4} \alpha\pi B(0, \tfrac{\alpha p}{2})$$
[R1]	$$\dfrac{\dfrac{\alpha^2 p}{2} C(0, \tfrac{\alpha p}{4}) + \alpha(1 - \tfrac{\alpha}{2}) p\gamma(\tfrac{\alpha p}{4}) + (1-\alpha)q\gamma(0)}{\alpha p + (1-\alpha)q} < \dfrac{1}{2} \alpha\pi B(0, \tfrac{\alpha p}{4})$$
[R2]	$$\gamma(0) < \dfrac{1}{4} \alpha\pi\beta(0)$$
[D0]	$$\dfrac{\dfrac{\alpha^2 p}{2}[2C(0,\alpha p)+C(0, \tfrac{\alpha p}{2})] + \alpha(1 - \tfrac{\alpha}{2}) p\gamma(\tfrac{\alpha p}{2}) + \alpha(1-\alpha)p\gamma(\alpha p) + 2(1-\alpha)q\gamma(0)}{2[\alpha p + (1-\alpha)q]}$$ $$< \dfrac{3}{4} \alpha\pi \dfrac{2B(0,\alpha p) + B(0, \tfrac{\alpha p}{2})}{3}$$
[D1]	$$\dfrac{\dfrac{\alpha^2 p}{2} C(0, \tfrac{\alpha p}{2}) + \alpha(1 - \tfrac{\alpha}{2}) p\gamma(\tfrac{\alpha p}{2}) + (1-\alpha)q\gamma(0)}{\alpha p + (1-\alpha)q} < \dfrac{1}{2} \alpha\pi B(0, \tfrac{\alpha p}{2})$$
[D2]	$$\dfrac{\dfrac{\alpha^2 p}{2} C(0, \tfrac{\alpha p}{2}) + \alpha p\gamma(0) + \alpha(1 - \tfrac{\alpha}{2})p\gamma(\tfrac{\alpha p}{2}) + 2(1-\alpha)q\gamma(0)}{2[\alpha p + (1-\alpha)q]} < \dfrac{1}{4} \alpha\pi B(0, \tfrac{\alpha p}{4})$$

Legend for Tables IVa and IVb. [R]: A_1 recessive to A_0. [D]: A_1 dominant to A_0. [0]: haplodiploid full sisters. [1]: diploid full sibs. [2]: diploid half-sibs. $\theta_0 = (1-\alpha)q$, $\theta_2 = \tfrac{\alpha}{2} p + (1-\alpha)q$, $\theta_3 = \tfrac{3}{4} p + (1-\alpha)q$, $\theta_1 = \alpha p + (1-\alpha)q$.

Table IVb. Conditions for Fixation of the altrusit gene A_1, in the six
models $[R,D] \times [0,1,2]$

$[F]$: Fixation

$[R0]$	$$\frac{\frac{\alpha^2 p}{2}[2C(\theta_0,\theta_1) + C(\theta_2,\theta_1)] + \alpha(1 - \frac{\alpha}{2})p\gamma(\theta_2) + \alpha(1-\alpha)p\gamma(\theta_0) + 2(1-\alpha)q\gamma(\theta_1)}{2[\alpha p + (1-\alpha)q]}$$ $$< \frac{3}{4}\alpha\pi\left\{\frac{2B(\theta_0,\theta_1) + B(\theta_2,\theta_1)}{3} + (1-\alpha)(q-p)\frac{2D(\theta_0,\theta_1) + D(\theta_2,\theta_1)}{3}\right\}$$
$[R1]$	$$\frac{\frac{\alpha^2 p}{2}C(\theta_2,\theta_1) + \alpha(1 - \frac{\alpha}{2})p\gamma(\theta_2) + (1-\alpha)q\gamma(\theta_1)}{p + (1-\alpha)q}$$ $$< \frac{1}{2}\alpha\pi\{B(\theta_2,\theta_1) + (1-\alpha)(q-p)D(\theta_2,\theta_1)\}$$
$[R2]$	$$\frac{\frac{\alpha^2 p}{2}C(\theta_2,\theta_1) + \alpha p\gamma(\theta_1) + \alpha(1 - \frac{\alpha}{2})p\gamma(\theta_2) + 2(1-\alpha)q\gamma(\theta_1)}{2[\alpha p + (1-\alpha)q]}$$ $$< \frac{1}{4}\alpha\pi\{B(\theta_2,\theta_1) + (1-\alpha)(q-p)D(\theta_2,\theta_1)\}$$
$[D0]$	$$\frac{\frac{3\alpha^2 p}{2}C(\theta_2,\theta_1) + \alpha(2 - \frac{3\alpha}{2})p\gamma(\theta_2) + 2(1-\alpha)q\gamma(\theta_1)}{2[\alpha p + (1-\alpha)q]}$$ $$< \frac{3}{4}\alpha\pi\{B(\theta_2,\theta_1) + (1-\alpha)(q-p)D(\theta_2,\theta_1)\}$$
$[D1]$	$$\frac{\frac{\alpha^2 p}{2}C(\theta_3,\theta_1) + \alpha(1 - \frac{\alpha}{2})p\gamma(\theta_3) + (1-\alpha)q\gamma(\theta_1)}{\alpha p + (1-\alpha)q}$$ $$< \frac{1}{2}\alpha\pi\{B(\theta_3,\theta_1) + (1-\alpha)(q-p)D(\theta_3,\theta_1)\}$$
$[D2]$	$$\gamma(\theta_1) < \frac{1}{4}\alpha\pi\{\beta(\theta_1) + (1-\alpha)(q-p)\gamma'(\theta_1)\}$$

newborn and are therefore unrelated. Altruism cannot evolve in such random groups, at least on both criteria of initial increase and fixation. In fact, the conditions of Table IV, in the limit, $\alpha \to 0$, reduce to

$$\gamma(0) < 0 \quad \text{for [I]} \quad \text{and} \quad \gamma(q) < 0 \quad \text{for [F]} .$$

But an altruist trait is characterized in that $\beta(\theta)$ and $\gamma(\theta)$ are positive for all $0 \le \theta \le 1$.

B. Constant Cost and Benefit

Let $\gamma(\theta) = \gamma_0$, $\beta(\theta) = \beta_0$ for all $0 \le \theta \le 1$, γ_0 and β_0 being two positive constants. This corresponds to the fitness determination $\phi_S(\theta) = 1 + \beta_0\theta$, $\phi_A(\theta) = 1 + \beta_0\theta - \gamma_0$ (cf. Section II). In this case all conditions of Table IV reduce to

$$\gamma_0 < \beta_0\alpha\pi s ,$$

where π is defined in (4) and s follows (6). This condition is precisely the Hamilton Rule in which relatedness is measured by $r_{Mem,Alt} = \alpha\pi s$ (cf. Eq. (5)). Thus, in the special case of constant cost and benefit functions, the Hamilton Rule indeed gives the correct conditions for initial increase and fixation of altruism.

C. Weak Penetrance of the Altruist Gene

When the altruist gene A_1 rarely expresses altruistic behavior, i.e., if both p and q are small, the Hamilton Rule describes an approximate condition ensuring initial increase or fixation which indeed becomes exact in the limit as $p,q \to 0$, provided that cost and benefit are then measured by $\gamma(0)$ and $\beta(0)$, respectively. For example, if the penetrances for residents and intruders are of the same order of magnitude, so that p/q

remains positive and finite in the limit as $p,q \to 0$, all conditions of Table IV converge to

$$\gamma(0) < \beta(0)\alpha s\pi^*$$

where $\pi^* = \lim_{p,q \to 0} \pi = \dfrac{\alpha p/q}{\alpha p/q + (1-\alpha)}$, is the limiting probability for a donor to be a resident. Hence $\alpha s\pi^*$ gives the correct value of $r_{\text{Mem,Alt}}$.

D. The General Case

The Hamilton Rule cannot provide a useful criterion for initial increase and fixation of altruism in the general case. In fact, it postulates measures of cost and benefit which do not depend on any of the factors which influence relatedness. Useful measures of this sort in general do not exist when cost and benefit vary with the proportion of altruists in a group.

Examination of Table IV reveals that all conditions for *initial increase* have the form

$$C < B \, r_{\text{Mem,Alt}} , \tag{7}$$

where the relatedness measure $r_{\text{Mem,Alt}}$ is given by (5), and C and B are weighted averages of the cost and benefit functions, β and γ, over an appropriate interval of [0,1]. Equation (7) formally resembles the Hamilton Rule, but by no means is equivalent to it. In fact, the "average" benefit and cost B and C are themselves functions of the very same parameters which affect $r_{\text{Mem,Alt}}$. In general

$$C = C(s,\alpha,p,q), \quad B = B(s,\alpha,p,q) ,$$

where s serves to identify the sibship model [0], [1], and [2]. In particular, depending on the shape of the cost and benefit functions, it is possible that condition (7) violates even the Hamilton Property, because a change in a parameter which *increases* $r_{Mem,Alt}$ might affect C and B to make (7) *more difficult* to achieve.

Similar considerations apply to the conditions for *fixation*. These have the form

$$C < [B + (1-\alpha)(q-p)D] \, r_{Mem,Alt} \tag{8}$$

where $D = D(s,\alpha,p,q)$ indicates a weighted average of the derivative of the cost function over an appropriate interval of $[0,1]$. Note that (8) achieves the same form as (7) in the case of pure sibling groups ($\alpha = 1$) or where intruders and residents have the same penetrance ($p = q$) or, finally, when the cost function is a constant.

E. Weak Selection

Several authors (e.g., Charlesworth, 1980; Michod, 1982) have argued that the Hamilton Rule provides a good approximation of the conditions for evolution of altruism when selection is weak. However, equations (7) and (8) are generally *independent of the scale of selection intensity*. To account for the magnitude of selection we may prescribe fitness functions as follows.

$$\phi_S(\theta) = 1 + \epsilon h_S(\theta), \quad \phi_A(\theta) = 1 + \epsilon h_A(\theta),$$

where h_S, h_A are general bounded functions such that $h_S(0) = 0$ and $h_A(\theta) < h_S(\theta)$ for all $\theta \in [0,1]$. The parameter ϵ scales the intensity of selection such that small values of ϵ entail small effects on the fitness of all group members, independent of the group composition. Corresponding to the above fitness assignment we have the cost and benefit functions

$$\gamma(\theta) = \epsilon[h_S(\theta) - h_A(\theta)] = \epsilon\gamma^*(\theta),$$

$$\beta(\theta) = \epsilon[\theta h_R'(\theta) + (1-\theta)h_S'(\theta)] = \epsilon\beta^*(\theta) \, ,$$

so that γ^* and β^* are scale independent. Substituting these specifications into the conditions of Table IV shows immediately that the results are independent of ϵ, the intensity of selection. In particular, the conditions (7) and (8) persist if the averages B, C and D are applied to β^* and γ^*. Furthermore, because (7) and (8) are valid, independently of ϵ, we see that the violation of the Hamilton Property may apply even when selection is weak.

F. Kinds of Altruistic Behavior with the Hamilton Property

The scope of validity of the Hamilton Property is more basic than the applicability of the Hamilton Rule, however attractive the simplicity of the Hamilton Rule might be. In this perspective we need to identify the kinds of altruistic behavior whose initial increase, or fixation, is facilitated by increasing relatedness. This depends on the properties of the cost and benefit functions γ and β. Since with the sib-intruder sampling scheme, relatedness depends on the four factors s, α, p and q we have to consider the Hamilton Property relative to each one of these separately and in combination. Thus it is relevant to determine properties of $\{\gamma,\beta\}$ such that initial increase, or fixation, is facilitated by increasing s or α or p or by decreasing q. For example, the case of constant γ and β has the Hamilton Property relative to each one of these factors for both criteria of evolution of altruism by initial increase and by fixation.

We now present results limited to the class of fitness assignments, spanned by the Generalized CCF-Model, introduced in Section II:

$$\phi_S(\theta) = 1 + bh(\theta), \quad \phi_R(\theta) = 1 + (1-a)bh(\theta) - c \, .$$

The associated cost and benefit functions are

$$\gamma(\theta) = c + abh(\theta), \quad \beta(\theta) = (1-a\theta)bh'(\theta) ,$$

subject to the following conditions:

$$b > 0, \; 0 \leq a \leq 1, \; 0 < c \leq 1, \; h(\theta) \uparrow \text{ for } \theta \in [0,1], \; h(0) = 0, \; h'(0) = 1 .$$

Also, we focus on the role of s, α and q. The influence of p is more subtle and will be considered elsewhere.

Table V. Sufficient conditions on the fitness function $h(\theta)$ such that initial increase [I] or fixation [F] is facilitated by the indicated changes of the components of relatedness s, α and q.

A. Increasing Siblings Relatedness s

	[R]	[D]
[I]	convex h	all h
[F]	all h	concave h

B. Increasing Proportion of Residents α

		[R]			[D]		
		[0]	[1]	[2]	[0]	[1]	[2]
[I]			all h			all h	
[F]	$q \leq p$		$\theta h'/h$ increasing with θ			$\theta h'/h$ increasing with θ	
	$q > p$		concave h			concave h	

C. Decreasing Intruder's Penetrance q

	[R]			[D]		
	[0]	[1]	[2]	[0]	[1]	[2]
[I]		all h			all h	
[F]		concave h			concave h	

Legend. [R]: altruist gene A_1 recessive to A_0. [D]: A_1 dominant to A_0. [0]: haplodiploid full sisters. [1]: diploid full sibs. [2]: diploid half-sibs. p: resident's penetrance.

The conditions for initial increase and fixation of the altruist gene A_1 for this class of fitness functions, are readily obtained from Table IV by substitution for γ and β. These are used to determine properties of $h(\theta)$ which are *sufficient* to guarantee that initial increase, or fixation, is facilitated by increasing s or α or by decreasing q. Recall that s is given by (6). The results of this analysis are presented in Table V.

The results convey a striking robustness of the Hamilton Property, at least within the limits of the Generalized CCF-Model of fitness assignments. *Any* fitness function h is guaranteed to have the Hamilton Property with respect to the parameters s, α, q, at least for one of the two criteria [I] or [F]. Relatively mild requirements on the shape of h extend considerably the domain of validity of the Hamilton Property. For example, part C of Table V asserts that, if h is concave, the Hamilton Property with respect to q holds at both initial increase and fixation for dominant and recessive genes and for the three kinds of sibships. The Hamilton Property with respect to α, if intruders are less likely than residents to commit to altruism (part B, $q \le p$, in Table V), holds at fixation if $\theta h'(\theta)/h(\theta)$ is an increasing function. Roughly speaking, this condition requires h to be strongly convex.

G. Kinds of Altruistic Behavior without the Hamilton Property

The Hamilton Property can be violated with respect to any one of the three parameters s, α and q with sufficiently strong departures from the shapes prescribed in Table V. Consider the event of altruist fixation with respect to increasing values of α, when $q \le p$. The Hamilton Property prevails if $\theta h'(\theta)/h(\theta)$ is increasing with θ. An example of its violation arises by specifying h to be the explicit concave function

$$h(\theta) = \frac{\theta}{1 + m\theta} \,, \qquad m > 0 \,, \tag{9}$$

and by taking $a = 0$ in the CCF-Model, which makes $\gamma(\theta) \equiv c$. Assume further that the altruist gene is always expressed among residents (i.e., $p = 1$) and completely repressed among intruders (i.e., $q = 0$). If m is sufficiently large, i.e., h is strongly concave, then *increasing* the proportion α of residents makes the fixation event *more difficult*, when α is near 1. This is illustrated in Figure 1 which gives a pictorial representation of the conditions for fixation of a recessive gene in the diploid sibs case (model [R1]), when $m = 2\sqrt{2}$. Fixation only obtains within the shaded area of the $\{\alpha, b/c\}$ plane. Thus, if b/c is too small, fixation cannot occur for any value of α. If b/c is sufficiently large, fixation occurs for an interval of α extending to 1. But if b/c is moderate, then fixation occurs for a moderate range of α but is prohibited when relatedness is further sharpened by increasing α.

The same choice of h (Eq. 9) provides examples of the violation of the Hamilton Property relative to s, with respect to the criterion of initial increase of a recessive gene (cf. part A of Table V). Moreover, if we choose h to be the explicit convex function

$$h(\theta) = \frac{(1+\theta)^m - 1}{m} \,, \qquad m > 1 \,, \tag{10}$$

and take $a = 0$ in the CCF-Model, then the Hamilton Property is violated relative to q at fixation, provided m is sufficiently large, i.e., h is strongly convex (cf. part C of Table V). In this case fixation is made easier by *increasing* the intruder's penetrance q.

The foregoing conclusions hold even when selection is weak, since h is independent of the scale of selection intensity, which is effectively

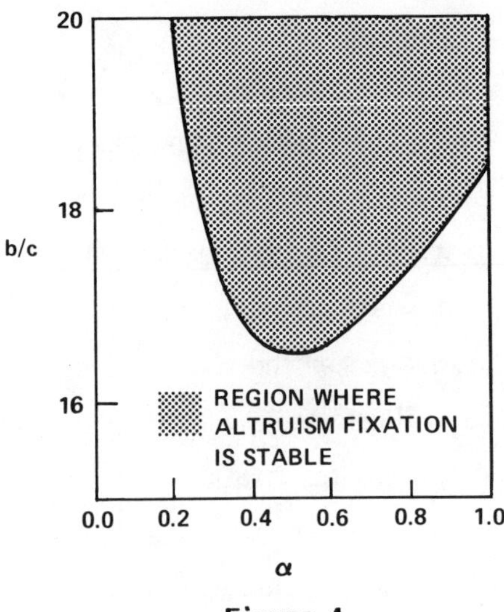

Figure 1

measured by the magnitude of b and c . For example, fixation in Figure 1 only depends on the ratio b/c. It is the shape of the fitness functions, not their scale, which accounts for the failure of the Hamilton Property.

V. REVIEW OF THE RESULTS IN RELATION TO THE CLASSICAL THEORY OF ALTRUISM

We have studied the problem of Initial Increase and Fixation of an altruist gene. Altruism is performed within local groups of fixed size comprised of residents, members of a common sibship, and a number of intruders who are unrelated individuals. Three kinds of sibships are considered (haplodiploid sisters, diploid sibs of both sexes, and diploid half-sibs). Individuals of the appropriate genotype express altruism with probabilities (penetrances) p and q for residents versus intruders, respectively. The average relatedness between a recipient and a donor of altruism is a function of four parameters: (i) the average relatedness among residents, s, depending on the kind of sibship; (ii) the fraction α of residents in the group; (iii) the intruders' penetrance, q; and (iv) the residents' penetrance, p. The relatedness increases with s, α and p, and decreases with q (Eq. 5).

The sib-intruder model compared to the pure sibs model, presents greater flexibility in studying the effects of changing levels of relatedness on the evolution of altruism. Moreover, altruistic interactions between unrelated or distantly related individuals seem to be a real possibility, particularly in vertebrate societies which are less rigidly structured than insect societies. For example, in a community of the social Florida Scrub Jays, where adult non-reproductives assist at nesting a reproductive pair, about 32% of the helpers were unrelated to at least one member of the pair assisted and 3.6% of them were unrelated to both members (based on Table 5.3 in Woolfenden and Fitzpatrick, 1984). Altruistic or cooperative interactions have been observed among ground squirrels, especially between mother and daughter and between littermate sisters, and these seem to be intrinsic to their social life. But also cases have been observed where a young from a different nest has joined a group of siblings and remained with the "foster" family (Sherman, 1980).

In our model, selection is expressed by a pair of local fitness functions which represent the viability of altruist and nonaltruist individuals respectively as a function of the number of altruists in a group. At the population level the local fitness functions are converted into a pair of global fitness functions, which depend on the expected proportion of altruists in a group (Sec. IIA,B). The change over successive generations of the genotypic frequencies in the population involves only the global fitness functions.

In order to assess the domain of validity of the classical theory of altruism, in terms of the properties of the fitness functions, we have analyzed the conditions for Initial Increase and Fixation of the altruist gene. The main results of the classical theory are summarized by two principles, namely the Hamilton Rule and the Hamilton Property (Sec. 1). The first asserts a well known condition for the evolution of altruism. The second indicates the more general and qualitative property that the conditions for the evolution of altruism, in particular the conditions for Initial Increase or Fixation, should be facilitated, that is become less restrictive, as the relatedness of the recipients to donors increases. The Hamilton Rule implies the Hamilton Property, but the Hamilton Property can be satisfied without the Hamilton Rule. In studying the validity of the Hamilton Property, we focused on a particular family of fitness regimes, the Generalized CCF-Model, where both global fitness functions depend on a single function (a study of more general fitness regimes will be presented elsewhere). The results of our analysis clearly suggest that the classical theory of altruism needs to be revised. The reasons and the possible directions of such a revision, as indicated by our results, are outlined in the following discussion.

The classical theory implicitly assumes constant cost and benefit of altruistic behavior. Equivalently, the fitness functions of the two phenotypes are linear and parallel (Sec. IIA,C). In this case the Hamilton Rule provides the exact condition for both Initial Increase and Fixation of an altruistic gene (Sec IVB).

The shape of the fitness functions resulting from any particular case of altruistic behavior is determined by the details of its ecology. For example, in some situations each additional altruist can only procure a diminishing fitness increment to the recipients so that the resulting fitness function is concave. A good example of this case is helping behavior among Florida Scrub Jays. Parents assisted by one or more helpers produce annually about 50% more fledglings than parents without helpers, but this increase is about the same irrespective of the number of helpers (based on Table 8.8 in Woolfenden and Fitzpatrick, 1984). By contrast, a convex fitness function corresponds to situations where each additional altruist contributes marginally more to the recipient's fitness.

The shape of the fitness functions can depend on many specifics. To illustrate, suppose that each altruist procures the same amount of a resource which is then distributed equally among the recipients. If all group members are recipients, the relevant fitness function is linear, but if only non-altruists are recipients the resulting fitness function is convex (cf. models I and II in Charlesworth, 1978). The group size may also significantly affect the shape of the fitness functions. For example, the convex function given by Eq. (10), noted as an example of violation of the Hamilton Property, for integer m is the global counterpart of the local function

$$g(x) = (2^x - 1)/m, \quad x = 0,1,...,m+1$$

for a group of $m+1$ individuals among which x are altruists. In this case, as the group size grows, the strength of convexity of the global function increases.

The foregoing discussion well justifies consideration of altruistic behaviors with nonlinear fitness functions. If the assumption of linearity is relaxed, the Hamilton Rule ceases to be useful as a condition for Initial Increase or Fixation of altruism (Sec. IVD,E). The breakdown of the Hamilton Rule, however, does not preclude the validity of the Hamilton Property, which appears to have much wider scope as attested to by the results of Sec. IVF. We have documented in Table V substantial classes of nonlinear fitness functions that enjoy the Hamilton Property, which moreover cannot be violated in any case at both Initial Increase and

Fixation (but cases of violations at both ends are known for more general fitness regimes). For example, with concave fitness functions, and certain additional conditions on the expression of the altruistic gene (see Table V for details), Initial Increase and Fixation are facilitated whenever the relatedness of recipients to donors is increased either by increasing the relatedness among residents, or by decreasing the proportion of intruders in a group or by decreasing the intruders' penetrance. Similar properties pertain to cases of convex fitness functions.

The Hamilton Property is the basis of two widely held empirical inferences. The first is that altruistic traits will be more frequently found in cases (e.g., species) where the relatedness in groups is higher, so that a decreasing incidence of altruistic traits is expected as relatedness declines. A celebrated application concerns the explanation that it offers for the much higher number of independent occurrences of eusociality among Hymenoptera compared to other insect orders. This expectation tends to influence the direction of related ethological studies. For example, if a seemingly altruistic trait were found to be associated with low levels of relatedness, one might doubt the altruistic nature of the trait and search for direct selective advantages.

The second prediction argues that whenever an altruistic trait has evolved, natural selection should favor inheritable modifications that in some way restrict the set of recipients to be those individuals who are most closely related to the donor. Obviously this inference supports the first. The frequency distribution of relatedness values associated with newly evolved altruistic traits is expected to acquire a stronger skewness in favor of the highest relatedness values by successive modifications.

In the context of the sib-intruder situation, these inferences suggest that most altruistic traits would be found where few or no intruders can ever enter a kin group. Furthermore, among the cases where some intruders are likely to invade kin groups, those where the expression of altruism is repressed in intruders, should be the most common. That is, the joint frequency distribution of the α and q values associated with independently evolved altruistic traits would concentrate on α values near one and q values near zero. These predictions are justified by the Hamilton Property. However, while our results indicate robustness of the Hamilton Property with respect to nonlinearities of the fitness functions, at the same time they also set forth the limits of its validity. As discussed in Section IVG, violations of the Hamilton Property are possible with respect to each of three parameters determining relatedness. In particular, strongly concave fitness functions produce violations relative to α when the intruders' penetrance q is smaller than the residents' penetrance p. Strongly convex fitness functions produce violations of the Hamilton Property with respect to q and, if $q > p$, also with respect to α. These violations can occur only at fixation, since the Hamilton Property is always satisfied at Initial Increase with respect to α and q. When the Hamilton Property is violated relative to α, fixation can occur most easily for intermediate values of α since fixation becomes increasingly difficult to achieve as α shrinks to zero (Sec. IVA). Violation of the Hamilton Property with respect to q may reverse its implication, so that Fixation is facilitiated by increasing rather than decreasing q as in the case of the fitness function given by Eq. (10).

These findings suggest that the standard empirical predictions discussed above are invalid in certain cases and different sorts of predictions should be made depending on the shape of the fitness functions. Thus, consider altruistic traits that have reached fixation and suppose that for all of them, p has some intermediate value such that some but not all residents express altruism. We can tentatively classify these traits into three categories. A different set of predictions applies to each.

(i) *Traits for which the resulting fitness function is strongly concave.* They fulfill the Hamilton Property relative to q. Hence most traits corresponding to such fitness regimes will be associated with values of q close to zero. But when q < p, the Hamilton Property relative to α is violated due to the strong concavity. Then we would expect that most of such traits are associated with intermediate values of α. Relative to these traits the joint frequency distribution of α and q is concentrated on some intermediate range of α and on values of q near zero.

(ii) *Traits for which the resulting fitness function is strongly convex.* The Hamilton Property relative to q is violated and reversed so that increasing q facilitates fixation. The majority of these traits will have q close to one. But q > p (p is assumed intermediate) entails that the Hamilton Property is also violated with respect to α as a consequence of strong convexity. For these traits the joint distribution of α and q is concentrated on some intermediate range of α and on values of q near one.

(iii) *Traits resulting in linear or moderately non-linear fitness functions.* They generally abide by the Hamilton Property relative to both α and q. The joint distribution of α and q conforms to the classical predictions with α near 1 and q close to 0.

This is still a tentative classification since our results have a qualitative nature and are limited to the Generalized CCF family of fitness regimes. More precise characterizations of the shape of fitness functions in relation to the Hamilton Property are certainly desirable. Nevertheless, our results clearly indicate that empirical research should consider accurate measurements of fitness functions since the nature of proper empirical predictions critically depends on the qualitative properties of these functions.

REFERENCES

Boorman, S. A., and Levitt, P. R. (1980). "The Genetics of Altruism." Academic Press, New York.

Cavalli-Sforza, L. L., and Feldman, M. W. (1978). Darwinian selection and altruism. *Theor. Pop. Biol.* **14**, 268-280.

Charlesworth, B. (1978). Some models of the evolution of altrustic behavior between siblings. *J. Theor. Biol.* **72**, 297-319.

Charlesworth, B. (1980). Models of kin selection. *In* "Evolution of Social Behavior: Hypotheses and empirical tests" (H. Markl, ed.), pp. 11-26. Verlag Chemie, Weinheim.

Crozier, R. H. (1970). Coefficients of relationship and the identity of genes by descent in the hymenoptera. *Amer. Natur.* **104**, 216-217.

Fisher, R. A. (1930). "The Genetical Theory of Natural Selection." Oxford University Press, Oxford.

Haldane, J. B. S. (1932). "The Causes of Evolution." Longmans Green, London.

Hamilton, W. D. (1964). The genetical evolution of social behavior. I,II. *J. Theor. Biol.* **7**, 1-52.

Karlin, S. (1977). Protection of recessive and dominant traits in a subdivided population with general migration structure. *Amer. Natur.* **111**, 1145-1162

Karlin, S., and Matessi, C. (1983). Kin selection and altruism. *Proc. R. Soc. Lond. B* **219**, 327-353.

Lessard, S., and Karlin, S. (1982). A criterion for stability-instability at fixation states involving an eigenvalue one with applications in population genetics. *Theor. Pop. Biol.* **22**, 108-126.

Levitt, P. R. (1975). General kin selection models for genetic evolution of sib altruism in diploid and haplodiploid species. *Proc. Natl. Acad. Sci. USA* **72**, 4531-4535.

Matessi, C., and Jayakar, S. D. (1976). Conditions for the evolution of altruism under Darwinian selection. *Theor. Pop. Biol.* **9**, 360-387.

Matessi, C., and Karlin, S. (1984). On the evolution of altruism by kin selection. *Proc. Natl. Acad. Sci. USA* **81**, 1754-1758.

Michod, R. E. (1982). The theory of kin selection. *Ann. Rev. Ecol. Syst.* **13**, 23-55.

Michod, R. E., and Hamilton, W. D. (1980). Coefficients of relatedness in sociobiology. *Nature* **288**, 694–697.

Sherman, P. W. (1980). The limits of ground squirrel nepotism. *In* "Sociobiology: Beyond Nature/Nurture?" (G. W. Barlow and J. Silverberg, eds.), pp. 505–544. Westview Press, Inc., Boulder, Colorado.

Woolfenden, G. E., and Fitzpatrick, J. W. (1984). "The Florida Scrub Jay." Princeton University Press, Princeton, N.J.

TOWARDS A THEORY FOR THE EVOLUTION OF LEARNING

Marcus W. Feldman[1]

Department of Biological Sciences
Stanford University
Stanford, California 94305

Luigi L. Cavalli-Sforza

Department of Genetics
Stanford University Medical School
Stanford, California 94305

ABSTRACT

Population genetic models for the transmission of a learned dichotomous trait are analyzed. The genetics is haploid and the different genotypes learn at different traits. Natural selection acts on the learned phenotype after transmission. It is shown that if the phenotypic transmission is vertical, no genetic polymorphism can be stable. If the learning is horizontal, however, genetic and phenotypic polymorphism is possible. If the haploids are sexual, then the larger set of vertical transmission parameters allows a genetic polymorphism under conditions which are derived.

I. INTRODUCTION

The ability to receive, store and transmit information that is not

strictly encoded in DNA evolved from a primeval state in which all the

[1] Research supported in part by NIH grants GM28016 and GM10452-22.

    725

information that determined survival ability was encoded in the genes. Although we have a vast quantitative literature on evolution under genetically determined natural selection, and a growing literature on quantitative aspects of behavioral evolution, theory pertaining to the passage from the state of completely innate transmission to one where learning of selectively important traits may occur has not seen substantial quantitative development.

In a series of recent studies, we have attempted to address this question of the evolution of learning using extremely simple genetic models. In Cavalli-Sforza and Feldman (1983a) we considered the evolution of communication for which the selection was intrinsically frequently dependent. This frequency dependence entailed that only if the population was in some way structured, for example, if communication occurred primarily between family members, or if mating occurred preferentially between communicators could this social trait increase when rare. The conditions of this model are probably most relevant to the learning of such group-specific behaviors as language. The frequency dependence arises from the fact that any advantage to the learned trait is manifest only when there are both "transmitters" and "receivers." (See also Eshel and Cavalli-Sforza, 1982.)

Another class of models was introduced by Cavalli-Sforza and Feldman (1983b) in order to compare the evolutionary consequences of genetic and nongenetic transmission. Two phenotypes, labeled 1 and 2 are considered. Genotypes are identified by the letters A and a while the subscripted letters A_1, A_2, a_1, a_2 are phenotypes 1 and 2 carrying genotypes A and a, respectively. These are called phenogenotypes. Initially there is a single gene a, and a_1 transmits phenotype 1 rate b to its asexually produced offspring, while all offspring of a_2 are a_2. Following this asexual vertical transmission there is oblique transmission

(Cavalli-Sforza and Feldman, 1981) from members of the *whole* parental generation to the offspring generation at rate f.   Darwinian viability selection then acts with phenotype 2 having fitness 1-s relative to 1 for phenotype 1. (The term "Darwinian" is used to differentiate this type of selection from that due to variation in modes and rates of transmission, which we have called cultural selection). When f < 1, b < 1   and $(1+s)(b+f) > 1$, we showed that a phenotypic polymorphism of a_1 and a_2 is stable. Now, a new mutant, A, arises near this phenotypic equilibrium such that A is necessarily A_1.   In other words, its phenotype is completely genetically determined. We showed that A will always increase; innate determination of the favored phenotype 1 "wins." Actually, under the assumption that f = 0,   a more general result concerning purely vertical transmission holds and is described next.

II. VERTICAL TRANSMISSION ALONE: ASEXUAL HAPLOIDS

Consider again the two allele (A,a) two phenotype (1,2) situation with asexual reproduction and where the transmission rule is specified by the following Table I.

Table I. Transmission coefficients for alleles A and a with phenotypes 1 and 2.

Probability of offspring type

		A_1	A_2	a_1	a_2
	A_1	α_1	$1 - \alpha_1$	0	0
parent	A_2	$1 - \alpha_2$	α_2	0	0
	a_1	0	0	δ_1	$1 - \delta_1$
	a_2	0	0	$1 - \delta_2$	δ_2

The transmission rule of Table I was one of those studied by Feldman and Cavalli-Sforza (1984) where again the viability of phenotype 2 was $1-s$, relative to 1 for that of phenotype 1. Set u_1, u_2, v_1, v_2 to be the frequencies of A_1, A_2, a_1, a_2, respectively. Then the recursion (with non-overlapping generations) specifying the frequencies u_1', u_2', v_1', v_2' in the next generation is

$$\bar{w}u_1' = \alpha_1 u_1 + (1-\alpha_2)u_2 \tag{1a}$$

$$\bar{w}u_2' = (1-s)[(1-\alpha_1)u_1 + \alpha_2 u_2] \tag{1b}$$

$$\bar{w}v_1' = \delta_1 v_1 + (1-\delta_2)v_2 \tag{1c}$$

$$\bar{w}v_2' = (1-s)[(1-\delta_1)v_1 + \delta_2 v_2] \tag{1d}$$

where

$$\bar{w} = 1 - s[(1-\alpha_1)u_1 + (1-\delta_1)v_1 + \alpha_2 u_2 + \delta_2 v_2] . \tag{1e}$$

In this model we showed that there always exists a single stable phenotypically polymorphic equilibrium on each of the two gene fixation edges. Call these $E_A = (\hat{u}_1, \hat{u}_2, 0, 0)$ and $E_a = (0, 0, \hat{v}_1, \hat{v}_2)$. We then showed that when E_A is locally stable to the introduction of a, E_a is unstable to the introduction of A. In other words, it is impossible for both gene fixations to be simultaneously either stable or unstable. The average fitnesses at E_A and E_a are

$$\bar{\hat{w}}_A = \hat{u}_1 s(\alpha_1 + \alpha_2 - 1) + 1 - s\alpha_2 \tag{2a}$$

$$\bar{\hat{w}}_a = \hat{v}_1 s(\delta_1 + \delta_2 - 1) + 1 - s\delta_2 , \tag{2b}$$

where $\hat{u}_1$ solves the quadratic

$$u_1^2 s(\alpha_1 + \alpha_2 - 1) + u_1(2 - \alpha_1 - \alpha_2 - s\alpha_2) - (1 - \alpha_2) = 0 , \tag{3}$$

and $\hat{v}_1$ solves an analogous quadratic with α_1 and α_2 replaced by δ_1 and δ_2.

The system (1) is linear in u_1, u_2, v_1 and v_2, and the matrix on the right partitions into blocks representing A (with u_1 and u_2) and a (with v_1 and v_2). The evolution is then controlled by the leading eigenvalues of the two 2×2 matrices. In fact, it is not difficult to show that the leading eigenvalue of the matrix for A is larger than that for a, so that A will fix, under the condition

$$\overline{\hat{w}_A} > \overline{\hat{w}_a} \, . \tag{4}$$

Only in the highly unlikely case that $\overline{\hat{w}_A} = \overline{\hat{w}_a}$ can a polymorphism result. Thus the marginal mean fitnesses of the alleles are sufficient to qualitatively describe the evolutionary outcome.

III. A MODEL WITH BOTH VERTICAL AND OBLIQUE LEARNING IN ASEXUAL HAPLOIDS

The model described in the introduction included oblique transmission, but was very restrictive in the parameters of vertical transmission. The model described in Section 2 allows general vertical transmission, but no oblique transmission. We proceed to a model that allows a fairly general system of both vertical and oblique transmission parameters.

The model is constructed in stages: Again there are two genotypes A, a, and four phenogenotypes A_1, A_2, a_1, a_2 in frequencies u_1, u_2, v_1, v_2 with $u_1 + u_2 + v_1 + v_2 = 1$. First vertical (parent-to-offspring) transmission takes place at rate α for A_1 and rate δ for a_1. Thus, after vertical transmission there are $\alpha u_1 A_1$ and $(1-\alpha)u_1 A_2$ offspring with $\delta u_3 a_1$ and $(1-\delta)u_3 a_2$ offspring. After vertical transmission oblique transmission occurs via contact between individuals of phenotype 1 in the parental generation and those of phenotype 2 (remaining after vertical transmission) in the offspring and allows the conversion of A and a into A_1 and a_1, respectively. In the most general case the rate of conversion through olibque transmission will be β_1 for an offspring of an A_1 parent who remains A_2 after vertical transmission, and γ_1 for an offspring of an a_1 parent who is still a_2 after vertical transmission. β_2 and γ_2 are the rates of conversion for the A and a offspring of A and a parents. Note that in terms of the parameters of Table 1 we have set $\alpha_2 = \delta_2 = 1$ with $\alpha_1 = \alpha$ and $\delta_1 = \delta$. Allowance is made here for phenotype 2 offspring of phenotype 1 parents to be "sensitized" so that their rate of conversion, by contact, to phenotype 1 differs from that of the offspring of phenotype 2, i.e., $\beta_1 \neq \beta_2$, $\gamma_1 \neq \gamma_2$. A single round of oblique transmission is assumed to occur prior to the action of natural selection.

The fitness of A_2 and a_2 are $1-s$ relative to 1 for A_1 and a_1, i.e., selection, is at the phenotypic level only. The resulting recursions for the frequencies are

$$\overline{w}u_1' = u_1\alpha + (1-\alpha)\beta_1 u_1(u_1+v_1) + \beta_2 u_2(u_1+v_1) \tag{4a}$$

$$\overline{w}u_2' = (1-s)\{u_1(1-\alpha)[1-\beta_1(u_1+v_1)] + u_2[1-\beta_2(u_1+v_1)]\} \tag{4b}$$

$$\overline{w}v_1' = v_1\delta + (1-\delta)\gamma_1 v_1(u_1+v_1) + \gamma_2 v_2(u_1+v_1) \tag{4c}$$

$$\overline{w}v_2' = (1-s)\{v_1(1-\delta)[1-\gamma_1(u_1+v_1)] + v_2[1-\gamma_2(u_1+v_1)]\} \tag{4d}$$

where $\overline{w}$ is the sum of the right hand sides. It will be assumed that all transmission rates, α, β_1, β_2, δ, γ_1, γ_2 are less than unity.

The first question we ask is, when a is fixed what are the conditions under which A will increase? This involves an equilibrium analysis first of the a_1, a_2 polymorphism in the absence of A_1 and A_2 (i.e., $u_1 = u_2 = 0$), and then the stability of such polymorphisms, in the higher dimensional (u_1,u_2,v_1,v_2) frequency space, to the introduction of A_1 and A_2.

For $u_1 = u_2 = 0$ the equilibria from (4c) and (4d) are $\hat{v}_1 = 0$ and the roots of the quadratic equation

$$1-\delta-\gamma_2-s+v_1[s(\gamma_2 + \delta)-\gamma_1(1-\delta)+\gamma_2] + v_1^2 s[\gamma_1(1-\delta)-\gamma_2] = 0. \quad (5)$$

If

$$\gamma_2+\delta+s > 1 , \quad (6)$$

then there exists exactly one polymorphic root v_1 in the interval $(0,1)$. On the other hand, if

$$\gamma_2 + \delta + s < 1 , \quad (7)$$

and

$$\gamma_1(1-\delta) < \gamma_2 \quad (8)$$

no polymorphic equilibrium is admissible.

When (7) holds and the opposite inequality to (8) is true, i.e.,

$$\gamma_1(1-\delta) > \gamma_2 , \quad (9)$$

then it is possible that both roots of (5) are admissible. In addition to (7), this requires

$$\gamma_1(1-\delta) - \gamma_2 > s(\gamma_2+\delta) \quad (10)$$

$$(1-2s)[\gamma_1(1-\delta) - \gamma_2] < s(\gamma_2+\delta) \quad (11)$$

and

$$[\gamma_1(1-\delta) - \gamma_2 + s(\gamma_2+\delta)]^2 > 4s(1-s)[\gamma_1(1-\delta) - \gamma_2] . \quad (12)$$

For example, if $s = 0.7$, $\gamma_2 = \delta = 0.1$ and $\gamma_1 = 0.8$ there are three equilibria $\hat{v}_1 = 0$, $\hat{v}_1 = 0.2784245$, and $\hat{v}_1 = 0.8275663$. Alternatively, when (7) and (9) hold there are no polymorphic equilibria if any of (10), (11), and (12) is violated.

The stability of the equilibria on the boundary where a is fixed is considered first with respect to perturbations in the boundary $u_1 = u_2 = 0$.

Clearly, $\hat{v}_1 = 0$ is locally stable if $\gamma_2 + \delta + s < 1$, and locally unstable if $\gamma_2 + \delta + s > 1$. When (9) holds, it can be shown that the transformation (4c) is monotone increasing and that both $\hat{v}_1 = 0$ and the larger root of (5) can be stable. When $\gamma_2 + \delta + s > 1$ and (8) holds, (4c) may not be monotone, and we have not been able to prove global convergence to the unique polymorphic a_1, a_2 equilibrium. However, numerical iteration indicates that this does occur. In what follows we shall assume that the a_1, a_2 polymorphism is stable in the boundary where $u_1 = u_2 = 0$.

IV. INITIAL INCREASE OF A WHEN a IS FIXED

When a is fixed ($u_1 = u_2 = 0$) it may be fixed either at $\hat{v}_1 = 0$ or at a root of (5) at which there is phenotypic polymorphism. Clearly, if $\hat{v}_1 = 0$, then A can initially increase only if $\alpha > 1-s$, i.e., if the rate of vertical transmission of phenotype 1 in A individuals exceeds the fitness of phenotype 2 in a. More interesting is the case where the mutation to A arises near $\hat{v}_1$, a root of (5). In this case, whether A succeeds in entering the population is determined by the eigenvalues of the 2×2 local stability matrix obtained by linearizing (4a) and (5b) near $\hat{v}_1$. These eigenvalues are the roots of the quadratic

$$\hat{\overline{w}}_a^2 \lambda^2 - \lambda \hat{\overline{w}}_a \{1 + \alpha - s + v_1[\beta_1(1-\alpha) - \beta_2(1-s)]\}$$

$$+ (1-s)\{\alpha + [(1-\alpha)\beta_1 - \beta_2]\hat{v}_1\} = 0 , \qquad (13)$$

where
$$\hat{\overline{w}}_a = 1 - s + s\delta\hat{v}_1 + s\gamma_1\hat{v}_1^2(1-\delta) + s\gamma_2\hat{v}_1(1-\hat{v}_1)$$

$$= \delta + \gamma_2 + \hat{v}_1[\gamma_1(1-\delta) - \gamma_2] . \qquad (14)$$

Again we have used the subscript a on $\hat{\overline{w}}$ to indicate that it is computed at the a_1, a_2 fixation. $\hat{\overline{w}}_A$ refers the corresponding quantity at A_1, A_2, fixation.

Now from (14)

$$\hat{\overline{w}}_a - (1-s) = s\hat{v}_1\hat{\overline{w}}_a ,$$

so that

$$\hat{v}_1 = [\hat{\overline{w}}_a - (1-s)]/s\hat{\overline{w}}_a . \qquad (15)$$

On substitution of (15) into (13), the condition for instability of $\hat{v}_1$ to invasion by A_1, A_2, namely that (13) has a root greater than unity, is

$$[\hat{\overline{w}}_a - (1-s)]\{\delta + \gamma_2 + \hat{v}_1[\gamma_1(1-\delta) - \gamma_2] - \alpha - \beta_2 - \hat{v}_1[\beta_1(1-\alpha) - \beta_2]\} < 0. \qquad (16)$$

Since $\hat{\overline{w}} > 1-s$, condition (16) reduces to

$$\alpha + \beta_2 + \hat{v}_1[\beta_1(1-\alpha) - \beta_2] > \delta + \gamma_2 + \hat{v}_1[\gamma_1(1-\delta) - \gamma_2] . \qquad (17)$$

The criterion (17) for success of A has an interesting qualitative interpretation. In the Appendix we show that (17) is equivalent to

$$\hat{\overline{w}}_A > \hat{\overline{w}}_a , \qquad (18)$$

Thus, the condition (17) for initial increase of A is equivalent to a greater average fitness at the (A_1, A_2) polymorphism than for (a_1, a_2) at the corresponding polymorphism. This condition is, of course, highly suggestive that no genetically polymorphic interior equilibria can be stable.

In the special case $\beta_1 = \beta_2 = \beta$; $\gamma_1 = \gamma_2 = \gamma$, the parental phenotype has no influence on the likelihood that an offspring will be converted by oblique transmission. In this case the equilibrium quadratics (5) and (A1) may only have zero or one valid root. (Two is impossible.) The transformation (4c) in this case is monotone increasing for all values of α, β, γ, δ and s, and the argument is algebraically simpler than the previous general case. The conclusion here is that A increases when rare if

$$\alpha + \beta - \alpha\beta\hat{u}_1 > \gamma + \delta - \gamma\delta\hat{v}_1 . \tag{19}$$

V. VERTICAL AND HORIZONTAL TRANSMISSION

Instead of contact between offspring and members of their parents' generation, we now assume that contact is with peers. Thus, individuals who fail to learn phenotype 1 from their parents contact their peers who did learn and are converted according to the parameters β_i and γ_i. In this case the recursion system (4) of the previous section is modified to become

$$\overline{w}u_1' = u_1\alpha + \beta_1(1-\alpha)u_1(\alpha u_1 + \delta v_1) + \beta_2 u_2(\alpha u_1 + \delta v_1) \tag{20a}$$

$$\overline{w}u_2' = (1-s)\{u_1(1-\alpha)[1 - \beta_1(\alpha u_1 + \delta v_1)] + u_2[1 - \beta_2(\alpha u_1 + \delta v_1)] \tag{20b}$$

$$\overline{w}v_1' = v_1\delta + (1-\delta)\gamma_1 v_1(\alpha u_1 + \delta v_1) + \gamma_2 v_2(\alpha u_1 + \delta v_1) \tag{20c}$$

$$\overline{w}v_2' = (1-s)\{v_1(1-\delta)[1 - \gamma_1(\alpha u_1 + \delta v_1)] + v_2[1 - \gamma_2(\alpha u_1 + \delta v_1)] \tag{20d}$$

where

$$\overline{w} = 1 - s\{v_1(1-\delta)[1 - \gamma_1(\alpha u_1 + \delta v_1)] + v_2[1 - \gamma_2(\alpha u_1 + \delta v_1)]$$

$$+ u_1(1-\alpha)[1 - \beta_1(\alpha u_1 + \delta v_1)] + u_2[1 - \beta_2(\alpha u_1 + \delta v_1)]\} . \tag{20e}$$

Again we consider the initial increase of A near a polymorphism E_a of a_1, a_2 and the dual problem of the initial increase of a near a polymorphism E_A of A_1, A_2. The quadratic corresponding to (5) that specifies E_a is

$$1 - s - \delta - \delta\gamma_2 + v_1[s\gamma_2\delta + s\delta + \gamma_2\delta - \delta(1-\delta)\gamma_1] + v_1^2[\gamma_1(1-\delta) - \gamma_2]s\delta = 0, \tag{21}$$

while the quadratic for E_A is obtained from (21) be replacing δ, γ_1 and γ_2 with α, β_1 and β_2, respectively. Corresponding to (14) the value of the normalizer $\overline{w}$ in (20) at the relevant root of (21) is

$$\hat{\overline{w}}_a = \delta[1 + \gamma_2 + \hat{v}_1[\gamma_1(1-\delta) - \gamma_2)] \, , \tag{22}$$

and at the A-fixation equilibrium

$$\hat{\overline{w}}_A = \alpha\{1 + \beta_2 + \hat{u}_1[\beta_1(1-\alpha) - \beta_2)] \, , \tag{23}$$

The local stability analysis of the gene fixation equilibria proceeds as in the previous section, but with rather different conclusion. We find that E_a is unstable if

$$\delta\{1 + \gamma_2 + \hat{v}_1[\gamma_1(1-\delta) - \gamma_2]\} < \alpha + \beta_2\delta + \hat{v}_1\delta[\beta_1(1-\alpha) - \beta_2], \tag{24a}$$

while E_A is unstable if

$$\alpha\{1 + \beta_2 + \hat{u}_1[\beta_1(1-\alpha) - \beta_2]\} < \delta + \gamma_2\alpha + \hat{u}_1\alpha[\gamma_1(1-\alpha) - \gamma_2]. \tag{24b}$$

In general, these two inequalities may both be valid.

We have carried out an extensive numerical analysis of the recursion system (20) by randomly choosing parameter sets $(s,\alpha,\beta_1,\beta_2,\delta,\gamma_1,\gamma_2)$ and initial vectors (u_1,u_2,v_1,v_2) and iterating (20) to equilibrium. In 15 of 1000 parameter sets chosen in this way, a full genotypic and phenotypic polymorphism was revealed. In no case were both E_A or E_a found to be stable, and, although we cannot rule this possibility out, it must be regarded as highly unlikely. In some 60 cases one of E_A or E_a was stable together with points on the phenotype fixation edge where $u_1 = v_1 = 0$. These points constitute a neutral boundary of equilibria and the precise point attained depends on the initial frequency vector. Clearly there is no corresponding boundary for fixation of phenotype 1 because the vertical transmission of this phenotype is assumed to be imperfect, unlike our assumption on phenotype 2.

A. Sexual Haploids under Uniparental Transmission

Consider again the same phenotypes 1 and 2 and alleles A and a as before, but now suppose that there is random mating between the phenogenotypes. One parent is a transmitter of phenotype 1; the other is not. The mechanism by which the choice of parent to be the transmitter is made is not specified. One can imagine that the genotype at some other set of loci might determine this role. Insofar as the following analysis is concerned, this issue is ignored. If the transmitting parent is A_1, then its offspring of genotype A are A_1 with probability $\beta_{1,1}$, and its

Table II. Uniparental transmission with sexual haploids.

Transmitting parent	Other parent	Mating frequency	Chance of offspring			
			A_1	A_2	a_1	a_2
A_1	A_1	u_1^2	$\beta_{1,1}$	$1-\beta_{1,1}$	0	0
A_1	A_2	$u_1 u_2$	$\beta_{1,1}$	$1-\beta_{1,1}$	0	0
A_1	a_1	$u_1 v_1$	$\beta_{1,1}/2$	$(1-\beta_{1,1})/2$	$\beta_{1,2}/2$	$(1-\beta_{1,2})/2$
A_1	a_2	$u_2 v_2$	$\beta_{1,1}/2$	$(1-\beta_{1,1})/2$	$\beta_{1,2}/2$	$(1-\beta_{1,2})/2$
A_2	A_1	$u_2 u_1$	0	1	0	0
A_2	A_2	u_2^2	0	1	0	0
A_2	a_1	$u_2 v_1$	0	1/2	0	1/2
A_2	a_2	$u_2 v_2$	0	1/2	0	1/2
a_1	A_1	$v_1 u_1$	$\beta_{2,1}/2$	$(1-\beta_{2,1})/2$	$\beta_{2,2}/2$	$(1\beta_{2,2})/2$
a_1	A_2	$v_1 u_2$	$\beta_{2,1}/2$	$(1-\beta_{2,1})/2$	$\beta_{2,2}/2$	$(1\beta_{2,2})/2$
a_1	a_1	v_1^2	0	0	$\beta_{2,2}$	$1-\beta_{2,2}$
a_1	a_2	$v_1 v_2$	0	0	$\beta_{2,2}$	$1-\beta_{2,2}$
a_2	A_1	$v_2 u_1$	0	1/2	0	1/2
a_2	A_2	$v_2 u_2$	0	1/2	0	1/2
a_2	a_1	$v_2 v_1$	0	0	0	1
a_2	a_2	v_2^2	0	0	0	1

offspring of genotype a are a_1 with probability $\beta_{1,2}$. $\beta_{2,1}$ and $\beta_{2,2}$ are corresponding parameters for a_1 parents. We then have the mating structure of Table II with u_1, u_2, v_1, v_2 as before, the frequencies of A_1, A_2, a_1, a_2.

The recursions for the phenotype frequencies after selection are

$$\overline{w}u_1' = \beta_{1,1}u_1[(u_1+u_2) + (v_1+v_2)/2] + \beta_{2,1}v_1(u_1+u_2)/2 \tag{25a}$$

$$\overline{w}u_2' = (1-s)\{(1-\beta_{1,1})u_1[(u_1+u_2) + (v_1+v_2)/2] + (1-\beta_{2,1})v_1(u_1+u_2)/2$$
$$+ [u_2(v_1+v_2) + v_2(u_1+u_2)]/2 + u_2(u_1+u_2)\} \tag{25b}$$

$$\overline{w}v_1' = \beta_{2,2}v_1[(v_1+v_2) + (u_1+u_2)/2] + \beta_{1,2}u_1(v_1+v_2)/2 \tag{25c}$$

$$\overline{w}v_2' = (1-s)\{(1-\beta_{2,2})v_1[(v_1+v_2) + (u_1+u_2)/2] + (1-\beta_{1,2})u_1(v_1+v_2)/2$$
$$+ [v_2(u_1+u_2) + u_2(v_1+v_2)]/2 + u_2(v_1+v_2)\} \tag{25d}$$

where $\overline{w}$ is the sum of the right sides.

When $\beta_{1,1} + s > 1$ there is an A-fixation equilibrium $\hat{u}_1 = (\beta_{1,1} + s - 1)/s\beta_{1,1}$ and if $\beta_{2,2} + s > 1$, the corresponding a-fixation point is $\hat{v}_1 = (\beta_{2,2} + s - 1)/s\beta_{2,2}$. These are the points called E_A and E_a in the previous sections. We again ask for the conditions under which E_A and E_a are stable. It is a relatively simple matter to show that under the existence conditions $\beta_{1,1} + s > 1$, $\beta_{2,2} + s > 1$, E_A is stable to the introduction of a if

$$\beta_{1,1} > (\beta_{1,2} + \beta_{2,2})/2 \tag{26}$$

and E_a is stable to the introduction of A if

$$\beta_{2,2} > (\beta_{2,1} + \beta_{1,1})/2 . \tag{27}$$

Thus, if

$$\beta_{1,2} - \beta_{1,1} > \beta_{1,1} - \beta_{2,2} > \beta_{2,2} - \beta_{2,1} , \tag{28}$$

there is protection of the genetic polymorphism. It is conceivable, however, that the system reaches the phenotypic fixation edge where $u_1 = v_1 = 0$ rather than an isolated polymorphism. This possibility is still under investigation.

VI. DISCUSSION

Aoki (1984) has considered models related to that of Cavalli-Sforza and Feldman (1983a), except that vertical transmission was absent and it was assumed that the transmission rule allowed the recursions to be written in terms only of the gene frequency and phenotype frequency rather than phenogenotype frequencies. Thus there were two free dimensions rather than three. Other treatments of epidemic dynamics with genetic variation by Gillespie (1975) and May and Anderson (1983) assume that within generations of the genotypes the epidemic runs its course to produce genotype dependent fitnesses which are functions of the numbers of each genotype that survive the epidemic.

The models proposed here are extremely simple in their view of the learning process. An offspring may copy an advantageous trait from its parent with the trait, or acquire the trait on contact with its carriers in the population. This combination of vertical and epidemic transmission ignores details of the learning process, in particular age at which learning takes place. In addition, the contact described here is "mass action." Other models involving more than one contact, or different ways of choosing the individuals to be contacted are under investigation.

In the model with oblique transmission, the successful allele is that which produces the highest "mean fitness" at equilibrium (condition (18)). Interpretation of this condition is somewhat complicated because the mean fitness involves the genetically determined contact parameters as well as the phenotypic fitness differences. Under restrictive conditions, for example, $\alpha = \delta$, $\beta_2 = \gamma_2 = 0$, it is true that $\beta_1 > \gamma_1$ entails $\hat{\overline{w}}_A > \hat{\overline{w}}_a$ but with more general parameter sets, this intuition is more difficult to confirm. Thus it is not clear that the winning allele produces the highest frequency of the advantageous phenotype at equilibrium, although this is an entirely reasonable conjecture.

The models do illustrate that alleles which allow increased learning of a behavior which increases fitness, either by imitation of parents, or of others in the population, can succeed and eventually fix. However, with asexual haploids we must be careful; with horizontal transmission, genetic and phenotypic polymorphism is possible. Intuitively, some kind of trade-off between better vertical transmission by one allele and better contagious transmission by the other seems to be essential for this outcome, but the phenomenon warrants further investigation.

With sexual haploids the possibility arises that A_1 parents produce more a_1 offspring than A_1 offspring. (This may be a function either of better parental transmission, or a higher rate of learning by the offspring, or both). When phenotype 1 is the fitter, this is tantamount to "heterozygote advantage" and the possibility of polymorphism is not unexpected.

The models involving asexual haploids described above have clear analogies in microbial ecology where the advantageous phenotypic state might represent antibiotic resistance which is plasmid based, and which may be acquired from the medium. In this event, vertical transmission is likely to be almost perfect, but other modes of transmission may be subject to genetic variation.

APPENDIX: Proof that (17) is equivalent to $\bar{\hat{w}}_A > \bar{\hat{w}}_a$ in Section IV.

From (4a) and (4b)

$$\bar{\hat{w}}_A = \alpha + \beta_2 + \hat{u}_1[\beta_1(1-\alpha) - \beta_2] ,$$

and $\hat{u}_1$ solves

$$1-\alpha-\beta_2-s+u_1[s(\beta_2+\alpha)-\beta_1(1-\alpha)+\beta_2]+u_1^2s[\beta_1(1-\alpha)-\beta_2] = 0. \quad (A1)$$

Here, (A1) is the analogous equilibrium quadratic for A_1, A_2 fixation to (5) which pertains to a_1, a_2 fixation. For $\bar{\hat{w}}_A > \bar{\hat{w}}_a$ we require

$$\alpha + \beta_2 + [\beta_1(1-\alpha) - \beta_2]\hat{u}_1 > \delta + \gamma_2 + \hat{v}_1[\gamma_1(1-\delta) - \gamma_2] , \quad (A2)$$

where $\hat{u}_1$ solves (A1) and $\hat{v}_1$ solves (5). Assume that $\beta_1(1-\alpha) > \beta_2$. (If the opposite inequality holds, a parallel argument works.) Then (A2) entails that

$$\hat{u}_1 > \frac{\delta + \gamma_2 - \alpha - \beta_2 + \hat{v}_1[\gamma_1(1-\delta) - \gamma_2]}{\beta_1(1-\alpha) - \beta_2} = k(\hat{v}_1) , \quad (A3)$$

say. If there is a single polymorphic root of (A1), it is obvious that for (21) to hold we require that the equilibrium quadratic (A1) evaluated at $u_1=k(\hat{v}_1)$ be negative. If there are two roots of (A1), the larger is stable and the requirement is the same. After considerable algebraic simplification the condition for this to occur reduces to (17).

ACKNOWLEDGMENT

The authors are grateful to Joel Peck for his computational assistance and comments on the draft.

REFERENCES

Aoki, K. (1984). A population genetic model of the evolution of oblique cultural transmission. *Proc. Japan Academy* **60, Ser. B**, 310-313.

Cavalli-Sforza, L. L., and Feldman, M. W. (1983b). Paradox of the evolution of communication and of social interactivity. *Proc. Natl. Acad. Sci. USA* **80**, 2017-2021.

Cavalli-Sforza, L. L., and Feldman, M. W. (1983a). Cultural versus genetic adaptation. *Proc. Natl. Acad. Sci. USA* **80**, 4993-4996.

Eshel, I., and Cavalli-Sforza, L. L. (1982). Assortment of encounters and evolution of cooperativeness. *Proc. Natl. Acad. Sci. USA* **79**, 1331-1335.

Feldman, M. W., and Cavalli-Sforza, L. L. (1984). Cultural and biological evolutionary processes: Gene-culture disequilibrium. *Proc. Natl. Acad. Sci. USA* **81**, 1604-1607.

Gillespie, J. H. (1975). Natural selection for resistance to epidemics. *Ecology* **56**, 493-495.

May, R. M., and Anderson, R. M. (1983). Epidemiology and genetics in the coevolution of parasites and hosts. *Proc. R. Soc. Lond. B.* **219**, 281-313.

GENETIC MODELS OF ENDOSPERM EVOLUTION IN HIGHER PLANTS

M. G. Bulmer

Department of Biomathematics
University of Oxford
5 South Parks Road
Oxford OX1 3UB, England

ABSTRACT

The evolutionary significance of the genetics of the endosperm (which is typically triploid in angiosperms, with two contributions from the mother and one from the father) has aroused considerable interest; several authors have attempted to explain it in terms of kin selection. This paper gives a critique of two recent population genetics models of the effect of endosperm genetics on resource allocation to seeds; it then presents a new model for a perennial plant. The main conclusion of the analysis of this model is that kin selection theory predictions are correct provided that the Michod-Hamilton definition of relatedness is used. Finally, the problem of the evolution of triploid endosperm is discussed.

I. INTRODUCTION

The seeds of seed-plants contain three genetically different tissues: the seed-coat (of maternal origin), the endosperm (which acquires nutrients from the mother-plant and stores them for later use by the embryo), and the embryo itself. In gymnosperms the endosperm (*sensu lato*) is of gametophytic origin and is thus a haploid tissue genetically identical with the maternal contribution to the embryo. In angiosperms the endosperm is usually triploid with two contributions identical with the maternal contribution to the embryo (as a result of a doubling of the maternal contribution to the endosperm) and a single contribution identical with the paternal contribution to the embryo (as a result of double fertilization, that is to say fertilization of the endosperm as well as the embryo by identical haploid pollen cells). This is the usual Polygonum type of angiosperm endosperm, but a number of variants exist (Maheshwari, 1950). In all cases there is double fertilization, and the variants differ in the maternal contribution to the endosperm, as shown in Table I. In this table, the four products of meiosis (an ordered tetrad) and their descendants are labelled 1 through 4, with the convention that the maternal contribution to the embryo is labelled 1. For a detailed description see Maheshwari (1950).

The evolutionary significance of endosperm genetics has aroused considerable interest, and several authors have attempted to explain it in terms of kin selection (Charnov, 1979; Westoby and Rice, 1982; Queller, 1983, 1984; Willson and Burley, 1983, pp. 75-93). Following Charlesworth (1980), the relatedness, r_{XY}, between an individual X and the endosperm Y is defined as the probability that X contains a gene identical by descent with a random gene at the same locus sampled from the endosperm. The relatedness is shown in Table II between three types of individual X and the endosperm: X_1, the endosperm's own embryo; X_2, another embryo on the same plant with a different father; X_3, the mother plant itself.

(In calculating $r_{X_1,Y}$ from the information in Table I, note that $1 \equiv 2$ and $3 \equiv 4$ if there is segregation at the first meiotic division whereas $1 \equiv 3$ and $2 \equiv 4$ for second division segregation, where $\equiv$ denotes identity by descent. Observe that $r_{X_1,Y}$ for Allium and Adoxa endosperm depends on

Table I. Maternal contribution to the endosperm of angiosperms.

Type of endosperm	Maternal contribution
Polygonum	11
Oenothera	1
Allium	12
Adoxa	13 or 14, probably with equal frequency
Drusa	12, 13 or 14, probably with equal frequency
Penaea, Fritillaria, Plumbagella, Plumbago	1234
Peperomia	8 cells, probably 11223344

Table II. Relatedness between three types of individual, X, and the endosperm; q is the frequency of second division segregation.

Type of endosperm	X_1 the endosperm's own embryo	X_2, another embryo on same plant	X_3, the maternal plant
Gymnosperm (gametophytic)	1	1/2	1
Polygonum	1	1/3	2/3
Oenothera (zygotic)	1	1/4	1/2
Allium	$1-q/3$	1/3	2/3
Adoxa	$2/3+q/6$	1/3	2/3
Drusa	7/9	1/3	2/3
Penaea etc.	3/5	2/5	4/5
Peperomia	5/9	4/9	8/9
Maternal plant	1/2	1/2	1

the frequency of second division segregation, which varies from 0 for loci near the centromere to 2/3 for distal loci (Fincham, 1983). Previous authors, for example, Law and Cannings (1984), have assumed first division segregation.)

If the endosperm can extract extra resources from the mother plant which benefit its own embryo by an increment in fitness b at a cost c to other embryos on the same plant, kin selection theory predicts that this strategy will succeed if

$$r_{X_1,Y} b > r_{X_2,Y} c .$$ (1)

If the cost is experienced as a decrement in fitness of the mother plant, the strategy will succeed if

$$r_{X_1,Y} b > r_{X_3,Y} c .$$ (2)

In this paper I first consider some population genetics models to determine the accuracy of these predictions of kin selection theory. I then extend one of these models to consider the related but separate question of the evolution of angiosperm endosperm types from the ancestral gymnosperm type.

II. GENETIC MODELS OF SEED PROVISIONING

A. Law and Cannings' Model

Law and Cannings (1984) consider the following model. A fixed resource, M, is to be divided amongst a fixed number of seeds S. There are two endosperm phenotypes, 'overconsumer' and 'underconsumer'. If the proportion of overconsumers on a particular plant is α, the fitnesses of over- and underconsumers on that plant are $w(\alpha)$ and $v(\alpha)$, respectively, with

$$w(\alpha) \geq v(\alpha) ,$$
$$w(1) = v(0) ,$$
$$w'(\alpha) < 0, \quad v'(\alpha) < 0 .$$ (3)

The two phenotypes are determined by a single locus with two alleles. Law and Cannings determine the conditions for the overconsumer allele to spread when rare under different assumptions about its mode of expression and for different endosperm types, assuming random mating, no selfing and a different paternity for all seeds on the same plant.

For example, for Polygonum type endosperm they find that a dominant overconsumer mutant (expressed in single, double or triple dose) will always spread, as will one with a dosage threshold (expressed in double or triple, but not in single dose). However, the condition for the spread of a recessive overconsumer mutant (expressed only in triple dose) is restrictive, being that

$$w(0) + v'(0)/4 > v(0) . \qquad (4)$$

This model is similar to Macnair and Parker's (1979) model for conflict between members of the same full-sib brood for parental resources. They suppose that brood members compete with each other for a fixed total resource, and that an individual's fitness will be an increasing function of his level of solicitation. If the population is fixed at a solicitation level x, a dominant mutant soliciting for level $x + \delta x$ will always spread if $\delta x > 0$, so that x will increase without limit, if there is no cost of solicitation. A recessive mutant, on the other hand, may or may not spread. Macnair and Parker remark that this difference between dominant and recessive mutants is more apparent than real because a recessive mutant will always spread if δx is small, so that x will increase without limit by a succession of small recessive mutations.

To see if this is true of Law and Cannings' model we must inject some more detail into it. Let m_1 and m_2 be the resources obtained by an individual over- and underconsumer, respectively, and suppose that the ratio $p = m_1/m_2$ is independent of α. Then

$$m_1 = p/(\alpha p + 1 - \alpha)$$
$$m_2 = 1/(\alpha p + 1 - \alpha) \qquad (5)$$

if we assume without loss of generality that $M/s = 1$. Suppose also that the fitness of a seed is an increasing function, $f(m)$, of the resource it receives so that $w(\alpha) = f(m_1)$, $v(\alpha) = f(m_2)$. The condition for a recessive mutant to spread from (4) is

$$f(\rho) - (\rho-1)f'(1)/4 > f(1) . \qquad (6)$$

If ρ is near unity, we can write $f(\rho) \simeq f(1) + (\rho-1)f'(1)$, so that the condition becomes $3(\rho-1)f'(1)/4 > 0$, which is certainly true.

The same result holds for other types of endosperm considered by Law and Cannings (1984, Table 4). An overconsumer mutant with constant efficiency ρ slightly greater than unity will always spread whether it is dominant or recessive or has a dosage threshold; the only exceptions occur when the rules of expression ensure that over- and underconsumer phenotypes do not occur on the same plant so that all mutants under this model are neutral.

Thus under this model one expects the endosperm to evolve to extract resources from the parental plant as efficiently as possible. Kin selection does not operate because an underconsumer benefits not a random sample of its half-sibs but a sample selected for not sharing its phenotype. When the most efficient possible endosperm has evolved, each seed will receive the same resource, $m = M/S$. For fixed M one would expect selection to operate on S, the number of seeds produced by the parent plant, to maximize the total reproductive output, $Sf(M/S)$, which leads to the equation in m:

$$f'(m) = f(m)/m . \qquad (7)$$

This can be interpreted as the resource allocation 'desired' by the maternal plant (see (8) and the last line in Table II), which is appropriate since the maternal plant controls S.

It should be mentioned that these conclusions depend on the special case of Law and Cannings' model embodied in Eq. (5). There are other ways of incorporating small effects in the context of their model, and it remains to be seen how general are the conclusions presented here.

B. Queller's Model

Law and Cannings (1984) argue that kin selection does not operate under their model because underconsumers do not benefit a random sample of their half-sibs. They suggest that this condition is most likely to be met, so that kin selection will lead to restraint in resource utilization by the endosperm, when underconsumers release extra maternal resources for future reproduction from which all subsequent seeds will benefit irrespective of their genotype. This echoes the conclusion of Parker and Macnair (1978) and Macnair and Parker (1978, 1979) about the distinction between intra-brood conflict and conflict between successive broods.

Queller (1984) considers a model based on the parent-offspring conflict model of Parker and Macnair when conflict is between successive broods. He supposes that a fixed resource M is to be divided amongst a large number of seeds. A plant producing seeds with over- and under-consumer phenotypes taking resources m_1 and m_2 respectively with frequencies α and $1-\alpha$ will have $S = M/(\alpha m_1 + (1-\alpha)m_2)$ seeds. In this model S is variable while m_1 and m_2 are fixed, so that the penalty for overconsumption is to reduce the total seed production; this penalty falls equally on both phenotypes. In Law and Cannings' model, by contrast, S is fixed and m_1 and m_2 are functions of α, leading to (5).

Queller considers Gymnosperm and Polygonum type endosperm, and finds that under additive gene action a rare mutant taking resource $m+dm$ will spread in a population taking resource m (dm small) if

$$r_{X_1,Y}\, f'(m) > r_{X_2,Y}\, f(m)/m \, . \tag{8}$$

An equilibrium (evolutionarily stable strategy) is attained when the equality holds. Queller notes that this accords with the kin selection argument (1), because the increase in fitness to seeds taking additional resource dm is $f'(m)dm$, while the total decrement in fitness to other seeds on the plant through decreasing their number is $f(m)dm/m$. However, this rule breaks down for Polygonum endosperm under non-additive gene action because of the interaction of the latter with the asymmetry of the parental contributions to the endosperm. As explained later, the rule can be saved by using a more sophisticated definition of relatedness.

Queller justifies this model by supposing that seeds acquire resources in succession (rather than simultaneously) until the total resource M is exhausted. Thus it may serve as a model of an annual plant with a long and continuous flowering season, if it is assumed that the resource for reproduction is present at the beginning of the season in a storage organ (though this is unlikely to be true of an annual plant). The model also assumes that all seeds take the same amount of resource until the total resource is exhausted, whereas the optimal solution would probably lead to seeds at the head of the queue taking more than their successors.

C. Model of a Perennial Plant

So far we have discussed models of an annual plant with either simultaneous flowering (Law and Cannings) or sequential flowering (Queller). I shall now develop a similar model for a perennial plant. Suppose that an adult plant produces S seeds each year until it dies; the total resource allocated to reproduction each year by this plant is $M = S\bar{m}$, where $\bar{m}$ is the mean allocation per seed. M cannot exceed M_0, the resource available for allocation each year, and the remaining resource, $M_0 - M$, is allocated to functions helping the plant to survive. The probability of a plant surviving to reproduce next year is $p(M)$, a decreasing function of M; the probability that a seed will germinate and survive to reproduce next year is $f(m)$, an increasing function of m, the resource allocated to that seed. For simplicity it will be assumed that the fecundity and mortality of adult plants are age-independent. The evolution of both S and m must be considered. I shall first consider the evolution of m for fixed S.

To illustrate the methodology I shall discuss in detail the evolution of m when it is controlled by a gametophytic (gymnosperm) endosperm. Suppose that resource allocation to endosperm is controlled by a locus with two alleles, so that there are two endosperm genotypes, C_1 and C_2, receiving resources m_1 and m_2, respectively. Let the frequencies of the genotypes C_1C_1, C_1C_2 and C_2C_2 among adult plants in a particular year be x, y and z, and write $P = x + \frac{1}{2} y$, $Q = \frac{1}{2} y + z$ for the gene frequencies of C_1 and C_2. With random pollination we can evaluate the genotypes of new plants flowering the following year as follows:

Maternal genotype	Endosperm/Embryo genotype	Frequency of surviving seedlings
C_1C_1	C_1/C_1C_1	$SxPf(m_1)$
	C_1/C_1C_2	$SxQf(m_1)$
C_1C_2	C_1/C_1C_1	$\frac{1}{2} SyPf(m_1)$
	C_1/C_1C_2	$\frac{1}{2} SyQf(m_1)$
	C_2/C_1C_2	$\frac{1}{2} SyPf(m_2)$
	C_2/C_2C_2	$\frac{1}{2} SyQf(m_2)$
C_2C_2	C_2/C_1C_2	$SzPf(m_2)$
	C_2/C_2C_2	$SzQf(m_2)$
Total		$SPf(m_1) + SQf(m_2)$

After adding in the existing adult plants which survive, the genotype frequencies the following year are

$$x' = \{SP^2 f(m_1) + xp(Sm_1)\}/\overline{w}$$

$$y' = \{SPQ[f(m_1) + f(m_2)] + yp(\tfrac{1}{2}Sm_1 + \tfrac{1}{2}Sm_2)\}/\overline{w}$$

$$z' = \{SQ^2 f(m_2) + zp(Sm_2)\}/\overline{w} \tag{9}$$

where

$$\overline{w} = SPf(m_1) + SQf(m_2) + xp(Sm_1) + yp(\tfrac{1}{2}Sm_1 + \tfrac{1}{2}Sm_2) + zp(Sm_2). \tag{10}$$

We now write $m_1 = m$, $m_2 = m + dm$ where dm is small. Under weak selection we may suppose that to order dm the population is in Hardy-Weinberg equilibrium, so that $x = P^2 + O(dm)$ and so on. Expanding the expression for $Q' = \tfrac{1}{2}y' + z'$ in a MacLaurin series, we find to order dm that

$$\Delta Q \equiv Q' - Q = \tfrac{1}{2}SPQ[Sf(m) + p(Sm)]^{-1}[f'(m) + p'(Sm)]dm \tag{11}$$

where f' and p' denote first derivatives.

The direction in which m evolves will depend on the sign of $f' + p'$. (Remember that $f' > 0$ and $p' < 0$.) Suppose first that m takes its maximum value M_0/S; if $f'(M_0/S) + p'(M_0) > 0$, a mutant with negative dm will not invade, and there is a stable situation with the plant devoting all its resources to reproduction. Such a plant will presumably be an annual. Otherwise the population will evolve to a value $m < M_0/S$ such that $f'(m) + p'(Sm) = 0$. (Note that the criterion for ΔQ to be positive does not depend on Q; a mutant which can invade will go to fixation.) This result agrees with the kin selection result (2) since the benefit to the seed of an additional resource dm is $f'dm$ and the cost to the parent plant is $-p'dm$ and since $r_{X_1,Y} = r_{X_3,Y}$ for gymnosperm endosperm.

Table III. Selection pressure on the endosperm in a perennial plant.

Type of endosperm	ΔQ (divided by $SPQ/(Sf+p)$)
Gymnosperm	$\frac{1}{2}(f'+p')d_1$
Maternal	$(\frac{1}{2}f'+p')[Pd_1 + Q(d_2-d_1)]$
Zygotic	$(f'+\frac{1}{2}p')[Pd_1 + Q(d_2-d_1)]$
Polygonum (general)	$[(\frac{1}{2}-Q)(d_1+d_2)+Qd_3]f'+\frac{1}{2}[-Qd_1+Pd_2+Qd_3]p'$
" (dominant, $d_1=d_2=d_3=d$)	$P(f'+\frac{1}{2}p')d$
" (recessive, $d_1=d_2=0,d_3=d$)	$Q(f'+\frac{1}{2}p')d$
" (threshold, $d_1=0,\ d_2=d_3=d$)	$\frac{1}{2}(f'+p')d$
" (additive, $d_i = id$)	$(1\frac{1}{2}f'+p')d$

This analysis was extended to other types of endosperm, and the results are given in Table III. For an endosperm with ploidy π it was assumed that an endosperm genotype with i C_2 and $\pi-i$ C_1 alleles would acquire resources $m+d_i$, with $d_0 = 0$ and d_i always small. The results closely parallel those of Queller (1984), bearing in mind that kin selection theory predicts the criterion (8) for the spread of a mutant taking increased resources under his model, while it predicts the criterion

$$r_{X_1,Y}\, f'(m) > -r_{X_3,Y}\, p'(Sm) \tag{12}$$

under the present model.

The kin selection prediction (12) is correct for haploid (gymnosperm) endosperm and for diploid endosperm (whether maternal or zygotic), and a mutant which can invade when rare will go to fixation, provided only that there is not overdominance in the diploid case. For triploid (Polygonum) endosperm, the kin selection prediction is only valid under additive gene action. With complete dominance or recessivity Polygonum behaves like zygotic endosperm; with a dosage threshold like gametophytic endosperm.

These results have an intuitive explanation. With complete dominance or recessivity the doubling of the maternal contribution to the endosperm does not affect the phenotype, so that it behaves like zygotic (Oenothera) endosperm. With a dosage threshold, the maternal contribution to the endosperm determines the phenotype, so that it behaves like gametophytic endosperm.

It is not difficult to show for any of the endosperm types in Table I that the kin selection prediction (12) is valid under additive gene action of small effect. Let p_{ij} be the probability that a mother of genotype C_1C_2 contributes i C_2 alleles to the endosperm and j C_2 alleles to the zygote ($i = 0,1,...,\pi-1$; $j = 0,1$), and that the last allele in the endosperm is contributed by the father through double fertilization. Then

$$r_{X_1.Y} = (1 + 2\Sigma i p_{11})/\pi$$
$$r_{X_3.Y} = (\pi-1)/\pi . \tag{13}$$

If endosperm with i C_2 and $\pi-i$ C_1 alleles takes resource $m+id$, where d is small, the argument used above shows that, to order d,

$$\Delta Q = \frac{\frac{1}{2}\pi\, SPQd}{Sf + p} (r_{X_1.Y}\, f' + r_{X_3.Y}\, p') , \tag{14}$$

in accordance with kin selection theory. Additive gene action seems the most likely type of action, at least for genes of small effect.

The kin selection prediction (12) is only valid for Polygonum endosperm under additive gene action if one uses the simple expression for relatedness shown in Table II, but it is valid in general for all the types of endosperm in Table III if one uses the more general definition of relatedness proposed by Michod and Hamilton (1980). (See Seger (1981) for an illuminating derivation and discussion of different definitions of relatedness using the covariance theorem of Price (1970, 1972).) The Michod–Hamilton definition is

$$R_{XY} = \text{Cov}(Y_p, X_g)/\text{Cov}(Y_p, Y_g) \,. \tag{15}$$

In this formula Y_p is the phenotypic value of the endosperm, which in the present model is d_i for an endosperm of ploidy π with i C_2 and $\pi - i$ C_1 alleles, Y_g is the genetic value of the endosperm, defined as i/π, and X_g is the genetic value of the diploid individual X defined as the relative frequency of C_2 alleles in his genotype $(0, {}^1\!/_2$ or $1)$. The covariances are calculated by writing down the joint distributions of the pairs of random variables in the absence of selection, so that the results will only be valid under weak selection.

For Polygonum endosperm endosperm we find

$$R_{X_1,Y} = \{({}^1\!/_2 - Q)(d_1 + d_2) + Qd_3\}/\{({}^1\!/_3 - Q)d_1 + ({}^2\!/_3 - Q)d_2 + Qd_3\}$$
$$R_{X_3,Y} = {}^1\!/_2 \{-Qd_1 + Pd_2 + Qd_3\}/\{({}^1\!/_3 - Q)d_1 + ({}^2\!/_3 - Q)d_2 + Qd_3\}. \tag{16}$$

Comparison with Table III shows that substitution of these expressions in (12) gives the correct criterion.

There is, however, a curious feature of the above results. Under additive gene action $(d_i \propto i)$, one would expect that R_{XY} from (15) should equal r_{XY} calculated in Table II, but this is not the case. For example, from (16) $R_{X_1,Y} = 9/10$, $R_{X_3,Y} = 3/5$ which disagree with the values for Polygonum endosperm in Table II, although the ratios of the two coefficients are the same. $(R_{X_1,Y}/R_{X_3,Y} = r_{X_2,Y}/r_{X_3,Y} = 3/2.)$ This paradox is not resolved by redefining the genetic value as the absolute rather than the relative gene frequency (e.g., 0, 1 or 2 rather than 0, ${}^1\!/_2$ or 1 for a diploid) since this gives $R_{X_1,Y} = 3/5$, $R_{X_3,Y} = 2/5$.

It should also be noted that the outcome of selection for Polygonum endosperm depends on the mode of gene action, so that strictly speaking no evolutionarily stable strategy exists. For example, if an equilibrium has been reached under additive gene action, an allele causing m to increase will invade and spread to fixation if it is dominant or recessive, while an allele causing m to decrease will spread if it has a threshold. As noted above, additive gene action seems most likely, at least for the genes of small effect considered here, but the different consequences of other types of gene action should be borne in mind.

Let us now consider the evolution of S for fixed m, which is presumably under the control of the mother plant. If the maternal genotypes C_1C_1, C_1C_2 and C_2C_2 produce S, $S+d_1$ and $S+d_2$ seeds, respectively, we find in the same way as before that

$$Q = \frac{PQ(\tfrac{1}{2} f + mp')}{Sf + p} (Pd_1 + Q(d_2 - d_1)), \tag{17}$$

if d_1 and d_2 are small. If there is not overdominance, then at equilibrium

$$\tfrac{1}{2} f + mp' = 0. \tag{18}$$

Furthermore, in a stationary population we must have

$$1 - p(M) = Sf(m) = \frac{M}{m} f(m)$$

so that M satisfies

$$p'(M) + (1 - p(M))/2M = 0 \tag{20}$$

provided that the solution of this equation is less than M_0. As an example, suppose that

$$p(M) = 1 - e^{-\beta(M_0 - M)}. \tag{21}$$

The equilibrium value of M from (20) is

$$M = 1/2\beta \tag{22}$$

provided that this is less than M_0. (The stationarity condition in Eq. (19) should not be thought of as an extra equation for m and M with the

Table IV. The equilibrium value of m from (23) and (24).

$r_{X_1,Y}/r_{X_3,Y}$ Type of endosperm	$1/2$ maternal	1 haploid (gametophytic)	$1\,1/2$ polygonum	2 zygotic (oenothera)
α				
1	1.49	3.05	3.67	4.08
2	2.92	4.15	4.72	5.11
4	5.54	6.51	7.02	7.37
8	10.22	11.05	11.51	11.82

functions f and p fixed. Rather, it expresses the fact that under density-dependent regulation with fixed m and M, the functions f and p will change through changes in population density until Eq. (19) is satisfied.)

Finally, consider the coevolution of m and S under additive gene action for the endosperm. The equations

$$r_{X_1,Y}\, f' + r_{X_3,Y}\, p' = 0 \tag{12 bis}$$

$$1/2\, f + mp' = 0 \tag{18 bis}$$

lead to

$$r_{X_1,Y}\, f' - r_{X_3,Y}\, f/2m = 0 . \tag{23}$$

As an example, suppose that

$$f(m) = p(1 - e^{-m})/(1 + e^{\alpha - m}) . \tag{24}$$

(This is about the simplest function which is increasing, convex for small m so avoiding the solution $m = 0$ and concave for large m.) The solution of (23) depends on the two parameters α and $r_{X_1,Y}/r_{X_3,Y}$, and is shown in Table IV. As expected, m increases as the embryo becomes more closely related to the endosperm, but the magnitude of the effect depends on the shape of the function f as measured by the parameter α.

III. HOW DID TRIPLOID ENDOSPERM EVOLVE?

In the last section we took the genetic constitution of the endosperm as given, and considered how that constitution might have affected resource allocation to seeds. I shall now consider how the peculiar genetic constitution of the endosperm might itself have evolved. The evolution of Polygonum endosperm, normal in angiosperms, from the primitive gametophytic type, normal in gymnosperms, requires two events, double fertilization and a doubling of the maternal contribution. Double fertilization is impossible in gymnosperms because the endosperm develops several weeks or even months before ferilization of the ovum, but it became possible in angiosperms when the time of fertilization of the ovum was brought forward to coincide with the beginning of endosperm development. A plausible evolutionary scenario suggests that double fertilization, once it became physiologically possible, arose as an attempt by the father to gain more resources for 'his' seed at the expense of other seeds on the mother plant, and that the doubling of the maternal contribution then evolved as the mother plant's response to regain control of the endosperm for herself (Charnov, 1979; Queller, 1983; Willson and Burley, 1983). This theory regards conflict between the interests of the two parents as the driving force behind the evolution of triploid endosperm. I shall now subject these verbal arguments to a population genetic analysis.

Consider first the evolution of double fertilization. Imagine a haploid, gametophytic endosperm with wild type allele C_1 and suppose that a mutant, C_2, expressed in pollen, makes diploid (zygotic) endosperm by double fertilization. There are four possible endosperm genotypes: C_1, C_2, C_1C_2 and C_2C_2. The first and third arise if the maternal contribution to the endosprem and zygote is C_1, the second and fourth if it is C_2.

Since the only function of the endosperm is to acquire resources for the embryo, the mutant allele will be selectively neutral unless the four genotypes differ in the resources they acquire, and we assume that these are m, $m+d_2$, $m+d_3$ and $m+d_4$, respectively. Since most gymnosperms are perennial, we use the model of a perennial plant developed in the last section, and by the same argument as before (assuming the d_i's to be small) we find that

$$\Delta Q = \frac{\tfrac{1}{2}\,SPQ}{(Sf + p)}\,\{[d_3 + 2Q(d_4 - d_3)]f' + [Pd_2 + Q(d_4 - d_3)]p'\}\,. \quad (25)$$

Assume now that resource allocation is at equilibrium for gametophytic endosperm so that $f' + p' = 0$ (see (11) and Table III). The expression inside the curly braces in (25) is $p(d_3-d_2) + Qd_4$. The conditions for the gene to invade when rare and then go to fixation are that (i) $d_3 > d_2$, (ii) $d_4 > 0$. If the d_i's are of the same sign, and hence positive if (ii) is satisfied, the first condition says that C_2 has a greater effect in a diploid heterozygote than in a hemizygote. This only seems plausible if the effect on resource allocation is largely due to diploidy itself rather than to an epistatic effect of C_2 or to a locus closely linked to it. Thus the physiological theory of Brink and Cooper (1947), which suggests that double fertilization evolved because it enhanced the efficiency of the endosperm through heterosis, is not at variance with the parental conflict theory. A pure effect of diploidy would be represented by putting $d_2 = 0$, $d_3 = d_4 > 0$. Substituting these values into (25) shows that the expression inside the braces becomes d_3f', which is always positive. Thus if diploid endosperm takes more resources than haploid endosperm, a gene causing double fertilization with no other side effect will always go to fixation. The intuitive reason for this is that it is always in the father's interest to increase investment in his own seed at the expense of any other seeds the mother plant may produce to whom he is unrelated.

Consider now the doubling of the maternal contribution as a response by the mother to regain control of the endosperm. Imagine a diploid, zygotic endosperm with wild type allele C_2, and suppose that a mutant C_3 expressed in the gametophyte makes triploid (Polygonum) endosperm by doubling the maternal contribution. There are four possible endosperm genotypes: C_2C_2, C_2C_3, $C_2C_3C_3$, $C_3C_3C_3$. The first two arise if the haploid gametophyte is C_2, the last two if it is C_3; the first and third arise if the pollen is C_2, the second and fourth if it is C_3. Suppose that the resources acquired by these four genotypes are m, $m+d_2$, $m+d_3$ and $m+d_4$, respectively, where the d_i's are small. If Q is the frequency of the C_3 allele, we find in the same way as before that

$$\Delta Q = \frac{\tfrac{1}{2} SPQ}{(Sf + p)} \{[(1-2Q(d_2+d_3)+2Qd_4]f' + [-Qd_2+Pd_3+Qd_4]\} . \quad (25)$$

Assuming that resource allocation is at equilibrium for zygotic endosperm so that $f' + \tfrac{1}{2} p' = 0$, the expression inside the braces is $(d_2-d_3)f'$, so that a mutant will invade and go to fixation if $d_2 > d_3$. If this is a ploidy effect (with $d_2 = 0$) then d_3 must be negative so that triploids are less efficient than diploids. This seems unlikely to be true of triploids in general, but might be true of triploids with two identical haplotypes which might be less heterotic than the corresponding diploid. If it is a pleiotropic effect of the C_3 allele independent of ploidy, then this allele must cause a reduction in endosperm function.

This analysis confirms the validity of the parental conflict theory of the evolution of triploid endosperm, while delimiting the conditions under which it can occur. One important caveat must however be discussed. The mother plant may be able to control the endosperm directly, either through sealing it off from digestible maternal tissue by the seedcoat (Westoby and Rice, 1982) or by imposing developmental arrest by releasing abscissic acid or withholding water (Cook, 1981). In this case resource allocation to the endosperm would be under maternal control. The kin selection theory of seed provisioning discussed in the last section would remain valid with the endosperm regarded as if it were a maternal tissue (see last line of Table II), but the parental conflict theory of endosperm would become inapplicable since there would be no conflict.

It is therefore important to know whether seed weight is a character under maternal or endosperm control. Tschermak (1922) describes Mendelian experiments on several species. There is evidence that seed size is under maternal control in *Linum* (Tammes, 1911) and *Pisum* (Tschermak, 1922), but that it is determined by the endosperm/zygote in *Nicotiana* (Goodspeed, 1912) and *Phaseolus* Tschermak, 1922). However, Sax (1923) failed to find evidence of endosperm inheritance in *Phaseolus,* perhaps because his experiments involved different different lines, segregating for different loci. Chandraratna and Sakai (1960) conclude that grain weight in rice is determined by a combination of maternal nuclear and cytoplasmic inheritance. Thus there is evidence for genes determining seed weight which act both in the mother and in the endosperm (or zygote). In reviewing the genetic control of seed size, Harper, Lovell and Moore (1970) conclude thus: "To what degree seed size is determined by the forces of supply from the parent and by the forces of demand from the developing embryo and endosperm is an interesting and still largely open question."

ACKNOWLEDGMENTS

I thank Alan Grafen, Richard Law and Oliver Mayo for valuable comments.

REFERENCES

Brink, R. A., and Cooper, D. C. (1947). The endosperm in seed development. *Bot. Rev.* **13**, 423–541.

Chandraratna, M. F., and Sakai, K. (1960). A biometrical analysis of matriclinous inheritance of grain weight in rice. *Heredity* **14**, 365–373.

Charlesworth, B. (1980). Models of kin selection. *In* "Evolution of Social Behavior: Hypotheses and Empirical Tests" (H. Markl, ed.), pp. 1–26. Verlag Chemie, Weinheim.

Charnov, E. L. (1979). Simultaneous hermaphroditism and sexual selection. *Proc. Natl. Acad. Sci.* **76**, 2480–2484.

Cook, R. E. (1981). Plant parenthood. *Nat. Hist.* **90**, 30–35.

Fincham, J. R. S. (1983). "Genetics." Wright, PSG, Bristol.

Goodspeed, T. H. (1912). Quantitative studies of inheritance in *Nicotiana* hybrids. I. The relation between the weights of hybrid tobacco seed and the inheritance of certain characters in F_2. *Univ. Calf. Pub. Bot.* **5**, 87–116.

Harper, J. L., Lovell, P. H., and Moore, K. G. (1970). The shapes and sizes of seeds. *Ann. Rev. Ecol. Syst.* **1**, 327–356.

Law, R., and Cannings, C. (1984). Genetic analysis of conflicts arising during development of seeds in the Angiospermophyta. *Proc. R. Soc. Lond. B* **221**, 53–70.

Macnair, M. R., and Parker, G. A. (1978). Models of parent-offspring conflict. II. Promiscuity. *Anim. Behav.* **26**, 111–122.

Macnair, M. R., and Parker, G. A. (1979). Models of parent-offspring conflict. III. Intra-brood conflict. *Anim. Behav.* **27**, 1202–1209.

Maheshwari, P. (1950). "An Introduction to the Embryology of the Angiosperms." McGraw-Hill, New York.

Michod, R. E., and Hamilton, W. D. (1980). Coefficients of relatedness in sociobiology. *Nature* **288**, 694–697.

Parker, G. A., and Macnair, M. R. (1978). Models of parent-offspring conflict. I. Monogamy. *Anim. Behav.* **26**, 97–111.

Price, G. R. (1970). Selection and covariance. *Nature* **227**, 520–521.

Price, G. R. (1972). Extension of covariance selection mathematics. *Ann. Hum. Genet.* **35**, 485–490.

Queller, D. C. (1983). Kin selection and conflict in seed maturation. *J. Theor. Biol.* **100**, 153–172.

Queller, D. C. (1984). Models of kin selection on seed provisioning. *Heredity* **53**, 151–165.

Sax, K. (1923). The association of size differences with seed-coat pattern and pigmentation in *Phaseolus vulgaris*. *Genetics* **8**, 552–560.

Seger, J. (1981). Kinship and covariance. *J. Theor. Biol.* **91**, 191–213.

Tammes, T. (1911). Das Verhalten fluktuierend variierender Merkmale bei der Bastardierung. *Rec. Trav. bot. neerl.* **8**, 201.

Tschermak, von E. (1922). Über die Vererbung des Samengewichtes bei Bastardierung verschiedener Rassen von *Phaseolus vulgaris*. *Ztsch. indukt. Abstamm. Vererb.* **28**, 23-52.

Westoby, M., and Rice, B. (1982). Evolution of the seed plants and inclusive fitness of plant tissues. *Evolution* **36**, 713-724.

Willson, M. F., and Burley, N. (1983). "Mate Choice in Plants: Tactics, Mechanisms, and Consequences." Princeton University Press.

THE SELECTION OPERATING ON THE EVOLUTION EQUILIBRIUM OF THE FREQUENCY OF SEXUAL REPRODUCTION IN PREDOMINANTLY ASEXUAL POPULATIONS

Dan Cohen

Department of Botany
Hebrew University
Jerusalem, Israel

Daniel Zohari

Department of Genetics
Hebrew University
Jerusalem, Israel

ABSTRACT

Most tested predominantly asexual plant species are characterized by highly multiline or polyclonal populations, representing high levels of ecological diversity. Almost all such plants maintain, however, the floral apparatus for cross-pollination, and consistently produce low levels of sexual seeds. This and other evidence seems to indicate a considerable turnover of lines and clones, and a recurrent production of new clones from the sexual offspring.

A model is formulated to explain the ESS level of sexuality in predominantly asexual populations. The optimal ESS level of sexuality maximizes the equilibrium number of clones in a population, reached when the rates of generation of new clones and extinction of existing clones are equal. The optimal sexuality maximizes the ratio between the functions which express the effects of sexuality on generation and on extinction, respectively, and is independent of the rates themselves and their dependence on the number of existing clones.

Optimal sexuality is expected therefore to be independent of the number or the turnover of clones in natural populations. It is expected to be an increasing function of the diversity of available new niches, and a decreasing function of the density independent component of the growth rate of the population.

I. INTRODUCTION

Modeling the evolution of sexual reproduction has been the subject of much active research, e.g., Maynard-Smith (1978). Much of the research concerns itself with the evolutionary advantage of 100% sexual reproduction versus 100% asexual reproduction. It seems more appropriate however to investigate the evolutionary forces which select for any degree of sexuality, assuming of course that there is genetic variance for this trait in the population.

Asexual reproduction by apomixis, i.e., production of offspring without meiosis and fertilization is quite common in many plant and animal species (Bell, 1983). Self-pollination, which produces offspring identical to their parents is also very common, mainly in annual and weedy plants (Jain, 1976). It has been noted that in most such cases there is a consistent low level of production of genetically diverse, sexually produced seeds. Its importance for the generation of genetic variety of the many clones or homozygous lines comprising such species has been repeatedly pointed out (e.g., Stebbins, 1957)

However, no explicit model has been suggested for the evolutionary dynamics and equilibrium for the degree of sexuality or of outcrossing in such poly-clone or multiline species.

We suggest a model for the evolution of sexuality in polyclonal or multiline populations, which is the optimal balance between the effects of sexuality or of outcrossing on the rate of generation of new clones or lines and on the rate of extinction of existing clones or lines.

II. The Model

Our model deals with the question of the ESS for the fraction of sexually produced seeds in a population with a relatively low frequency of sexual reproduction.

We assume that the population consists of a number of asexual clones, or pure homozygous lines, where each individual plant produces sexual offspring at some frequency X which is genetically determined. Assuming that there exist mutants with different levels of sexuality, we want to find the ESS level of sexuality in the population.

We assume that the sexual offspring are produced by random matings within the common gametic pool of any local population, and that the sexuality alleles are distributed with equal probabilities among all the sexual offspring.

The selection on the sexuality gene operates by the association by "hitchhiking" of different alleles to the clones generated by them. Successful clones carry with them the associated sexuality alleles, and increase their frequency in the population, relative to sexuality alleles associated with less successful clones (Maynard Smith, 1978).

The essential characteristics of the model are similar for apomictic polyclonal populations and for self-pollinated multiline populations. Therefore, for the sake of simplicity of the presentation, we shall use the term clone to indicate both an asexual clone and a pure homozygous line. Also, we shall use the terms sexual and asexual to denote out-crossing and selfing, respectively.

Each existing clone may become extinct with a probability P_E per reproduction season. P_E is an increasing function of the number of established clones in the population, K, because of the increased competition. P_E is also an increasing function of the fraction X of sexual offspring produced by the clone, because the sexual offspring do not contribute to the reproduction output of the clone.

The rate of extinction of existing clones, $E(X,K)$ is therefore an increasing function of X, and an increasing concave function of K.

New clones are generated from sexual offspring of existing clones at a rate $G(X,K)$. $G(X,K)$ is an increasing function of the fraction of sexual offspring, X. It is proportional to X at low X. It increases less then proportionally as X increases at higher levels of X, because when there are many new clones, an increasing fraction of them are ecologically equivalent. $G(X,K)$ is a decreasing function of K, because the number of available niches decreases as K increases.

It should be noted that the effect of sexuality on the generation and extinction of clones is in addition to the effect of the overall mutation rate m. Mutation in a clone generates genetically different offspring which increases the probability of extinction of the clone and generates new clones similar to genetically diverse sexual offspring.

For any given monomorphic level of sexuality X in the population, the number of clones will reach an equilibrium level $\hat{K}$ where the rate of generation and establishment of new clones becomes equal to the extinction rate of existing clones, i.e., $G(X,K) = E(X,K)$.

An ESS for sexuality is defined as the level of monomorphic X^* in a population which cannot be invaded by a rare $X_i = X^*$.

We can show that X^* is the level of sexuality which generates the maximum level of $\hat{K}$ in monomorphic populations, and that for any invading X_1, its frequency in the population will increase if and only if $\hat{K}(X_1) > \hat{K}(\tilde{X})$, where $\tilde{X}$ is the X of the resident type.

Let us assume two types of clones in the population with X_1 and X_2 levels of sexuality, at fractions f and $1-f$, respectively, of the total number of clones. For each X level there coresponds an equilibrium number of clones, $\hat{K}_1$ and $\hat{K}_2$, respectively. Since the ecological characteristics of the clones are independent of their level of sexuality, we may assume that both $E_1(X_1,K_1)$ and $G_1(X_1,K_1)$ are proportional to f. Noting that $K_1 + K_2 = K$, we get therefore: $E(X_1,K_1) = fE(X_1,K)$, and $G(X_1,K_1) = fG(X_1,K)$. Similarly, $E(X_2,K_2) = (1-f)E(X_2,K)$, and $G(X_2,K_2) = (1-f)G(X_2,K)$. Thus, at any intermediate level of f, if $\hat{K}_1 > \hat{K}_2$, $E(X_2,K) > G(X_2,K)$, so that the number of clones with X_2 decreases, and X_1 displaces X_2 from the population. The opposite effect of X_2 displacing X_1 will occur if $\hat{K}_1 < \hat{K}_2$. (See Fig. 1.)

The equilibrium level of X^* in the population which is an ESS that cannot be displaced, is that for which the equilibrium number of clones $\hat{K}$ is maximized. (See Fig. 2.)

The ESS of sexuality depends therefore on the exact forms of the dependence of the rates of generation and extinction of clones on the sexuality and on the number of clones. Since $\hat{K}$ is the solution to the equation $G(K,X) = E(K,X)$, $\hat{K}$ is given by some function $f(x)$. X^* maximizes $\hat{K}$, i.e., at X^*, $f'(X^*) = 0$, $f''(x^*) < 0$.

In general, $G(X,K)$ is a decreasing function of K, and $E(X,K)$ is an increasing function of K. $G(X,K) - E(X,K) = H(X,K)$ is therefore a decreasing function of K. $\hat{K}$ is the solution for the equation $H(X,K) = 0$. $\hat{K}$ is maximized when $H(X,K)$ has a maximum with respect to X at $\hat{K}$. Thus, at $K = \hat{K}$, and $X = X^*$, $G(X,K) = E(X,K)$ and $\partial G(X,K)/\partial X = \partial E(X,K)/\partial X$.

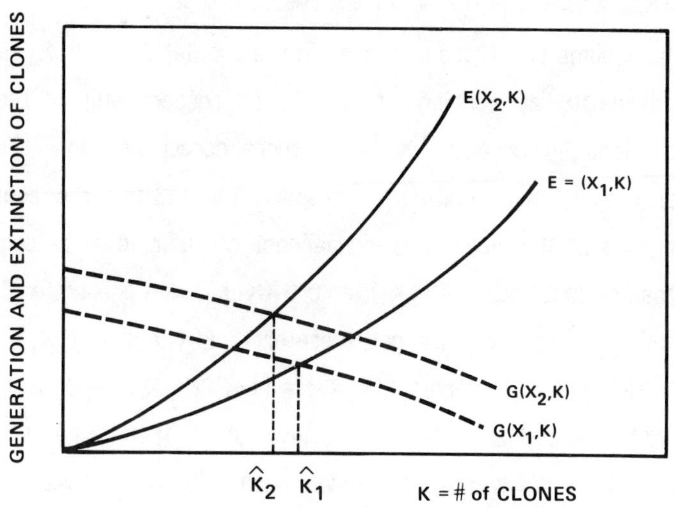

Figure 1

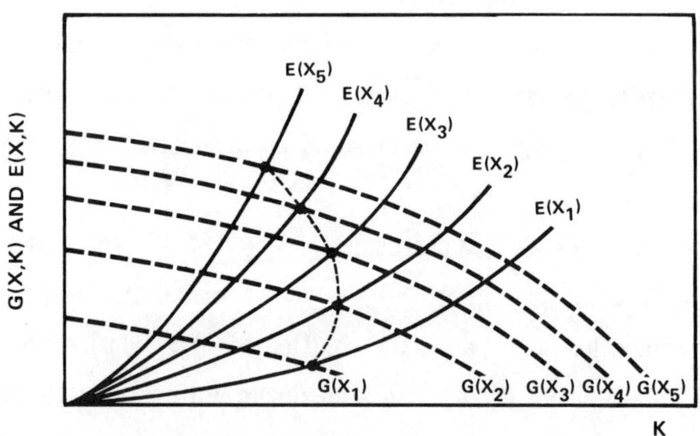

Figure 2

G(X,K) may be reasonably expressed as the product of two functions which give the effects of X and of K on G, i.e.,

$$G(X,K) = g(X)h(K) . \qquad (1)$$

Similarly, E(X,K) may be expressed as a product,

$$E(X,K) = u(X)v(k) . \qquad (2)$$

In this case, $\hat{K}$ is the solution for the equation $g(x)h(K) = u(x)v(K)$, i.e.,

$$g(x)/u(x) - v(k)/h(K) = 0 \qquad (3)$$

$v(K)/h(K)$ is a monotonic increasing function of K, because $v(K)$ increases and $h(K)$ decreases.

$g(X)$ an increasing function with a decreasing first derivative, while $u(X)$ is an increasing function with a non-decreasing or with increasing first derivative. Since $g(X)$ tends to level off at high X, and $u(X)$ always increases, $g(X)/u(X)$ must decrease at high X.

At low X, $g(X)$ usually increase at a higher rate than $u(X)$, so that $g(X)/u(X)$ increases at low X. Therefore, $g(X)/u(X)$ has an intermediate maximum at X*.

The derivative $\partial[g(X)/u(X)]/\partial X = g'(X)u(X) - g(X)u'(X)/[u(X)]^2$.

Since $g'(X)u(X)$ is high at low x and decreasing, while $g(X)u'(X)$ is increasing, $g(X)/u(X)$ is decreasing at low X and increasing at high X, with a maximum at X*, at which:

$$g'(X)u(X) - g(X)u'(X) = 0 . \qquad (4)$$

Since $v(K)/h(K)$ is monotonic increasing, $\hat{K}$ from equation (3) is *maximized* at the maximal level of $g(X)/u(X)$, i.e., at X*. *This is independent of the exact forms of* $v(K)/h(K)$ *or of* $g(X)/u(X)$. X* *is therefore the ESS level of sexuality in the population.*

u(X) may be an almost linear function of X, because the probability of extinction of a clone is expected to be almost proportional to the loss of sexual offspring from its populations. Thus, X* depends mainly on the curvature of g(X) and the slope of u(X).

It should be noted, however, that if both g(X) and u(X) are linear functions of X, $\hat{K}$ is maximized either at X* = m or at X* = 1, i.e., at zero sexuality or maximal sexuality.

X* = m when g'(m)u(m) < g(m)u'(m), and X* = 1 when g'(1)u(1) > g(1)u'(1).

A high degree of sexuality completely changes, however, the structure of the model, which is then no longer applicable, because clones do not persist at a high degree of sexuality.

II. APPROXIMATE FORMS FOR E(X,K) AND G(X,K)

A. Extinction

A fairly realistic form for E(X,K) takes into account the concavity with increasing K, and assumes a linear effect of X, in addition to the effect of mutation.

Thus, we may choose the form

$$E(X,K) = [1 + r(x+m)](aK+bK^2) . \tag{5}$$

B. Generation

The realistic representation of G(X,K) has to take into account a small rate of generation by mutation, although the generation of new clones by mutation or by sexual reproduction are very similar in the principle, and may be combined as one factor.

An important property of $G(X,K)$ is its decreasing rate of increasing as X is increased, because of a multiple generation of ecologically equivalent clones.

$G(X,K)$ is a decreasing function of K, probably by a higher power. Thus, we may choose the form

$$G(X,K) = [g(X+m)/(c+X+m)](1-sK-hK^2) . \qquad (6)$$

We shall now redefine X as $X+m$, assuming therefore that X cannot take values less than m.

We now get

$$E(X,K) = (1+rX)(aK+bK^2) \qquad (7)$$

$$G(X,K) = [gX/(C+X)](1-sK-hK^2) . \qquad (8)$$

$\hat{K}$ is the solution for $G(X,K) = E(X,K)$ which gives the quadratic equation

$$K^2[b(1+rX)+hgx/c+X]+K[a(1+rX)+sgx/(c+X)]-gX/(c+X) = 0. \qquad (9)$$

An approximate simplification by neglecting the K^2 terms in both $E(X,K)$ and $G(X,K)$ gives $\hat{K}$ as the solution to the linear form

$$K[a(1+rx) + sgX/(C+X)] - gX/(c+X) = 0 . \qquad (10)$$

from which we get the approximation

$$\hat{K} = gX/[a(1+rX)(C+X) + sgX] . \qquad (11)$$

In this case, $\hat{K}$ is an increasing function of g and a decreasing function of a, c, s and r.

Rewriting $H(X,K) = 0$ as

$$(1+rX)(c+X)/gX = (1-sK-hK^2)/(aK+bK^2) \qquad (12)$$

we find X^* by calculating the minimum of the L.H.S. of (12), because R.H.S. of (12) is a decreasing function of K.

L.H.S. of (12) can be simplified to $[c/x + rc + 1 + rx]/g$, the minimum of which is at

$$X^* = \sqrt{c/r} . \tag{13}$$

$X^* = \sqrt{c/r}$ is thus the ESS solution for sexuality, which maximizes the number of coexisting clones at the steady state equilibrium between generation and extinction of clones.

In this case, X^* is completely independent of the parameters which express the effects of K on the rates of generation and extinction. This is because the effects of X and K on $G(X,K)$ and $E(X,K)$ are by independent multiplicative functions, which can be separated at $\hat{K}$, when

$$G(X,K) = E(X,K) .$$

In this case, X^* is determined only by the ratio between c, the coefficient of linearity of X in $G(X,K)$, and r, the linear coefficient of X in $E(X,K)$.

c is the level of X at which $G(X,K)$ is half saturated with respect to X. It indicates the degree to which the fraction of ecologically distinct clones increases linearly with the total number of available ecological niches in the environment. If this number is large, new clones will most likely differ in their ecological requirements, so that saturation of $G(X,k)$ will occur at higher levels of X. This will lead to a higher optimal X in the population.

The effect of sexuality on increasing the rate of extinction is expressed by the factor $1 + rX$. The coefficient r is expected to increase with the degree of saturation of the habitat by the existing clones. The degree of saturation is expected to increase when the number of existing clones increases. When the extinction coefficients a and b

are small, $\hat{K}$ is large. In this case, the coefficient r is expected to be large, and X^* is expected to decrease. The opposite effect is expected when the extinction coefficients are large. In this case, $\hat{K}$ is small, and the coefficient r is expected to be small, with an increase in X^*.

Since X is bounded between the mutation rate $\hat{m}$ and 1, there are two possible equilibrium levels for X^*.

1. $X^* = m$, i.e., no sexuality. This will occur if $\hat{K}$ is a decreasing function of X at $X = m$, i.e., $\sqrt{c/r} < m$.

2. $m < X^*$, at $\partial\hat{K}/\partial K = 0$, i.e., $X^* = \sqrt{c/r}$, i.e., an intermediate optimal sexuality.

IV. DISCUSSION

Natural populations with a predominantly asexual mode of reproduction are usually made up of a large number of distinct clones in predominantly apomictic (parthenogenetic) species, or a large number of homozygous lines in selfing species.

The number of such clones or lines may be very large. Their spatial distributions often suggest that at least some such lines or clones are adapted to some subsets of the overall range of ecological dimensions for the whole species (Brown, 1979, Nevo *et al.*, 1979).

In addition, there are strong indications that there is a continuous turnover of clones or lines in most species. Some clones or lines appear to have persisted for many hundreds or even thousands of generations. Others seem to have appeared fairly recently indicating a fairly high rate of turnover, e.g., Hebert (1978).

Our model of a balance between generation and extinction of clones seems therefore to represent reasonably well the population structure of many species with a predominantly asexual mode of reproduction.

The level of sexuality in many such species may be very low. Levels of 1% or less are commonly reported, for example, in Wild Barley (Nevo *et al.*, 1979).

Yet, the maintenance of the apparatus of sexual reproduction in perfect order, with a considerable investment of resources to attract mates in animals or pollinators in plants, suggests very strongly that there is a strong selection for the maintenance of the observed low level of sexuality. In such cases, it is clear that the low level of sexuality is not accidental or a residual relic of a historical sexual ancestor. In fact, in plants which produce 1% or less of their seeds by cross-pollination, the rare sexual seeds must carry the cost of the whole floral apparatus. In many such plants, the investment in pollen, in the attractivity of the flowers and in reward for pollinators, does not seem to be much less than in similar related 100% sexual species (Charnov, 1982).

The fitness of rare sexual seeds must therefore be very high. This fitness must however, become much lower as the frequency of sexual seeds increases. At th ESS level of sexuality, the fitness of sexual seeds per unit of invested resources has to become equal to twice the fitness of the asexual seeds, to compensate for the loss of 50% of the genes of the parent of sexual offspring. In our model, the fitness of an allele for sexuality is expressed as the long term rate of increase of its frequency in the whole population of clones.

A. Tests and Predictions of the Model

The rate of generation of new clones and of the extinction of existing clones may be measured or estimated in natural populations. It is very difficult however to estimate the effect of the number of existing clones on these rates. It is practically impossible to estimate the effect of the degree of sexuality on these rates.

Our model allows us to predict that the distribution of equilibrium levels of sexuality will probably represent different combinations of $g(x)$ and $u(x)$, or of c and r in the effects of sexuality on the generation and extinction of clones. This is because the same levels of X^* may be reached at very different levels of $g(X)$ and $u(X)$ or of r and c, provided that the maxima of their ratios remain the same.

A high r population represents a highly saturated ecosystem, with strong competition between the clones. Any increase in sexuality of a clone increases very strongly the probability of its extinction.

This selects for a low level of X^*. However, the level of X^* may increase as the result of a high level of c, which indicates a more linear $g(X)$ representing the degree of non-overlapping between the different ways by which new clones may get established.

A low r type of population represents a less saturated environment, such that an increase in sexuality has a small effect on increasing the probability of extinction of a clone. This selects for a higher X^*. Low levels of X^* may be the result, however, of low c, which represents a degree of overlapping between newly generated clones, so that the effect of X on the generation rate is highly convex.

Our model may be considered as a special case of the evolution of modifying features under a regime of clonal selection. According to Eshel (1973), the stable equilibrium level of the modifying character, which may include mutation, or sex, or some other character affecting clonal survival, maximizes the average survival probability of the clones in the population. Our model predicts the maximixation of the equilibrium number of clones, rather than their survival probabilities. The model may also be considered as a further extension of the model of Williams (1975) of clonal selection and the equilibrium level of sexuality in a population.

Our model is essentially a steady state model for the turnover of clones or lines. It does not assume any long term changes in the environment. It does assume however that the complexity and diversity of habitats and niches may be such that the equilibrium number of clones may vary over a wide range, from a low level close to zero up to a fairly high level. If the equilibrium number of clones is very high, the population becomes ecologically equivalent to a completely sexual population. If such a higher number of clones is maintained at a low level of optimal sexuality, we may expect that selection will maintain the clonal structure of the population.

The assumptions of our model become invalid however at a high level of sexuality. In this case, the purely clonal structure of the population breaks down, and the population reverts to a mixture of clonal and sexual individuals, which requires a different model for its analysis.

Our model suggests therefore that a clonal mode of reproduction may be optimal even in very complex habitats with many diverse biological interactions which would normally tend to favor a sexual model of reproduction according to the "Tangled Bank" model of Ghiselin (1974). This will be the case in situations where a very large number of ecologically very diverse clones can be maintained by an optimal low level of sexuality. Since the optimal level of sexuality in polyclonal populations depends on the balance between the effects of sexuality on the generation of new clones, and on the exinction of existing clones, a thorough study of both effects is necessary in order to be able to understand the distribution of sexual and asexual models of reproduction in natural populations.

An environment with a high diversity of micro-habitats and micro-niches can be occupied at equilibrium either by a polyclonal population with a large number of ecologically diverse clones, or by a sexual population with a large diversity of genotypes generated by the sexual process. Both genetic systems provide a diversity of rare genotypes which are maintained in equilibrium by their adaptation to rare micro-habitats. They are essentially almost equivalent ecologically.

If all the micro-habitats are occupied by appropriately adapted clones in a polyclonal population, there are no micro-habitats in which the sexual genotypes may have a selective advantage. In this case we expect that the polyclonal mode of reproduction will dominate.

If, however, some fraction of the micro-habitats are not occupied by clones adapted to them, then these unoccupied micro-habitats will provide niches in which the sexual offspring may have an advantage over the asexual offspring. A sufficiently large fraction of unoccupied micro-habitats will provide an overall advantage for the sexual versus the asexual mode of reproduction.

The fraction of unoccupied niches will increase when the equilibrium number of clones in the population decreases. An important process which reduces the number of clones is random extinction caused by fluctuations in the availability of the micro-niches together with a stochastic process of extinction of clones which have reached low levels of population size. Thus, a more diverse environment will have a large number of micro-niches occupied by a large number of small clones, which are more likely to become extinct. This means that a larger fraction of all the niches will be unoccupied by specifically adapted clones in more diverse environments, which will confer a greater advantage for the sexual mode of reproduction.

We suggest that this may be an important factor in explaining the observed trend of increased prevalence of sexual reproduction in more complex and diverse environment (Ghiselin, 1974; Bell, 1984).

Such an explanation allows us to make another yet untested prediction. Random extinction decreases when population sizes increases. We may, therefore, predict that the fraction of niches unoccupied by clones decreases in species with large and dense populations, and that in these species there would be a higher probability for a polyclonal asexual mode of reproduction compared with species with less dense populations. We may therefore expect species with very small and restricted populations to be more sexual than species with large and dense populations.

For example, bacteria have huge polyclonal populations characterized by a very small fraction of genetically diverse offspring. On the other hand, populations of large mammals are relatively small in absolute numbers, and are always 100% sexual.

B. The Effects of Local Extinctions and Recolonizations, and of Dispersal

Local extinctions of clones create temporary open micro-niches which remain open until they are reoccupied by migration from neighboring areas. These temporarily available micro-niches provide a niche space in which the sexual offspring have an advantage over clones which are not specifically adapted to this niche space. Effective long distance dispersal will lead to rapid recolonization of locally vacated micro-niches, and will decrease the overall amount of available niche space for sexual offspring.

It would be possible to predict therefore that effective long dispersal or mixing in a species will lead to a decrease of the prevalence of sexual reproduction. On the other hand, dispersal and mixing expose individuals of any clone to a much greater variety of possible micro-niches, and would thus favor a higher degree of sexuality among the dispersing individuals.

In plants, the distance traveled by outcrossing pollen is often much greater than the dispersal distance of seeds. In this case, cross pollination between plants adapted to different ecological conditions in adjacent patches will produce seeds which are not well adapted to the local conditions. Selfing or apomixis in a parent plant will increase therefore the overall fitness of its offspring relative to an outcrossing plant. Strong spatial heterogeneity and limited seed dispersal will therefore favor a higher fraction of asexual seeds. Such, indeed, was found to be the case in grasses growing on extremely patchy distribution of heavy metal contamination in soils in Britain (Jain and Bradshaw, 1966).

In fact, both selfing and apomixis act as isolating mechanisms between ecologically differentiated subpopulations, caused by spatial heterogeneity or resource specialization, or both. Selfing and/or apomixis will evolve therefore whenever there is a strong selection favoring genetic isolation between sexually accessible subpopulations.

Sib competition, on the other hand, will favor the sexual offspring among non-dispersing locally competing offspring.

Both genetic isolation and wide range environmental tolerance have been pointed out by Lynch (1984) as important characteristics of parthenogenetic species.

Since all these effects are expected in natural populations, the overall effect of dispersal and heterogenetiy is difficult to predict. Available data suggest that in some snails, selfing is associated with limited dispersal and local ecological differentiation (Selander and Hudson, 1976). The same is true of quite a number of selfers and apomicts in plants (Jain, 1976). On the other hand, the dandelion, *Taraxacum officinalis* is a typically apomixtic plant with widely dispersing seeds, with many specialized clones occupying different micro-habitats.

REFERENCES

Bell, G. (1982). "The Masterpiece of Nature: The Evolution and Genetics of Sexuality." Croom Helm Ltd., London.

Brown, A. H. D. (1979). Enzyme polymorphism in plant populations. *Theor. Pop. Biol.* **15**, 1-42.

Charnov, E. R. (1982). "The Theory of Sex Allocation." Princeton University, Princeton, N.J.

Eshel, I. (1973). Clone selection and the evolution of modifer genes. *Theor. Pop. Biol.* **4**, 196-208.

Ghiselin, M. T. (1974). "The Economy of Nature and the Evolution of Sex." University of California Press, Berkeley, Calif.

Hebert, P. D. N. (1978). The population biology of Daphnia (Crustacea, Daphnidae). *Biol. Rev.* **53**, 387-426.

Jain, S. K. (1976). The evolution of inbreeding in plants. *Ann. Rev. Ecol. Syst.* **7**, 469-495.

Jain, S. K., and Bradshaw, A. D. (1966). The evolutionary divergence among adjacent plant populations. I. The evidence and its theoretical analysis. *Heredity* **21**, 407-441.

Lynch, M. (1984). Destabilising hybridisation, general purpose genotypes and geographic parthenogenesis. *Quart. Rev. Biol.* **59**,. 257-290.

Maynard Smith, J. (1978). "The Evolution of Sex." Cambridge Universtiy Press.

Nevo, E., Zohary, D., Brown, A. D. H., and Haber, M. (1979). Genetic diversty and environmental associations of wild barley, *Hordeium spontaneum,* in Israel. *Evolution* **33**, 815-833.

Selander, R. K., and Hudson, R. O. (1976). Animal population structure under close inbreeding; the land snail in Rumina in Southern France. *Amer. Natur.* **110**, 695-718.

Stebbins, G. L. (1957). Self fertilisation and population variability in higher plants. *Amer. Natur.* **91**, 337-354.

Williams, G. C. (1975). "Sex and Evolution." Princeton University Press, Princeton, N.J.